ORBIT
A FOUNDATION SERIES

SPACE NUCLEAR POWER

by
Joseph A. Angelo, Jr., Ph.D.
and
David Buden

ORBIT BOOK COMPANY, INC.

MALABAR, FLORIDA
1985

Original Edition 1985

Printed and Published by
ORBIT BOOK COMPANY, INC.
2005 Township Road
Malabar, FL 32950

Printed in the United States of America

This book represents the views of the authors and does not necessarily reflect the official opinion of the United States Department of Defense, the Department of the Air Force, the Department of Energy, or the National Aeronautics and Space Administration.

Library of Congress Cataloging in Publication Data

Angelo, Joseph A.
Space nuclear power.

1. Space vehicles—Nuclear power plants. I. Buden, David. II. Title.
TL1102.N8A54 1985 629.47′443 84-16701
ISBN 0-89464-000-3

Dedicated to all individuals who endeavor to make humanity's movement into space a near-term reality: and especially to the young people who will be the explorers and settlers of new worlds

Contents

Foreword

For more than three decades, larger and more powerful rockets have inspired awe among those privileged to observe their launching. But the very factor which makes the firing of a large rocket so spectacular—the consumption of tons of fuel in just seconds—also suggests practical limits for chemical propulsion. And the limits apply not only to propulsion. Any application which requires high power for more than the few minutes of a chemical rocket's firing, or power levels greater than can be provided by collecting the sun's diffuse radiation, is a candidate for nuclear power. The reason for nuclear power in space is simple and compelling: an increase by a factor of as much as one million in the energy which is contained in a mass of nuclear fuel compared with the same mass of chemical fuel.

With nuclear power as the only source for the most ambitious future space applications, *Space Nuclear Power* is a welcome and timely addition to the literature. Starting with the fundamentals of nuclear physics, the authors develop the essentials of space nuclear power—radioactivity, nuclear engineering, thermodynamics, energy conversion, heat transport, and space heat rejection. With physics and engineering as the foundation, they review past and current nuclear power experience, including nuclear power sources on highly successful explorations of the Moon, Mars, and beyond. In fact, *Space Nuclear Power* provides a concise history of space nuclear energy sources, both radioisotope and reactor. Of course, no treatment of nuclear power could be complete without addressing the safety issue. This is done in the context of power plant design, and both United States and Soviet policies are reviewed.

Energy is the keystone of man's extraterrestrial development. *Space Nuclear Power* is an indispensable guide to the realization of that energy.

JOHN T. MAY
Vice Dean of Faculty
United States Air Force Academy.

Introduction

Human development at the start of the third millennium will be highlighted by the establishment of man's extraterrestrial civilization. Three critical cornerstones for this exciting development are currently perceived: (1) the availability of compact energy sources for power and propulsion; (2) the ability to process materials anywhere in the Solar System; and (3) the creation of permanent human habitats in outer space [1, 2].

Energy—reliable, abundant, and portable—is a key factor in developing and sustaining man's permanent presence in outer space. Space nuclear power systems, in turn, represent an enabling technology that must be effectively incorporated in future space programs, if imaginative and ambitious space applications and utilization programs are to occur in the next few decades [3]. For example, the movement of large quantities of cargo from low-Earth-orbit (LEO) to high-Earth-orbit (HEO) or lunar destinations, the operation of very large space platforms throughout cislunar space, and the startup and successful operation of lunar bases and settlements can all benefit from the creative application of advanced space nuclear power technology. Very imaginative future space projects, such as asteroid movement and mining, planetary engineering and climate control, and human visitations to Mars and the planets beyond cannot even begin to be considered without the availability of compact energy supplies in the megawatt (MW) and, ultimately, hundreds of megawatt class.

Since the start of the space age, a range of nuclear power supply options has been developed by the United States to support the civilian and military space program. Tomorrow's space exploration and exploitation activities will require even larger quantities of reliable, long-lived power. Nuclear energy judiciously applied in future space missions offers several distinctive advantages over traditional solar and chemical space power systems. These advantages include: compact size; low to moderate mass; long operating lifetimes; operation in hostile environments (e.g., trapped radiation belts, surface of Mars, moons of the outer planets); operation

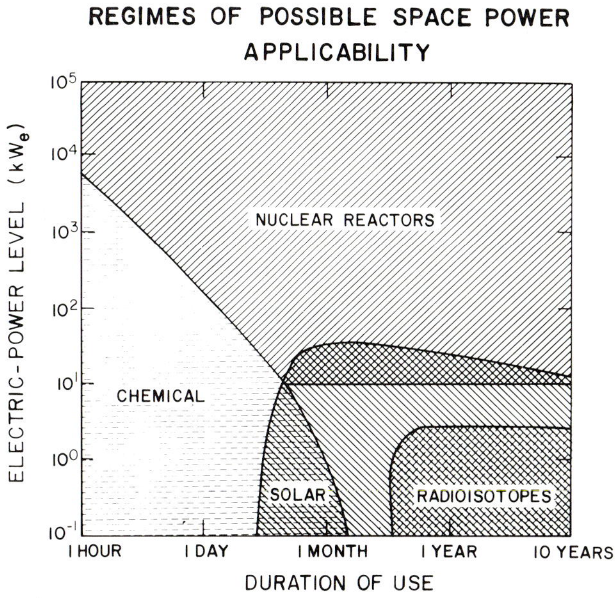

Fig. I.1 Regimes of possible space power applicability. *Courtesy of Los Alamos National Laboratory.*

independent of the distance from the Sun or of the orientation to the Sun; and high system reliability and autonomy [4, 5, 6]. In fact, as power requirements approach the hundreds of kilowatts and megawatt regime, nuclear energy appears to be the only realistic power option [6] (see Fig. I.1).

Space nuclear power technology involves the use of the thermal energy liberated by nuclear processes. This heat is released in the spontaneous, yet predictable decay of radioisotopes; in the controlled fission of heavy nuclei (such as uranium-235) in a sustained neutron chain reaction; or in the fusion or joining together of light nuclei such as deuterium and tritium. The thermal energy so liberated may then be employed directly in space system processes demanding large quantities of heat or else converted into electric power. Until controlled thermonuclear fusion is actually achieved, nuclear energy applications will be based on the use of radioisotope decay or the nuclear fission process.

A generic space nuclear power system is depicted in Fig. I.2. Here, the primary system output is electrical energy, which is produced by converting radioisotope decay heat or the thermal energy released in nuclear fission to electrical energy, using static conversion (e.g., thermoelectric) or dynamic conversion (e.g., the Rankine or Brayton cycle) principles. Contemporary options for nuclear energy heat sources and companion power conditioning subsystems are shown in Fig. I.3. While these options are, of course, not all-inclusive, they nevertheless form the basis for intelligent planning of nuclear power source applications for space missions in the next few decades. Radioisotope and reactor power systems are further classified in Fig. I.4 [7].

For space power applications in the megawatt (electric) regime and beyond, there are a number of interesting advanced nuclear reactor technology options. These designs include the solid core nuclear reactor (a derivative of the nuclear rocket technology (Rover) program), the fluidized bed reactor, and the gaseous core nuclear reactor [8].

Since 1961, the United States has launched over twenty National Aeronautics and Space Adminstration (NASA) and military space systems which derived all or part of their power requirements from nuclear energy sources. These systems are summarized in Table I.1. As can be seen in this table, all but one of the previous missions used plutonium-238 as the fuel in various radioisotope thermoelectric generator (RTG) systems. The SNAP-10A was a compact nuclear reactor which used fully enriched uranium-235 as the fuel. The acronym SNAP stands for Systems for Nuclear Auxiliary Power, with the odd-numbered units representing radioisotope heat sources and the even-numbered units nuclear reactors.

Since its inception, the United States space nuclear power program has placed great emphasis on safety. Contemporary policy and practices for all space nuclear power sources promote designs which ensure that the levels of radioactivity and the probability of nuclear fuel release will not provide any significant risk to the Earth's population or to the terrestrial environment [7]. For radioisotope heat sources, this aerospace nuclear safety policy essentially consists of providing containment that is not prejudiced under any circumstances, including launch accidents, reentry, or impact on land or water. For nuclear reactors, the safety mechanism consists of maintaining subcriticality under all conditions, normal and otherwise, in the Earth's atmosphere or on the Earth's surface. After the reactor has experienced power operation in space, the reactor will be prevented from reentering the terrestrial biosphere until the fission products and other radioactive materials no longer represent a radiological risk. Orbital lifetimes in excess of three hundred years support this safety approach.

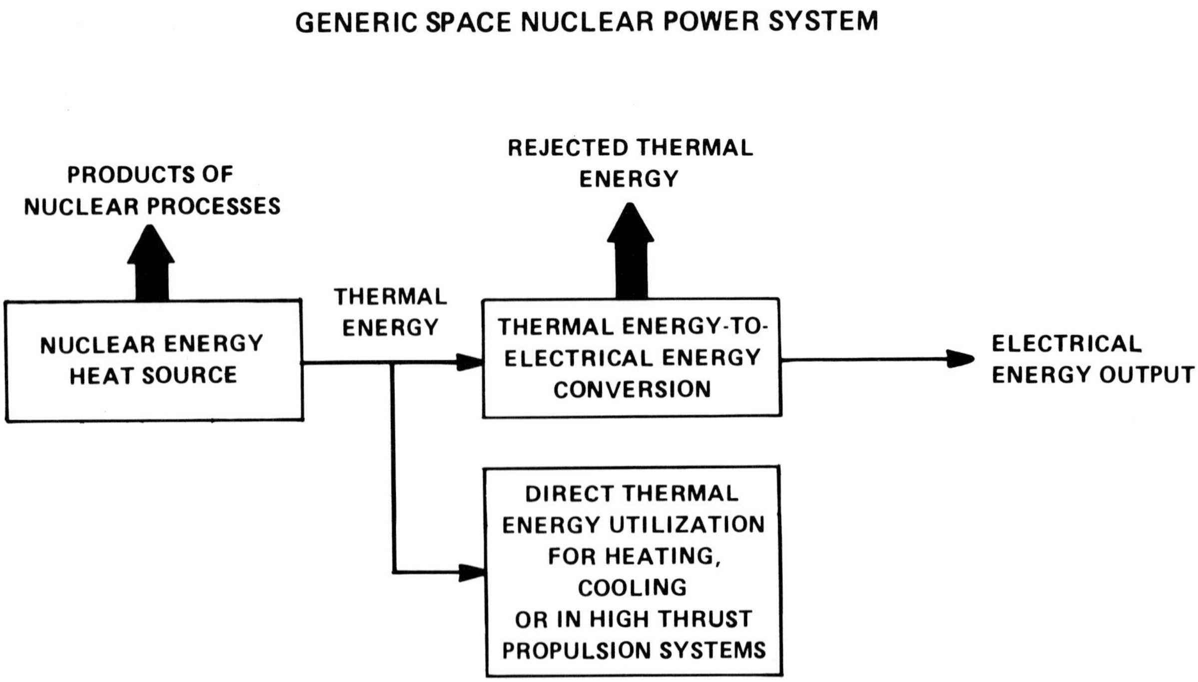

Fig. I.2 Generic space nuclear power system. *Courtesy of Los Alamos National Laboratory.*

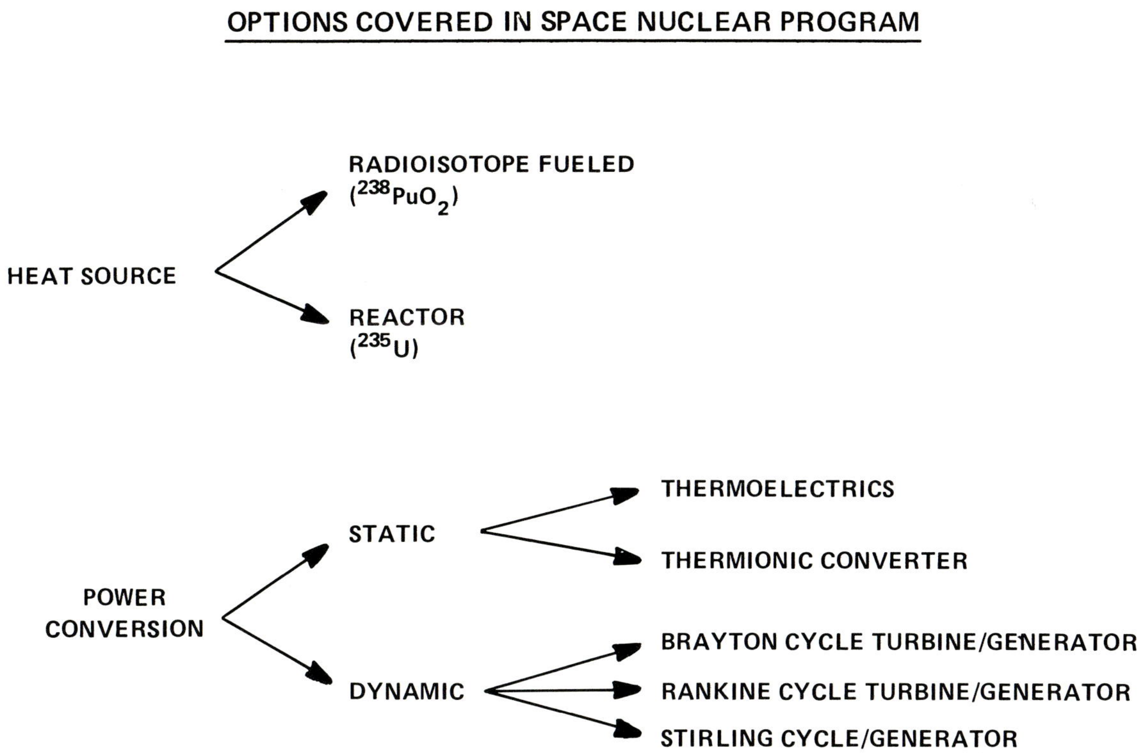

Fig. I.3 Options covered in space nuclear program. *Courtesy of Los Alamos National Laboratory.*

CLASSIFICATION OF NUCLEAR POWER SYSTEM TYPES BEING CONSIDERED FOR SPACE APPLICATION

NUCLEAR POWER SYSTEM TYPE	ELECTRIC POWER RANGE (MODULE SIZE)	POWER CONVERSION
RADIOISOTOPE THERMOELECTRIC GENERATOR (RTG)	UP TO 500 We	STATIC: THERMOELECTRIC
RADIOISOTOPE DYNAMIC CONVERSION GENERATOR	0.5 kWe – 10 kWe	DYNAMIC; BRAYTON OR ORGANIC RANKINE CYCLES
REACTOR SYSTEMS HEAT PIPE SOLID CORE THERMIONICS	10 kWe – 1000 kWe	STATIC: THERMOELECTRIC, THERMIONICS DYNAMICS: BRAYTON, RANKINE, OR STIRLING
REACTOR SYSTEM HEAT PIPE SOLID CORE	1 – 10 MWe	BRAYTON CYCLE RANKINE CYCLE STIRLING CYCLE
REACTOR SOLID CORE PELLET BED FLUIDIZED BED GASEOUS CORE	10 – 100 MWe	BRAYTON CYCLE (OPEN LOOP) STIRLING MHD

Fig. I.4 Classification of nuclear power system types being considered for space application. *Courtesy of Los Alamos National Laboratory.*

Table I.1 Summary of space nuclear power systems launched by the United States (1961–1984).

POWER SOURCE	SPACECRAFT	MISSION TYPE	LAUNCH DATE	STATUS
SNAP-3B	TRANSIT 4A	Navigational	June 29, 1961	Successfully achieved orbit
SNAP-3B	TRANSIT 4B	Navigational	November 15, 1961	Successfully achieved orbit
SNAP-9A	TRANSIT-5BN-1	Navigational	September 28, 1963	Successfully achieved orbit
SNAP-9A	TRANSIT-5BN-2	Navigational	December 5, 1963	Successfully achieved orbit
SNAP-9A	TRANSIT-5BN-3	Navigational	April, 21, 1964	Mission aborted: burned up on reentry
SNAP-10A (REACTOR)	SNAPSHOT	Experimental	April 3, 1965	Successfully achieved orbit
SNAP-19B2	NIMBUS-B-1	Meteorological	May 18, 1968	Mission aborted: heat source retrieved
SNAP-19B3	NIMBUS III	Meteorological	April 14, 1969	Successfully achieved orbit
SNAP-27	APOLLO 12	Lunar	November 14, 1969	Successfully placed on lunar surface
SNAP-27	APOLLO 13	Lunar	April 11, 1970	Mission aborted on way to moon. Heat source returned to South Pacific Ocean.
SNAP-27	APOLLO 14	Lunar	January 31, 1971	Successfully placed on lunar surface
SNAP-27	APOLLO 15	Lunar	July 26, 1971	Successfully placed on lunar surface
SNAP-19	PIONEER 10	Planetary	March 2, 1972	Successfully operated to Jupiter and beyond
SNAP-27	APOLLO 16	Lunar	April 16, 1972	Successfully placed on lunar surface
TRANSIT-RTG	"TRANSIT" (TRIAD-01-1X)	Navigational	September 2, 1972	Successfully achieved orbit
SNAP-27	APOLLO 17	Lunar	December 7, 1972	Successfully placed on lunar surface
SNAP-19	PIONEER 11	Planetary	April, 5, 1973	Successfully operated to Jupiter, Saturn, and beyond
SNAP-19	VIKING 1	Mars	August 20, 1975	Successfully landed on Mars
SNAP-19	VIKING 2	Mars	September 9, 1975	Successfully landed on Mars
MHW	LES 8/9	Communications	March 14, 1976	Successfully achieved orbit
MHW	VOYAGER 2	Planetary	August 20, 1977	Successfully operated to Jupiter and Saturn
MHW	VOYAGER 1	Planetary	September 5, 1977	Successfully operated to Jupiter and Saturn

References

1. J. A. Angelo, Jr. and D. Buden, "Space Nuclear Reactors—Energy Gateway Into the Next Millennium," 32nd Congress, International Astronautical Federation, Rome, Italy, September 6–12, 1981.
2. K. A. Ehricke, "Space Stations—Tools of New Growth in an Open World," *Proceedings of the 25th International Astronautical Congress*, Amsterdam, The Netherlands, 1974.
3. J. A. Angelo, Jr. and D. Buden, "Nuclear Power—Key to Man's Extraterrestrial Civilization," 17th Intersociety Energy Conversion Engineering Conference, Los Angeles, CA, August 8–13, 1982.
4. H. M. Dieckamp, *Nuclear Space Power Systems*, Atomics International, Canoga Park, CA, September 1967.
5. Atomic Industrial Forum, *Guidebook for the Application of Space Nuclear Power Systems*, New York, January 1969.
6. D. Buden et al., "Selection of Power Plant Elements for Future Reactor Space Electric Power Systems," LA-7858, Los Alamos National Laboratory, NM, September 1979.
7. U.S. Department of Energy, "Environmental Development Plan (EDP)—Space Applications," DOE/EDP-0026, April 1978.
8. D. Buden and J. A. Angelo, Jr., "The Role of Nuclear Reactors in Future Military Satellites," 8th DARPA Strategic Space Symposium, Naval Postgraduate School, Monterey, CA, June 22–25, 1982.

Chapter 1

Radioactivity

To understand the use of radioisotopes in space nuclear power systems, it is first necessary to explore some of the fundamental aspects of radioactivity and the principles governing radioactive decay. Radioactivity can be defined as the process involving the spontaneous decay or disintegration of an unstable atomic nucleus and is accompanied by the emission of nuclear radiations, such as alpha particles (α), beta particles (β), gamma rays (γ), neutrons (n), or protons (p).

Alpha Particles

An alpha particle is a positively charged nuclear particle consisting of two protons and two neutrons. It is identical to the nucleus of the helium atom. Alpha radiation is the least penetrating of the three common types of radiation (alpha, beta, and gamma), being stopped by a mere sheet of paper. It is not generally considered dangerous to living things unless the alpha-emitting substance has been ingested.

Beta Particles

The beta paticle is an elementary charged particle emitted from the nucleus during radioactive decay. It has the mass and charge of the electron (e). The electron is an elementary particle that has a rest mass $(m_0)_e$ of 9.10956×10^{-31} kg and an electric charge (e) of 1.60219×10^{-19} coulomb (C). A negatively charged beta particle is identical to an electron, while a positively charged beta particle is called a positron or positive electron. Beta radiation may cause skin burns and beta-emitters are harmful if they enter the body. However, they are easily stopped by a thin sheet of metal.

Gamma Rays

Similar in nature to X-rays, gamma rays are high frequency, short wavelength electromagnetic radiation which are emitted by nuclei during transitions from higher to lower nuclear energy states. Gamma radiation frequently accompanies alpha and beta emissions and always accompanies nuclear fission. Gamma rays are very penetrating and are best shielded against by dense materials such as lead or tungsten.

Neutrons

A neutron is an uncharged, elementary particle with a mass (m_n) of 1.67492×10^{-27} kg, which is slightly larger than the mass of the proton. It is found in the nucleus of every atom heavier than hydrogen. A free neutron (i.e., one outside the nucleus) is unstable and changes into a proton, beta particle, and antineutrino ($\bar{\nu}$) in a decay process with a half-life ($T_{1/2}$) of about twelve minutes (see Fig. 1.1). Half-life will be defined shortly. The lifetime of such a neutron in a typical space nuclear reactor is less than 0.001 second. Consequently, the instability of a free neutron is of no real significance in the operation of such systems. Neutrons sustain the fission chain reaction in a nuclear reactor and impose special shielding requirements.

DECAY OF A FREE NEUTRON

$$^{1}_{0}\mathrm{n} \xrightarrow[T_{1/2} \sim 12 \text{ min}]{} {}^{1}_{1}\mathrm{p} + {}^{0}_{-1}\mathrm{e} + \bar{\nu}_e$$

Fig. 1.1 Decay of a free neutron.

Neutrinos

The antineutrino ($\bar{\nu}$) is the antiparticle or antimatter form of the neutrino (ν). The neutrino is a zero rest mass, electrically neutral particle, appearing in the decay of certain nuclei and possessing great penetrating power. For example, solar neutrinos (those originating in the Sun) generally pass right through the Earth without interacting. At present, six types of neutrinos have been identified. However, only two types: electron neutrinos (ν_e) and electron antineutrinos ($\bar{\nu}_e$) are of any interest here. For the purposes involved with space nuclear power systems, it is not essential to make a precise distinction between the two types of electron neutrinos and we will refer to both as neutrinos throughout the text.

Protons

The proton is an elementary particle with a single positive electric charge (equal in magnitude to the charge on the electron) and a rest mass $(m_0)_p$ of 1.67261×10^{-27} kg. Protons are constituents of all atomic nuclei and the atomic number (Z) of an atom is equal to the number of protons in the nucleus (see Appendix A).

Nuclide Designations

In accordance with the recommendations of the International Union of Pure and Applied Chemistry, the following designations are used for nuclides:

The *mass number* of a nuclide is placed as a superscript to the left of the symbol for the chemical element of the nuclide, rather than to its right, as formerly designated; for example, ^{238}Pu, rather than Pu^{238} for plutonium-238.

The *atomic number* is placed as a left subscript; for example, $^{235}_{92}U$ or uranium-235.

The *state of ionization* is shown as a right superscript; for example, Ca^{++}.

The *number of neutrons* in the nucleus is shown as a right subscript; for example, $^{238}_{94}Pu_{144}$ for the isotope plutonium-238 which contains 94 protons and 144 neutrons.

Excited nuclear states are shown either as part of the left superscript, or sometimes the right superscript; for example, ^{110m}Ag of $^{110}Ag^{m}$ indicates an excited state of a silver-110 nucleus.

Fig. 1.2 depicts these nuclide designation relationships.

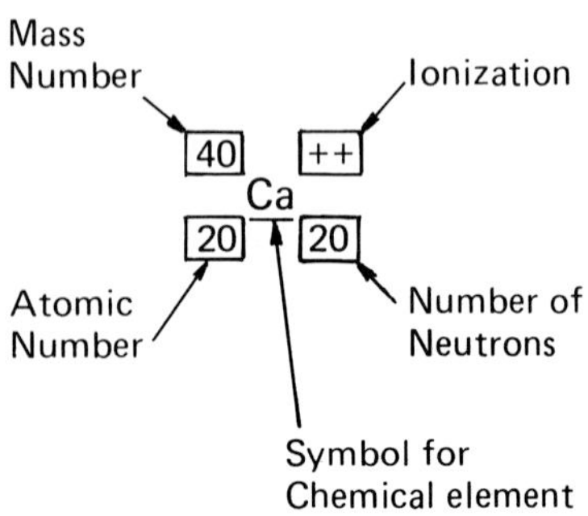

Fig. 1.2 Nuclide designation. *Courtesy of U.S. Dept. of Energy.*

Atomic Weight

The atomic weight is the mass of an atom relative to other atoms. The current basis of the scale of atomic weights is the most common isotope of the element carbon, namely carbon-12, which has arbitrarily been assigned an atomic weight of 12 (see Appendix A). The atomic mass unit (amu) is defined as one-twelfth the mass of a neutral atom of the most abundant carbon isotope, $^{12}_{6}C$. That is 1 amu = 1.66053×10^{-27} kg. Consequently, the mass of any atom is equal to the numerical value of its atomic weight. The atomic weight of any element is approximately equal to the total number of protons and neutrons in its nucleus.

Photons

A photon is a zero rest mass particle which serves as the carrier of a quantum or packet of electromagnetic energy. Photons have no charge and travel in a vacuum at the speed of light (c), namely, 2.9979×10^8m/s. The energy of a photon whose frequency is ν (Hertz) is given by eq. 1.6.

Mass and Energy Relationships

Einstein's theory of relativity allows us to describe the equivalence of mass and energy, in the well-known formula

$$E_0 = m_0 c^2 \tag{1.1}$$

where E_0 is the energy equivalent to the rest mass (m_0) and c is the speed of light. This deceptively simple equation describes the enormous amounts of energy liberated in nuclear reactions when quantities of matter are annihilated and appear in a variety of energy forms (e.g., kinetic, electromagnetic radiations). Consider, for example, the complete conversion of just one gram of matter into energy. This reaction would release some 9×10^{13} joules (J) of energy. In comparison, one kiloton (kT) of nuclear explosion yield corresponds to some 4.2×10^{12} joules or 2.61×10^{25} million electron volts (MeV) of energy [1, 2]. The electron volt (eV) is the kinetic energy gained by an electron when it is accelerated through an electric potential difference of one volt (V). The eV is a unit of energy frequently encountered in nuclear physics and engineering and is equal to 1.60219×10^{-19} joule. The atomic mass unit has an energy equivalence of 931.481 million electron volts. That is, 1 amu = 931.481 MeV. Before we continue our discussion of radioactivity, it is also important to recognize that if a particle is in motion, according to the theory of relativity, its mass will increase relative to an observer as follows:

$$m = \frac{m_0}{\sqrt{1 - \frac{v^2}{c^2}}} \tag{1.2}$$

where v is the particle speed. The total energy of the particle (E_T) is simply the sum of the rest mass energy (E_0) and its kinetic energy (E_{KE}). Using the mass in eq.

1.2, the expression for the total energy becomes

$$E_T = mc^2. \tag{1.3}$$

The particle's kinetic energy is given by

$$\begin{aligned} E_{KE} &= E_T - E_0 \\ &= mc^2 - m_0c^2 \\ &= \left[\frac{1-\sqrt{1-(v^2/c^2)}}{\sqrt{1-(v^2/c^2)}}\right] m_0c^2. \end{aligned} \tag{1.4}$$

If, however, $v \lesssim 0.2c$, then the particle's kinetic energy can be reasonably approximated by the formula from classical Newtonian mechanics, namely

$$E_{KE} = (1/2)m_0v^2. \tag{1.5}$$

Example 1.1. To illustrate several of these basic concepts, let us calculate the rest mass energy of the electron, $(E_0)_e$.

$$\begin{aligned} (E_0)_e &= (m_0)_e c^2 \\ &= (9.10956 \times 10^{-31}\ \text{kg})(2.9979 \times 10^8\ \text{m/s})^2 \\ &= 8.18713 \times 10^{-14}\ \text{joules} \\ &= 0.511\ \text{MeV} \end{aligned}$$

Zero Rest Mass Particles

It is very important to understand that the previously described equations for kinetic energy, eq. 1.4 and eq. 1.5, apply only to particles that have a nonzero rest mass. Zero rest mass particles travel at the speed of light and therefore are wavelike in nature. Due to this fact, their total energy is identical to that of a photon; namely,

$$E = h\nu \tag{1.6}$$

where h is Planck's constant (6.626196×10^{-34} joulesec) and ν is the frequency of the electromagnetic radiation. A more convenient form of this equation for total energy in electron volts is

$$E(\text{eV}) = (4.136 \times 10^{-15}\ \text{eV-sec})\nu(\text{sec}^{-1}). \tag{1.7}$$

The deBroglie equation which relates the momentum p of a particle of mass m moving at velocity v and its characteristic wavelength λ is given as

$$\lambda = \frac{h}{mv} = \frac{h}{p}. \tag{1.8}$$

For particles of nonzero rest mass, the momentum is given by the equation

$$p = mv. \tag{1.9}$$

For nonrelativistic velocities, the particle wavelength becomes

$$\lambda = \frac{h}{\sqrt{2m_0E}} \tag{1.10}$$

while for particles traveling at relativistic velocities, the wavelength is given by

$$\lambda = \frac{hc}{\sqrt{E_T^2 - E_0^2}}. \tag{1.11}$$

The momentum of a zero rest mass particle, such as a photon, can be expressed as

$$p = E/c \tag{1.12}$$

where E is the photon energy described by eq. 1.6. The deBroglie wavelength of the zero rest mass particle is then

$$\lambda = \frac{hc}{E}. \tag{1.13}$$

Using convenient units, we can express the wavelength in meters of a photon as

$$\lambda = \frac{1.24 \times 10^{-6}}{E(\text{eV})}. \tag{1.14}$$

For example, a one MeV gamma ray has a wavelength of 1.24×10^{-12} meters.

Radioactive Decay

A radioisotope is an unstable isotope of an element that decays, spontaneously emitting nuclear radiation(s). More than 1,300 natural and manmade radioisotopes have been identified.

The French scientist, Henri Becquerel, is credited with the discovery of natural radioactivity. While working with uranium salts and photographic plates in 1896, Becquerel discovered that the uranium emitted a penetrating radiation which seemed similar to the radiation that the German physicist, Wilhelm Konrad Roentgen, had produced a year earlier in experiments with a gas discharge tube. Actually, Roentgen had discovered X-rays in 1895—a discovery for which he was awarded the first Nobel Prize in Physics in 1901. Further investigation by Becquerel and by Pierre and Marie Curie showed that this radioactivity is a specific property of the element and that certain quantitative relationships existed. In particular, the atoms producing these radiations are unstable and decay at characteristic rates to form new atoms.

The actual existence of isotopes was discovered about 1913 after more than a decade of pioneering experimentation with naturally occurring radioactive materials. Isotopes are atoms of a given element that have the same number of protons in the nucleus but differing numbers of neutrons. For example, uranium-235 ($^{235}_{92}U$) and uranium-238 ($^{238}_{92}U$) are both naturally occurring radioisotopes of the element uranium. Because isotopes have the same number of protons and electrons, they are identical in chemical behavior. However, because they also have different numbers of neutrons, they have different atomic masses. As mentioned previously, certain isotopes, called radioisotopes, are unstable and

1.2, the expression for the total energy becomes

$$E_T = mc^2. \tag{1.3}$$

The particle's kinetic energy is given by

$$\begin{aligned} E_{KE} &= E_T - E_0 \\ &= mc^2 - m_0c^2 \\ &= \left[\frac{1 - \sqrt{1 - (v^2/c^2)}}{\sqrt{1 - (v^2/c^2)}} \right] m_0c^2. \end{aligned} \tag{1.4}$$

If, however, $v \lesssim 0.2c$, then the particle's kinetic energy can be reasonably approximated by the formula from classical Newtonian mechanics, namely

$$E_{KE} = (1/2)m_0v^2. \tag{1.5}$$

Example 1.1. To illustrate several of these basic concepts, let us calculate the rest mass energy of the electron, $(E_0)_e$.

$$\begin{aligned} (E_0)_e &= (m_0)_e c^2 \\ &= (9.10956 \times 10^{-31}\ \text{kg})(2.9979 \times 10^8\ \text{m/s})^2 \\ &= 8.18713 \times 10^{-14}\ \text{joules} \\ &= 0.511\ \text{MeV} \end{aligned}$$

Zero Rest Mass Particles

It is very important to understand that the previously described equations for kinetic energy, eq. 1.4 and eq. 1.5, apply only to particles that have a nonzero rest mass. Zero rest mass particles travel at the speed of light and therefore are wavelike in nature. Due to this fact, their total energy is identical to that of a photon; namely,

$$E = h\nu \tag{1.6}$$

where h is Planck's constant (6.626196×10^{-34} joulesec) and ν is the frequency of the electromagnetic radiation. A more convenient form of this equation for total energy in electron volts is

$$E(\text{eV}) = (4.136 \times 10^{-15}\ \text{eV-sec})\nu(\text{sec}^{-1}). \tag{1.7}$$

The deBroglie equation which relates the momentum p of a particle of mass m moving at velocity v and its characteristic wavelength λ is given as

$$\lambda = \frac{h}{mv} = \frac{h}{p}. \tag{1.8}$$

For particles of nonzero rest mass, the momentum is given by the equation

$$p = mv. \tag{1.9}$$

For nonrelativistic velocities, the particle wavelength becomes

$$\lambda = \frac{h}{\sqrt{2m_0E}} \tag{1.10}$$

while for particles traveling at relativistic velocities, the wavelength is given by

$$\lambda = \frac{hc}{\sqrt{E_T^2 - E_0^2}}. \tag{1.11}$$

The momentum of a zero rest mass particle, such as a photon, can be expressed as

$$p = E/c \tag{1.12}$$

where E is the photon energy described by eq. 1.6. The deBroglie wavelength of the zero rest mass particle is then

$$\lambda = \frac{hc}{E}. \tag{1.13}$$

Using convenient units, we can express the wavelength in meters of a photon as

$$\lambda = \frac{1.24 \times 10^{-6}}{E(\text{eV})}. \tag{1.14}$$

For example, a one MeV gamma ray has a wavelength of 1.24×10^{-12} meters.

Radioactive Decay

A radioisotope is an unstable isotope of an element that decays, spontaneously emitting nuclear radiation(s). More than 1,300 natural and manmade radioisotopes have been identified.

The French scientist, Henri Becquerel, is credited with the discovery of natural radioactivity. While working with uranium salts and photographic plates in 1896, Becquerel discovered that the uranium emitted a penetrating radiation which seemed similar to the radiation that the German physicist, Wilhelm Konrad Roentgen, had produced a year earlier in experiments with a gas discharge tube. Actually, Roentgen had discovered X-rays in 1895—a discovery for which he was awarded the first Nobel Prize in Physics in 1901. Further investigation by Becquerel and by Pierre and Marie Curie showed that this radioactivity is a specific property of the element and that certain quantitative relationships existed. In particular, the atoms producing these radiations are unstable and decay at characteristic rates to form new atoms.

The actual existence of isotopes was discovered about 1913 after more than a decade of pioneering experimentation with naturally occurring radioactive materials. Isotopes are atoms of a given element that have the same number of protons in the nucleus but differing numbers of neutrons. For example, uranium-235 ($^{235}_{92}U$) and uranium-238 ($^{238}_{92}U$) are both naturally occurring radioisotopes of the element uranium. Because isotopes have the same number of protons and electrons, they are identical in chemical behavior. However, because they also have different numbers of neutrons, they have different atomic masses. As mentioned previously, certain isotopes, called radioisotopes, are unstable and where $\bar{\nu}_e$ is the antineutrino. In this transformation, a neutron in the nucleus essentially changes into a proton and an electron. The daughter nucleus has experienced no change in atomic mass (A), but its atomic number (Z) has now increased by one. That is, in beta-minus decay

$$Z_{\text{daughter}} = Z_{\text{parent}} + 1$$

and

$$A_{\text{daughter}} = A_{\text{parent}}. \tag{1.20}$$

Consequently, there is a reduction in the neutron-to-proton ratio, and in general the daughter nuclide will tend to be more stable than the parent nuclide. If the daughter nuclide is not completely stable, another beta decay may occur, creating an isotope of still another element. A typical beta decay chain is shown in Fig. 1.5 [3].

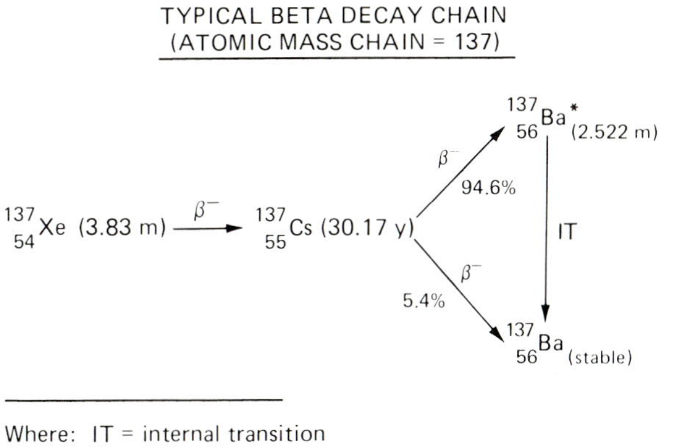

Fig. 1.5 Typical beta decay chain (atomic mass chain = 137). *Courtesy of Los Alamos National Laboratory.*

Beta decay is a so-called three body process. Since two different radiations are emitted from the nucleus, the energy given up in a single beta-minus transition must be divided between the beta particle and the antineutrino in a statistical manner. Consequently, when a large number of transitions between the same two energy levels in the parent nucleus and daughter nucleus is considered, the negative beta particles and antineutrinos have a continuous kinetic energy distribution from zero to a maximum value [$E_{\max}(\beta^-)$] called the endpoint energy. Considering the conservation of energy principle, this endpoint energy for a beta-minus transition is given by the equation

$$E_{\max}(\beta^-) = Q_\beta^- + E_p - E_d \tag{1.21}$$

where Q_β^- is the energy difference between the ground states of the parent nucleus and daughter nucleus

E_p is the excitation energy of the beta-emitting level in the parent nucleus

E_d is the excitation energy of the level in the daughter nucleus fed by beta-minus decay.

If a nucleus has too few neutrons, or too many protons, nuclear stability is achieved when a proton is converted into a neutron and a positive beta particle (β^+) or positron is emitted along with a neutrino. This reaction is simply

$${}^1_1p \rightarrow {}^1_0n + {}^{\;\,0}_{+1}\beta + \nu_e. \tag{1.22}$$

Here, the daughter nucleus experiences no change in atomic mass, but its atomic number now decreases by one. That is, in beta-plus decay

$$Z_{\text{daughter}} = Z_{\text{parent}} - 1$$

and

$$A_{\text{daughter}} = A_{\text{parent}}. \tag{1.23}$$

As in the case of beta-minus decay, the positrons are emitted during transitions between particular energy levels in the parent nucleus and daughter nucleus and have a continuous distribution of energies that can be characterized by an endpoint energy [$E_{\max}(\beta^+)$]. Again invoking the conservation of energy principle, this endpoint energy for a beta-plus transition is expressed as

$$E_{\max}(\beta^+) = Q_\beta^+ + E_p - E_d - 2m_0c^2 \tag{1.24}$$

where Q_β^+ is the energy difference between the ground states of the parent nucleus and daughter nucleus

E_p is the excitation energy of the beta emitting level in the parent nucleus

E_d is the excitation energy of the level in the daughter nucleus fed by beta-plus decay

m_0c^2 is the rest mass energy of the electron (0.511 MeV).

It should be noted that positron decay cannot occur unless the energy difference between the parent and daughter nuclear energy levels is greater than $2m_0c^2$ (or 1.022 MeV). That part of the total transition energy which "disappears" in the formation of two electron rest masses is normally transformed back to energy when the emitted positron experiences an annihilation reaction with an electron, producing two 0.511 MeV gamma rays.

A third beta decay process is called electron capture (EC) decay. In this reaction an orbiting atomic electron (e^-) is captured by the nucleus and combines with a proton to form a neutron. That is,

$${}^1_1p + {}^{\;\,0}_{-1}e \rightarrow {}^1_0n + \nu_e. \tag{1.25}$$

Here, the nucleus captures one of its own atomic electrons and emits a neutrino. This process competes with beta-plus decay whenever the transition energy is greater than $2m_0c^2$ (1.022 MeV) [3, 4]. As in beta-plus decay, EC decay decreases the atomic number of the parent nucleus by one, while the atomic mass remains the same. The capture by the nucleus of an atomic electron leaves the daughter atom with a vacancy in one of its atomic shells or atomic energy levels. The total energy available for electron capture decay is expressed as

$$E_{EC} = Q_{EC}^{+} + E_p - E_d - \Delta_X^E \tag{1.26}$$

where Q_{EC}^{+} is the energy difference between ground states of the parent nucleus and daughter nucleus
E_p is the excitation energy of the parent nucleus
E_d is the excitation energy of the level in the daughter nucleus fed by the decay process
Δ_X^E is the electron binding energy for shell X in the daughter atom.

All three beta decay processes (namely, β^-, β^+, and EC) appear in the decay chain shown in Fig. 1.6 [3].

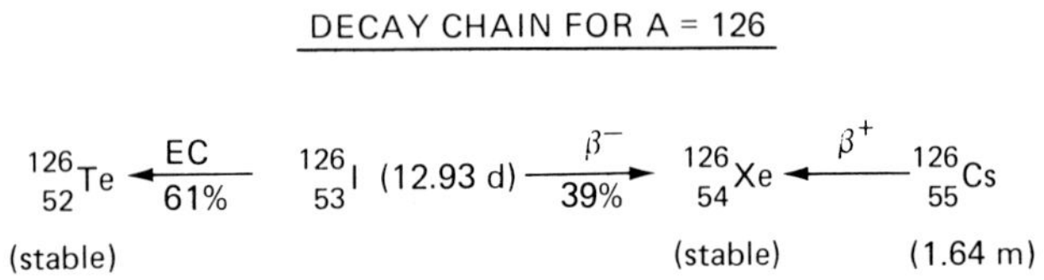

Fig. 1.6 Decay chain for $A = 126$. *Courtesy of Los Alamos National Laboratory.*

During alpha or beta decay processes, the majority of the excited states of a daughter nucleus decay very rapidly to lower energy states and eventually to the ground state by means of electromagnetic processes. This de-excitation of the daughter nucleus is accompanied by the emission of either gamma rays or internal conversion electrons. Long-lived isomeric states may also experience decay to lower energy states in the same nucleus through such electromagnetic transitions.

When a gamma ray is emitted by an excited daughter nucleus, the gamma ray energy is equal to the energy difference between the higher energy state and the lower energy state minus the energy of nuclear recoil (E_r). The energy of nuclear recoil is generally negligible, except for high energy transitions in light nuclei. It is given by the expression [3]

$$E_r \approx \frac{(5.4 \times 10^{-7})(E_\gamma)^2}{A} \text{ (keV)} \tag{1.27}$$

where E_r is the nuclear recoil energy (keV)
E_γ is the gamma ray energy (keV)
A is the atomic mass of the nucleus.

In the internal conversion process, the energy difference between the initial and final nuclear states is transferred directly to a bound atomic electron which is then ejected from the atom. The energy of an internal conversion electron (*ce*) emitted from an atomic shell X is given in terms of the corresponding gamma ray energy (E_γ) by the equation

$$E_{ce}^{x} = E_\gamma - \Delta_x^E \tag{1.28}$$

where Δ_x^E is the electron binding energy in shell X [3].

The electron capture and internal conversion processes always create a vacancy in an inner atomic electron shell. When an electron from an outer atomic electron shell fills this vacancy, an X-ray or an Auger electron is also emitted. The X-ray and Auger electrons are called atomic radiations. Vacancies that are produced when the initial vacancy is filled will, in turn, create other X-rays or Auger electrons. This cascade of atomic radiation continues until the only vacancies are those found in the outermost electron shell. A characteristic X-ray is simply a photon that is emitted when an electron from a higher atomic shell fills a lower shell vacancy. The energy of the emitted X-ray is equal to the energy difference between the two atomic electron shells. Auger electron emission competes with X-ray emission as a way in which energy is released when an outer shell electron fills an inner atomic shell vacancy. In the Auger electron process, the filling of an inner atomic shell vacancy is accompanied simultaneously by the ejection of an outer atomic shell electron. The residual atom is consequently left with two shell vacancies.

Law of Radioactive Decay

Radioactive decay is the decrease in the number of radionuclides present as a result of their spontaneous nuclear transmutation. It is an exponential process governed by the fundamental expression

$$dN/dt = -N\lambda \tag{1.29}$$

where N is the number of atoms of a particular radionuclide
λ is the decay constant (time^{-1})
t is time.

The decay constant is defined as the probability that a particular radionuclide will disintegrate in unit time. Equation 1.29 is the fundamental relationship describing radionuclide behavior and is called the law of radioactive decay. This relationship was first postulated by Rutherford and Soddy in 1902, as a result of their pioneering work on thorium and its daughter species [4]. In 1910, Bateman [5] created more general forms of the radioactive decay and growth equations, called the Bateman equations. These equations describe not only the decay processes of naturally occurring uranium, actinium, and thorium, but also the decay processes of the many fission product decay chains created in nuclear reactors. Today, modern computer-based computational techniques are used to study the production and decay of nuclides in complex decay schemes. An expanded decay chain diagram for plutonium-238 is presented in Fig. 1.7.

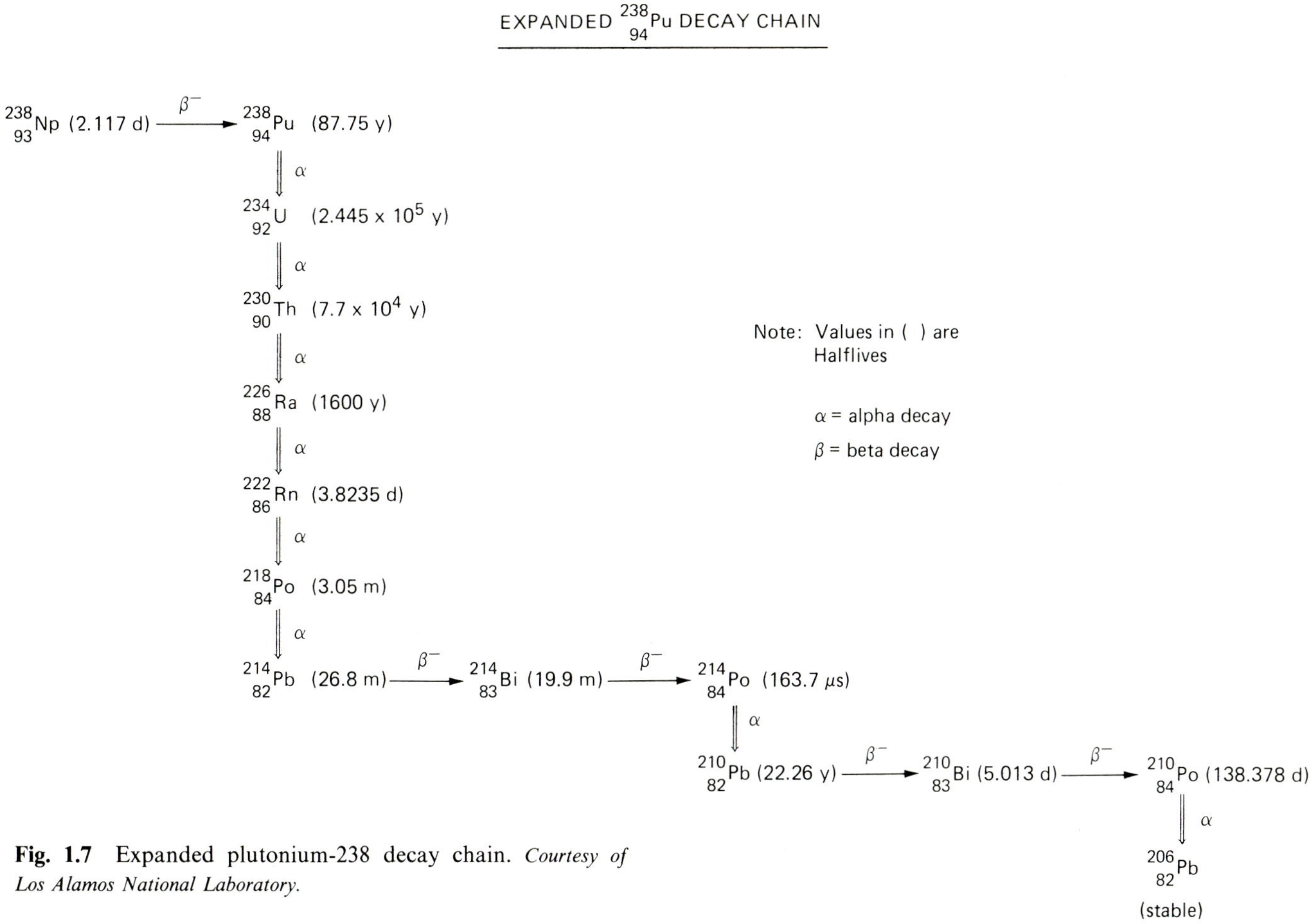

Fig. 1.7 Expanded plutonium-238 decay chain. *Courtesy of Los Alamos National Laboratory.*

Equation 1.29 describes the time rate of change of the radionuclide population. Here, the product $N\lambda$ is called the activity of the radioactive substance and corresponds to the total number of disintegrations per unit time. The negative sign in eq. 1.29 indicates that the activity level decreases with time. The curie (Ci) is a unit of activity commonly used to describe the intensity of radioactivity in a sample of material, and is defined as

$$1 \text{ Ci} = 3.7 \times 10^{10} \text{ disintegrations per second (dps)}. \quad (1.30)$$

The curie is approximately the rate of decay or activity of one gram of pure radium (indeed, this was the basis upon which the curie was originally defined—Ed.) and is named after Pierre and Marie Curie who discovered the element, radium, in 1898.

The SI (metric), or Systèms International d'Unites, unit of activity is called the becquerel (Bq). It is defined as exactly one disintegration per second which, in terms of the curie, can be expressed as

$$1 \text{ Bq} = 2.7027 \times 10^{-11} \text{ Ci}. \quad (1.31)$$

Both units are commonly found in nuclear engineering and physics and consequently both will appear in this text.

Equation 1.29 can be rewritten in integral form as

$$\int \frac{dN}{N} = -\int \lambda \, dt \quad (1.32)$$

using the separation of variables technique. Next we make two basic assumptions. First, we assume that the decay constant λ is indeed constant and independent of the age of the radionuclide. Second, we assume that λ is independent of the number of radionuclides present. With these assumptions eq. 1.32 can be integrated to yield

$$\ln\left(\frac{N}{N_0}\right) = -\lambda t. \quad (1.33)$$

The above important result is more commonly expressed in its exponential form as

$$N = N_0 e^{-\lambda t} \quad (1.34)$$

where N_0 is the number of atoms of a particular radionuclide present at time $t = 0$.

This universal law of radioactive decay applies to all radionuclides. The term, $e^{-\lambda t}$, can be interpreted simply as the probability that a particular nuclide will not decay in time t. The product, $N_0 e^{-\lambda t}$ represents the total number of radionuclides N that have survived from $t = 0$ to time t.

The half-life ($T_{1/2}$) is a convenient way of representing the rate of decay of a particular radionuclide and is defined as the time in which half the number of radionuclides of a particular isotope disintegrate to another nuclear form. The half-life is related to the decay constant as follows

$$T_{1/2} = (\ln 2)/\lambda \approx 0.693/\lambda. \qquad (1.35)$$

This result is obtained from eq. 1.33 by setting $N = \frac{1}{2} N_0$ and solving for $t = T_{1/2}$. The universal law of radioactive decay is extremely useful for calculating the temporal behavior of typical radioisotope fuels used in space nuclear power systems. The total thermal power level as a function of time, $\dot{Q}(t)$, for a particular (simple) decay chain is given by

$$\dot{Q}(t) = \dot{Q}_0 e^{-\lambda t} \qquad (1.36)$$

where $\dot{Q}_0$ is the initial thermal power level (in watts) at the start of the mission.

A simple decay chain is one in which the decay of the daughter nuclides and beyond may be ignored as a first approximation due to their very long half-lives. For example, the daughter nuclide of plutonium-238 is uranium-234 which has a half-life of 244,500 years. For all practical purposes in calculating the rate of decay of $^{238}_{94}Pu$ (the parent nuclide), we can treat its very long-lived daughter ($^{234}_{92}U$) as effectively being stable—at least with respect to its parent. Thus, eq. 1.34 or 1.35 is a valid representation of the decay dynamics of a plutonium-238 fueled radiosotope system.

Example 1.2. Consider a radioisotope system fueled by plutonium-238. Given a fuel capsule containing one kilogram of plutonium dioxide (PuO_2) with a specific thermal power at the beginning of the mission of 0.4 watt/gram for the fuel compound. What is the total thermal power output of the capsule after ten years of space flight?

Step 1. For $^{238}_{94}Pu$, the half-life is 87.75 years. We first compute the decay constant (λ).

$$\lambda = \frac{\ln 2}{T_{1/2}} = \frac{0.69315}{(87.75\text{ y})} = 7.9 \times 10^{-3}/\text{year}$$

Step 2. We next calculate the thermal energy generation rate in watts at the start of the mission.

$$\dot{Q}_0 = (1000\text{g})(0.4\text{ W/g})$$

$$\dot{Q}_0 = 400\text{ W}_{\text{thermal}}$$

Step 3. Then, to find $\dot{Q}(t)$ after ten years of space flight we use eq. 1.36 as follows:

$$\dot{Q}(t) = \dot{Q}_0 e^{-\lambda t}$$

$$\dot{Q}(10\text{y}) = (400\ W_{\text{th}}) e^{-(7.9\times 10^{-3}\text{y}^{-1})(10\text{y})}$$

$$\dot{Q}(10\text{y}) = (400)\ (0.924)$$

$$\dot{Q}(10\text{y}) = 369.6\ W_{\text{th}}$$

This example illustrates the value of using plutonium-238 as the radioisotope fuel in certain space nuclear power systems. After ten years of continuous operation, the radioisotope fuel is still supplying over 92 percent of its original thermal power level to the energy conversion systems.

Decay Chains

Daughter nuclides are the nuclides formed by the radioactive disintegration of the parent radionuclide (or first nuclide) in a particular decay chain. Consider now a very common situation in which a parent nuclide (W) decays into a daughter nuclide (X) which is also radioactive. The decay series can be represented as shown in

$$\text{W} \xrightarrow{\lambda_W} \text{X} \xrightarrow{\lambda_X} \text{Y (stable)}$$

(parent nuclide) (daughter nuclide) (granddaughter nuclide)

Fig. 1.8 Parent-daughter-granddaughter decay scheme. *Courtesy of Los Alamos National Laboratory.*

Fig. 1.8, where λ_W is the decay constant for the parent nuclide; and λ_X is the decay constant for the daughter nuclide. The general differential equation describing the growth and decay of the daughter nuclide X is

$$\frac{dX}{dt} = \lambda_W W - \lambda_X X \qquad (1.37)$$

where dX/dt is the time rate of change of the number of daughter nuclides X.

$\lambda_W W$ is the number of daughter nuclides being created per unit time

$\lambda_X X$ is the number of daughter nuclides decaying per unit time.

Some simplifying assumptions can now be applied to eq. 1.37. First, we assume that the only source of nuclide W is the initial supply, W_0, at time $t = 0$. Second, we also assume that at time $t = 0$, there are no daughter nuclides (i.e., $x = 0$, initially). With these assumptions, eq. 1.37 has the solution [3]

$$X(t) = W_0 \frac{\lambda_W}{\lambda_X - \lambda_W} \left[e^{-\lambda_W t} - e^{-\lambda_X t} \right]. \qquad (1.38)$$

This equation is the general solution to the growth and decay of a daughter nuclide, subject to the initial conditions $W = W_0$ and $X = 0$ at $t = 0$. If $W = W_0$ and $X = X_0$ at $t = 0$ (that is, there are also daughter nuclides initially present) then eq. 1.37 has the solution

$$X(t) = X_0 e^{-\lambda_X t} + W_0 \frac{\lambda_W}{\lambda_X - \lambda_W} \left[e^{-\lambda_W t} - e^{-\lambda_X t} \right]. \qquad (1.39)$$

We now consider a very short decay chain (i.e., parent to daughter nuclide series as shown in Fig. 1.9).

$$\text{W} \xrightarrow{\lambda_W} \text{X (stable)}$$

(parent nuclide) (daughter nuclide)

Fig. 1.9 Parent-daughter decay scheme. *Courtesy of Los Alamos National Laboratory.*

Table 1.1 Radiological decay data for plutonium-238 [*3, 6*].

half-life ($T_{1/2}$): 87.75 years

decay constant (λ): $7.899 \times 10^{-3}\ y^{-1}$

typical heat source mixture:

Pu-isotope	Weight-(%)
238	90.0
239	9.1
240	0.6
241	0.03
242	<0.01

major radiations:

Type	energy (MeV)	Yield (%)
α_1	5.36	0.1
α_2	5.46	28.3
α_3	5.50	71.6

alpha activity: 3.9×10^7 α/min-μg

spontaneous fission rate (metal): 3420 neut/s-g

Neutron production
[(α, n) reaction]

PuO_2	1.4×10^4	neut/s-g
PuF_4	2.1×10^6	neut/s-g

For this decay scheme, the accumulation of stable nuclide X is given by

$$X = W_0(1 - e^{-\lambda_W t}) \tag{1.40}$$

where W_0 is the initial population of the parent nuclide.

Helium Gas Buildup

When plutonium-238 decays not only is a daughter nuclide, uranium-234, formed but an energetic, ionized helium atom is also released. One area of concern in the use of alpha-emitting radioisotopes in space power systems is helium gas generation and subsequent pressure buildup in a closed fuel capsule. Equation 1.40 may be rewritten to express the helium gas generation associated with the decay of plutonium-238. Thus,

$$N_\alpha(t) = N_0[1 - e^{-\lambda t}] \tag{1.41}$$

where $N_\alpha(t)$ is the number of helium atoms in the fuel capsule at any time t

N_0 is the original number of plutonium-238 atoms in the fuel capsule at $t = 0$ (beginning of mission).

λ is the decay constant for plutonium-238 ($7.899 \times 10^{-3} y^{-1}$).

To complete our discussion of radioactivity, Table 1.1 provides some very useful nuclear data for plutonium-238 [3, 6].

References

1. John R. Lamarsh, *Introduction to Nuclear Engineering* (2nd Edition), Addison-Wesley Publishing Co., 1983.
2. E. Teller et al., *The Constructive Use of Nuclear Explosives*, McGraw-Hill Book Co., New York, 1968.
3. D. C. Kocher, *Radioactive Decay Data Tables*, U.S. Department of Energy, DOE/TIC-11026, 1981.
4. R. D. Evans, *The Atomic Nucleus*, McGraw-Hill Book Co., New York, 1955.
5. H. Bateman, Cambridge Philosophical Society Proceedings, **15**, pp. 423–427, 1910.
6. Health, Safety and Environmental Manual, "Plutonium," Technical Bulletin 505, Los Alamos Scientific Laboratory Manual, LA-7940-M, October 1979.

Chapter 2

Radiation Interaction With Matter

The design and operation of radioisotope power supplies, nuclear reactors, and radiation shields are basically dependent on the way that nuclear radiation interacts with matter. In this chapter we explore some of the fundamental principles that describe the passage of alpha particles, beta particles, gamma rays, and neutrons through different media.

Ionization, Excitation, and Absorption

An ion is an electrically charged particle of either sign. Ionization is the process of forming ions—that is, electrons are added or taken away from previously neutral atoms, causing the total electric charge on the atom to be numerically different than the positive charge within its nucleus. In general, the passage of radiation through matter produces ions by imparting energy to orbital electrons and by stripping them from their parent atom with a resultant loss of energy by the incident radiation. Some atoms have loosely bound orbital electrons that are easily removed by radiation interaction. Usually, the energy necessary to initiate an ionizing event will decrease as the atomic number (Z) of the medium increases. When ionized, an atom can lose more than one electron. If two electrons are lost, the atom is said to be doubly ionized.

Different forms of radiation have varying abilities to cause ionization. The specific ionization (SI) is defined as the number of ion pairs created per unit path length.

$$\text{Specific Ionization (SI)} = \frac{\text{Number Ion Pairs}}{\text{Unit Path Length}} \text{(ip/cm)} \qquad (2.1)$$

It is a measure of the ability to cause ionization in a particular medium. Specific ionization is a function of the material, type of radiation, and energy of the radiation.

In many instances, the energy imparted by the radiation to the orbital electron is insufficient to ionize the atom. However, it will cause the electron to jump to a higher energy orbit in a process called excitation. An electron in a higher energy orbit falls back into a lower energy orbital vacancy by emitting high energy photons (i.e., X-rays). This process is called deexcitation.

The final basic interaction between radiation and matter is absorption. Absorption can be defined as the transformation of radiation energy into some other form, such as internal energy (heat) of the medium.

Directly and Indirectly Ionizing Radiations

Directly ionizing radiations include all charged particles such as alpha particles, beta particles, and heavy ions. These undergo energy loss by interaction with orbital electrons (excitation and ionization) or atomic nuclei. Indirectly ionizing radiations consist of neutrons and electromagnetic radiations (X-rays and gamma rays). When these radiations interact with matter, they create secondary radiation that is ionizing. Neutrons, X-rays, and gamma rays lose energy by collisions with atomic nuclei or electrons. The charged particles liberated in such collisions then travel through the medium and interact with other atomic nuclei or orbital electrons [2, 3].

Quantities Used to Describe Radiation Interactions

In describing the transport of radiation through matter, we must consider the interaction that various particles or photons may have with the nuclei of the absorbing medium. The probability that a particular interaction occurs varies with the type of particle or electromagnetic radiation, its energy, and the medium through which it travels. This interaction probability is usually defined in terms of a microscopic cross section σ or macroscopic cross section Σ. The microscopic cross section can be considered as the effective area that a nucleus (or atom) presents to an incident particle, regardless of the particle's direction. Simply stated, a particle will interact with a nucleus with a probability of one, when the particle is incident upon this effective area; otherwise, the interaction probability is zero [4]. By convention, the microscopic cross section has units of cm^2/atom or barns/atom, where

$$\text{one barn (b)} \equiv 10^{-24}\ \text{cm}^2. \qquad (2.2)$$

According to nuclear physics folklore, the term "barn" originated when a scientist made the humorous statement that 10^{-24}cm^2 looks as big as a barn to a neutron.

Because of the wavelike behavior of particles and the very complex nature of the atomic nucleus, the magnitude of the microscopic cross section generally bears little relation to the actual physical size of the nucleus; rather σ is really a function of the properties of the atom and the energy and type of incident particle (or photon) [4].

As shown in Fig. 2.1a, if a nucleus has a cross section σ and is contained within some characteristic volume V_c of material which has a projected area A_c, then the probability that a particle incident on the volume will interact with that particular nucleus is the ratio of the effective area σ to the projected area A_c. We can thus consider that the ratio σ/A_c represents the reaction probability per nucleus associated with the area A_c and interpret it as the average number of reactions that occur in V_c per incident particle [4].

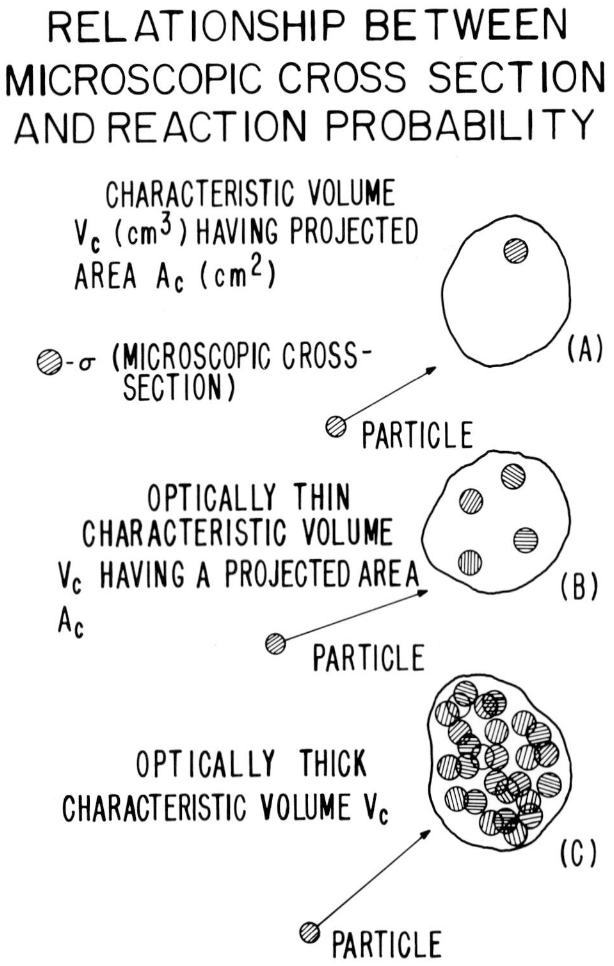

Fig. 2.1 Relationship between microscopic cross section and reaction probability. *Courtesy of Oak Ridge National Laboratory [4].*

If the nuclide density, N (nuclides/cm^3), is assumed uniform throughout the characteristic volume, then the total number of nuclei contained within V_c is given by the product NV_c. The reaction probability associated with the total number of nuclides is equal to the linear sum of the reaction probabilities corresponding to the individual nuclei—assuming that the characteristic volume is "optically thin" (i.e., there is no self-shielding). The reaction probability for V_c is thus

$$\text{Reaction Probability} = NV_c(\sigma/A_c). \qquad (2.3)$$

The summation of interaction probabilities for an optically thin medium is depicted in Fig. 2.1b. The optically thin assumption holds if the characteristic volume is differential (very small) in size. The projected area A_c will also be differential in size in this assumption. In the "optically thick" situation, shown in Fig. 2.1c, the effective cross section per nucleus is actually decreased because of the self-shielding effect.

By regrouping the parameters in eq. 2.3, we can rewrite the reaction probability for an optically thin characteristic volume as

$$\text{Reaction Probability} = \sigma N(V_c/A_c). \qquad (2.4)$$

But the term V_c/A_c is simply the characteristic differential distance dS_c traveled by particles within V_c. If we define the macroscopic cross section for neutrons as Σ (μ for gamma rays), then

$$\Sigma \equiv N\sigma. \qquad (2.5)$$

Thus the product $\Sigma \cdot dS_c$ is the reaction probability associated with the differential track length dS_c. It follows that Σ represents the number of reactions per particle track length (typical units are cm^{-1}).

There is an alternate interpretation of Σ, the macroscopic cross section. In penetrating a distance dx, a particle will experience on the average $\Sigma\, dx$ interactions per unit of time. Consider a beam of particles incident normal to an infinite slab of some material. We define $\Phi(x)$ as the uncollided flux at thickness x. The number of particles being removed in differential distance dx is

$$-d\Phi(x) = \Phi(x)\Sigma\, dx. \qquad (2.6)$$

Integrating eq. 2.6 yields

$$\Phi(x) = \Phi_0 \cdot e^{-\Sigma x} \qquad (2.7)$$

where Φ_0 is the incident flux (particles/cm^2-s)
x is the distance into the medium (cm).

The probability that a particle experiences an interaction in a distance less than L is

$$P(x \leqslant L) = \int_0^L \frac{\Sigma\Phi(x)\,dx}{\Phi_0}$$
$$= \int_0^L \Sigma e^{-\Sigma x}\,dx$$
$$= 1 - e^{-\Sigma L} \tag{2.8}$$

where $e^{-\Sigma x}$ is the probability density function for an interaction.

The mean distance a particle travels before it undergoes an interaction is called the mean free path and is given by

$$\lambda \equiv \int_0^{\infty} x\Sigma e^{-\Sigma x}\,dx = 1/\Sigma. \tag{2.9}$$

The macroscopic cross section is a function of the properties of the nuclide (properties of nucleus only in case of neutrons) and the type and energy of the incident particle or electromagnetic radiation. The microscopic cross section σ in terms of cross-sectional area per nucleus seen by the incident particle is then defined as

$$\sigma \equiv \Sigma/N \tag{2.10}$$

where N is the nuclear density (particles/cm^3)
σ is the cross-sectional area (cm^2 or barns) for a particular interaction per nucleus.

The reaction rate per unit volume at the space point ($\bar{r}$) is

$$R(\bar{r}) = \Sigma\Phi(\bar{r}) \tag{2.11}$$

where $R(\bar{r})$ is the number of reactions per cm^3 per second
Σ is the macroscopic cross section (cm^{-1}) or the reactions per particle per cm
$\Phi(\bar{r})$ is the particle flux (particles/cm^2-s); also the total track length traced by particles per unit volume and time at the space point $\bar{r}$.

The differential reaction rate is expressed as [4]

$$R(\bar{r}, E) = \Sigma(E) \cdot \Phi(\bar{r}, E) \tag{2.12}$$

where $R(\bar{r}, E)dE$ represents the number of reactions at space point $\bar{r}$ per unit volume and time that involve particles with differential energy dE about the energy E
$\Sigma(E)$ is the macroscopic cross section for the reaction (It depends on the incident particle's energy as well as the nuclear characteristics of the medium)
$\Phi(\bar{r}, E)$ is the differential particle flux density (particles/cm^2-s-MeV).

The reaction rate for an energy interval or group ($\Delta E_k = E_k - E_{k+1}$) is related to the differential reaction rate by the expression

$$R_k(\bar{r}) = \int_{E_{k+1}}^{E_k} R(\bar{r}, E)\,dE$$
$$= \int_{E_{k+1}}^{E_k} \Sigma(E) \cdot \Phi(\bar{r}, E)\,dE. \tag{2.13}$$

This expression leads to a definition of the reaction rate in terms of an energy-averaged macroscopic cross section Σ_k (group cross section) and the group flux $\Phi_k(\bar{r})$

$$R_k(\bar{r}) = \Sigma_k \Phi_k(\bar{r}) \tag{2.14}$$

where the group cross section is defined as

$$\Sigma_k \equiv \frac{\int_{E_{k+1}}^{E_k} \Sigma(E)\Phi(\bar{r}, E)\,dE}{\int_{E_{k+1}}^{E_k} \Phi(\bar{r}, E)\,dE} \tag{2.15}$$

and the group flux is

$$\Phi_k(\bar{r}) \equiv \int_{E_{k+1}}^{E_k} \Phi(\bar{r}, E)\,dE. \tag{2.16}$$

The reaction rate concept, briefly introduced here, is essential in the formulation of the various equations used to describe radiation transport (e.g., the Boltzmann transport equation). Reaction rates are also used in the determination of energy deposition and biological dose from known or calculated values of particle flux density [4].

Alpha Particle Interactions

Alpha particles are ejected from radionuclides with speeds on the order of one-twentieth that of the speed of light. They typically have energies ranging from 4 to 9 MeV. Alpha particles lose energy by ionization and excitation as they pass through matter. There are two basic alpha particle-induced ionization processes: (1) direct collision with an orbital electron; and (2) an electrostatic field interaction between the alpha particle and the orbital electrons. Excitation occurs, usually at the end of an alpha particle's range (described shortly) when the energy transferred to the orbital electron is insufficient to remove it from the atom.

Let W be defined as the average energy needed to create an ion pair in a given substance. This value varies for alpha particles in different gases, ranging from $W_\alpha = 22$ eV per ion pair (ip) in xenon to $W_\alpha = 46$ eV/ip in pure helium; but the value of W is relatively constant in a given gas for different alpha particle energies. For air, $W_\alpha = 35$ eV/ip.

Because of its double positive charge and large mass, the alpha particle creates a rather large number of ion pairs per unit path length. The actual value of alpha particle specific ionization (SI_α) is a function of particle velocity and the atomic number and mean ionization potential of the absorbing medium. The specific ionization for alpha particles in air goes from approximately 10,000 ip/cm to 70,000 ip/cm [1]. If the ionization produced by the alpha particle is plotted as a function of penetration into the absorbing medium, we find that the specific ionization gradually increases as the alpha particle loses energy (see Fig. 2.2) [1, 2]. A plot of this type is called a Bragg curve. The specific ionization reaches a peak value and then falls to zero as the alpha particle loses all of its energy, acquires first one, then two electrons, and becomes a neutral helium atom.

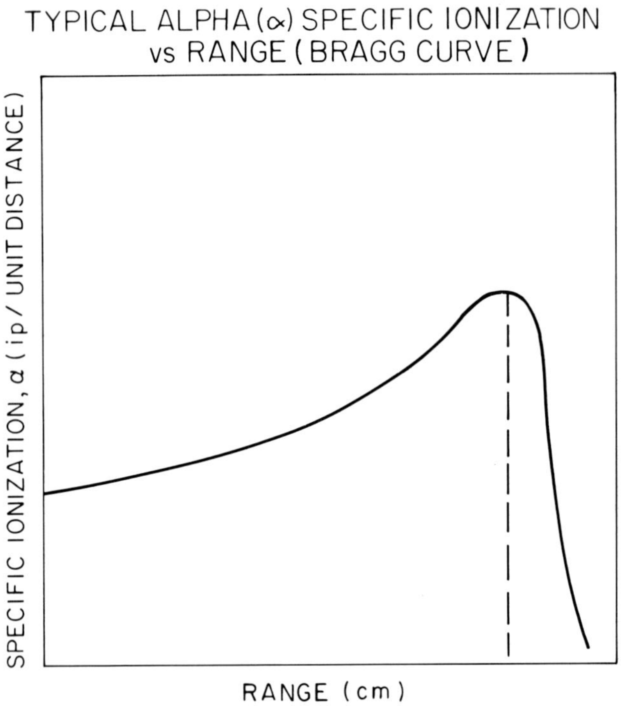

Fig. 2.2 Typical alpha specific ionization versus range (Bragg Curve).

The specific ionization well within the range of the alpha particle is given by

$$SI_\alpha = \frac{dE/dx}{W_\alpha} \tag{2.17}$$

where dE/dx is the linear stopping power of the medium which is often expressed in units of MeV/cm. The linear stopping power describes energy transfer as ionization and excitation along the path of a charged particle. However, the mathematical expression for linear stopping power collapses for very low energy alpha particles, and we cannot express a value for SI_α near the end of an alpha particle path.

The average specific ionization is given by

$$\overline{SI}_\alpha = \frac{\text{\# of ion pairs formed}}{\text{length of path}} = \frac{E_\alpha}{W_\alpha \cdot R}\ (\text{ip/cm}) \tag{2.18}$$

where E_α is the alpha particle energy (eV)
R is the range of the alpha particle in the substance (cm).

The linear stopping power for heavy particles of charge $(Z \cdot e)$ is given as [1]

$$dE/dx = \frac{4\pi e^4 Z^2}{m_0 v^2} NB \tag{2.19}$$

where N is the atom density (number of atoms/cm^3) for the medium
m_0 is the rest mass of the particle
v is the particle velocity
Z is the atomic number (or number of charges) associated with the particular radiation
e is the electron charge
B is the atomic stopping power.

The atomic stopping power is a function of the particle velocity, the atomic number, and the mean ionization potential (I) of the medium. For a particle whose velocity v is very much less than the speed of light, the nonrelativistic expression for the atomic stopping power is [1]

$$B = Z \ln\left[\frac{2m_0 v^2}{I}\right]. \tag{2.20}$$

The mass stopping power is defined as the linear stopping power S divided by the density ρ of the medium. That is,

$$\text{Mass Stopping Power} \equiv (S/\rho) = (1/\rho)\, dE/dx\ (\text{MeV/g-cm}^2) \tag{2.21}$$

In radiation protection, the ratio of two stopping powers, or the relative stopping power, is a useful quantity. For alpha radiation, air is frequently taken as the reference medium, and other absorbing media are compared to it. The relative mass stopping power $(S/\rho)_m$ of a medium m with respect to air is given by

$$(S/\rho)_m \equiv \frac{(S/\rho)\text{ of medium}}{(S/\rho)\text{ of air}} = \frac{(B/\bar{A})_{\text{medium}}}{(B/\bar{A})_{\text{air}}} \tag{2.22}$$

where B is the atomic stopping power
$\bar{A}$ is the effective atomic weight.

Because of their large specific ionization, alpha particles use up their energy in very short distances. In fact, most alphas from a given source lose their energy in about the same distance or range (see Fig. 2.3). The range (in cm) of alpha particles in air at 15°C and one atmosphere pressure can be expressed as a function of their energies (in MeV) as [1]

$$R_{\text{air}}(\text{cm}) = 0.56\, E \qquad (\text{for } E < 4 \text{ MeV})$$
$$= 0.318\, E^{3/2} \qquad (\text{for } 4 < E < 7 \text{ MeV}). \tag{2.23}$$

For other media, the alpha particle range can be approximated using the Bragg–Kleeman rule [1, 2]

$$R_m = \frac{\rho_{\text{air}} R_{\text{air}}}{\rho_m (S/\rho)_m} \tag{2.24}$$

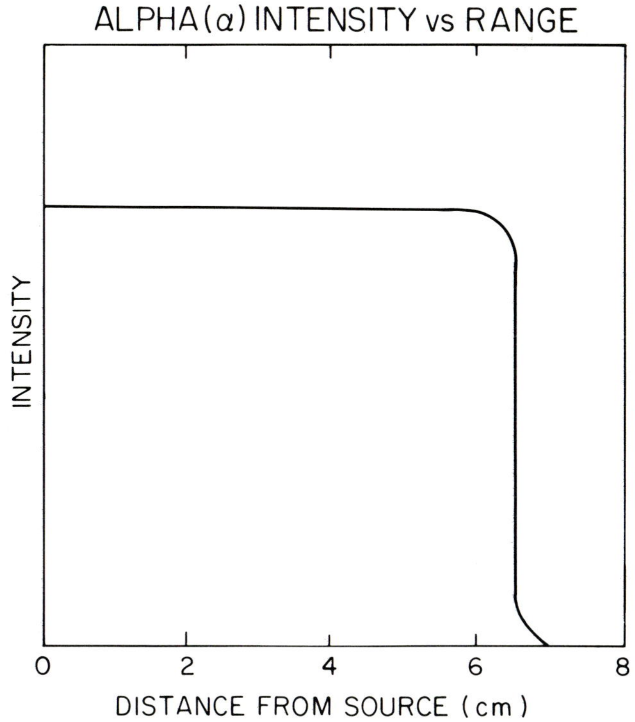

Fig. 2.3 A typical plot of alpha intensity versus range.

where $(S/\rho)_m$ is the relative mass stopping power of the medium with respect to air. For a number of substances

$$B_m / B_{air} \approx \sqrt{\bar{A}_m} \,/\, \sqrt{\bar{A}_{air}}\,. \qquad (2.25)$$

Using the Bragg–Kleeman rule, we find the range in any medium (R_m) is given by

$$R_m(\text{cm}) = \frac{3.2 \times 10^{-4} \sqrt{\bar{A}_m}}{\rho_m} R_{air} \qquad (2.26)$$

where $\bar{A}_m$ is the average atomic weight of the medium
R_{air} is the alpha particle range in air (cm)
ρ_m is the density of the medium (g/cm^3).

The following values for air at 15° C (288 K) and one atmosphere pressure (101.3 kPa) have been used: $\bar{A}_{air} = 14.6$ and $\rho_{air} = 1.226 \times 10^{-3} g/cm^3$. The range relationship expressed in eq. 2.26 is usually valid within ± 15 percent. For a compound or mixture, the quantity $\sqrt{\bar{A}_m}$ can be found from

$$\sqrt{\bar{A}_m} = \frac{n_1 A_1 + n_2 A_2 + n_3 A_3 + \ldots n_i A_i}{n_1\sqrt{A_1} + n_2\sqrt{A_2} + n_3\sqrt{A_3} + \ldots n_i\sqrt{A_i}} \qquad (2.27)$$

where n_i is the atomic fraction of the element i of atomic weight A_i.

Because there is such a large difference in density between a solid medium and a gas, the range of alpha particles in solids is very small.

Example 2.1. It is interesting here to determine the range of uranium-238 alpha particles ($E_\alpha = 4.195$ MeV) in uranium dioxide (UO_2) fuel for which the density is given as 10.9 g/cm^3.

Step 1. Starting with eq. 2.27,

$$\sqrt{\bar{A}_m} = \frac{n_1 A_1 + n_2 A_2}{n_1\sqrt{A_1} + n_2\sqrt{A_2}}$$

$$\sqrt{\bar{A}_m} = \frac{(1/3)(238) + (2/3)(16)}{(1/3)\sqrt{238} + (2/3)\sqrt{16}} = 11.526$$

Step 2. Using eq. 2.23,

$$R_{air} = 0.318\, E^{3/2} = 0.318\,(4.195)^{3/2} = 2.732 \text{ cm}$$

Step 3. Using eq. 2.26, we obtain the range of alpha particles in UO_2

$$R_m = \frac{3.2 \times 10^{-4} \sqrt{\bar{A}_m}}{\rho_m} R_{air}$$

$$R_m = \frac{3.2 \times 10^{-4}(11.526)}{(10.9)}(2.732) = 9 \times 10^{-4} \text{ cm}.$$

Since alpha particles have short ranges in most substances, they are much less of a hazard to human beings than other external radiations. In fact, alpha particles with energies up to 7.5 MeV cannot even penetrate the protective layer of skin found on most parts of the human body! However, once inside the body, the extremely limited range of the alpha particle becomes very significant. In this case, the alpha source is surrounded by living tissue. This means that radiation induced cell damage will be highly localized near the point of origin of the alphas. Thus, if an alpha source is lodged in them, great biological damage can be done to small, essential body organs because all of the alpha particle energy will be absorbed in that particular organ and not spread out over a larger volume of tissue. For this reason, alpha particles represent a very real concern as an internal radiation hazard.

Beta Particle Interactions

Unlike alpha particles, beta particles are not emitted with discrete energies, but display a continuous energy spectrum (see Fig. 2.4). Radionuclides emit beta particles at all energies, up to some maximum value E_{max}. This maximum beta particle energy value is characteristic of a particular radionuclide and can be found in appropriate nuclear data tables [5, 6]. The average energy of a beta source is about (1/3) E_{max} in the case of beta-minus (β^-) particles and 0.4 E_{max} for positrons (β^+). Beta particles lose energy in a number of ways as they pass through matter. Loss by collisions with orbital electrons, leading to ionization and excitation, is the most frequent mechanism. But for energetic beta particles, interactions with the nucleus also occur and lead

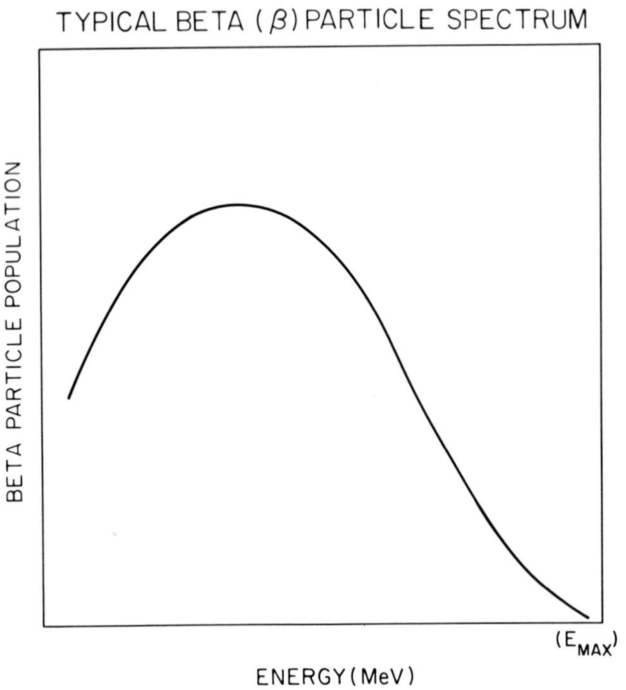

Fig. 2.4 Typical beta particle spectrum.

to the emission of X-rays in a process called Bremsstrahlung—a German word meaning braking radiation.

In a given substance the beta particle will undergo a large number of collisions, until it eventually loses all its kinetic energy. The processes leading to ionization in the substance are the same as those for alpha particles. However, because of its smaller mass and lower electric charge (one half that of the alpha particle), these interactions take place at less frequent intervals. In fact, for a given energy, beta particles have a much greater range than alphas. Consequently, beta particles do not produce as many ions per unit path length as do alpha particles.

The W value for beta particles in air is 34 eV per ion pair. The specific ionization values for beta particles are much lower than those for alpha particles. As an approximation, the beta particle specific ionization (SI_β) varies inversely as the square of the particle's speed for energies up to 10 MeV. In air the specific ionization values range from 60 to 7000 ip/cm.

Well within the range of a beta particle, the specific ionization is defined as

$$SI_\beta = (dE/dx)/W_\beta \tag{2.28}$$

where, as before, dE/dx is the linear stopping power of the substance (eV/cm).

The average specific ionization for beta particles ($\overline{SI}_\beta$) is defined, as it was for alpha particles in eq. 2.18, as

$$\overline{SI}_\beta = \frac{\#\text{ of ion pairs}}{\text{path length}} = \frac{E_\beta}{W_\beta R}\,(\text{ip/cm}) \tag{2.29}$$

where E_β is the energy of the beta particle (eV) and R is the range in the substance (cm). The specific ionization of betas as a function of their energies is approximately given by

$$SI_\beta \approx 33 + 63\,E_{\max}^{-0.9} \quad \text{when} \quad 0.05 < E_{\max} < 2\text{ MeV}. \tag{2.30}$$

The linear stopping power due to ionization and excitation can be expressed as [1]

$$\frac{dE}{dx} = \frac{2\pi e^4 NZ}{m_0 v^2} B^* \tag{2.31}$$

where N is the atom density (number of atoms/cm^3)
B^* is the electron (or beta) stopping power
Z is the atomic number of the medium
e is the charge on the electron (beta particle)
m_0 is the mass of the beta particle (g)
v is the velocity of the beta particle (cm/s).

The electron stopping power is a function of the speed of the electron and the mean ionization potential of the substance. This expression gives the rate of energy loss along the actual path travelled. Since beta particles are greatly scattered in passing through a given substance, the actual path traversed is often much greater than the straight-line path. (Typically the actual path is about 1.2 to 4 times the straight-line path.)

In addition to ionization and excitation, some of the beta particle's energy may appear as X-rays that are produced as a result of interactions with the nuclei of the absorbing substance. The resulting continuous X-ray spectrum is associated with Bremsstrahlung or braking radiation. Processes leading to the production of Bremsstrahlung are inefficient for electron energies below 1 MeV; however, as the electron energy increases above 1 MeV, the production of X-rays becomes more significant.

When a charged particle is either accelerated or decelerated in an electric field, electromagnetic radiation may be given off. If an electron passes close to an atomic nucleus while travelling through a substance, the charge on the nucleus will exert a force on the electron and this will cause a deflection in the particle's path. During this acceleration process, the electron may radiate energy in any amount from zero up to its total kinetic energy $E_{KE}(\beta)$. The total Bremsstrahlung per atom is roughly proportional to $(Z/m)^2$, where Z is the atomic number of the absorbing material and m is the mass of the incident charged particle. Because of this $1/m^2$ dependence, the amount of Bremsstrahlung is almost completely negligible for all particles except electrons.

For a beta source of maximum beta energy $E_{\max}(\beta)$,

expressed in MeV, the fraction of beta particle energy converted to X-rays is approximately

$$F_x(\beta) \approx 3.33 \times 10^{-4} Z E_{max}(\beta). \qquad (2.32)$$

For example, a one MeV beta particle will lose about 3 percent of its energy as Bremsstrahlung when absorbed in lead. The spectral distribution will contain X-rays of energy from zero up to $E_{max}(\beta)$. In most cases, we can assume that the X-rays that are emitted correspond to the maximum energy of the beta particle for the purposes of shielding design. This will normally result in a safety factor in radiation protection calculations.

Another useful relationship is the following which describes the relative magnitudes of beta energy loss via Bremsstrahlung to total beta energy loss [3]

$$\frac{\text{Bremsstrahlung Energy Loss}}{\text{Total Energy Loss}} = \frac{Z \cdot \bar{E}}{(Z \cdot \bar{E}) + 800} \qquad (2.33)$$

where $\bar{E}$ is the average beta particle energy ($\bar{E} \approx E_{max}(\beta)/3$). In designing a shield for pure beta emitters, it is important to consider the possibility that Bremsstrahlung could be produced within the shield material itself.

The stopping power for electrons decreases as the atomic number Z of the absorber increases. This occurs because substances of high Z have fewer electrons per unit mass and these electrons are more tightly bound to the parent atom. But as Z increases, the multiple scattering of electrons also increases. The effect of multiple scattering is to increase the actual path of the electron in a substance. These two effects then tend to balance each other, so that the density of a material is a good indicator of its relative ability to stop electrons. It is common practice to express the range of electrons or beta particles in terms of a density-thickness. That is, $t(\text{cm}) \cdot \rho\ (\text{g/cm}^3) = \text{range}\ (\text{g/cm}^2)$. The range is then nearly independent of the type of the absorbing substance. The relationship between the maximum range and the maximum energy of the electron has been expressed as [2, 7]

$$R_{max}(\text{mg/cm}^2) = 412\ E_{max}^{[1.265 - 0.0954 \ln E_{max}]}$$
$$(\text{for } 0.01 < E_{max} < 2.5 \text{ MeV}), \qquad (2.34)$$

and

$$R_{max}(\text{mg/cm}^2) = 530\ E_{max} - 106$$
$$(\text{for } E_{max} < 2.5 \text{ MeV}). \qquad (2.35)$$

These expressions can be used to find the range of beta particles in any substance in terms of the density-thickness (mg/cm^2 or g/cm^2).

Example 2.2. What is the range of a 1.5 MeV beta particle in mg/cm^2 in any substance? Starting with eq. 2.34,

$$R\ (\text{mg/cm}^2) = 412\ E^{[1.265 - 0.0954 \ln E]}$$

$$R = 412(1.5)^{[1.265 - 0.0954 \ln(1.5)]}$$

$$R \approx 677 \text{ mg/cm}^2$$

Beta particle absorption is an exponential process described by the equation

$$I = I_0 e^{-\mu t} \qquad (2.36)$$

where I_0 is the initial intensity (particles/cm$^2 \cdot$ s)
I is the intensity after passing through a thickness (t) of absorber
t is the thickness of the absorber (cm)
μ is the absorption coefficient (cm^{-1}).

Experiments have shown that the mass absorption coefficient μ/ρ, where ρ is the density of the substance in g/cm^3, is almost independent of the atomic weight of the absorber. We can express the mass absorption coefficient as [2]

$$\frac{\mu}{\rho} = 17(E_{max})^{-1.14}(\text{cm}^2/\text{g}) \qquad (2.37)$$

for most substances, where E_{max} is given in MeV. We can rewrite eq. 2.36 in terms of the mass absorption coefficient as follows

$$I = I_0 e^{-(\mu/\rho)\psi} \qquad (2.38)$$

where ψ is expressed in g/cm^2 (i.e., $\psi = t \cdot \rho$). Fig. 2.5 displays a typical beta intensity versus range curve.

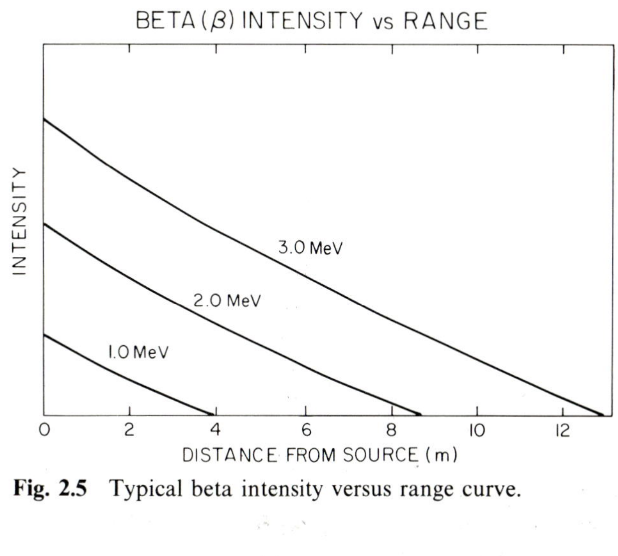

Fig. 2.5 Typical beta intensity versus range curve.

A beta source may pose an external radiation hazard, depending upon the energy of the particles. For example, a 70 keV beta is needed to penetrate the body's protective layer of skin and do damage to living tissue. Normally, beta radiation is considered only a slight external hazard. A beta source can be fully shielded by a small thickness of aluminum. However, an external radiation hazard may also result from the production of

X-rays in the shield material itself which is used to stop the energetic beta particles. The amount of this Bremsstrahlung can be reduced by choosing a shielding substance of low atomic number Z.

The beta particle is not as great an internal hazard as the alpha particle. Although the beta particle has a much greater range in tissue, its specific ionization is less. Thus, it is not as effective as the alpha in causing damage in a small volume of tissue.

Gamma Ray Interaction With Matter

Traditionally in nuclear physics high energy electromagnetic radiation has been classified according to its origin, independent of its actual energy level. As discussed in Chapter 1, X-rays are considered atomic in origin, coming from the atomic transitions of bound electrons and from Bremsstrahlung processes. Gamma rays arise from processes in the atomic nucleus, such as the fission reaction, fission product decay, neutron inelastic scattering, and neutron radiative capture [4]. The first two gamma ray production processes are discussed in Chapter 3, while the latter two will be discussed shortly in this chapter. The interaction mechanisms of these electromagnetic radiations are independent of their origin and dependent only on their actual energy [3, 4]. As a convention in this book, we will treat high energy X-ray and gamma ray interactions with matter as gamma ray interactions.

The interaction of gamma rays in materials (e.g., shields) of interest in nuclear engineering is completely dominated by the following three processes: the photoelectric effect, Compton scattering, and pair production. All three processes involve electrons that then ionize or excite atoms of the substance.

Photoelectric Effect

The photoelectric effect is characterized by the complete absorption of the incident gamma ray, which interacts directly with a bound orbital electron. Momentum is conserved by the recoil of the entire atom, thereby permitting a total loss of energy by the gamma ray. As a result, the more tightly bound electrons have the greatest probability of absorbing the incident gamma ray. Approximately 80 percent of the photoelectric absorptions involve K-shell electrons [4]. In this process, the orbital electron (called a photoelectron) is ejected from the atom with a kinetic energy given by

$$(E_{\mathrm{KE}})_e = E_\gamma - \phi_e \tag{2.39}$$

where E_γ is the energy of the incident gamma ray (eV)
and ϕ_e is the binding energy (or work function) of the bound electron (eV).

The photoelectron then causes ionization and excitation in the absorbing medium. The remainder of the energy, equal to ϕ_e, appears as very soft characteristic X-rays that are usually referred to as fluorescence radiation. This fluorescence radiation is significant only when the photoelectric interaction occurs in heavy metals. Even then, the range of these emitted X-rays is comparable to the range of the photoelectrons, and their contribution to the overall penetrating radiation dose is so small that they can be assumed to be absorbed at their point of origin [4].

Table 2.1 Gamma ray energies at which photoelectric effect contributes one-half of the total absorption coefficient.

Atomic No., Z	Element	Gamma-Ray Energy (MeV)	K-Shell Ionization Energy (MeV)
1	Hydrogen (H)	10^{-4}	1.4×10^{-5}
4	Beryllium (Be)	0.011	2.2×10^{-4}
6	Carbon (C)	0.016	2.8×10^{-4}
8	Oxygen (O)	0.025	5.2×10^{-4}
13	Aluminum (Al)	0.046	1.5×10^{-3}
20	Calcium (Ca)	0.079	4.0×10^{-3}
26	Iron (Fe)	0.11	6.9×10^{-3}
42	Molybdenum (Mo)	0.195	2.0×10^{-2}
50	Tin (Sn)	0.25	2.9×10^{-2}
74	Tungsten (W)	0.42	6.06×10^{-2}
82	Lead (Pb)	0.50	8.8×10^{-2}
92	Uranium (U)	0.62	1.16×10^{-1}

The photoelectric effect is important when the energy of the gamma ray is low, that is, less than 1 MeV. However, the incident photon energy must be greater than ϕ_e for the process to occur at all. For small values of E_γ, the photons will interact with electrons in the outer shells of the atom. As E_γ increases, more of the inner electrons are released. The photoelectric effect is also more likely to occur in materials with a high atomic number. For elements of interest in shielding, Table 2.1 presents the gamma ray energies at which the photoelectric effect provides one-half the total absorption coefficient. The ionization energy of the K-shell is also given [4, 8]. The qualitative dependence of the probability of the photoelectric effect occurring is given by the approximation

$$\sigma_{\mathrm{pe}} \propto Z^4 / E_\gamma^3 \, . \tag{2.40}$$

Compton Effect

The Compton effect, or Compton scattering process, is named after the American physicist, Arthur Holly Compton (1892–1962) who discovered this phenomenon in 1923 [9]. It involves an elastic collision between a gamma ray and a free (unbound) electron. However, as the energy of the incident gamma ray increases, even the tightly bound electron begins to behave as a free electron. In Compton scattering, kinetic energy and momentum are conserved. The result is that a photon of lesser energy (i.e., longer wavelength) is scattered at an angle θ with respect to the initial direction of the gamma ray (see Fig. 2.6).

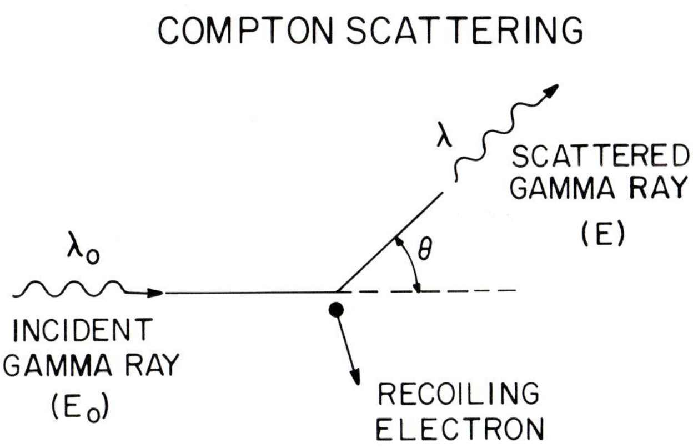

Fig. 2.6 Compton scattering.

The functional relationship between the loss of gamma ray energy and its change in direction is given by

$$E = \frac{E_0}{1 + E_0(1 - \cos\theta)/m_0c^2} \tag{2.41}$$

where E_0 is the initial gamma ray energy (MeV)
E is the final gamma ray energy (MeV)
θ is the change in the angle experienced by the gamma ray
m_0c^2 is the rest mass energy of the electron (0.511 MeV).

Fig. 2.7 presents a plot of eq. 2.41, the Compton equation [4]. We define the Compton wavelength as [1, 2, 4]

$$\lambda_c = \frac{h}{m_0c^2} = 2.426 \times 10^{-10} \text{ cm}. \tag{2.42}$$

The Compton wavelength is the wavelength of a photon whose energy is equal to the rest mass energy of the electron ($m_0c^2 = 0.511$ MeV). The Compton equation can then be rewritten as

$$\lambda - \lambda_0 = \lambda_c(1 - \cos\theta) \tag{2.43}$$

where λ is the photon wavelength after Compton scattering (cm)
λ_0 is the initial gamma ray wavelength (cm).

Since the gamma ray interacts with individual electrons in Compton scattering, we can define a differential cross section $\sigma_T(\theta)$ for this process as the cross section per electron for scattering through a given angle θ per unit solid angle Ω. Expressed in Thomson units, this differential cross section $\sigma_T(\theta)$ is given by the Klein–Nishina formula [1, 4, 10] (neglecting polarization effects)

$$\sigma_T(\theta) = \frac{d\sigma}{d\Omega} = \left(\frac{3}{16\pi}\right)\frac{E^2}{E_0^2}\left[\frac{E_0}{E} + \frac{E}{E_0} - \sin^2\theta\right]$$

(Thomson Units) (2.44)

where

$$\text{one Thomson unit} \equiv \left(\frac{8\pi}{3}\right)\left(\frac{e^2}{m_0c^2}\right)^2 = 0.665 \text{ barn.} \tag{2.45}$$

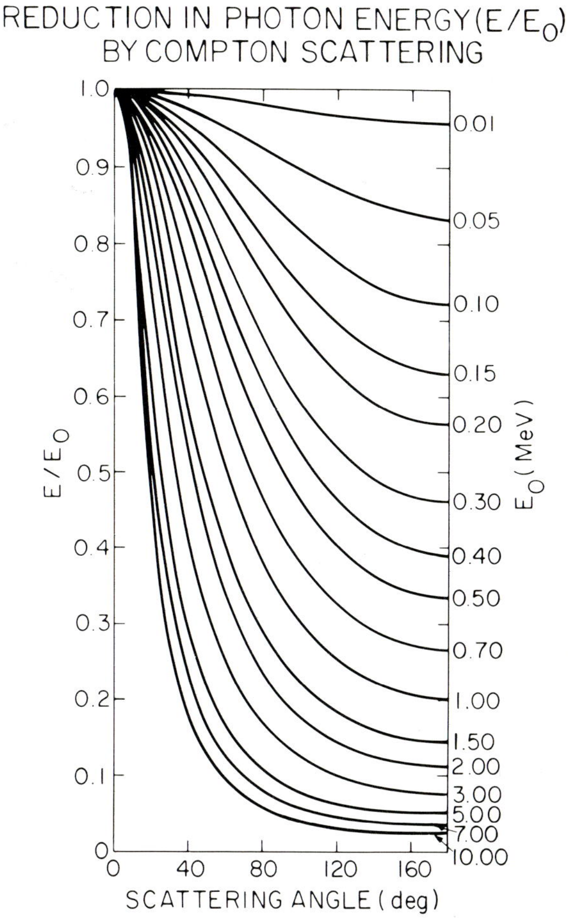

Fig. 2.7 Reduction in photon energy (E/E_0) by Compton scattering. *Courtesy of Oak Ridge National Laboratory* [4].

Fig. 2.8 depicts the dependence of the differential cross section ($d\sigma/d\Omega$) on the scattering angle θ for various initial gamma ray energies E_0 [4]. It should be noted that $\sigma_T(\theta)$ is only symmetric for the nonrelativistic limit $E_0 \to 0$. As the initial gamma ray energy increases, the differential cross section becomes more and more sharply peaked in the forward direction. Consider an initial gamma ray energy, $E_0 = 2.5$ MeV. For this case, $\sigma_T(\theta)$ decreases to one half its initial value when $\theta \approx 20°$ and then falls to approximately 3 percent of its initial value for $\theta = 180°$. This characteristically forward scattering of the high energy gamma rays strongly influences the nature of the deep penetration problem. Because the gamma ray does not disappear in Compton scattering, given a wide beam of gamma radiation passing through a thick shield, some of the photons initially scattered out of the beam may later be scattered back into the beam. This leads to a greater dose of radiation reaching a point beyond the shield than would normally be expected. For this reason a correction or buildup factor is used in shielding calculations (see Chapter 7).

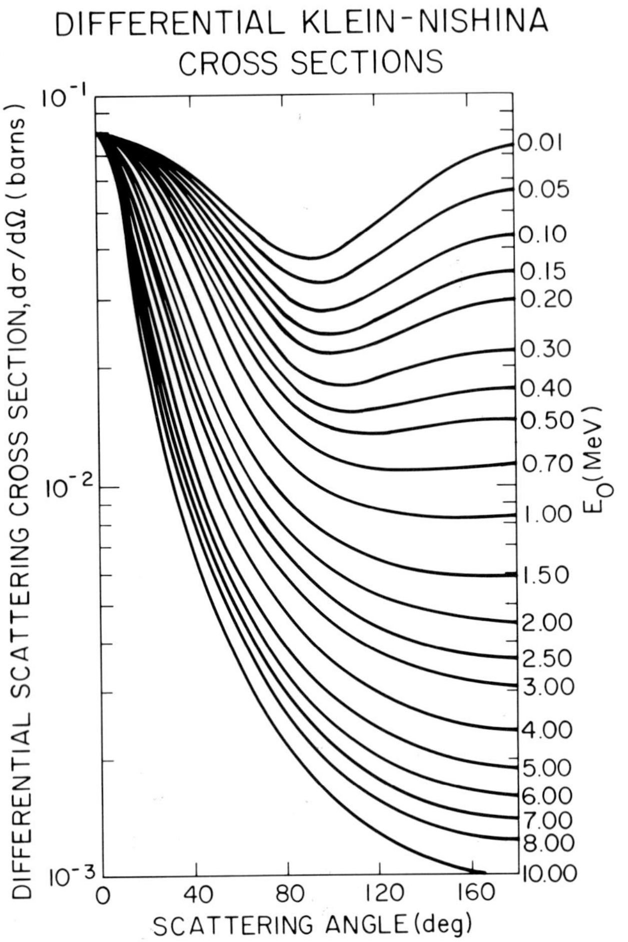

Fig. 2.8 Differential Klein–Nishina cross sections. *Courtesy of Oak Ridge National Laboratory* [4].

The total cross section for Compton scattering per electron can be obtained by integrating the differential cross section over 4π solid angle [1,4]

$$\sigma_c(E) = \int_{4\pi} \sigma_T(\theta)\, d\bar{\Omega} \tag{2.46}$$

where

$$d\bar{\Omega} = 2\pi \sin\theta\, d\theta. \tag{2.47}$$

Substituting eq. 2.44 into eq. 2.46 and integrating, we obtain the following expression for the total cross section [4]

$$\sigma_C(E) = \frac{3}{4}\left\{ \frac{(1+E)}{E^3}\left[\frac{2E(1+E)}{(1+2E)} - \ln(1+2E)\right] + \frac{1}{2E}\ln(1+2E) - \frac{(1+3E)}{(1+2E)^2} \right\} \tag{2.48}$$

where the units of gamma ray energy E are given in terms of the electron rest mass energy ($m_0c^2 = 0.511$ MeV). For small gamma ray energies (that is, $E \ll m_0c^2$) the value of $\sigma_C(E)$ approaches one Thomson unit and for large gamma ray energies it approaches $(3E/8)\cdot(\ln 2E + 1/2)$.

The cross section related to the average fractional energy carried off by the scattered gamma ray is defined as

$$\sigma_{C_s}(E) \equiv \int_{4\pi} \frac{E}{E_0}\sigma(\theta)\, d\bar{\Omega} \tag{2.49}$$

which yields on integration

$$\sigma_{C_s}(E) = \frac{3}{8}\left[\frac{\ln(1+2E)}{E^3} + \frac{2(1+E)(2E^2-2E-1)}{E^2(1+2E)^2} + \frac{8E^2}{3(1+2E)^3}\right]. \tag{2.50}$$

The other cross section of significance relates to the average fractional energy lost by a gamma ray suffering a Compton collision. The fraction increases monotonically as the gamma ray energy increases. This cross section represents the energy absorption or energy deposition cross section and is defined by [4]

$$\sigma_{C_a}(E) = \int_{4\pi} \frac{(E-E')}{E}\sigma(\theta)\, d\Omega \tag{2.51}$$

where E and E' represent the photon energy before and after interaction, respectively (MeV). Eq. 2.51 can also be written as

$$\sigma_{C_a}(E) = \sigma_C(E) - \sigma_{C_s}(E). \tag{2.52}$$

As previously mentioned, the various Compton cross sections are usually expressed on a per electron basis, since the Compton effect occurs with free electrons. To obtain a Compton cross section per atom $(\sigma_C)_{atom}$, we simply multiply σ_C per electron by the atomic number Z. That is,

$$(\sigma_C)_{atom} = Z \cdot \sigma_c\,. \tag{2.53}$$

Pair Production

The process in which a gamma ray is completely absorbed and an electron-positron pair appears is called pair production. It must occur within the electrostatic (Coulomb) field of a charged particle and requires an incident gamma ray energy of at least 1.02 MeV, since the total rest mass energy of two electrons is $2m_0c^2 =$ 1.02 MeV [2,4]. Momentum is conserved by the recoil of the charged particle, thereby making the complete absorption of the gamma ray energy possible. Either a nucleus or an atomic electron can provide the necessary electrostatic field. However, since it has a greater electric charge, most pair production reactions for gamma ray energies of interest occur in the electrostatic field of the nucleus.

A simple energy balance indicates that the sum of the kinetic energies of the electron-positron pair ($KE_{\beta+}$, $KE_{\beta-}$) is equal to the incident gamma ray energy (E_γ) less the rest mass equivalent of the electron pair ($2m_0c^2$) and the kinetic energy of the recoil nucleus (KE_{recoil}). That is,

$$KE_{\beta+} + KE_{\beta-} = E_\gamma - 2m_0c^2 - KE_{recoil}. \quad (2.54)$$

Because of its relatively large mass, the kinetic energy acquired by the recoil nucleus in pair production can frequently be neglected in eq. 2.54. If, on the other hand, the pair production process occurs in the Coulomb field of an atomic electron, a considerable quantity of energy is transferred to the recoil electron and a triplet is formed. This triplet is composed of the recoil electron and the newly created electron-positron pair. Because of momentum considerations, the gamma ray threshold energy for triplet production is 2.04 MeV [4].

Every positron created in pair production gives up its kinetic energy by ionization and then ultimately interacts with an ordinary electron, converting the electron and itself into electromagnetic radiation called annihilation radiation. In this process, two 0.511 MeV gamma rays are emitted in opposite directions. These secondary gamma rays are very important for high energy ($E_\gamma > 6$ MeV) gamma ray sources and medium-to-high atomic number materials [4].

Except for the effects of screening, the nuclear pair production cross sections $(\sigma_{pp})_n$ are theoretically and experimentally proportional to Z^2 [1,2,4]. Screening occurs when the positive charge of the nucleus is effectively reduced by the surrounding cloud of negatively charged orbital electrons. If screening is neglected and $m_0c^2 \ll E_\gamma \ll 137(m_0c^2)Z^{-1/3}$, then the pair production cross section σ_{pp} is given by [4]

$$\sigma_{pp} = \sigma_0 Z^2\left[\frac{28}{9}\ln\frac{2E_\gamma}{m_0c^2} - \frac{218}{27}\right] \quad (2.55)$$

where

$$\sigma_0 = \frac{1}{137}\left(\frac{e^2}{m_0c^2}\right)^2 = 5.80 \times 10^{-28}\ \text{cm}^2/\text{nucleus} \quad (2.56)$$

and m_0c^2 is the rest mass equivalent of the electron (0.511 MeV)
E_γ is the incident gamma ray energy (MeV)
Z is the atomic number of the target nucleus.

If complete screening is assumed and $E_\gamma \gg 137(m_0c^2)Z^{-1/3}$, then the pair production cross section is given by

$$\sigma_{pp} = \sigma_0 Z^2\left[\frac{28}{9}\ln(183Z^{-1/3}) - \frac{2}{27}\right] \quad (2.57)$$

Table 2.2 presents the gamma ray energies at which pair production provides one-half the total absorption coefficient for elements of interest in shielding calculations [4, 8].

Table 2.2 Gamma ray energies at which pair production provides one-half of the total absorption coefficient.

Atomic No., Z	Element	Gamma-Ray Energy (MeV)
1	Hydrogen (H)	78
4	Beryllium (Be)	35
6	Carbon (C)	28
8	Oxygen (O)	20
13	Aluminum (Al)	15
20	Calcium (Ca)	12
26	Iron (Fe)	9.5
42	Molybdenum (Mo)	7.5
50	Tin (Sn)	6.5
74	Tungsten (W)	5.2
82	Lead (Pb)	5.0
92	Uranium (U)	4.8

Pair production by interaction with an atomic electron's Coulomb field can occur when $E_\gamma > 2.04$ MeV. The ratio of the total cross section for Z electrons of a given atom to the nuclear pair production cross section $(\sigma_{pp})_n$ can be expressed as [4]

$$\frac{Z \cdot (\sigma_{pp})_e}{(\sigma_{pp})_n} = \frac{1}{C \cdot Z} \quad (2.58)$$

where C is a constant that depends on E_γ. If $E_\gamma = 6.5$ MeV, $C \approx 2.6$; while if $E_\gamma = 100$ MeV, $C \approx 1.2$.

Finally, the total pair production cross section σ_{pp} can be written as

$$\sigma_{pp} = (\sigma_{pp})_n + Z \cdot (\sigma_{pp})_e = (\sigma_{pp})_n\left(1 + \frac{1}{CZ}\right) \quad (2.59)$$

where $1/CZ = 0$, if $E < 2.04$ MeV.

Total Gamma Ray Cross Sections

Three processes—the photoelectric effect, the Compton effect, and pair production—account for the main gamma ray interactions with matter. As shown in Fig. 2.9, both the photoelectric and Compton effects decrease with increasing gamma energy, while pair production increases with increasing gamma energy.

The total microscopic gamma ray interaction cross

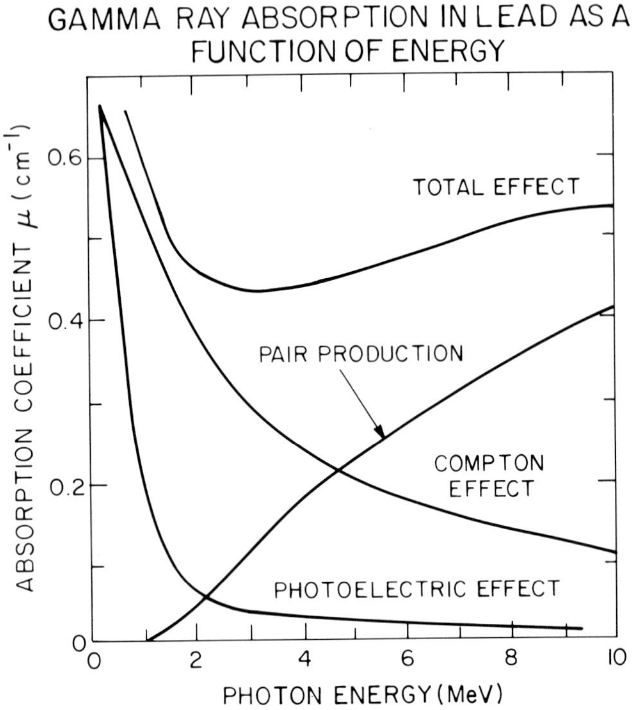

Fig. 2.9 Gamma ray absorption in lead (Pb) as a function of energy.

section for typical shielding calculations is expressed as [2, 4]

$$\sigma_{tot} = \sigma_{pe} + \sigma_{pp} + Z \cdot \sigma_c \tag{2.60}$$

where σ_{tot} is the total gamma ray interaction cross section per atom (cm^2/atom)

σ_{pe} is the photoelectric cross section per atom (cm^2/atom)

σ_{pp} is the pair production cross section per atom (cm^2/atom)

σ_c is the total Compton cross section per electron (cm^2/electron)

Z is the atomic number (electrons/atom).

The linear attenuation coefficient μ, typical units cm^{-1}, provides a macroscopic measure of the total gamma ray interaction cross section. It is defined as [2, 4]

$$\mu \equiv \sigma_{tot} \frac{\rho \cdot N_{AV}}{A} \tag{2.61}$$

where ρ is the density (g/cm^3)

N_{AV} is Avogadro's number (6.022×10^{23} atoms/g-atom)

A is the gram atomic weight (g/g-atom).

The mass attenuation coefficient is designated as μ/ρ and has the common units, cm^2/g [2, 4].

The gamma ray energy deposition coefficient can be defined as

$$\mu_a = \frac{\rho N_{AV}}{A} \{\sigma_{pe} + \sigma_{pp} + Z \cdot \sigma_{C_a}\} \tag{2.62}$$

where σ_{C_a} is the Compton energy deposition cross section per electron (see eq. 2.51 and eq. 2.52). Equation 2.62 assumes that the photoelectric and pair production reactions are pure absorptions.

Gamma Ray Absorptions

Consider now a narrow, monoenergetic beam of photons sent through a thin substance (see Fig. 2.10). Gamma ray interactions may occur in any of the three ways previously discussed. Energy will be removed from this beam by Compton scattering and by absorption (i.e., photoelectric effect and pair production). Since we have assumed a narrow beam and thin geometry, a Compton scattering process will be the equivalent to an absorption, i.e., the scattered photon will be removed from the beam. For this narrow beam of monoenergetic photons (or thin geometry situation), gamma ray attenuation is described by

$$I = I_0 e^{-\mu x} \tag{2.63}$$

where I_0 is the incident gamma ray intensity (photons/cm^2 − s)

I is the intensity after passing through a thickness x (photons/cm^2 − s)

x is the thickness of the material (cm)

μ is the total linear attenuation coefficient (cm^{-1}).

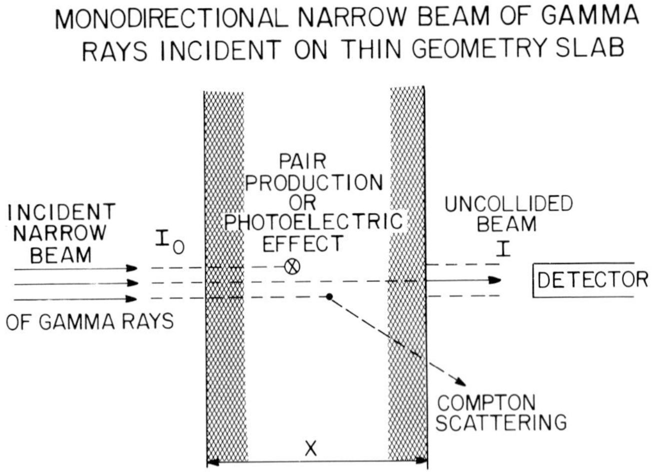

Fig. 2.10 Monodirectional narrow beam of gamma rays incident on thin geometry slab.

In the case of a compound shield or attenuating medium, the total linear attenuation coefficient for the compound can be found from

$$\mu = \sum_{i=1}^{n} (\mu/\rho)_i \cdot \rho_i \tag{2.64}$$

where $(\mu/\rho)_i$ is the mass attenuation coefficient of the ith element of the compound (cm^2/g) and ρ_i is the density of the ith element (g/cm^3) as it appears in the mixture.

ρ_i is given by the product of the weight fraction ψ_i of the ith element and the density of the compound ρ_{mix}.

$$\rho_i = (\psi)_i \cdot \rho_{mix} \qquad (2.65)$$

The weight fraction of an element ψ_i in a compound is found from

$$\psi_i = \frac{k_i A_i}{A_{mix}} \qquad (2.66)$$

where k_i is the number of ith atoms of atomic mass A_i
A_{mix} is the atomic or molecular weight of the mixture.

Example 2.3. Find the linear attenuation coefficient for one MeV gamma rays in water. The mass attenuation coefficients for hydrogen (H) and oxygen (O) are 0.126 and 0.0635 cm^2/g, respectively. Starting with eq. 2.64,

$$\mu = \sum_{i=1}^{2} (\mu/\rho)_i \cdot \rho_i = (\mu/\rho)_H \cdot \rho_H + (\mu/\rho)_O \cdot \rho_O$$

$$= 0.126 \frac{k_H A_H}{A_{H_2O}} \rho_{H_2O} + 0.0635 \frac{k_O A_O}{A_{H_2O}} \rho_{H_2O}$$

$$= 0.126 \frac{(2)(1)}{(18)} (1) + 0.0635 \frac{(1)(16)}{(18)} (1)$$

$$= 0.0704 \text{ cm}^{-1}.$$

Special Terms

A quantity of usefulness in shielding estimations is the half-value layer $X_{1/2}$. This quantity is the thickness of a substance that reduces the intensity of a beam of radiation to one half its initial value. The half-value layer is found from

$$X_{1/2} = (\ln 2)/\mu \qquad (2.67)$$

The mean free path λ is the average distance that a photon travels before interacting. The mean free path is equal to the reciprocal of the linear attenuation coefficient. That is,

$$\lambda = 1/\mu. \qquad (2.68)$$

If the thickness of an absorber is equal to one mean free path, then the intensity of radiation will be reduced by a factor of e.

The buildup factor (here for slab geometry) can be defined as the ratio of actual gamma ray intensity to that predicted by narrow-beam (thin geometry) calculations [1, 2]

$$B(\mu, x) = \frac{\text{true intensity}}{\text{“thin geometry” intensity}} = \frac{I}{I_0 e^{-\mu x}} \qquad (2.69)$$

or

$$\phi = \phi_0 B(\mu, x) e^{-\mu x}.$$

It corrects for an underestimation of scattered radiation reaching a point beyond an absorbing medium or shield (see Fig. 2.11 and Chapter 7).

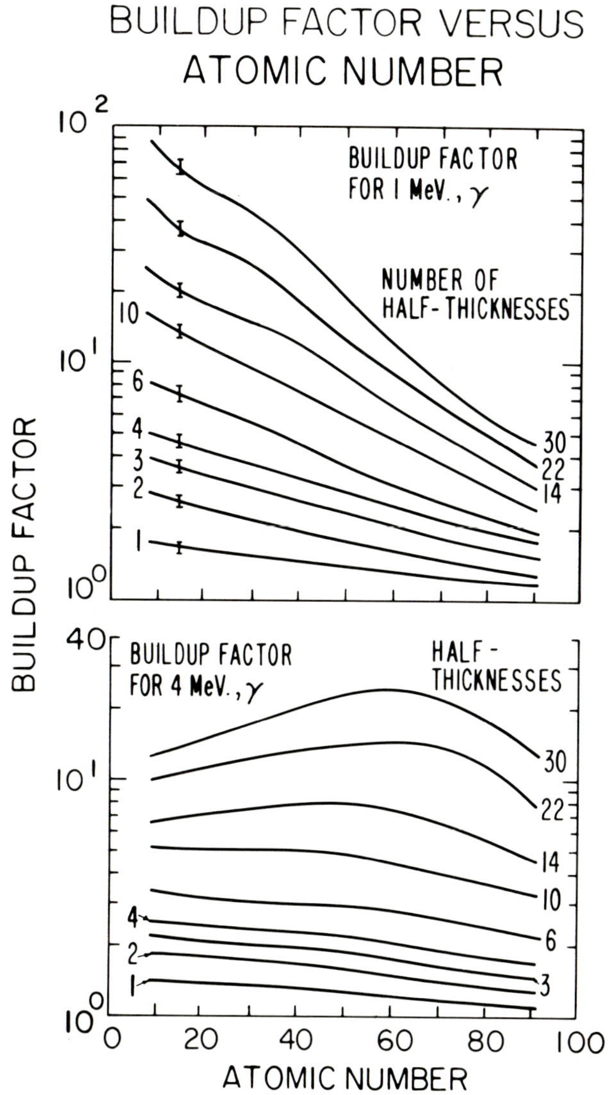

Fig. 2.11 Buildup factor versus atomic number.

Relative Hazard

Gamma rays are of most concern as external radiation hazards. They have great ranges in air and other substances and pose a significant health hazard at fairly large distances from a source. In almost all cases, shielding is needed to minimize this potential hazard. Since gamma radiation is so penetrating, the damage done to tissue extends throughout the entire body. Because of Compton scattering, care must also be taken in shield design to prevent scattered radiation from exceeding intended radiation levels beyond the shield. Furthermore, any cracks or breaks in a gamma ray shield (as might occur with a stack of lead bricks) could promote radiation streaming.

Neutron Interactions

Neutron Energies

The presence of a highly penetrating, electrically neutral radiation was experimentally demonstrated between 1930 and 1932 by scientists who observed that certain light elements, such as beryllium, when bombarded by alpha particles emitted an unknown radiation that manifested itself by ejecting protons from paraffin. The English physicist, Sir James Chadwick, resolved the mystery in 1932 by establishing the existence of the neutron (see Chapter 1) [9]. The classic reaction leading to this discovery is

$$^9_4\mathrm{Be} + ^4_2\alpha \rightarrow ^{13}_6C \rightarrow ^{12}_6C + ^1_0n. \quad (2.70)$$

Neutrons, like protons, are constituents of the atomic nucleus and both are frequently referred to as nucleons. Outside of the nucleus the free neutron is not stable and decays to a proton and an electron in approximately 12 minutes. In reactor core calculations, such neutron decay phenomena can be neglected because of the extremely short transport lifetimes (typically much less than 1 second). This may not always be the case in space nuclear power applications, however, where we might be concerned about neutron radiations from a shadow or partially shielded reactor that is a few tens or even hundreds of kilometers away. In such situations, neutron decay and time of flight factors should be taken into account. However, for the vast majority of nuclear power work, the neutrons are considered stable with respect to the transport and diffusion phenomena they experience in the reactor.

The theory of wave mechanics suggests that all moving particles, including neutrons, exhibit wavelike properties. If its kinetic energy is less than the energy equivalent of its rest mass (i.e., $E_n < 939.5$ MeV), the neutron's nonrelativistic deBroglie wavelength λ_n is given by [4]

$$\lambda_n = (h/p) \approx \frac{2.860 \times 10^{-12}}{\sqrt{E_n}} \quad (2.71)$$

where λ_n is the deBroglie wavelength (cm)
E_n is the neutron energy (MeV)
h is Planck's constant
p is the momentum of the neutron.

The types of interactions that neutrons undergo when passing through matter are quite strongly dependent on their energies. As can be seen in eq. 2.71, the neutron's wavelength is inversely proportional to its velocity. This fact provides useful insight into the general nature of neutron interactions with matter. For example, thermal neutrons (defined shortly) have wavelengths on the same order as the atomic spacings in crystals. Consequently, when thermal neutrons interact with a crystal, diffraction patterns in accordance with Bragg's law are created. It should be noted that the principal difference between the diffraction of neutrons and of X-rays and gamma rays is that neutrons interact primarily with nuclei whereas energetic photons interact primarily with orbital electrons. Neutrons with energies ranging from 1 to 100 MeV exhibit wavelengths on the order of nuclear dimensions, while relativistic neutrons (e.g., 10 GeV) have wavelengths comparable to the spacing of nucleons within the nucleus.

Neutrons are classified according to energy, with the groupings or divisions occurring somewhat naturally on the basis of behavior generic to particular energy ranges.

Neutrons with energies from 0 to about 1 or 2 keV are sometimes referred to as *slow neutrons*. This is a broad classification and does not serve as a very precise description of the energy state of low energy neutrons. We shall avoid use of this designation.

Thermal neutrons are in thermal equilibrium with the atoms (or molecules) of the medium. A particular thermal neutron may gain or lose energy in any one collision with the nuclei of the medium, but there is no net energy exchange when a large number of such thermal neutrons diffuse through a nonabsorbing medium. For this situation the kinetic energy of the neutrons is distributed statistically according to the Maxwell–Boltzmann equation as derived from the kinetic theory of gases. The equation for spectral distribution is [4]

$$n(E) = \frac{dn}{dE} = \frac{2\pi n}{(\pi kT)^{3/2}} e^{-(E/kT)} E^{1/2} \quad (2.72)$$

where $dn = n(E)dE$ is the number of neutrons having energies in the interval dE about E
n is the total number of neutrons in the system
E is the neutron energy (eV)
k is the Boltzmann constant (8.61×10^{-5} eV/K)
T is the absolute temperature (K).

When the Maxwell–Boltzmann distribution applies, the energy corresponding to the most probable thermal neutron velocity (e.g., 2200 m/s at 20° C) is given by kT (0.0252 eV); the average kinetic energy of the thermal neutrons is (3/2)kT; and the most probable neutron energy is (1/2)kT. For absorbing media the true energy distribution of the neutrons deviates from the Maxwell–Boltzmann distribution because neutron absorptions tend to increase with decreasing energy. This causes an increase in the average energy. The spectral distribution depicted in Fig. 2.12 shows very little distortion, as characteristic of poorly absorbing media, while the distribution shown in Fig. 2.13 clearly depicts the effects of preferential absorptions at the lower neutron energies [4]. This phenomenon is called hardening.

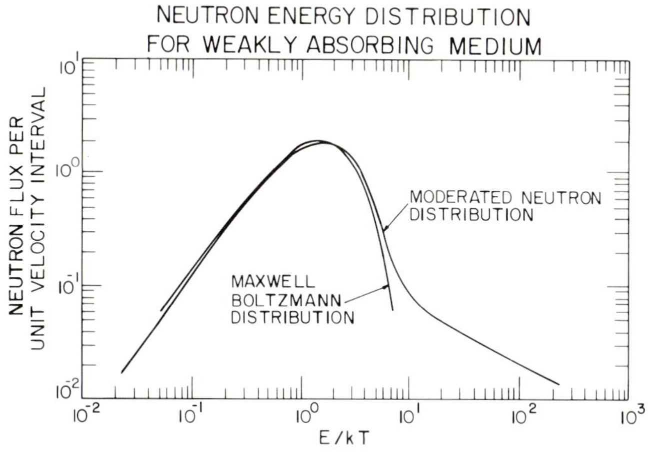

Fig. 2.12 Neutron energy distribution for weakly absorbing medium. *Courtesy of Oak Ridge National Laboratory* [*4*].

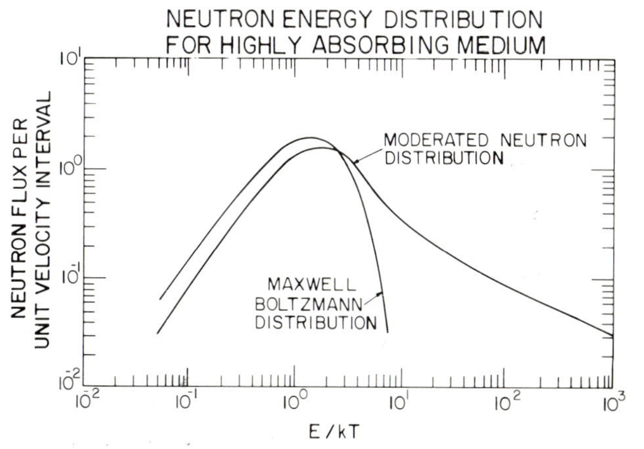

Fig. 2.13 Neutron energy distribution for highly absorbing medium. *Courtesy of Oak Ridge National Laboratory* [*4*].

We can partially compensate for it by using an effective neutron temperature T_n defined as the number that, when used in the Maxwell–Boltzmann distribution, gives the best least-squares fit to the true distribution. The effective neutron temperature T_n has been found to be approximately related to the environmental temperature T_m in the following fashion [4]

$$T_n = T_m\left[1 + \frac{1.84\Sigma_a(\mathrm{kT})}{\xi \cdot \Sigma_s}\right] = T_m(1 + 0.46\Delta) \quad (2.73)$$

where $\Sigma_a(\mathrm{kT})$ is the macroscopic absorption cross section evaluated at the energy kT (cm^{-1})

Σ_s is the macroscopic scattering cross section for thermal neutrons (cm^{-1})

ξ is the average logarithmic energy decrement per collision

$$\Delta = 4\Sigma_a(\mathrm{kT})/(\xi \cdot \Sigma_s).$$

Neutrons with an average energy less than that of thermal neutrons are sometimes referred to as *cold neutrons*.

Epithermal neutrons are the neutrons in a system whose energies exceed those of a Maxwell–Boltzmann distribution for the effective temperature of the neutron-supporting medium.

Neutrons having energies in the range between 1 and 100 eV are frequently called *resonance neutrons*. Many nuclides exhibit strong absorption of neutrons at well-defined energies in this energy regime, and these absorptions are commonly referred to as resonance absorptions.

Neutrons in the energy range from 1 keV to approximately 0.5 MeV are frequently called *intermediate neutrons*. The slowing down and removal aspects of neutron transport primarily involve intermediate neutrons. In addition, the behavior of intermediate neutrons strongly affects the design of high performance shields.

Neutrons having energies between 0.5 MeV and approximately 15 MeV are called *fast neutrons*. Most neutrons produced in nuclear fission or fusion reactions are fast neutrons.

Neutrons referred to as having very high energy usually fall in the 15 to 50 MeV regime, while those referred to as being ultra high energy neutrons fall in an energy range greater than 50 MeV. These extremely energetic neutrons are either cosmic rays or the by-products of high energy accelerator operations. The term *relativistic neutrons* is sometimes given to these particularly energetic neutrons.

Microscopic Cross Sections

Neutron interactions can include: elastic scattering, inelastic scattering, radiative capture, charged particle reactions, and fission reactions. The relative importance of these interactions is established by their respective cross sections, the magnitude of which varies with the nature of the medium and the energy of the neutron. As previously discussed for gamma rays, the term cross section is also used to describe the probability that a neutron will interact with a given medium. If the probability of interaction is high for a given process, its cross section will also be large. The microscopic cross section σ for neutron interactions and reactions expresses the probability per atom (or more properly per nucleus). As illustrated in Fig. 2.14, the total microscopic cross section σ_{total} is the sum of the separate cross sections for all the specific processes that can occur. One perception from Fig. 2.14 is that the total cross section is the sum of the elastic scattering cross section $\sigma_{\mathrm{elastic}}$ and the nonelastic scattering cross section $\sigma_{\mathrm{nonelastic}}$. That is,

$$\sigma_{\mathrm{total}} = \sigma_{\mathrm{elastic}} + \sigma_{\mathrm{nonelastic}}. \quad (2.74)$$

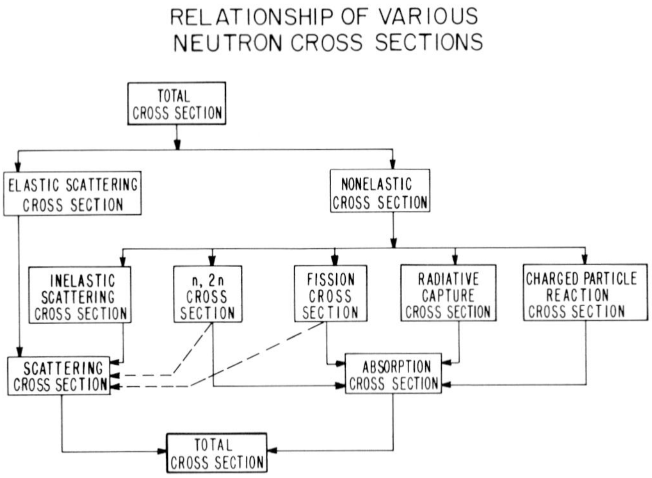

Fig. 2.14 Relationship of various neutron cross sections. *Courtesy of Oak Ridge National Laboratory* [*4*].

This classification distinguishes between interactions in which the neutron and the target nucleus undergo an elastic collision and all other interactions.

The total cross section may also be taken as the sum of the total scattering cross section $\sigma_{\text{scattering}}$ and the total absorption cross section $\sigma_{\text{absorption}}$.

$$\sigma_{\text{total}} = \sigma_{\text{scattering}} + \sigma_{\text{absorption}} \,. \tag{2.75}$$

In this interpretation the scattering cross section includes both elastic and nonelastic scattering processes and the absorption cross section includes all processes that cause the product nucleus to differ in either atomic mass (A) or atomic number (Z) from the original target nucleus. In some neutron transport calculations, all neutron-nucleus interactions that have neutrons leaving the target nucleus are considered by convention to be scattering processes. Thus, the $(n, 2n)$ and fission processes shown in Fig. 2.14 could be considered in this interpretation as scattering processes, noted by the dashed lines [4]. Again by common practice, absorption processes are generally called nuclear reactions rather than interactions, since there is a change in the target nucleus.

Elastic Scattering

Elastic scattering between neutrons and their target nuclei may be viewed as the collisions of perfectly elastic spheres (see Fig. 2.15). Such processes are characterized by the conservation of kinetic energy and momentum. Classical mechanics analyses are generally performed in the center-of-mass (CM) system or reference frame, which assumes that the center of mass of the neutron-nucleus system remains at rest. An external observer, for example, a person performing a physical or laboratory measurement, would view this same collision process in the laboratory system or frame of reference (L). Experimental results are more useful when expressed in the L-system, while theoretical relationships are more readily developed in the CM frame of reference system.

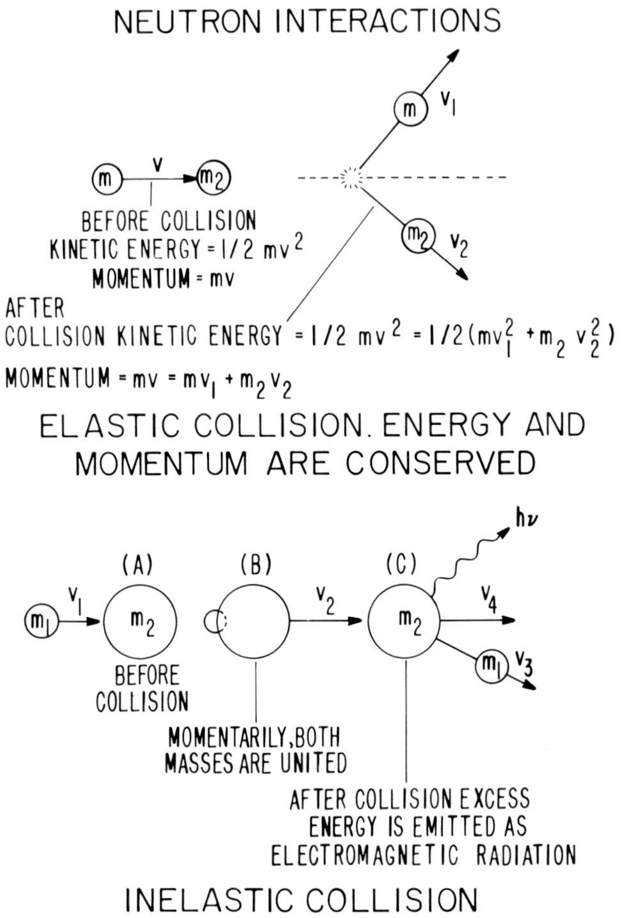

Fig. 2.15 Neutron interactions.

If we assume that the target nucleus in the laboratory system is stationary, then the change in the kinetic energy of the neutron in an elastic collision is uniquely related to the change in its direction by the following relation

$$E = \frac{E_0}{(A+1)^2}(A^2 + 2\eta A + 1) \tag{2.76}$$

where E_0 is the initial neutron energy in the laboratory system (MeV)
E is the final neutron energy in the laboratory system (MeV)
A is the ratio of the atomic mass of the nuclide to the mass of the neutron
$\eta = \cos\theta$ is the cosine of the scattering angle in the center of mass system.

The minimum value of E/E_0, corresponding to the maximum loss in energy by a neutron in a single collision, occurs for a head-on collision (i.e., $\theta = \pi$ or

$\eta = -1$). The minimum final neutron energy is given by

$$E_{\min} = \alpha E_0 \tag{2.77}$$

where

$$\alpha = \left[(A-1)/(A+1)\right]^2. \tag{2.78}$$

The cosine of the scattering angle in the laboratory system, μ, is related to the cosine of the scattering angle in the center of mass system, η, by [2, 4]

$$\mu = \frac{1+\eta A}{\left(A^2+2\eta A+1\right)^{1/2}}. \tag{2.79}$$

If we assume further that scattering in the CM system is isotropic—a good assumption for hydrogen, for example—then the following relationships are valid for the elastic scattering process.

a. The probability that a neutron will experience a change of direction into a differential solid angle $d\overline{\Omega}$ corresponding to a conical element defined by the scattering angles θ and $\theta + d\theta$ in the CM system is

$$p(\theta)\,d\theta = \frac{d\overline{\Omega}}{4\pi} = \frac{1}{2}\sin\theta\,d\theta. \tag{2.80}$$

b. The fraction of all elastic scattering collisions that involve neutrons of energy E_0 scattered into the energy interval dE about the final energy E is given by

$$p(E \to E_0)\,dE = \frac{dE}{E_0(1-\alpha)}. \tag{2.81}$$

c. The average final energy of all neutrons of initial energy E_0 is given by

$$\overline{E} = \int_{\alpha E_0}^{E_0} E \cdot p(E \to E_0)\,dE = \left[(1+\alpha)/2\right]E_0. \tag{2.82}$$

d. The average value of the cosine of the scattering angle in the laboratory system is given by

$$\overline{\mu} = 2/(3\mathrm{A}). \tag{2.83}$$

This is an important parameter in neutron transport calculations.

e. The average logarithmic energy decrement per collision is defined by

$$\xi \equiv \overline{\ln\frac{E_0}{E}} = \int_{\alpha E_0}^{E_0} \ln\left(\frac{E_0}{E}\right) p(E \to E_0)\,dE = 1 + \frac{\alpha \ln\alpha}{1-\alpha}. \tag{2.84}$$

An interesting property of the parameter ξ is its independence from the initial energy of the neutron. This quantity can also be interpreted as the average change in lethargy per collision, where the lethargy is defined as

$$u \equiv \ln(E_0/E) \tag{2.85}$$

Table 2.3 Typical neutron collision parameters.

Material	Atomic Mass (A)	Normal Density (g/cm^3)	ξ	α
Hydrogen (H)	1	8.9×10^{-5} (gas)	1.000	0
Lithium (Li)	7	0.534	0.268	0.563
Beryllium (Be)	9	1.85	0.209	0.640
Carbon (C)	12	1.60	0.158	0.716
Oxygen (O)	16	0.0014 (gas)	0.120	0.779
Sodium (Na)	23	0.971	0.083	0.840
Iron (Fe)	56	7.86	0.0357	0.931
Tungsten (W)	184	19.2	0.0108	0.979
Uranium (U)	238	19.1	0.0084	0.983
Plutonium (Pu)	239	19.6	0.0083	0.983

and the average change in lethargy per collision is given by

$$\overline{u-u_0} = \overline{\ln(E_0/E)} = \xi. \tag{2.86}$$

The lethargy is a useful alternate energy variable in the description of the slowing down process for neutrons (see Chapter 3). Table 2.3 provides some typical neutron collision parameters [2, 11, 12].

For most nuclides the isotropic assumption generally provides a good description of the elastic scattering process for neutrons in the thermal-to-keV energy range. However, it is not valid for high energy (MeV) neutrons or for interactions in which chemical binding energies are significant. Therefore, in shielding calculations that generally involve fast neutrons, the isotropic scattering assumption cannot be safely made, except when hydrogen is the scattering medium. With hydrogen the scattering process is virtually isotropic in the center-of-mass system for neutron energies up to 14 MeV [4].

In a more general treatment of the scattering process that applies to inelastic as well as to elastic scattering, the cross sections for scattering are represented by an expansion in a series of Legendre polynomials

$$\sigma_e(E,\eta) = \frac{\sigma_e(E)}{2\pi}\sum_m \frac{2m+1}{2} A_m(E) P_m(\eta) \tag{2.87}$$

where $\sigma_e(E,\eta)$ is the energy-dependent differential cross section for elastic scattering of neutrons of energy E through an angle in the CM system whose cosine is η per unit solid angle

$A_m(E)$ are the energy-dependent Legendre coefficients (where A_0 is always $= 1$)

$P_m(\eta)$ are the Legendre polynomials.

The energy-dependent elastic scattering cross section $\sigma_e(E)$ is obtained by integrating the differential elastic scattering cross section $\sigma_e(E,\eta)$ over all possible values of η

$$\sigma_e(E) = 2\pi \int_{-1}^{+1} \sigma_e(E,\eta)\, d\eta. \qquad (2.88)$$

The representation of the cross section as an expanded series of orthogonal functions is necessary in certain solutions of the Boltzmann equation. The elastic collision process is an important energy degradation mechanism and is most important for the light elements, especially hydrogen. Efficient neutron shields always contain hydrogeneous materials.

Nonelastic Reactions

All neutron reactions that are not elastic scatterings can be called nonelastic reactions. The sum of their cross sections is often referred to as the nonelastic cross section. We will use the Bohr compound nucleus model here for describing nonelastic neutron reactions with target nuclei. In this model the interaction begins when the neutron comes within the range of the nuclear forces of the nucleus and ends when the products of the interaction depart from this region. In the interim the combination of the neutron and the target nucleus is called the compound nucleus. It is convenient to consider the interaction as happening in two distinct stages. First, a compound nucleus is formed when the incident neutron loses its identity and gets incorporated into this unstable new system. This brings to the compound nucleus an excitation energy that consists of a portion of the neutron's kinetic energy and its binding energy with respect to the compound nucleus. (Consistent with the conservation of momentum principle, kinetic energy of the neutron appears as translational energy of the compound nucleus.) The binding energy per nucleon is defined as the average energy needed per nucleon to separate the nucleus into its constituent parts. This excitation energy is statistically shared by all the nucleons. Second, the unstable compound nucleus separates into its products when one or more of the nucleons acquire sufficient energy to escape. This happens when the nucleons acquire energy that is equal to or greater than their binding energy. In some cases the unstable compound nucleus may emit excitation energy in the form of gamma radiation. The resultant product nucleus may also be unstable and undergo radioactive transformation. Eventually a group of stable end products are formed (see Fig. 2.16).

The nonelastic neutron reactions are divided into scattering processes and absorption processes. As previously mentioned, the absorption processes are often termed nuclear reactions. They include radiative (n,γ) capture, charged particle reactions such as (n, p) and (n,α), $(n,2n)$ reactions, and fission [4].

Inelastic Scattering

During inelastic scattering, abbreviated (n,n'), the inci-

SCHEMATIC REPRESENTATION OF COMPOUND NUCLEUS FORMATION (INELASTIC SCATTERING)

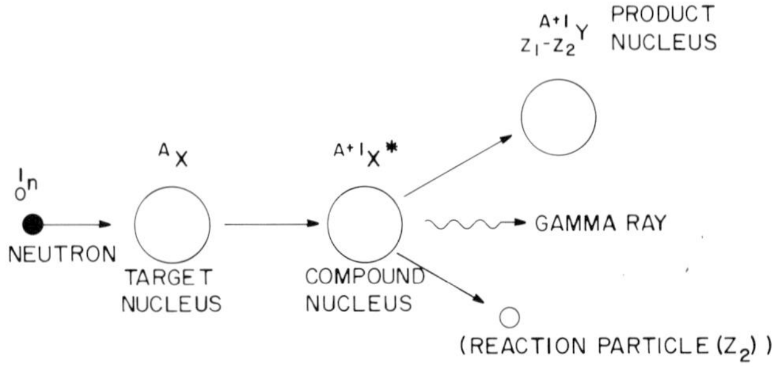

Fig. 2.16 Schematic representation of compound nucleus formation (inelastic scattering).

dent neutron n and the target nucleus form a compound nucleus from which a lower energy neutron n' is subsequently emitted. If the kinetic energy of the incident neutron is greater than the lowest energy level of the target nucleus, the nucleus may be left in an excited state following neutron emission. (If not, a form of capture scattering called elastic scattering may occur. This particular process is very much like inelastic scattering in that it involves the formation of a compound nucleus and the subsequent emission of a neutron with nearly all its original kinetic energy.) Inelastic scattering results in the emitted neutron having much less energy than the incident neutron. This process, therefore, contributes strongly to the energy degradation process and the attenuation of neutrons in shields. The excited product nucleus returns to the ground state by gamma ray emission. An energy level diagram of this process is provided in Fig. 2.17 [4].

The theoretical or experimental determination of detailed inelastic scattering cross sections is a complicated and challenging task. The following generalizations, however, can be made:

a. The inelastic scattering cross section usually increases with increasing neutron energy.

b. The threshold for the interaction in light elements is generally several MeV, and the gamma rays emitted are consequently harder and more energetic than those emitted in interactions with lower energy thresholds.

c. The thresholds for inelastic scattering in heavy elements are lower, sometimes considerably so, than for the reaction in light elements [4].

Nuclear Reactions

The term nuclear reaction is often reserved for those neutron-nucleus interactions in which the product nucleus differs in either atomic number or atomic mass from the target nucleus. The type of reaction that actually occurs depends on several factors. Two especially significant factors are the nuclear structure of the target nucleus and the energy of the incident neutron.

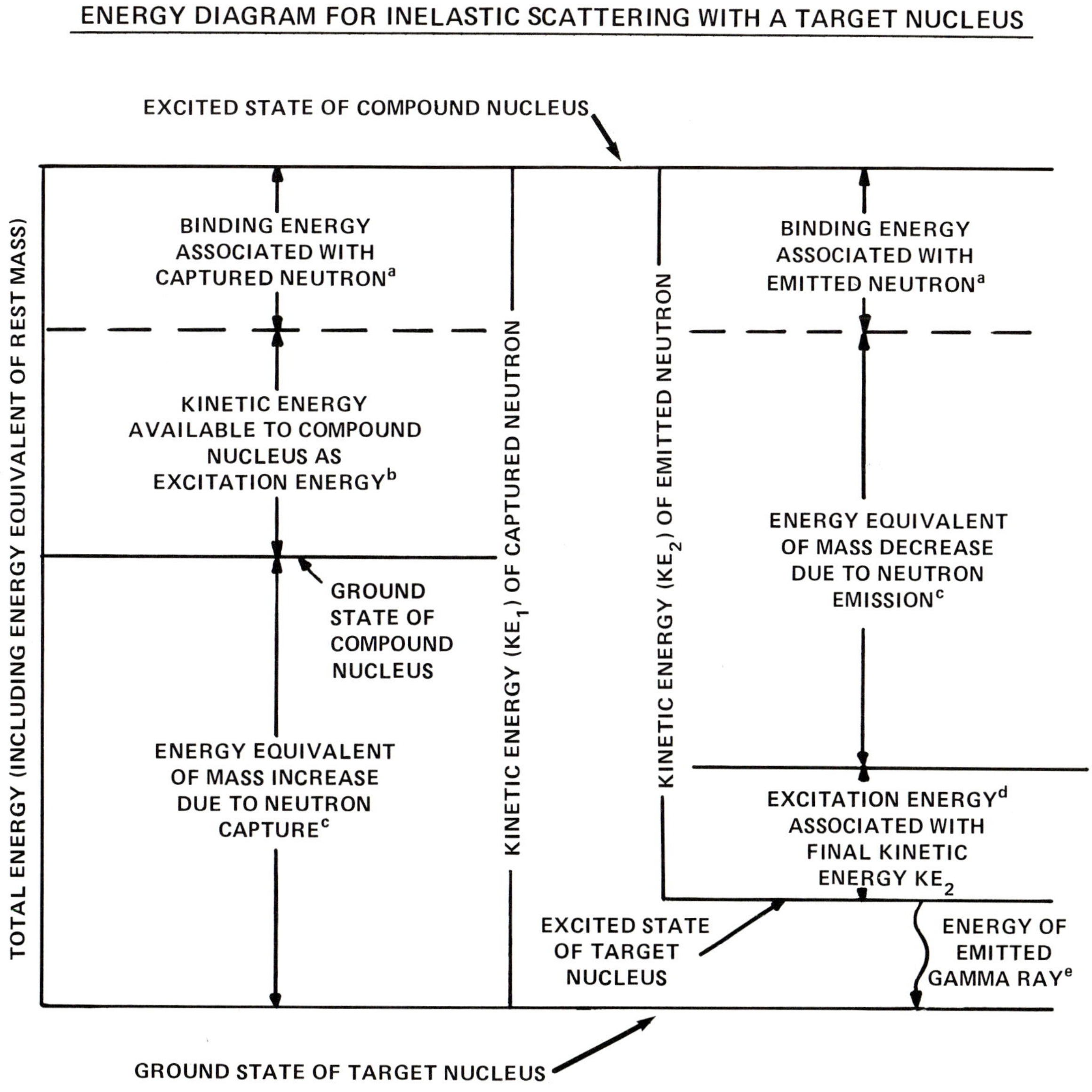

[a]Given by 931 MeV $[(M_X + M_n) - M_y] \approx 7$ MeV, where M_X, M_n and M_y are the masses, in atomic mass units, of the target nucleus, the neutron, and the compound nucleus, respectively.

[b]Given by $KE_1\ [M_X/(M_X + M_n)]$. Due to the conservation of momentum, some of the kinetic energy of the incident neutron must appear as translational energy of the compound nucleus and is therefore unavailable as excitation energy.

[c]Given by 931 $(M_y - M_X) \approx 931$ MeV.

[d]Given by $KE_2\ [(M_X + M_n)/M_X]$.

[e]Given by $KE_1\ [M_X/(M_X + M_n)] - KE_2\ [M_X + M_n)/M_X]$.

Fig. 2.17 Energy diagram for inelastic neutron scattering with a target nucleus. *Courtesy of Oak Ridge National Laboratory* [4].

Some reactions are possible for neutrons of any energy, while others have threshold energies. On occasion, more than one de-excitation process is possible, with the relative probabilities of each being given by their respective reaction cross sections. A reaction between a target nucleus and a neutron that has zero kinetic energy is expressed by the following equation [4]

$$ {}^{A}_{Z_1}X + {}^{1}_{0}n \rightarrow {}^{A+1}_{Z_1}X \rightarrow {}^{(A+1)-b}_{(Z_1-Z_2)}Y + {}^{b}_{Z_2}Y + Q \quad (2.89) $$

where ${}^{A}_{Z_1}X$ is the target nucleus with atomic number (Z_1) and mass number (A)
${}^{1}_{0}n$ is the incident neutron (assumed to have zero kinetic energy)
${}^{A+1}_{Z_1}X$ is the compound nucleus in its excited state
${}^{(A+1)-b}_{(Z_1-Z_2)}Y$ is the product nucleus
${}^{b}_{Z_2}Y$ is the emitted particle with atomic number Z_2
Q is the nuclear reaction energy.

The nuclear reaction energy Q is determined by the change in mass of the nuclear reactants. Taking one atomic mass unit (amu) as 931 MeV, eq. 2.89 can be written in the form of a total energy equation. That is, an explicit expression for Q is given by

$$ Q(\text{MeV}) = 931\left[M_x + M_n - (M_y + M_p) \right] \quad (2.90) $$

where Q is the nuclear reaction energy (MeV)
M_x is the mass of the target nucleus (amu)
M_n is the mass of the incident neutron (amu)
M_y is the mass of the product nucleus (amu)
M_p is the mass of the emitted particle (amu).

If Q is positive, then the nuclear reaction is exothermic and neutrons with essentially zero kinetic energy can cause it to occur. If, on the other hand, Q is negative, the reaction is endothermic, and the neutron (or other incident particle) must possess sufficient kinetic energy to compensate for the difference. In exothermic reactions, a small amount of nuclear mass is transformed into the kinetic energy of the products; while in endothermic reactions, kinetic energy of the incident particle is converted into mass.

In the radiative capture or (n, γ) reaction, the target nucleus captures the incident neutron, and one or more gamma rays, frequently called capture gamma rays, are emitted as the excited compound nucleus returns to ground state. This is an exothermic reaction. The parasitic capture (non-fission capture) of a neutron by a uranium-235 nucleus is an example of radiative capture.

$$ {}^{235}_{92}\text{U} + {}^{1}_{0}n \rightarrow {}^{236}_{92}\text{U}^* \rightarrow {}^{236}_{92}\text{U} + \gamma \quad (2.91) $$

Inelastic absorption reactions of the $(n, 2n)$ and $(n, 3n)$ type occur with energetic neutrons. For example, the $(n, 2n)$ reaction involves the emission of two neutrons. It occurs with incident neutrons of kinetic energy greater than 10 MeV. For certain very light and very heavy nuclides, the thresholds for the $(n, 2n)$ reaction are sometimes much less than 10 MeV. Consider, for example, ${}^{2}_{1}\text{H}$, ${}^{6}_{3}\text{Li}$, and ${}^{9}_{4}\text{Be}$ which have thresholds of 3.34 MeV, 6.2 MeV, and 1.85 MeV respectively; and ${}^{209}_{83}\text{Bi}$, ${}^{232}_{90}\text{Th}$, and ${}^{238}_{92}\text{U}$ have thresholds of 7.4 MeV, 6.44 MeV, and 6.0 MeV, respectively [4]. These reactions are endothermic in nature. The $(n, 2n)$ reaction could be especially significant in space nuclear power systems containing beryllium, since ${}^{9}_{4}\text{Be}$ nuclei have loosely bound neutrons that can easily be ejected [2, 4].

In charged particle reactions, the compound nucleus achieves de-excitation by emitting a charged particle such as a proton (${}^{1}_{1}p$) or an alpha particle (${}^{4}_{2}\alpha$). Such reactions may either be endothermic or exothermic; but they are most probable for incident neutrons in the MeV energy regime [2, 4]. One example of the (n, p) reaction is the production of tritium [2]

$$ {}^{3}_{2}\text{He} + {}^{1}_{0}n \rightarrow {}^{4}_{2}\text{He}^* \rightarrow {}^{3}_{1}\text{T} + {}^{1}_{1}p + 0.764 \text{ MeV}. \quad (2.92) $$

An (n, α) reaction is described below as

$$ {}^{10}_{5}\text{B} + {}^{1}_{0}n \rightarrow {}^{7}_{3}\text{Li}^* + {}^{4}_{2}\alpha + 2.30 \text{ MeV} \quad (2.93) $$

$$ {}^{7}_{3}\text{Li}^* \rightarrow {}^{7}_{3}\text{Li} + \gamma(0.478 \text{ MeV}). $$

The isotope ${}^{10}_{5}\text{B}$ can serve as an important component in shields designed to suppress high energy capture gamma production [4].

In the fission reaction, the absorption of a neutron by certain heavy nuclei can cause the compound nucleus to split or fission into two separate fission fragments. This very energetic exothermic reaction is accompanied by the release of gamma rays and additional neutrons. The fission reaction is described in Chapter 3.

Neutron Absorption

In treating neutron transport through matter, we may wish to express the probability of a reaction occurring in terms of the thickness of the medium. The probability per unit path length, called the macroscopic cross section Σ, is used. The total macroscopic cross section Σ_{tot} is related to the total microscopic cross section σ_{tot} by the equation

$$ \Sigma_{\text{tot}} = N \cdot \sigma_{\text{tot}} (\text{cm}^{-1}) \quad (2.94) $$

where N is the number of atoms per unit volume of substance (atoms/cm^3).

We now consider a cubic centimeter of matter in terms of a cross-sectional area of 1 cm^2 and a thickness of 1 cm. Each atom of the medium (assumed homogeneous and neglecting screening) has a cross section σ_{tot}. The total reaction cross section will then be the sum of the contributions from each atom. But N is the total number of atoms in the volume of material under consideration. Therefore, $N \cdot \sigma_{tot}$ represents the total reaction cross section.

The macroscopic cross section for neutrons is similar to the attentuation coefficient μ used for gamma ray interactions, since both quantities express a reaction probability per unit path length. By analogy, the mean free path λ for neutrons is given by

$$\lambda_{tot} = 1/\Sigma_{tot}. \tag{2.95}$$

In this case, λ_{tot} expresses the average distance a neutron travels before interacting in the medium.

We now consider a narrow, well-collimated beam of monoenergetic neutrons passing through a substance. Let n be defined as the number of neutrons per cubic centimeter in the beam. We further assume that each neutron has the same speed, v. Then, the monoenergetic neutron flux ϕ is given by

$$\phi = nv(\text{neutrons/cm}^2 - \text{s}). \tag{2.96}$$

The number of reactions per unit time in a cubic centimeter of substance is expressed as

$$\frac{\text{number of reactions}}{\text{cm}^3 - \text{s}} = \phi \cdot \Sigma_{tot}. \tag{2.97}$$

The decrease in flux as the beam moves through a thickness of material Δx is given by

$$\Delta\phi = -\phi\Sigma_{tot}\Delta x. \tag{2.98}$$

If we assume thin slab geometry (that is, if a neutron is absorbed or scattered it is removed from the beam) then eq. 2.98 may be written as a differential equation

$$d\phi/dx = -\Sigma_{tot} \cdot \phi. \tag{2.99}$$

Removal Cross Section

The removal cross section Σ_R is a parameter that is used in fast neutron attenuation calculations. It is a measure of the ability of a material to remove a fast neutron from further consideration as a fast neutron for shielding calculation purposes. This involves not only the interaction of the fast neutron with a nucleus of the material, but also its subsequent interaction with nuclei of the hydrogeneous material. Fast neutrons that are either degraded in energy by inelastic scatterings or wide-angle deflected by elastic scatterings are unlikely to penetrate large distances into a hydrogen-rich medium. Table 2.4 lists selected, experimentally derived values of the removal cross section [12]. At least as far as fast neutrons are concerned, we might consider this removal cross section as an equivalent to an absorption cross section. However, we must emphasize that this concept is valid only for a substance that is sufficiently thick and followed by a sufficiently hydrogen-rich medium in order to guarantee that a fast neutron which interacts in the substance is indeed removed or lost from the fast flux. The macroscopic removal cross section can be expressed as

Table 2.4 Removal cross sections.

Material	N_0 at 20° C (atom/cm^3)	σ_r (barns)	Σ_r (cm^{-1})
Hydrogen (H)	(—gas—)	1.00	—
Lithium (Li)	0.460×10^{23}	1.01	0.046
Beryllium (Be)	1.200×10^{23}	1.07	0.132
Boron (B)	1.390×10^{23}	0.97	0.135
Carbon (C) (graphite)	1.130×10^{23}	0.81	0.065
Aluminum (Al)	0.603×10^{23}	1.31	0.079
Iron (Fe)	0.848×10^{23}	1.98	0.168
Nickel (Ni)	0.913×10^{23}	1.89	0.173
Zirconium (Zr)	0.423×10^{23}	2.36	0.101
Tungsten (W)	0.631×10^{23}	3.13	0.198
Lead (Pb)	0.330×10^{23}	3.53	0.116
Uranium (U)	0.473×10^{23}	3.60	0.170

$$\Sigma_R = \sigma_R \cdot N \tag{2.100}$$

where σ_R is the macroscopic removal cross section (cm^{-1})
N is the atom density (atoms/cm^3).

If we have a mixture of elements that make up the shielding material, then the macroscopic removal cross section for the mixture is bound from [2]

$$(\Sigma_R)_{mix} = \sum_i N_i \cdot (\sigma_R)_i \tag{2.101}$$

where N_i is the atom density of the ith element in the mixture (atoms/cm^3)
$(\sigma_R)_i$ is the microscopic removal cross section of the ith element.

Neutron Activation

Many manmade radioactive substances are produced when parent materials are exposed to the neutron flux in a reactor. Absorption occurs mainly for thermal neutrons. The thermal activation cross section σ_{act} expresses the probability that a certain substance will be activated if it absorbs a thermal neutron.

We first assume that a sample containing N atoms/cm^3 is exposed to a thermal neutron flux ϕ. If σ_{act} is the probability of activation per atom, the total probability is $N \cdot \sigma_{act}$. The rate of formation of radioactive atoms is then

$$\text{Formation Rate} = \phi\sigma_{act}N. \tag{2.102}$$

During the time that radioactive atoms are being formed in this neutron activation process, some of them are also decaying. The decay rate of the newly formed radioactive nuclei N^* is given by

$$\text{Decay Rate} = \lambda N^* \tag{2.103}$$

where N^* is the number of radioactive nuclei present and λ is the decay constant (time^{-1}). The net rate of growth of the activated species is thus

$$\text{Rate of Growth} = \text{Formation Rate} - \text{Decay Rate}. \quad (2.104)$$

The corresponding differential equation is

$$dN^*/dt = \phi\sigma_{act}N - \lambda N^*. \quad (2.105)$$

If the flux is constant for the time that the material or sample is irradiated by thermal neutrons, then the equation can be integrated to give

$$N_t^* = \left[(\phi\sigma_{act}N)/\lambda\right]\left[1 - e^{-\lambda t}\right] \quad (2.106)$$

where N_t^* gives the number of radioactive nuclei present after an irradiation time t. The activity A_t (disintegrations/second) after an irradiation time t (assuming material is now removed from neutron flux) is

$$A_t = \lambda N_t^* = \phi\sigma_{act}N(1 - e^{-\lambda t}). \quad (2.107)$$

The activity A after the sample has been removed from the neutron flux for a time t_r is expressed as

$$A = A_t e^{-\lambda t_r} = \phi\sigma_{act}N(1 - e^{-\lambda t})e^{-\lambda t_r} \quad (2.108)$$

where t is the time the sample or material has been irradiated (s)

t_r is the time *after* irradiation (s) or the time since removal from the neutron flux.

The activity A_t of the activated material in the neutron flux will increase until the formation rate is equal to the decay rate. At this point, no further increase in activity will occur. For a particular material, the maximum value that the activity will reach is $\phi\sigma_{act}N$, which is called the saturation activity, A_{sat}. The term $(1 - e^{-\lambda t})$ varies from 0 for $t = 0$ to 1 for $t = \infty$. For any time t, this term gives the fraction of A_{sat} that the neutron irradiated material has reached. Eq. 2.108 is valid if all the target nuclei receive the same flux. In effect this means that the activated material is thin; also implied is that the number of target atoms remains approximately constant (i.e., there is no significant decrease in the number of target nuclei).

Example 2.4. A 1 gram tungsten sample is exposed for 3 days at an average thermal neutron flux of 10^{11} neutrons/cm^2-s in a reactor. We know that σ_{act} is 40 barns and that the half life of tungsten-187 is 24 hours. What is the activity level of the tungsten sample 12 hours after it has been removed from the reactor? We also wish to find the saturated activity (A_{sat}) for this sample.

$$A = \phi\sigma_{act}N(1 - e^{-\lambda t})e^{-\lambda t_r}$$

Given: $\sigma_{act} = 40b = 40 \times 10^{-24}\ \text{cm}^2$

$\lambda = 0.693/\text{day} = 0.693\ \text{days}^{-1}$

$\phi = 10^{11}\ \text{neutrons/cm}^2\text{-s}$

$N = (m/A)N_{AV} = (1\ \text{g}/186\ \text{g/g-atom})(6.022\ \text{atoms/g-atom})$

$N = 3.238 \times 10^{21}$ atoms

$A = \phi N\sigma_{act}(1 - e^{-\lambda t})e^{-\lambda t_r}$

$A = (10^{11}\ \text{neutrons/cm}^2\text{-s})(3.238 \times 10^{21}\ \text{atoms}) \cdot (40 \times 10^{-24}\text{cm}^2/\text{atom}) \cdot (1 - e^{-(0.693/\text{day})(3\ \text{days})})(e^{-(0.693/\text{d})(0.5\ \text{d})})$

$A = 8.0 \times 10^9$ disintegrations/s = 0.22 curie

$A_{sat} = \phi\sigma_{act}N$

$= (10^{11}\ \text{neutrons/cm}^2\text{-s})(40 \times 10^{-24}\ \text{cm}^2/\text{atom})(3.238 \times 10^{21}\ \text{atoms})$

$= 1.295 \times 10^{10}$ disintegrations/s = 0.35 curie

Relative Hazard

Because neutrons move through matter rather freely until they interact, the damage done in living tissues extends throughout the body. Just as for gamma rays, the more radiosensitive body organs will receive damage from an external neutron field. The damage done in tissue is a function of neutron energy. Fast neutrons, for example, produce many collisions in tissue. Since almost 67 percent of the atoms in the body are hydrogen atoms, about 80 to 95 percent of the energy deposited in tissue by fast neutrons occurs in hydrogen scattering reactions. The recoil protons so produced will then lose their kinetic energies by ionization. For this reason, fast neutrons are actually more hazardous than thermal neutrons. Thermal neutrons, however, can produce damage in tissue through (hydrogen) scattering collisions and through absorption reactions. Two capture reactions are of special interest: H(n,γ)D and $^{14}N(n, p)^{14}C$. The first represents a radiative capture reaction

$$^1_0n + {}^1_1\text{H} \rightarrow {}^2_1\text{D} + \gamma. \quad (2.109)$$

The second is a charged particle reaction

$$^1_0n + {}^{14}_7\text{N} \rightarrow {}^{14}_6\text{C} + {}^1_1p. \quad (2.110)$$

Since some of the neutron interactions can produce heavy recoil particles, recoil protons, and gamma rays, the transfer of energy by the neutrons to tissue will vary. In general, however, the relative effectiveness of neutrons in producing damage in living matter is from 2.5 to 10 times that of gamma radiation. Nonetheless, each type of radiation, neutron and gamma, represents a significant external health hazard and proper shielding configurations will be required in space nuclear power applications to protect both the astronaut crew and delicate electronic equipment.

References

1. R. D. Evans, *The Atomic Nucleus*, McGraw-Hill Book Co., New York, 1955.
2. John R. Lamarsh, *Introduction to Nuclear Engineering* (2nd Edition), Addison-Wesley Publishing Co., Inc., Reading, Mass., 1983.
3. International Atomic Energy Agency (IAEA), *Radiation Protection Procedures*, Safety Series No. 38, Vienna, Austria, 1973.
4. P. Stevens and H. C. Claiborne, *Weapons Radiation Shielding Handbook*, Chapter 2—"Basic Concepts of Radiation Shielding Analysis," Defense Atomic Support Agency, DASA-1892-5, June 1970.
5. D. C. Kocher, *Radioactive Decay Data Tables*, U.S. Department of Energy, DOE/TIC-11026, 1981.
6. Knolls Atomic Power Laboratory, "Chart of the Nuclides," 12th edition, revised to April 1977.
7. L. Katz and A. S. Penfold, *Rev. Mod. Phys.*, **24**, 28, 1952.
8. H. Goldstein, *Fundamental Aspects of Reactor Shielding*, Addison-Wesley Publishing Co., Reading, Mass., 1959.
9. I. Asimov, *Asimov's Biographical Encyclopedia of Science and Technology*, (revised edition), 1972.
10. O. Klein and Y. Nishina, *Z. Physik*, **52**, 853, 1929.
11. John R. Lamarsh, *Introduction to Nuclear Reactor Theory*, Addison-Wesley Publishing Co., Reading, Mass., 1972.
12. Argonne National Laboratory, *Reactor Physics Constants* (2nd Edition) ANL-5800, 1963.

Chapter 3

Nuclear Reactor Theory

Introduction

A nuclear reactor can be defined as a device containing fissile nuclear material, such that a chain reaction of fission events can be maintained and controlled. A fissile nuclide is a nuclide (e.g., $^{233}_{92}U$, $^{235}_{92}U$, or $^{239}_{94}Pu$) capable of being split or fissioned upon absorbing neutrons of any energy. A thermal reactor is one in which thermal neutrons are the predominant cause of fission reactions, while a fast reactor is one designed so that the majority of fissions occur at fast neutron energies, for instance, > 100 keV.

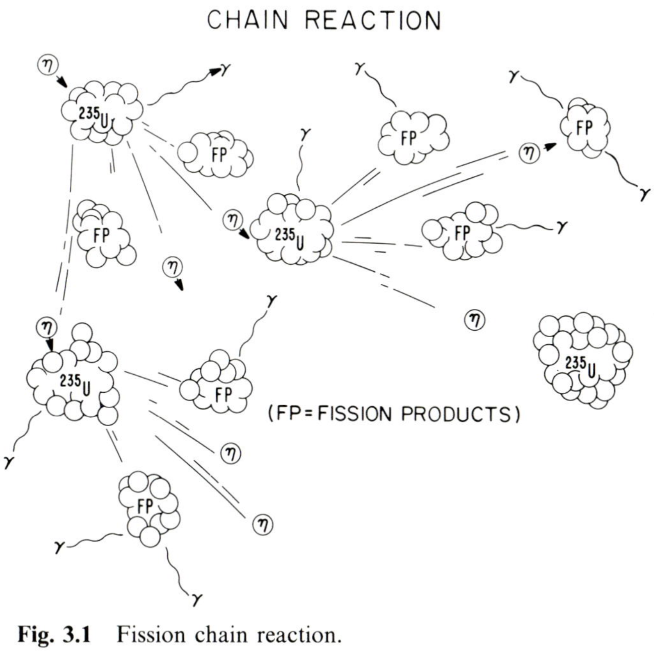

Fig. 3.1 Fission chain reaction.

A chain reaction is a reaction that stimulates its own repetition. For a fission chain reaction, a fissile nucleus absorbs a neutron, splits, and releases additional neutrons (see Fig. 3.1). A fission chain reaction is self-sustaining when at least one neutron per fission event survives to create another fission reaction. The two lighter elements produced in the splitting of the heavy nucleus are called fission products. The multiplication factor k is used to describe the fission chain reaction. It is defined as [1–3]

$$k \equiv \frac{\text{number of nuclear fissions (or neutrons) in one generation}}{\text{number of nuclear fissions (or neutrons) in the immediately preceding generation}} . \quad (3.1)$$

When $k = 1$, the fission reaction is critical or self-sustaining and power production occurs at a steady rate. If $k < 1$, the chain reaction is called subcritical and the number of fissions occurring per generation (or the neutron population) will eventually reach zero. If $k > 1$, the chain reaction is supercritical and the number of fission reactions (or the neutron population) increases each generation (see Fig. 3.2).

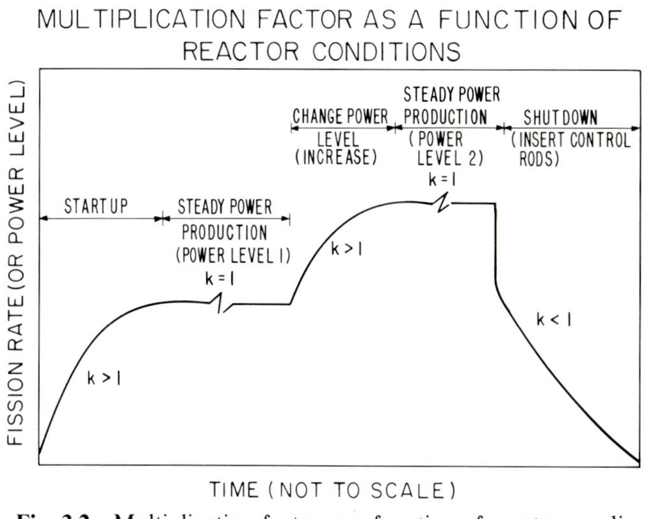

Fig. 3.2 Multiplication factor as a function of reactor conditions.

In general, a space nuclear reactor will consist of a core, nuclear fuel, coolant, a reflector, control drums or rods, and a radiation shield [4–7] (see Fig. 3.3). A moderator is found only in thermal reactors. The core is the central reactor region. It contains the nuclear fuel, a moderator (thermal reactors only), suitable structural materials, and coolant passages or heat removal devices, such as heat pipes. The nuclear fuel is a fissile

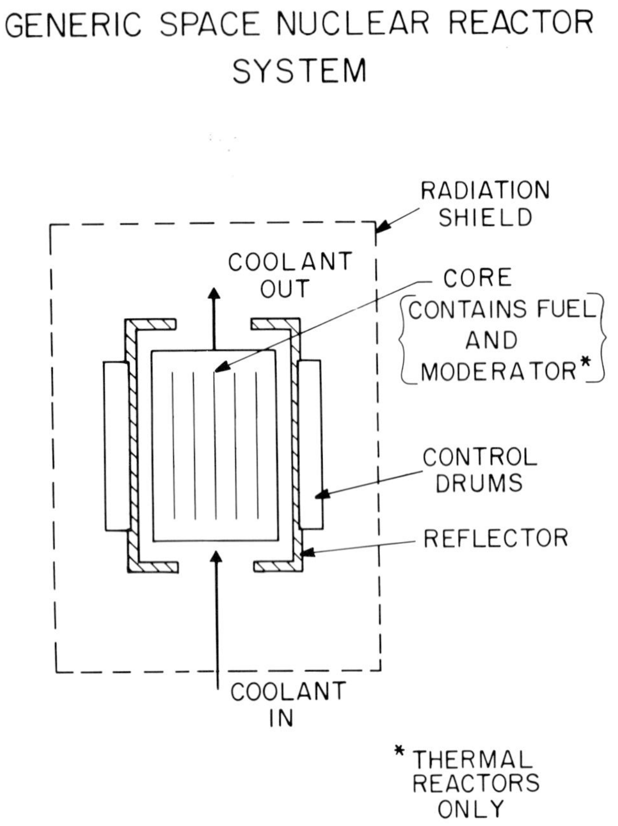

Fig. 3.3 Generic space nuclear reactor system.

material, like highly enriched (93.5 percent) uranium-235. It sustains the fission chain reaction, is responsible for the criticality of the reactor, and provides for the release of large quantities of energy in the nuclear fission reaction. This nuclear fuel may be in solid, liquid, or gaseous form. We will, however, focus our attention on solid fuel reactors in this chapter. The moderator is a low-mass material, such as hydrogen, that slows down or moderates neutrons from a fission energy spectrum to a thermal energy spectrum, or perhaps an epithermal spectrum. (Epithermal neutrons are neutrons with energies above thermal values.) This moderating material is found only in thermal or epithermal reactors and is not used in fast reactors.

The coolant removes thermal energy from the reactor core. This heat removal function may be accomplished by using a pumped-loop working fluid, such as a liquid metal or gaseous mixture of xenon and helium, or by using a self-contained heat pipe system (see Chapter 4). In fast reactors, care must be taken in the choice of coolants to avoid using low atomic mass fluids that would also undesirably slow down or moderate fission spectrum neutrons in the core.

The reflector is a material, like beryllium, located adjacent to the reactor core that scatters neutrons back into the core. It reduces neutron leakage and permits a more uniform power production within the core through flux flattening. Control rods or drums, containing neutron absorbers such as boron carbide (B_4C), are used to regulate the nuclear reactor. Movement of these rods or drums adjusts the multiplication factor, thereby controlling the reactor's power level (see Fig. 3.2).

The radiation shield protects the payload (astronauts and equipment) from the neutrons and gamma rays originating in the reactor core. This shield may either completely surround the reactor system (4π shield) or may only be a partial or shadow shield, if the reactor is located at one extreme of the spacecraft or space platform. Typical space reactor shield materials include lithium hydride (LiH) for neutron attenuation and tungsten (W) for gamma ray absorption.

In this chapter we will explore the nuclear fission process, examine the fundamental concepts governing the operation of a nuclear reactor, and discuss the major engineering characteristics of space nuclear reactors. Detailed treatment of specific space reactor systems occurs in subsequent chapters.

The Fission Process

Nuclear fission begins with the formation of a compound nucleus in an excited state. If the excitation energy exceeds a critical energy, then fission may occur. The alternative to the fission mode of deexcitation is the emission of gamma radiation that returns the compound nucleus to its ground state. Calculated values of the fission threshold energy are shown in Table 3.1 for a wide range of atomic masses [8, 9]. As indicated in this

Table 3.1 Neutron fission thresholds as a function of nuclear mass (calculated).

Mass No., A	Fission Threshold (MeV)
16	18.5
60	48
100	47
140	62
200	40
236	~5

table, only the very heavy nuclides ($A > 230$) have reasonably low threshold energies. Experimental values of the fission thresholds for selected heavy nuclides are presented in Table 3.2 [8, 9]. The significance of a negative threshold energy (i.e., $^{233}_{92}U$, $^{235}_{92}U$, and $^{239}_{94}Pu$) is that neutrons with essentially zero kinetic energy (thermal neutrons) can cause these nuclides to undergo "thermal" fission. That is why we have previously called them "fissile" nuclides. All the other heavy nuclides shown (i.e., $^{232}_{90}Th$, $^{234}_{92}U$, $^{236}_{92}U$, $^{238}_{92}U$, and $^{237}_{93}Np$) can only undergo fast fission.

Table 3.2 Neutron fission thresholds of heavy nuclides (experimental).

Target Nucleus	Compound Nucleus	Fission Threshold (MeV)
^{232}Th	^{233}Th	1.3
^{233}U	^{234}U	< 0
^{234}U	^{235}U	0.4
^{235}U	^{236}U	< 0
^{236}U	^{237}U	0.8
^{238}U	^{239}U	1.2
^{237}Np	^{238}Np	0.4
^{239}Pu	^{240}Pu	< 0

When fission occurs, the highly unstable compound nucleus almost always splits into two fission fragments. Splitting into three or more fragments is theoretically possible, but such an event is a rare occurrence and may subsequently be ignored here. Since the nucleons are more tightly bound within the fission fragments than they were in the original heavy nucleus, a very large amount of energy (typically 200 MeV per fission) is released in this nuclear reaction. (The difference in binding energy between the original heavy nucleus and its fission products is related to a quantity known as the mass defect.) Table 3.3 presents a typical energy distribution for uranium-235 fission [1–3]. Generally, nuclear

Table 3.3 Typical energy distribution for Uranium-235 fission.

Energy Form	Energy Released (MeV)	Energy Potentially Recoverable (MeV)
Kinetic Energy of Fission Fragments	168	168
Decay of Fission Products		
• Beta Radiation	8	8
• Gamma Radiation	7	7
• Neutrinos	12	—
Prompt (Fission) Gamma Radiation	7	7
Kinetic Energy of Fission Neutrons	5	5
Capture Gamma Radiation	—	3–12
TOTALS	207	198–207

fission results in the prompt emission of several neutrons. The average number of neutrons released per fission (ν) is a function of the neutron energy and the fissile nuclide. For example, thermal fission of uranium-235 releases about 2.5 neutrons per fission, while thermal fission of plutonium-239 is accompanied by the release of about 2.9 neutrons per fission. In addition, some of the fission products themselves are nuclides which possess more neutrons than necessary for nuclear stability. These unstable fission fragments (with half-lives of up to approximately one minute) emit delayed neutrons. For thermal fission of uranium-235 and plutonium-239, the delayed neutron fractions (β) are about 0.7 percent and 0.2 percent, respectively [8] (see also Table 3.4 [10]). Because the fission fragments are neutron-rich, they undergo radioactive decay, emitting mainly beta and gamma radiations.

Table 3.4 Delayed neutron fraction (β) for fast and thermal fission [*10*].

Nuclide	Fast Fission	Thermal Fission
^{233}U	0.0027 ± 0.0002	0.00264 ± 0.0002
^{235}U	0.0065 ± 0.0003	0.0065 ± 0.0003
^{238}U	0.0157 ± 0.0012	—
^{239}Pu	0.0021 ± 0.0002	0.0021 ± 0.0002
^{240}Pu	0.0026 ± 0.0003	—

Figures 3.4 and 3.5 illustrate the fission product mass numbers as a function of fission yield for the thermal and fast (14 MeV) fission of uranium-235 and plutonium-239, respectively [8]. The fission yield is defined as

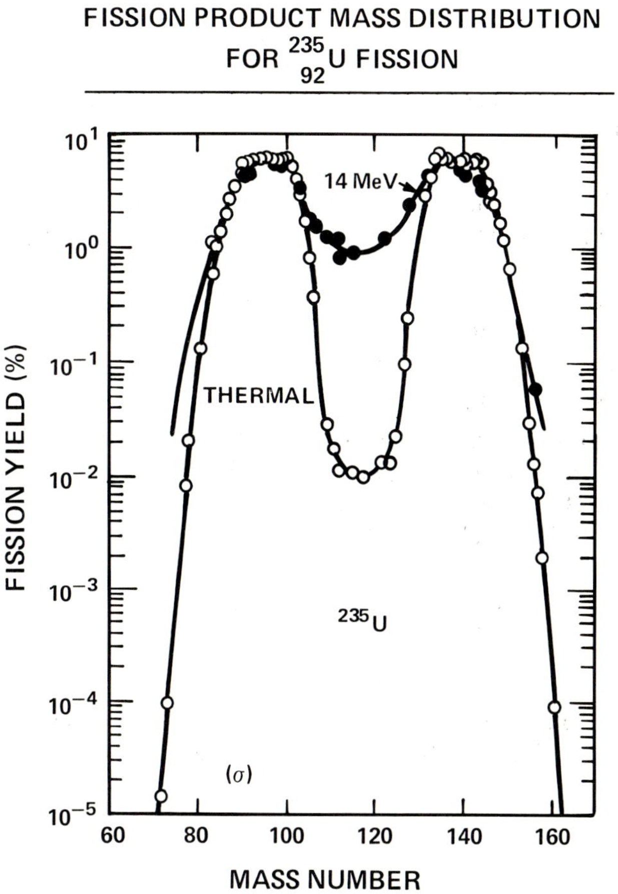

Fig. 3.4 Fission product mass distribution for uranium-235 fission [*8*].

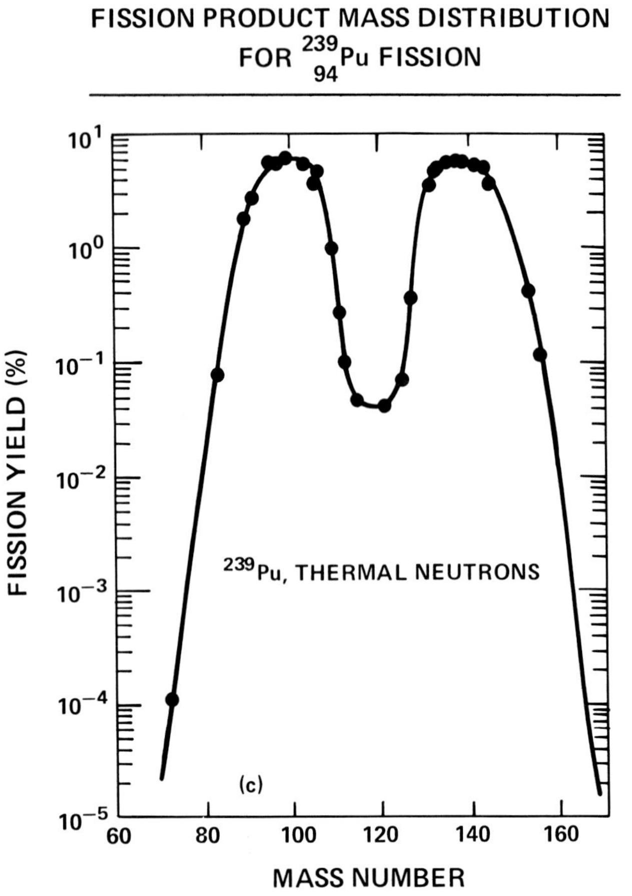

Fig. 3.5 Fission product mass distribution for plutonium-239 fission [8].

the percent of the total number of fissions that produces fission products of a given mass number. As the energy of the incident neutron increases, the probability of producing two fission products of nearly the same size increases and the characteristic dip that is present in the thermal fission-yield plots at approximately mass number 120 disappears.

The fission neutron energy spectrum has been experimentally determined for uranium-235, the nuclide used most extensively as a space nuclear reactor fuel, and is well described by the formula [8, 11]

$$N(E) = 0.453 e^{-(E/0.965)} \sinh\sqrt{2.29E} \qquad (3.2)$$

where $N(E)$ is the fraction of neutrons emitted per fission with energy E(MeV) per unit energy range. Table 3.5 presents values of $N(E)$ together with two other useful quantities: $F(E)$ and $G(E)$. $F(E)$ is the fraction of neutrons with energies greater than the energy E and is given by

$$F(E) = \int_E^\infty N(E')\,dE'. \qquad (3.3)$$

Table 3.5 Prompt neutron spectrum due to $^{235}_{92}$U fission by thermal neutrons.

E (MeV)	N(E) (MeV^{-1})	F(E)	G(E) (MeV)
0.00	0.000	1.000	1.981
0.25	0.290	0.948	1.973
0.50	0.347	0.867	1.942
0.75	0.358	0.779	1.887
1.00	0.347	0.690	1.810
1.25	0.325	0.606	1.715
1.50	0.298	0.528	1.608
1.75	0.268	0.457	1.493
2.00	0.239	0.394	1.374
2.50	0.184	0.2884	1.138
3.00	0.138	0.2082	0.919
3.50	0.102	0.1486	0.726
4.00	0.0738	0.1050	0.563
4.50	0.0528	0.0736	0.430
5.00	0.0375	0.0512	0.324
5.50	0.0263	0.03544	0.241
6.00	0.0184	0.02439	0.178
6.50	0.0127	0.01669	0.130
7.00	8.77(−3)*	0.01138	9.41(−2)
7.50	6.01(−3)	7.72(−3)	6.76(−2)
8.00	4.10(−3)	5.22(−3)	4.83(−2)
8.50	2.79(−3)	3.52(−3)	3.43(−2)
9.00	1.88(−3)	2.364(−3)	2.42(−2)
9.50	1.27(−3)	1.583(−3)	1.70(−2)
10.00	8.56(−4)	1.058(−3)	1.187(−2)
10.50	5.74(−4)	7.05(−4)	8.25(−3)
11.00	3.84(−4)	4.686(−4)	5.72(−3)
11.50	2.56(−4)	3.108(−4)	3.947(−3)
12.00	1.70(−4)	2.058(−4)	2.714(−3)
13.00	7.48(−5)	8.97(−5)	1.271(−3)
14.00	3.26(−5)	3.88(−5)	5.88(−4)
15.00	1.41(−5)	1.67(−5)	2.69(−4)
16.00	6.07(−6)	7.10(−6)	1.22(−4)
17.00	2.59(−6)	3.00(−6)	5.48(−5)
18.00	1.10(−6)	1.26(−6)	2.44(−5)

*read as: 8.77×10^{-3}

$G(E)$ is the energy per fission carried by neutrons having energies greater than E and is given by

$$G(E) = \int_E^\infty E' N(E')\,dE'. \qquad (3.4)$$

As mentioned previously, the average number of neutrons (both prompt and delayed) released per fission is given the symbol, ν. We also define a factor, η, as the number of neutrons released by fission per neutron absorbed in the fissile fuel [1–3].

Since radiative (n,γ) capture competes with the fission process for neutrons, η is always less than ν. In fact,

$$\eta = \frac{\sigma_f}{\sigma_a} = \frac{\sigma_f}{\sigma_\gamma + \sigma_f} \tag{3.5}$$

where σ_f is the microscopic fission cross section (b)
σ_a is the microscopic absorption cross section (b)
σ_γ is the microscopic cross section for radiative capture (b).

By defining the capture-to-fission ratio as

$$\alpha \equiv \sigma_\gamma / \sigma_f, \tag{3.6}$$

eq. 3.5 becomes

$$\eta = \nu\left[\frac{1}{1+\alpha}\right]. \tag{3.7}$$

If we have a mixture of fissile nuclides or a mixture of fissile and nonfissile nuclides, then, η is defined as the average number of neutrons emitted per neutron absorbed in the mixture. Thus,

$$\eta_{\text{mix}} = \frac{1}{\Sigma_a} \sum_i \nu_i (\Sigma_f)_i \tag{3.8}$$

where $(\Sigma_f)_i$ is the macroscopic fission cross section for the ith nuclide (cm^{-1})
Σ_a is the macroscopic absorption cross section for the mixture (cm^{-1})
ν_i is the average number of prompt and delayed neutrons released per fission by the ith nuclide.

Representative nuclide data related to these terms can be found in Figures 3.6 to 3.8 and Tables 3.6 and 3.7 [1–3, 12].

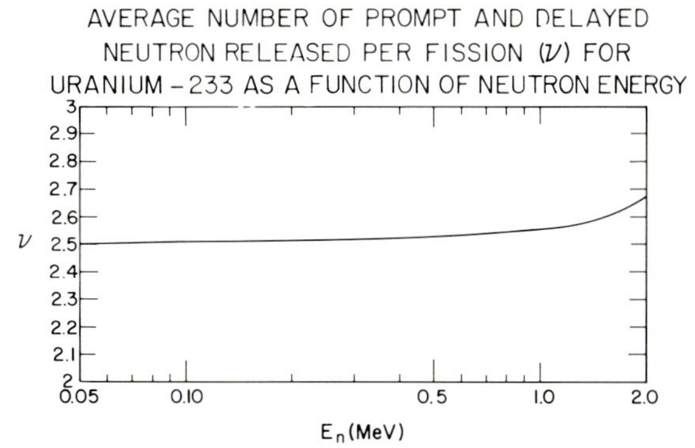

Fig. 3.6 Average number of prompt and delayed neutrons released per fission (ν) for uranium-233 as a function of neutron energy [12].

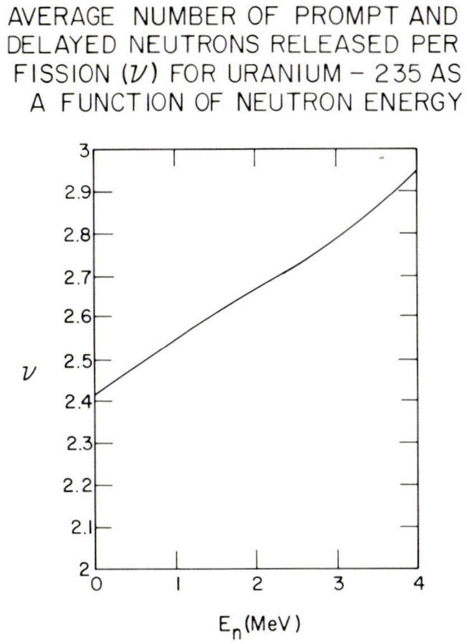

Fig. 3.7 Average number of prompt and delayed neutrons released per fission (ν) for uranium-235 as a function of neutron energy [12].

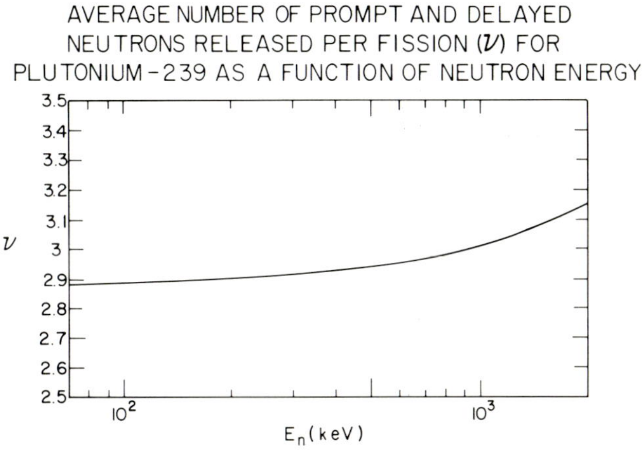

Fig. 3.8 Average number of prompt and delayed neutrons released per fission (ν) for plutonium-239 as a function of neutron energy [12].

Fundamentals of Nuclear Reactor Theory

Neutron Interactions

To understand the design and operation of a nuclear reactor we must first be able to describe how neutrons are distributed throughout the reactor system. As discussed in Chapter 2, a complete description of neutron behavior is a very complex undertaking and requires the solution of the Boltzmann transport equation. An approximate representation of the neutron distribution can be found by using the diffusion approximation. We will introduce some of the essential equations associated with this diffusion approximation and leave the more rigorous developments of both diffusion theory and transport theory to more formal treatments of nuclear

Table 3.6 Fissile nuclide data for thermal neutrons (0.0253 eV) [12].

Fissile Nuclide	σ_a* (barns)	σ_f (barns)	ν	α**	η
$^{233}_{92}U$	578.8	531.1	2.492	0.090	2.287
$^{235}_{92}U$	680.8	582.2	2.418	0.169	2.068
$^{239}_{94}Pu$	1011.3	742.5	2.871	0.362	2.108

$*\sigma_a \equiv \sigma_f + \sigma_\gamma$
$**\alpha \equiv \sigma_\gamma / \sigma_f$

Table 3.7 Selected fission neutron emission ($\bar{\nu}$) data (experimental) [1, 10].

Nuclide	Neutron Energy (MeV)	Number of Neutrons Released for Fission [$\bar{\nu}$] (both prompt & delayed)
$^{233}_{92}U + ^{1}_{0}n$	0.0	2.49 ± 0.02
	4.0	3.00 ± 0.12
	15.0	4.33 ± 0.17
$^{235}_{92}U + ^{1}_{0}n$	0.0	2.42 ± 0.02
	0.7	2.48 ± 0.10
	1.0	2.79 ± 0.35
	2.0	2.80 ± 0.15
	5.0	3.18 ± 0.35
	15.0	4.42 ± 0.19
$^{239}_{94}Pu + ^{1}_{0}n$	0.0	2.87 ± 0.03
	2.1	3.12 ± 0.15
	4.0	3.36 ± 0.11
	15.0	4.61 ± 0.20

reactor theory, e.g., [2, 3, 9].

Neutron flux can be expressed simply as

$$\phi = nv \tag{3.9}$$

where ϕ is the monoenergetic neutron flux (neutrons/cm^2-sec)
n is the neutron density (neutrons/cm^3)
v is the (monoenergetic) neutron speed (cm/sec)

Eq. 3.9 is especially significant because it defines a monoenergetic neutron flux. A collision or interaction density F can now be defined in terms of this flux [1–3]; that is,

$$F \equiv \Sigma_t \cdot \phi \text{ (neutrons/cm}^3\text{-sec)} \tag{3.10}$$

where Σ_t is the total macroscopic interaction cross section (cm^{-1}) (again at one particular energy). We now expand these concepts to incorporate neutron energy distributions by writing

$$dF = \Sigma_t(E) \cdot n(E)\,dE \cdot v(E) \tag{3.11}$$

where $n(E)\,dE$ represents the number of neutrons per cm^3 with energies between E and $E + dE$
$\Sigma_t(E)$ is the energy dependent total macroscopic cross section
$v(E)$ is the energy dependent neutron speed.

The total interaction rate becomes

$$F = \int_0^\infty \Sigma_t(E) v(E) n(E)\,dE. \tag{3.12}$$

If we next define an energy dependent neutron flux $\phi(E)$ as

$$\phi(E) \equiv n(e)v(E) \tag{3.13}$$

then, the total neutron interaction rate is given by

$$F = \int_0^\infty \Sigma_t(E)\phi(E)\,dE. \tag{3.14}$$

If we are interested in a certain interaction, eq. 3.14 can be rewritten in the appropriate form. For example, the number of fission reactions per cm^3 is given by

$$F_f = \int_0^\infty \Sigma_f(E)\phi(E)\,dE. \tag{3.15}$$

Diffusion Theory

Fick's law is the fundamental basis of diffusion theory. As originally developed, it represents a description of molecular diffusion under the influence of chemical concentrations. In essence, molecules can be observed to diffuse from regions of high concentration to regions of lower concentration. This chemical diffusion process as described by Fick's law is simply [13]

$$J_{z_1} = -D\frac{\partial n_1}{\partial z} \tag{3.16}$$

where J_{z_1} is the molecular current density for species 1 (molecules/cm^2-sec passing in the z-direction)
n_1 is the molecular concentration of species 1 (molecules/cm^3)
D is the diffusion coefficient (cm^2/sec).

If we assume that neutrons behave in a similar manner, moving from regions of higher concentration to regions of lower concentration, then Fick's law applied to neutron motion becomes [1–3]

$$J_z = -D\frac{d\phi}{dz}\,. \tag{3.17}$$

Here, J_z (in neutrons/cm^2-sec) is the neutron current density, that is—the net number of neutrons that pass through a unit area perpendicular to z-direction per unit time (see Fig. 3.9) and D (in cm) is the diffusion coefficient. The neutron current density vector **J** is

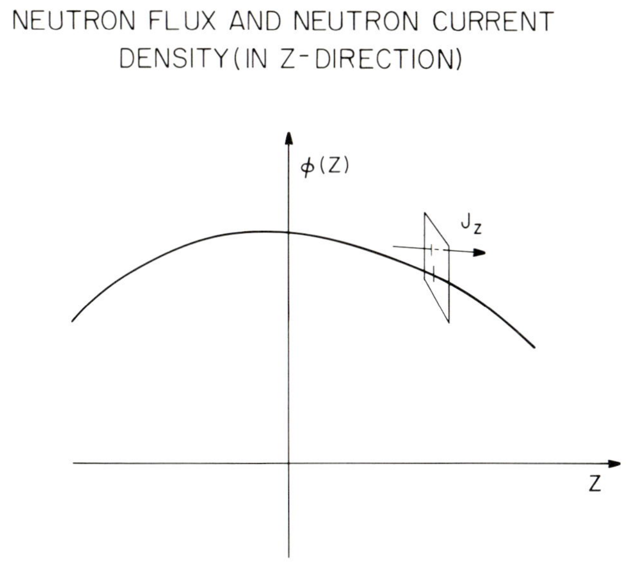

Fig. 3.9 Neutron flux and neutron current density (in z-direction).

expressed in terms of the gradient of the flux as [1–3, 14]

$$\mathbf{J} = -D \operatorname{grad} \phi = -D \nabla \phi \tag{3.18}$$

where $\nabla = \hat{i}\partial/\partial x + \hat{j}\partial/\partial y + \hat{k}\partial/\partial z$ is a vector differential operator. In this approximation, the diffusion coefficient D is given by [1–3, 14]

$$D = \tfrac{1}{3}\lambda_{tr} \tag{3.19}$$

where λ_{tr} is the transport mean free path (cm)

$$\lambda_{tr} = 1/\Sigma_{tr} = 1/\left[\Sigma_s(1-\bar{\mu})\right] \tag{3.20}$$

where Σ_{tr} is the macroscopic transport cross section (cm^{-1})
Σ_s is the macroscopic scattering cross section of the medium (cm^{-1})
$\bar{\mu}$ is the average cosine of the neutron scattering angle in the medium.

Here $\bar{\mu} = 2/(3A)$ where A is the atomic mass of the scattering medium nuclides.

There are several limitations in this Fick's law model of neutron diffusion. It is not valid in a medium that is a strong neutron absorber. It is not valid when neutron scattering is strongly anisotropic. Finally, it is not valid within approximately three mean free paths of either a boundary of the medium or a neutron source. Despite such limitations, the diffusion approximation can be used to provide a first estimate of reactor properties.

Neutron Continuity Equation

The fundamental principle of nuclear reactor theory is the neutron continuity equation or neutron conservation equation [2, 3, 14]. For a given volume, this equation states that the time rate of change of neutron density, $\partial n/\partial t$, is equal to the neutron production rate minus the rate of neutron absorption and minus the rate of neutron leakage. The neutron continuity equation can therefore be written as

$$\partial n/\partial t = (\text{Neutron Production}) - (\text{Neutron Absorption}) - (\text{Neutron Leakage}). \tag{3.21}$$

If our reactor system is in steady state, then $\partial n/\partial t = 0$ and the neutron continuity equation becomes

$$(\text{Neutron Production}) = (\text{Neutron Absorption}) + (\text{Neutron Leakage}). \tag{3.22}$$

The Boltzmann transport equation is a fundamental equation for expressing this continuity principle. Full treatment of the neutron continuity principle, however, is beyond the scope or purpose of this book. It is sufficient here to explore the last two equations above using diffusion theory. Thus, in the monoenergetic neutron model, the diffusion theory continuity equation is [1–3, 14]

$$\frac{\partial n}{\partial t} = S - \Sigma_a\phi - \operatorname{div}\mathbf{J} \tag{3.23}$$

where $\partial n/\partial t$ is the time rate of change of neutron density (neutrons/cm^3-sec)
S is the rate at which neutrons are emitted from sources within the medium (neutrons/cm^3-sec)
$\Sigma_a\phi$ is the rate at which neutrons are consumed by absorption within the medium (neutrons/cm^3-sec)
$\operatorname{div}\mathbf{J}$ is the divergence of the neutron current density (neutrons/cm^3-sec).

In rectangular coordinates

$$\operatorname{div}\mathbf{J} = \frac{\partial J_x}{\partial x} + \frac{\partial J_y}{\partial y} + \frac{\partial J_z}{\partial z}.$$

It should be noted that the divergence of the current density represents the term that describes neutron leakage in the diffusion approximation. For $\partial n/\partial t = 0$, eq. 3.23 becomes

$$S = \Sigma_a\phi + \operatorname{div}\mathbf{J}. \tag{3.24}$$

Equation 3.24 is called the steady state neutron continuity equation.

We now invoke the diffusion approximation (i.e., Fick's law) to describe neutron leakage by substituting eq. 3.18 into eq. 3.23 and by assuming that the diffusion coefficient is independent of location. Thus, we obtain

$$\frac{\partial n}{\partial t} = S - \Sigma_a \phi - \text{div}(-D\nabla\phi). \quad (3.25)$$

But the divergence of the gradient of a scalar function is an operator called the Laplacian. That is, div(grad ϕ) $= \nabla^2\phi$. In rectangular coordinates, $\nabla^2\phi = \partial^2\phi/\partial x^2 + \partial^2\phi/\partial y^2 + \partial^2\phi/\partial z^2$. Recalling eq. 3.9 ($\phi = nv$), we can write eq. 3.25 as

$$\frac{1}{v}\frac{\partial \phi}{\partial t} = S + D\nabla^2\phi - \Sigma_a\phi. \quad (3.26)$$

This is the time-dependent (monoenergetic) diffusion equation. The steady state form of the diffusion equation is

$$D\nabla^2\phi - \Sigma_a\phi + S = 0. \quad (3.27)$$

We next define a parameter, L^2, called the diffusion area, such that

$$L^2 \equiv D/\Sigma_a. \quad (3.28)$$

With this parameter introduced into eq. 3.27, the steady state diffusion equation becomes

$$\nabla^2\phi - (\phi/L^2) + (S/D) = 0. \quad (3.29)$$

The diffusion length is defined as [2, 3]

$$L \equiv \sqrt{D/\Sigma_a}\ (\text{cm}). \quad (3.30)$$

Physically, the diffusion length squared represents one-sixth of the mean square distance ($\bar{r}^2$), called the, crow flight distance, that a neutron travels from the point where it is emitted to the point where it is absorbed (see Fig. 3.10). That is, [1, 3]

$$L^2 = (1/6)\bar{r}^2. \quad (3.31)$$

In an analogous definition, the thermal diffusion length squared (L^2_{th}) represents one-sixth the mean square distance, crow flight distance, that a thermal neutron travels from the point where it just becomes thermal to the point where it is absorbed [2, 3, 14, 15].

There are several general boundary conditions that are used in solving the monoenergetic diffusion equation. First, the neutron flux must be finite and nonnegative in all regions of the medium, except perhaps for certain source regions. Second, the flux ϕ and the neutron current density **J** are continuous functions at the interface between two media that have different diffusion properties. Third, at the outer boundary of a reactor system (i.e., the vacuum interface) there can be no neutrons returning to the medium from this vacuum in any direction; or, stated differently, the flux must vanish at some finite distance, called the extrapolation distance d, outside the actual geometric boundary [2, 3, 14] (see Fig. 3.11).

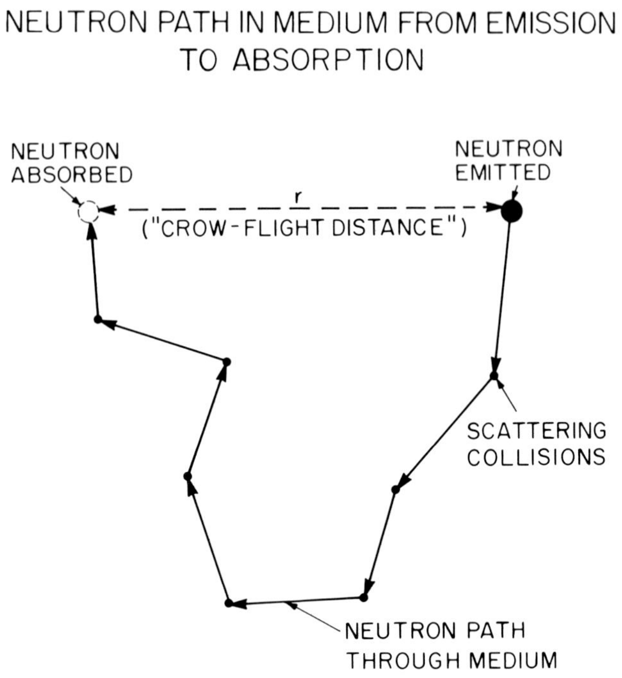

Fig. 3.10 Neutron path in medium from emission to absorption.

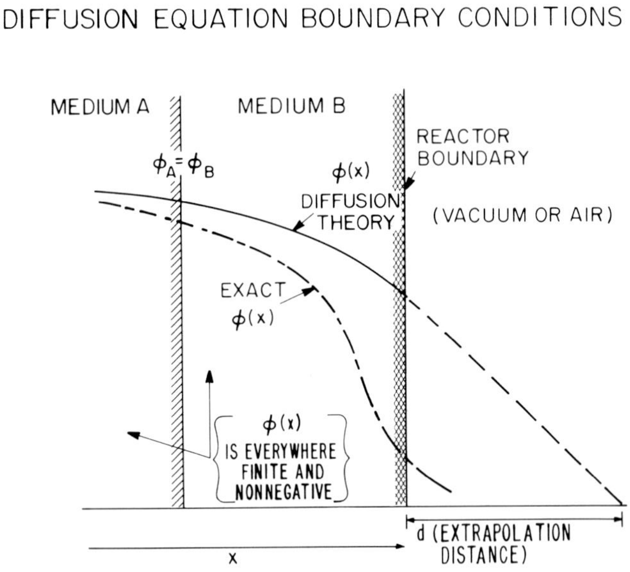

Fig. 3.11 Diffusion equation boundary conditions.

For many practical cases in diffusion theory, this extrapolation distance is given by [2, 3, 14]

$$d = 0.7104\lambda_{\text{tr}} \quad (3.32)$$

where λ_{tr} is the transport mean free path (cm) for the medium, and

$$\lambda_{\text{tr}} = 3D. \quad (3.33)$$

Before we investigate other aspects of reactor theory,

it is interesting to examine the solution of the steady state diffusion equation for an isotropic point source, S (neutrons/sec), in an infinite, homogeneous diffusion medium. If we assume that this point source is at the center of our coordinate system (spherical coordinates used with radius r), then the diffusion equation (eq. 3.26) becomes [1–3, 14] for $r > 0$

$$\frac{1}{r^2}\frac{d}{dr}r^2\left(\frac{d\phi(r)}{dr}\right) - \frac{1}{L^2}\phi(r) = 0. \tag{3.34}$$

The flux distribution (in spherical coordinates with radius r) is given by (1–3, 14]

$$\phi(r) = \frac{Se^{-r/L}}{4\pi Dr}. \tag{3.35}$$

Another one-dimensional diffusion theory problem of interest is an isotropic line source in an infinite diffusing medium. For convenience, cylindrical geometry assuming axial symmetry is invoked. The steady state diffusion equation for $r > 0$ becomes

$$\frac{1}{r}\frac{d}{dr}\left[r\frac{d}{dr}\phi(r)\right] - \frac{1}{L^2}\phi(r) = 0. \tag{3.36}$$

The flux distribution due to this isotropic line source in an infinite diffusing medium is given by [14]

$$\phi(r) = \frac{SK_0(r/L)}{2\pi D} \tag{3.37}$$

where $K_0(r/L)$ is the modified Bessel function of order zero of second kind. Table 3.8 lists values for $K_0(x)$ as a function of x [1, 16]. These data are useful in solving eq. 3.37.

Finally, consider an isotropic plane source of S neutrons per unit area per unit time in an infinite diffusing medium. The diffusion equation in rectangular coordinates is given by

$$\frac{d^2\phi(x)}{dx^2} - \frac{1}{L^2}\phi(x) = 0 \tag{3.38}$$

for $x \neq 0$. The spatial distribution of the neutron flux in the infinite diffusing medium is given by

$$\phi(x) = \frac{SLe^{-|x|/L}}{2D}. \tag{3.39}$$

Bare Reactors

As discussed earlier, the multiplication factor (k) is used to describe the fission chain reaction. This parameter is an indicator of the neutron balance in a particular nuclear reactor; that is, k is a measure of the balance between the number of neutrons lost from the reactor by either leakage or absorption and the number produced by the fission reaction. One of the main engineering challenges facing the designers of space nuclear power systems is the determination of the material composition and geometric dimensions necessary to achieve this neutron balance at various power levels.

Table 3.8 Modified Bessel function of order zero of second kind $[K_0(x)]$ [16].

x	$K_0(x)$
0.0	∞
0.1	2.427
0.2	1.753
0.3	1.372
0.4	1.115
0.5	0.9244
0.7	0.6605
1.0	0.4210
2.0	0.1139
3.0	0.0347
4.0	0.0112

We will introduce the concept of criticality calculations using the one group diffusion approximation. To start this investigation, we will first consider a bare, critical fast reactor. This reactor is called bare, since it consists of only one region, the core, and does not have a reflector. The core is assumed to consist of a uniform (homogeneous) mixture of fuel and coolant and all neutrons are treated as one energy group, fast. This bare, homogeneous fast reactor can be described by the following one group diffusion equation [1–3, 10, 14]

$$D\nabla^2\phi - \Sigma_a\phi + S = 0 \tag{3.40}$$

where D is the group diffusion coefficient (cm)
Σ_a is the group macroscopic absorption cross section for the fuel and coolant mixture (cm^{-1})
ϕ is the monoenergetic flux of fast neutrons (neutrons/cm^2-sec)
S is the neutron source density (neutrons/cm^3-sec).

In the critical reactor under consideration here, the neutron source density can be found from

$$S = \eta(\Sigma_a)_{fuel}\phi \tag{3.41}$$

where η is the average number of fission neutrons emitted per neutron absorbed in the nuclear fuel.
$(\Sigma_a)_{fuel}$ is the group macroscopic absorption cross section for the nuclear fuel (cm^{-1}).

The fuel utilization, f, is then defined as

$$f \equiv (\Sigma_a)_{fuel}/\Sigma_a. \tag{3.42}$$

Table 3.9 Typical fast reactor-one group data.

Nuclide or Element	ν	σ_f (barns)	σ_γ (barns)	σ_{tr} (barns)
$^{235}_{92}U$	2.6	1.4	0.25	6.8
$^{238}_{92}U$	2.6	0.095	0.16	6.9
$^{239}_{94}Pu$	2.98	1.85	0.26	6.8
Fe	—	—	0.006	2.7
Al	—	—	0.002	3.1
Na	—	—	0.0008	3.3

Note: $\sigma_a = \sigma_f + \sigma_\gamma$

Physically, the term f describes the fraction of absorbed neutrons that are absorbed by the fuel in the reactor. Table 3.9 presents typical one-group data for a fast reactor [1, 10].

The neutron source density can then be expressed as

$$S = \eta f \Sigma_a \phi. \tag{3.43}$$

We will now consider the multiplication factor, k_∞, for a critical fast reactor that is infinite in extent—that is, neutron leakage from this system does not occur. In this hypothesized infinite fast reactor, all neutrons are either absorbed in the fuel or in the coolant. Furthermore, the flux is constant and independent of position. If we recall the definition of the multiplication factor (eq. 3.1) as the number of fissions or neutrons in one generation divided by the number of fissions or neutrons in the preceding generation, then the absorption of $\Sigma_a\phi$ neutrons in one generation in this bare, infinite homogeneous fast reactor leads to absorption of $\eta f \Sigma_a \phi$ neutrons in the next generation. That is [1–3, 10, 14]

$$k_\infty = \frac{\eta f \Sigma_a \phi}{\Sigma_a \phi} = \eta f. \tag{3.44}$$

The value of k_∞ is the same for an infinite reactor as for a bare reactor of the same composition, since the terms f and η depend only on the material composition of the reactor [1–3]. Thus, eq. 3.43 and eq. 3.44 can be combined to form

$$S = k_\infty \Sigma_a \phi. \tag{3.45}$$

If we substitute eq. 3.45 into eq. 3.40, we obtain the one-group reactor equation [1–3]; namely,

$$D\nabla^2\phi - \Sigma_a\phi + k_\infty \Sigma_a \phi = 0. \tag{3.46}$$

By rearranging eq. 3.46 and defining, respectively, the one group diffusion area and the buckling as

$$L^2 \equiv D/\Sigma_a \tag{3.47}$$

and

$$B^2 \equiv \frac{k_\infty - 1}{L^2} \tag{3.48}$$

we obtain the one-group reactor equation [1–3]

Table 3.10 Critical bare reactor flux and buckling values.

Geometry*	Flux (ϕ)	Buckling (B^2)
Infinite Slab [thickness (x) = A]	$\cos\left(\frac{\pi x}{A}\right)$	$(\pi/A)^2$
Infinite Cylinder [radius (r) = R]	$J_0\left(\frac{2.405\, r}{R}\right)$**	$(2.405/R)^2$
Finite Cylinder [radius (r) = R] [height (z) = H]	$J_0\left(\frac{2.405\, r}{R}\right)\cos\left(\frac{\pi z}{H}\right)$	$(2.405/R)^2 + (\pi/H)^2$
Rectangular Parallelepiped [length (x) = A] [width (y) = B] [height (z) = C]	$\cos\left(\frac{\pi x}{A}\right)\cos\left(\frac{\pi y}{B}\right)\cos\left(\frac{\pi z}{C}\right)$	$(\pi/A)^2 + (\pi/B)^2 + (\pi/C)^2$
Sphere [radius (r) = R]	$\left(\frac{1}{r}\right)\sin\left(\frac{\pi r}{R}\right)$	$(\pi/R)^2$

* Dimensions are extrapolated values.
** $J_0(x)$ is Bessel function of the first kind of zero order.

$$\nabla^2\phi + B^2\phi = 0. \tag{3.49}$$

Combined with appropriate flux boundary conditions, eq. 3.49 provides the shape of the flux in a bare reactor and identifies conditions for criticality. The buckling parameter, B^2, is a measure of the extent to which the flux curves or buckles in the reactor. Of course, in an infinite reactor the flux is constant and exhibits no buckle, so that $B^2 = 0$. Mathematically, the B^2 term in eq. 3.49 is the lowest eigenvalue of the one group reactor equation for a critical reactor subject to the boundary condition that the flux go to zero at the extrapolated boundary. Table 3.10 summarizes the values of buckling and flux for critical bare reactors of various geometries [1–3]. All dimensions shown are extrapolated dimensions.

Fermi Theory of Bare Thermal Reactors

Consider now an infinite, homogeneous mixture of fuel and moderator. In this case, we assume that the ratio of moderator atoms to fuel atoms is very large, so that the majority of fission neutrons slow down to thermal energies before they initiate fission reactions. In other words, we now have a one-group model where the predominant neutron energy is thermal. Fermi age theory, named after the great Italian-American physicist, Enrico Fermi, is used to provide a slowing down source term to describe neutrons moderating from fast or fission energies to thermal energies. This simple, but powerful, model is completed by introducting fast (P_f) and thermal (P_t) nonleakage probabilities to compensate for leakage from finite-size reactor systems. Because we have assumed that the reactor is homogeneous and infinite, the thermal flux ϕ_t and the slowing down density q are independent of location.

Table 3.11 Selected slowing down data.

Material	A	α	ξ
Hydrogen (H)	1	0	1.000
Beryllium (Be)	9	0.640	0.209
Carbon (C)	12	0.716	0.158
Sodium (Na)	23	0.840	0.0825
Iron (Fe)	56	0.931	0.0357
Tungsten (W)	184	0.978	0.0108
Uranium (U)	238	0.983	0.0084
Plutonium (Pu)	239	0.983	0.0083

The slowing down density $q(\bar{r}, E)$ can be defined as the number of neutrons per unit volume (neutrons/cm^3) at the space point $\bar{r}$ whose energies fall below E per unit time (sec). However, since neutrons lose energy discretely, rather than continuously, by means of individual collisions, the slowing down density $q(\bar{r}, E)$ is equivalent to the total number of collisions per unit volume at $\bar{r}$ in which the energy of the colliding neutron falls below E [1–3].

It is now convenient to express the collision density, $F(E)$, in terms of a parameter called lethargy which can be written as

$$u \equiv \ln(E_0/E) \tag{3.50}$$

where E_0 is an arbitrary neutron energy datum (usually chosen as the most energetic). With every collision, a neutron's lethargy increases, since its energy decreases. Thus, as a result of a single collision

$$\Delta u = \ln(E_0/E') - \ln(E_0/E) = \ln(E/E') \tag{3.51}$$

where E is the neutron energy before collision (eV) and E' is the neutron energy after collision (eV).

The average increase in lethargy per collision, denoted ξ, can be obtained from the integral [1–3]

$$\overline{\Delta u} = \xi = \int_{\alpha E}^{E} \ln(E/E')P(E \to E')\,dE' \tag{3.52}$$

where here α is a "scattering" parameter defined as

$$\alpha \equiv \left[\frac{(A-1)}{(A+1)}\right]^2 \tag{3.53}$$

and $P(E \to E')$ is the probability distribution function, such that $P(E \to E')\,dE'$ is the probability that a neutron with energy E will emerge from a collision with an energy between E' and $E' + dE'$. In terms of mass number A, ξ is given by

$$\xi = 1 - \frac{(A-1)^2}{2A}\ln\left[\frac{(A+1)}{(A-1)}\right]. \tag{3.54}$$

Table 3.11 presents slowing down data for selected materials [1–3, 10]. In the Fermi slowing down model, all neutrons are treated as average neutrons, so that they all gain exactly ξ in lethargy at every collision. Figure 3.12 illustrates this continuous slowing down approximation for a single neutron. The slowing down of neutrons from fission energy E_0 to thermal energy E_{th} appears as a series of steps of approximately ξ height but gradually increasing time. The horizontal lines in Fig. 3.12 represent a constant neutron energy level, while it is traveling between collisions; the length of these horizontal lines depicts the time elapsing between consecutive collisions. This Fermi model, or continuous slowing down model, is more appropriate for heavier moderators (as opposed to hydrogen), since the variations in average behavior decrease with increasing moderator mass number. In the Fermi model, it is also assumed that diffusion theory is valid at all neutron energies.

To illustrate the application of Fermi age theory, we

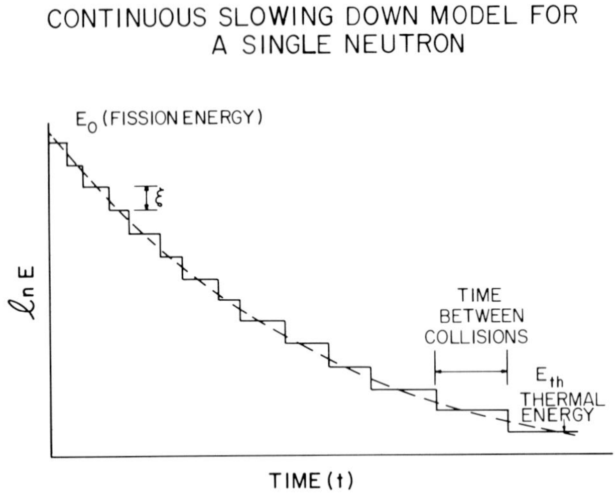

Fig. 3.12 Continuous slowing down model for a single neutron.

now consider an infinite slab of moderator that contains at its center an infinite planar source of S fast neutrons/cm^2-sec at the (fission) energy E_0. The slowing down density is now given by

$$q(x,\tau) = \frac{S}{\sqrt{4\pi\tau}} e^{-x^2/(4\tau)} \tag{3.55}$$

where τ is the Fermi age (cm^2). The Fermi age can be interpreted as being physically equal to one-sixth of the mean square (crow-flight) distance traversed by a neutron from the time of its emission (that is, age = 0) to the time that its age is τ. It should be noted, however, that τ has dimensions of length squared rather than time. While the concept of Fermi age applies to neutrons of any energy, it is of particular interest in the consideration of thermal neutrons. Consequently, $\sqrt{\tau_{th}}$ can be considered as the slowing down length of the thermal neutrons. The Fermi age (τ) is an important parameter in establishing the extent of neutron leakage during slowing down in a finite-dimensioned thermal reactor. A typical value of Fermi age for thermal neutrons from a fission source for beryllium at 20° C is 98 cm^2 [2].

Criticality in Bare Thermal Reactors

As discussed previously, the minimum condition for a sustained chain reaction is that on the average each fission event produces at least one neutron that survives to cause a fission reaction in the next generation. The size of the multiplication factor k in any reactor system is dependent upon the relative magnitude of four competing processes. These are: (1) leakage of neutrons out of the system; (2) nonfission or parasitic capture of neutrons by the fuel; (3) nonfission or parasitic capture of the neutrons by moderator, coolant, structural materials, etc.; and (4) fission capture by the fuel. Of course, only the last process releases neutrons that sustain the chain reaction; the number of neutrons released must offset neutron losses that occur in the first three processes.

Consider an infinite, homogeneous reactor system, consisting of fuel and moderator. The quantity f introduced in eq. 3.42 as the fuel utilization, is known as the "thermal utilization" in thermal reactors. It is simply the fraction of thermal neutrons absorbed in the fuel. That is,

$$f = \frac{\text{number of thermal neutrons absorbed in fuel}}{\text{total number of thermal neutrons absorbed in the reactor}} . \tag{3.56}$$

We can also state the thermal utilization as [1–3, 7]

$$f = (\bar{\Sigma}_a)_{\text{fuel}}/\bar{\Sigma}_a \tag{3.57}$$

where $\bar{\Sigma}_a$ is the macroscopic absorption cross section of the fuel-moderator mixture (i.e., $\bar{\Sigma}_a = (\bar{\Sigma}_a)_{\text{fuel}} + (\bar{\Sigma}_a)_{\text{moderator}}$)

$(\bar{\Sigma}_a)_{\text{fuel}}$ is the macroscopic absorption cross section of the fuel.

Recalling eq. 3.5, we next define η_t as the average number of fission neutrons emitted per thermal neutron absorbed in the fuel. Consequently, $\eta_t f \bar{\Sigma}_a \phi_t$ fast neutrons are emitted per cm^3/sec for $f\bar{\Sigma}_a\phi_t$ neutrons absorbed per cm^3/sec in the fuel, where ϕ_t is the thermal flux.

Since some fast neutrons in a thermal reactor will induce fission reactions in both fissile and fertile materials, we take these fast fissions into account by means of the fast fission factor ϵ which is defined as

$$\epsilon \equiv \frac{\text{number of fission neutrons from both fast and thermal fission}}{\text{number of fission neutrons generated by thermal fission alone}} . \tag{3.58}$$

Let us recap up to this point that the term $\eta_t \epsilon f \bar{\Sigma}_a \phi_t$ now represents the number of fission neutrons generated per cm^3/sec in our infinite, homogeneous thermal reactor.

The majority of neutrons in an infinite, homogeneous reactor are absorbed after they have slowed down to thermal energies. However, some of these neutrons are, in fact, absorbed while slowing down by certain nuclides that possess absorption resonances (for example, uranium-238). We, therefore, define a factor p called the resonance escape probability as follows [1–3, 7]

$$p \equiv \frac{\text{number of fission neutrons that escape resonance capture during slowing down}}{\text{total number of fission neutrons produced}} . \tag{3.59}$$

Consequently, in our infinite homogeneous thermal reactor, if $\overline{\Sigma}_a\phi_t$ thermal neutrons are absorbed, this will lead to the production of $p\eta_t f\epsilon\phi_t\overline{\Sigma}_a$ new thermal neutrons. Placing this in the context of the multiplication factor for an infinite system yields

$$k_\infty = \frac{p\eta_t f\epsilon\phi_t\overline{\Sigma}_a}{\overline{\Sigma}_a\phi_t} \tag{3.60}$$

or simply,

$$k_\infty = \eta_t\epsilon p f. \tag{3.61}$$

Equation 3.61 is known as the four-factor formula for an infinite homogeneous thermal reactor.

Of course, our infinite medium multiplication factor must be modified for a reactor's finite size. If we let P be defined as the probability that a neutron will not escape or leak out of the reactor during slowing down or as a thermal neutron, then the condition for a self-sustained chain reaction becomes

$$k_\infty P = 1 \tag{3.62}$$

where k_∞ is given by eq. 3.61. The nonleakage probability term P can be divided into two parts: P_f is the fast neutron nonleakage probability and P_{th} is the thermal nonleakage probability, such that

$$P = P_f P_{\text{th}}\,. \tag{3.63}$$

The proportion of neutrons leaking out or escaping from a finite size reactor can be reduced by either increasing the size of the reactor or by using a reflector. The ratio of neutrons leaking out to those absorbed in the reactor system is proportional to the external surface-to-volume ratio of the system. The nonleakage probability is determined by the reactor's shape. For a given volume, for example, a spherical reactor will have the minimum surface-to-volume ratio and, consequently, the smallest critical volume.

Considering a two-group (fast and thermal) reactor equation and invoking Fermi age theory, the solution of the continuity equation (eq. 3.21) becomes [1–3]

$$1 = \frac{k_\infty}{(1 + B^2\tau_{\text{th}})(1 + B^2L^2_{\text{th}})} \tag{3.64}$$

where B^2 is the buckling (cm^{-2})
L^2_{th} is the thermal diffusion area (cm^2)
τ_{th} is the Fermi or neutron age to thermal energies (cm^2).

It is important to recognize that eq. 3.64 is the two-group (fast and thermal) equation for a bare thermal reactor. The fast nonleakage probability can be expressed as

$$P_f = \frac{1}{(1 + B^2\tau_{\text{th}})} \tag{3.65}$$

while the thermal nonleakage probability is given by

$$P_{\text{th}} = \frac{1}{(1 + B^2L^2_{\text{th}})}\,. \tag{3.66}$$

Therefore, the multiplication factor for a bare, finite thermal reactor can be expressed in terms of these probabilities as

$$k = k_\infty P_f P_{\text{th}} \tag{3.67}$$

and the critical equation evolves from setting $k = 1$ for the particular system.

Reflected Reactors

The neutron economy of a reactor is improved considerably if the core is surrounded by a reflector or a blanket of material such as beryllium or graphite (carbon) that has a high scattering cross section and low neutron absorption characteristics. Neutrons leaking out of the core scatter in the reflector material and some of these return to the core. The use of a reflector helps reduce the critical mass of the system and sup-

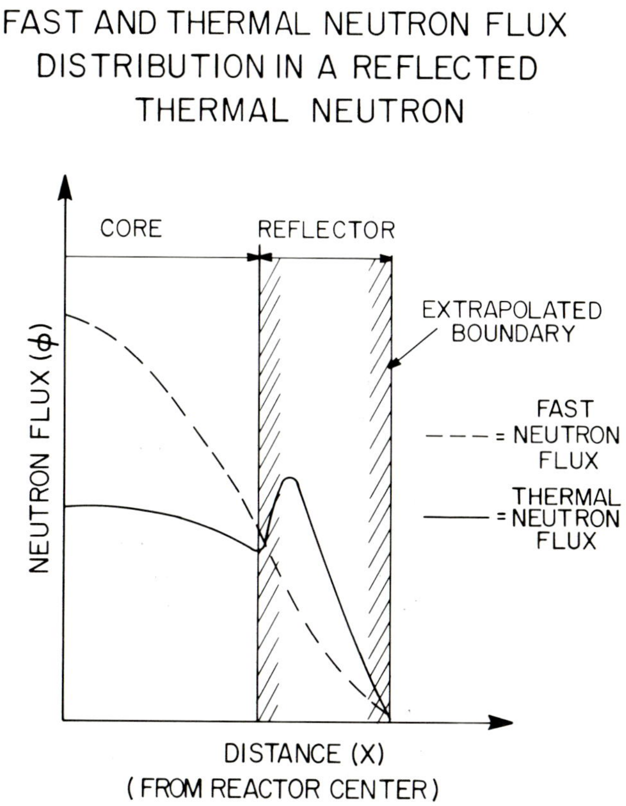

Fig. 3.13 Fast and thermal neutron flux distribution in a reflected thermal reactor.

Table 3.12 Power-dependent flux distributions for various bare critical reactors.

Reactor Geometry	Dimension(s)	Power Dependent Flux (ϕ)	Maximum-to-Average Flux Ratio (ϕ_{max}/ϕ_{ave})
Sphere	R (radius)	$\phi = \frac{P}{4R^2\Lambda\Sigma_f}\left(\frac{1}{r}\right)\sin\left(\frac{\pi r}{R}\right)$	3.29
Infinite Cylinder	R (radius)	$\phi = \frac{0.738\,P}{R^2\Lambda\Sigma_f}J_0\left(\frac{2.405\,r}{R}\right)$	2.32
Finite Cylinder	R (radius) H (height)	$\phi = \frac{3.63\,P}{V\Lambda\Sigma_f}\cos\left(\frac{\pi z}{H}\right)J_0\left(\frac{2.405\,r}{R}\right)$	3.64
Infinite Slab	T (thickness)	$\phi = \frac{1.57\,P}{T\Lambda\Sigma_f}\cos\left(\frac{\pi x}{T}\right)$	1.57

where P is the power density (W/cc), H is the height (cm), R is the radius (cm), V is the volume (cm^3), Σ_f is the macroscopic fission cross section (cm^{-1}), Λ is the recoverable energy per fission ($\approx$ 200 MeV $\approx 3.2 \times 10^{-11}$ joule)

ports flux flattening and a more uniform power generation in the core.

The reflector savings is a term used to describe the decrease in the critical dimensions of a reactor core, resulting from the use of a reflector. For a slab reactor of unreflected core thickness x_0 and reflected core thickness x, the reflector savings is defined as

$$\delta \equiv 1/2(x_0 - x). \tag{3.68}$$

Note that the (1/2) factor in eq. 3.68 is used because the slab reactor has two sides. Similarly, for a spherical reactor [1–3]

$$\delta \equiv R_0 - R \tag{3.69}$$

where R_0 is the bare core radius (cm)
R is the reflected core radius (cm).

Figure 3.13 depicts the fast and thermal neutron flux distributions in the core and reflector of a typical thermal reactor in which the moderator in the core and the reflector are the same type of material. The peaking of the thermal neutron flux in the reflector describes the "leaking" fast flux from the core slowing down in the reflector material with very little absorption. As we approach the extrapolated boundary, both the fast and thermal fluxes fall to zero [1–3].

Power Generation in a Reactor

The maximum flux (ϕ_{max}) in a bare, homogeneous reactor can always be found at the center of the reactor. The power density q'''(W/cm^3) is also highest at the reactor's center, since power density is proportional to flux. In addition, the flux in a reactor is established by the power level P at which the system is operated. Table 3.12 describes the power-dependent flux and the maximum-to-average flux ratios for selected critical bare reactors [1–3]. It should be noted that the energy recovered per fission (Γ) is approximately 200 MeV or 3.2×10^{-11} joule.

The rate of thermal energy generation $q'''(\bar{r})$ at the point $\bar{r}$ per unit volume in a reactor is given by [1–3, 17]

$$q'''(\bar{r}) = \Gamma \int_0^\infty \Sigma_f(E)\phi(\bar{r}, E)\,dE \tag{3.70}$$

where Γ is the energy recovered (locally) in the fuel per fission
$\Sigma_f(E)$ is the energy-dependent macroscopic fission cross section of the fuel (cm^{-1})
$\phi(\bar{r}, E)$ is the energy-dependent flux (neutrons/cm^2-sec)
$q'''(\bar{r})$ is the power density at point $\bar{r}$ (MeV/cm^3-s or W/cm^3).

In a thermal reactor, the majority of the fission reactions are caused by thermal neutrons, thus eq. 3.70 becomes

$$q'''(\bar{r}) = \Gamma\bar{\Sigma}_{f_{th}}\phi_t(\bar{r}) \tag{3.71}$$

where $\bar{\Sigma}_{f_{th}}$ is the average macroscopic fission cross section of the fuel for thermal neutrons (cm^{-1})
$\phi_t(\bar{r})$ is the thermal neutron flux (neutrons/cm^2-sec) at point $\bar{r}$.

In a fast reactor, most of the fission reactions are caused by fast neutrons, so that

$$q'''(\bar{r}) = \Gamma\Sigma_{ff}\phi_f(r) \tag{3.72}$$

where $\bar{\Sigma}_{f_f}$ is the average macroscopic fast neutron fission cross section for the fuel (cm^{-1})
$\phi_f(\bar{r})$ is the fast neutron flux at point $\bar{r}$.

In multigroup calculations, eq. 3.70 takes on the form

$$q'''(\bar{r}) = \Gamma \sum_{i=1}^{k} \Sigma_{f_i} \phi_i(\bar{r}) \tag{3.73}$$

where Σ_{f_i} is the macroscopic fission cross section of the fuel for the ith neutron energy group
$\phi_i(\bar{r})$ is the ith neutron group flux.

The spatial dependence of the neutron flux is a function of the structure and geometry of the reactor. Flux distributions were presented for various bare reactor configurations and these can be used in conjunction with equations 3.70 through 3.73 to describe heat generation within the core. A more detailed examination of heat production within a reactor core, including reflected cores and those containing complex fuel element-coolant passage geometries, is outside the scope or purpose of this particular book. Such considerations generally require sophisticated, computer-assisted analyses of flux behavior and heat transport processes within the core.

Afterheat

Afterheat or decay heat is the thermal energy released by the continuing radioactive decay of nuclides (mostly fission products) in a reactor core after the fission chain reaction has stopped. Once a reactor has operated at power for a few days, the decay of the accumulated fission products, primarily gamma and beta emitters, represents almost 10 percent of the total thermal power output of the system. After shutdown, a reactor will continue to generate thermal power P_s as a function of time. The amount of this postoperational thermal energy release is a function of the power level at which the reactor operated before shutdown (P_0) and the length of time (t_0) it operated at this power level, since both parameters establish the fission product inventory. We now consider a reactor which has operated at the constant power level P_0 for a sufficient period of time, so that the fission products are in equilibrium. We further assume that the afterheat or decay thermal power P_s at any time t_s after shutdown is proportional to the operating power level P_0, since fission product production during reactor operation is proportional to the power level. Figure 3.14 shows the ratio of decay heat to operating power $[P_s(t_s)/P_0]$ as a function of time after shutdown t_s for a thermal reactor fueled with uranium-235. This plot also assumes that 200 MeV per fission are recoverable [1–3, 17–19]. Figure 3.14 also provides an estimate of the core cooling requirement following shutdown. In space nuclear power plants either an auxiliary power supply is required to maintain the flow of the primary coolant through the core or devices, such as heat pipes, are needed to provide a passive postoperational cooling mechanism as an integral part of the design.

Reactor Kinetics

The study of the time-dependent behavior of the neutron population in a nuclear reactor is called reactor kinetics. Included in this area are the effects of prompt fission and delayed neutrons on reactor dynamics; the concept of reactivity; the effect of control rods, control drums, and chemical shims; and the influence of fission product poisoning on reactor performance.

The majority of neutrons released in the fission process appear essentially at the moment of nuclear fission. These neutrons are called prompt fission neutrons or simply prompt neutrons. However, a small fraction of these fission neutrons, designated β and called the delayed neutrons, are released long after the fission event has occurred. These delayed neutrons appear as a

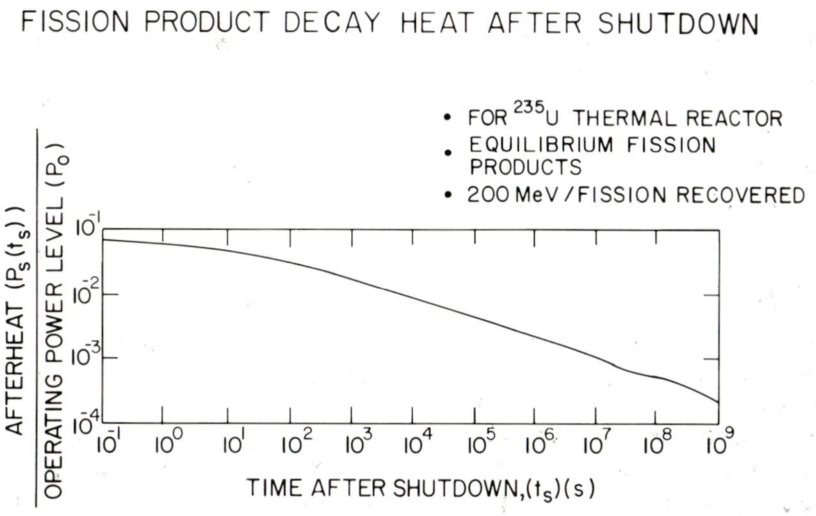

Fig. 3.14 Fission product decay heat as a function of operating power and time after shutdown.

Table 3.13 Delayed neutron data for thermal fission.

Fissile Nuclide	Number of Delayed Neutrons per Fission (total)	Delayed Neutron Fraction [β] (total)	Delayed Group Index (i)	Delayed Group Half-Life (sec) $[T_{1/2}]_i$	Relative Abundance (%)
$^{233}_{92}U$	0.0066	0.0026	1	55.00	8.6
			2	20.57	29.9
			3	5.00	25.2
			4	2.13	27.8
			5	0.615	5.1
			6	0.277	3.4
$^{235}_{92}U$	0.0158	0.0065	1	55.72	3.3
			2	22.72	21.9
			3	6.22	19.6
			4	2.30	39.5
			5	0.61	11.5
			6	0.23	4.2
$^{239}_{94}Pu$	0.0061	0.0021	1	54.28	3.5
			2	23.04	29.8
			3	5.60	21.1
			4	2.13	32.6
			5	0.618	8.6
			6	0.257	4.4

result of the radioactive decay of certain fission product nuclides and their fission-generated precursors. It turns out that the transient behavior of a nuclear reactor can be a function of both types of neutrons: the prompt fission fraction $1 - \beta$ and the delayed fraction β.

The prompt neutron lifetime l_p is defined as the average time that occurs between the emission of prompt neutrons during the fission process and their absorption inside the reactor. Similarly, the mean diffusion time t_d is the time a neutron spends as a thermal neutron before it is eventually absorbed. It corresponds to a neutron's average lifetime in an infinite system. In fast reactors, prompt neutron lifetimes are approximately 10^{-6} to 10^{-8} second, since fast neutrons in such systems are not permitted to reach thermal energies. On the other hand, in a thermal reactor, fast neutrons slow down rapidly and spend the majority of their lifetimes diffusing as thermal neutrons. Thus, for an infinite thermal reactor, $l_p \approx t_d$. In beryllium, the mean diffusion time is approximately 3.9 milliseconds, while in graphite it is approximately 17 milliseconds [1–3].

Reactors Without Delayed Neutrons

We will first consider reactor kinetics for a reactor system in which all neutrons are prompt fission neutrons—that is, an infinite thermal reactor in which we ignore for the moment the existence of delayed neutrons. In the absence of delayed neutrons, the prompt neutron lifetime l_p corresponds to the time between successive neutron generations. Recalling the definition of the multiplication factor k, we can write the following expression for an infinite thermal reactor system:

$$n(t + l_p) = k_\infty n(t) \tag{3.74}$$

where $n(t)$ is the number of fissions occuring at time t
$n(t + l_p)$ is the number of fissions occurring at a time $(t + l_p)$ seconds later.

With some manipulation we find that [1–3]

$$n(t) = n_0 e^{[(k_\infty - 1)/l_p]t}. \tag{3.75}$$

For the case in which the delayed neutrons are ignored, the reactor period T can be expressed as

$$T \equiv (l_p)/(k_\infty - 1). \tag{3.76}$$

Equation 3.75 can now be rewritten as

$$n(t) = n_0 e^{t/T} \tag{3.77}$$

Equation 3.77 is the classic kinetic equation for the prompt neutron only reactor model. Example 3.1 will help illustrate some of these concepts.

Example 3.1. Assume we have a large scale space nuclear power plant that has $k_\infty = 1.0000$ at some initial time $t = 0$ and then has its multiplication factor

(k_∞) increased from 1.0000 to 1.0005 in one prompt neutron lifetime ($l_p \approx 10^{-7}$ sec). What is the response of this reactor to such a change in multiplication factor? Assume the system is infinite in size and neglect the effect of delayed neutrons.

Step 1. The controlling equation for this particular situation is eq. 3.77

$$n(t) = n_0 e^{t/T}.$$

We must first calculate the reactor period T.

$$T = (l_p)/(k_\infty - 1) = 10^{-7}/(1.0005 - 1.0000)$$

$$T = 2 \times 10^{-4} \text{ sec.}$$

Step 2. Using eq. 3.77, we obtain

$$n(t)/n_0 = e^{t/T} = e^{t/(2\times 10^{-4})} = e^{5000t}.$$

Conclusion. The number of fissions and therefore the power level would increase by a factor of e^{5000t} for this fast reactor—even for a very small change in multiplication factor!

Delayed Neutron Groups

As previously mentioned, although delayed neutrons typically represent less than 1 percent of the neutrons liberated in the nuclear fission process, they play a very significant role in reactor control. The delayed neutrons originate from the decay (by neutron emission) of certain fission product nuclides. These delayed neutron precursors have been divided up into six distinct groups, each with a characteristic half-life. Delayed neutron data for selected fissile nuclides are presented in Tables 3.13 and 3.14. The decay for one such delayed neutron precursor is presented in Fig. 3.15 [1–3, 10].

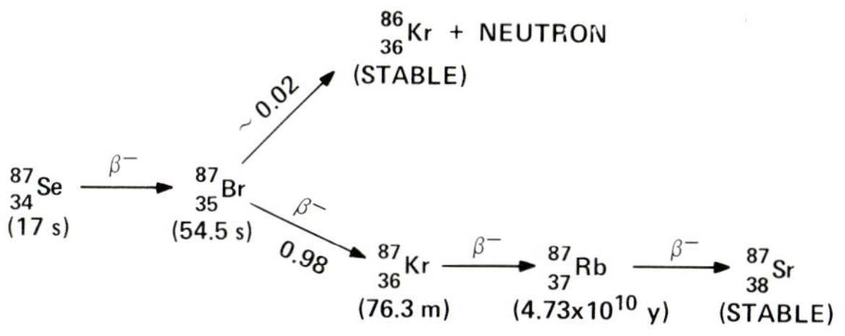

Fig. 3.15 Decay chain for delayed neutron precursor $^{87}_{35}$Br.

To perform credible reactor kinetic calculations, we must consider the contribution of each of the six delayed neutron groups. One way to simplify these calculations is to treat all the delayed neutrons as one single (hypothetical) delayed neutron precursor group—at least as a first approximation [1–3].

The governing concentration equation for this one-group precursor X is

$$\frac{dX}{dt} = \frac{\beta k_\infty \overline{\Sigma}_a \phi_t}{p} - \lambda X \qquad (3.78)$$

Table 3.14 Delayed neutron data for fast fission.

Fissile Nuclide	Number of Delayed Neutrons per Fission (total)	Delayed Neutron Fraction [β] (total)	Delayed Group Index (i)	Delayed Group Half-Life (sec) $[T_{1/2}]_i$	Relative Abundance (%)
$^{233}_{92}$U	0.0070	0.0026	1	55.11	8.6
			2	20.74	27.4
			3	5.30	22.7
			4	2.29	31.7
			5	0.546	7.3
			6	0.221	2.3
$^{235}_{92}$U	0.0165	0.0064	1	54.51	3.8
			2	21.84	21.3
			3	6.00	18.8
			4	2.23	40.7
			5	0.496	12.8
			6	0.179	2.6
$^{239}_{94}$Pu	0.0063	0.0020	1	53.75	3.8
			2	22.29	28.0
			3	5.19	21.6
			4	2.09	32.8
			5	0.549	10.3
			6	0.216	3.5

where X is the number of delayed neutron precursor nuclides (atoms/cm^3)
β is the delayed neutron fraction
λ is the decay constant for delayed neutron precursor group (sec^{-1})
ϕ_t is the thermal flux
$\bar{\Sigma}_a$ is the macroscopic absorption cross section (cm^{-1})
p is the resonance escape probability.

For an infinite, homogeneous thermal reactor we can write the time-dependent diffusion equation as [1–3, 10]

$$(1-\beta)k_\infty\phi_t + \left[(p\lambda X)/\bar{\Sigma}_a\right] - \phi_t = l_p\left[\frac{d\phi_t}{dt}\right] \quad (3.79)$$

where l_p is the prompt neutron lifetime (sec).

Consider now an infinite reactor that is critical ($k_\infty = 1$) at some initial time $t = 0$. We now introduce a step change in the multiplication factor k_∞ such that the reactor then becomes either subcritical ($k_\infty < 1$) or supercritical ($k_\infty > 1$). If we are interested in observing the flux (ϕ_t) as a function of time, we assume the solution has the following form

$$\phi_t(t) = Ae^{\omega t} \quad (3.80)$$

and

$$X(t) = X_0 e^{\omega t} \quad (3.81)$$

where A and X_0 are constants and ω is a root of the equation. Equation 3.79 can be rewritten as

$$\omega l_p = (1-\beta)k_\infty + \frac{\lambda\beta k_\infty}{(\omega+\lambda)} - 1. \quad (3.82)$$

Equation 3.82 can then be formed into

$$\frac{k_\infty - 1}{k_\infty} = \frac{\omega l_p}{(1+\omega l_p)} + \frac{\omega}{(1+\omega l_p)}\left[\frac{\beta}{\omega+\lambda}\right]. \quad (3.83)$$

For an infinite reactor the reactivity ρ is defined as

$$\rho \equiv \frac{k_\infty - 1}{k_\infty} \quad (3.84)$$

and for a reactor of finite size

$$\rho \equiv \frac{(k-1)}{k}. \quad (3.85)$$

The quantity, $k - 1$, is sometimes called the excess reactivity (k_{excess}). We now write eq. 3.83 in terms of reactivity, so that

$$\rho = \frac{l_p\omega}{(l_p\omega + 1)} + \frac{1}{(l_p\omega + 1)}\left[\frac{\beta\omega}{\omega+\lambda}\right]. \quad (3.86)$$

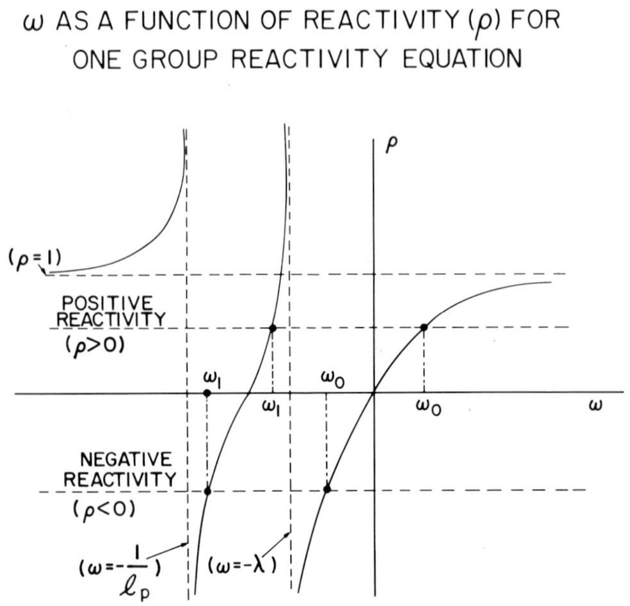

Fig. 3.16 Plot of ω as a function of reactivity (ρ) for one group reactivity equation.

Equation 3.86 is often called the reactivity equation, here expressed for the one-group delayed neutron approximation. The reactivity equation associates the reactivity ρ of the reactor with the appropriate values of ω by which equations 3.80 and 3.81 are solutions of equations 3.78 and 3.79. To help solve for the proper values of ω, we can plot eq. 3.86 for specific values of reactivity ρ (see Fig. 3.16). By inspection of Fig. 3.16, we observe two roots for eq. 3.86, ω_0 and ω_1 for either positive or negative reactivity values. Consequently, the time-dependent flux behaves as

$$\phi_t(t) = C_1 e^{\omega_0 t} + C_2 e^{\omega_1 t} \quad (3.87)$$

where C_1 and C_2 are constants. If we again inspect Fig. 3.16, we note that for either $\rho > 0$ or $\rho < 0$, the root ω_0 dominates eq. 3.87 after a short period of time. That is

$$\phi_t(t) \rightarrow e^{\omega_0 t}. \quad (3.88)$$

The reactor period T is defined as the time required for the neutron flux or number of neutrons to change by a factor e, therefore

$$T \equiv 1/\omega_0. \quad (3.89)$$

This relationship is sometimes called the stable reactor period, while the quantity $1/\omega_1$ is called the transient reactor period. Thus, the time-dependent behavior of the reactor flux can be described as

$$\phi_t(t) \rightarrow e^{t/T}. \quad (3.90)$$

Of course, our treatment here addressed only the rather

simplified one-delayed group model. More precise reactor kinetic calculations require the use of all six delayed neutron groups. The general reactivity equation then becomes [1–3]

$$\rho = \frac{l_p\omega}{(l_p\omega + 1)} + \frac{1}{(l_p\omega + 1)} \sum_{i=1}^{6} \frac{\beta_i\omega}{\omega + \lambda_i} \qquad (3.91)$$

where β_i is the delayed neutron fraction associated with the ith delayed group

λ_i is the decay constant associated with the ith delayed neutron group (sec^{-1}).

When a nuclear reactor remains a critical chain reaction due to prompt neutrons alone, it is said to be prompt critical. For this condition

$$(1 - \beta)k = 1. \qquad (3.92)$$

Using eq. 3.84 or 3.85, we find that for the prompt criticality conditon, $\rho = \omega$. At this point and beyond, the delayed neutron fraction no longer exerts any control influence on the transient behavior of the reactor's neutron population. Consequently, in normal reactor operating procedures, positive reactivity additions are usually limited to values less than β to avoid the creation of very short reactor periods.

A unit of reactivity, called the dollar ($), is defined as the amount of reactivity needed to make a reactor prompt critical. A cent (¢) of reactivity is the reactivity equivalent of one-hundreth of a dollar of reactivity. It should be noted that the value of β varies as a function of fissile material (see Tables 3.13 and 3.14). Therefore, a dollar of reactivity for a thermal reactor fueled with uranium-233 corresponds to $1\$(\rho) = \beta = 0.0026$; while a dollar of reactivity for a fast reactor fueled with uranium-235 corresponds to $1\$(\rho) = \beta = 0.0064$.

Example 3.2. Let us now assume that the reactor described in example 3.1 is a fast reactor fueled with uranium-235. What is the overall change in reactivity (in $), if the multiplication factor increases from 1.0000 to 1.0005?

Step 1. $\rho \equiv (k - 1)/k = (1.0005 - 1)/1.0005$

$$\rho \approx 0.0005$$

Step 2. But for ^{235}U (fast fission), $\beta = 0.0064$. Therefore,

$$\rho(\$) = \frac{\Delta\rho}{\beta} = \frac{0.0005}{0.0064} = 0.0781\$ \approx 7.8¢.$$

Again, the influence of delayed neutrons in determining transient reactor behavior is clearly illustrated. Without delayed neutron influence, even this relatively small positive reactivity change would have caused enormous neutron population increases.

Reactor Control

Generally, the level of criticality in a nuclear reactor is controlled by using control rods or drums and possibly chemical shims or burnable poisons. A control rod contains neutron absorbing materials or neutron poisons. When a reactor is critical (i.e., $k = 1$), if we move the control rod out of the core, the overall reactivity is increased and the reactor becomes supercritical. Conversely, if we insert a control rod into the core, the reactivity is decreased and the system becomes subcritical with the neutron population and power level decreasing in time. A chemical shim involves the use of different concentrations of neutron absorbers in a reactor's liquid coolant system. A burnable poison is a nuclide that possesses a large neutron absorption cross section. When this poison absorbs a neutron, it then becomes a low absorption cross section nuclide.

The control rods are generally used to change reactivity for the following purposes: (1) to permit the raising or lowering of a reactor's power level; (2) to compensate for long term reactivity changes in a reactor core, e.g., fuel burnup and fission product buildup; (3) to compensate for temperature-induced reactivity changes in the system; and (4) to permit rapid shutdown.

Chemical shims are usually employed to compensate for long term reactivity changes brought about by fissile material burnup or fission product buildup. Burnable poisons are frequently used to provide additional control reactivity at the startup of a cold clean (no fission product inventory) reactor core.

It is also interesting to note here that the reactor's temperature is one of the most significant parameters affecting its degree of criticality. The temperature coefficient of reactivity α_T is defined as

$$\alpha_T \equiv d\rho/dT \qquad (3.93)$$

where $d\rho$ is the reactivity change ($ or ¢)

dT is the temperature change (K).

If the multiplication factor k is close to unity, we can express eq. 3.93 as [1–3]

$$\alpha_T \approx \frac{1}{k}\frac{dk}{dT}. \qquad (3.94)$$

We frequently denote the fuel temperature coefficient of reactivity as the prompt temperature coefficient (α_{prompt}), since it is the fuel temperature that corresponds most rapidly to changes in power level. Thus, the value of α_{prompt} establishes the reactor's response to changes in power level or temperature. For control and safety purposes, reactors should have an α_{prompt} value

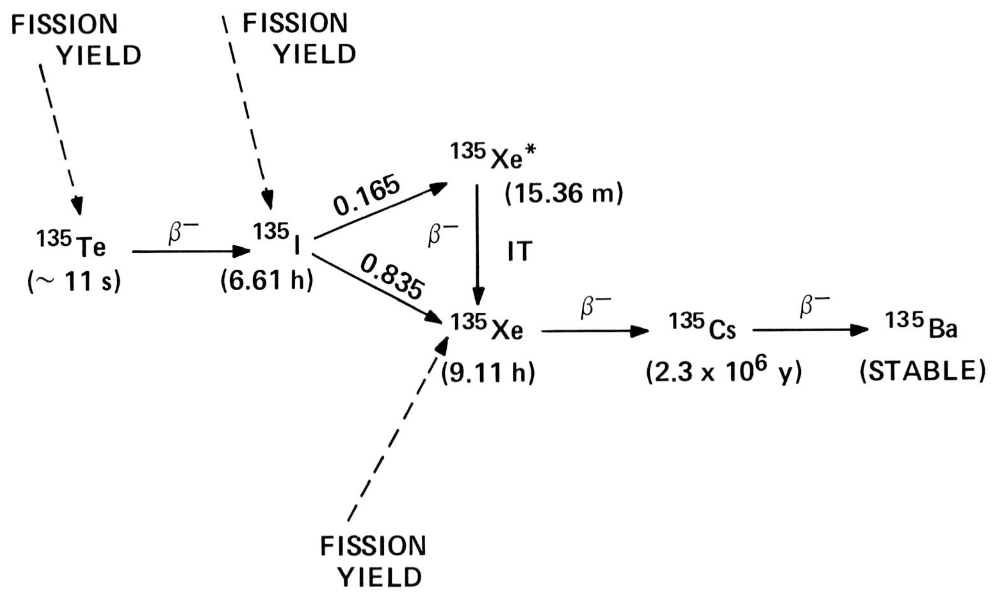

Fig. 3.17 Decay chain dynamics for fission product poison, xenon-135.

that is negative. This guarantees inherent temperature stability. If α_{prompt} is negative, dk/dT in eq. 3.94 is also negative. Therefore, an increase in temperature results in an inherent decrease in the multiplication factor. This decrease in the multiplication factor is accompanied by a subsequent decrease in power level, resulting eventually in a corresponding decrease in temperature and a return of the reactor to its original operating state. Conversely, a decrease in temperature will result in a positive dk, leading to a rise in power level back to the original operating state. Thus, if α_T is negative (especially α_{prompt}) then the reactor design is inherently stable with respect to temperature-induced reactivity changes. Since a reactor is designed with a negative temperature coefficient of reactivity, as it goes from a cold (ambient temperature) condition to its hot or elevated-temperature operating condition, it will experience a decrease in reactivity. This particular decrease in reactivity is called the temperature defect of the reactor.

Fission Product Influence

Since at least to some extent all fission products absorb neutrons, their gradual accumulation or buildup in a reactor's core produces a long term decrease in reactivity. Of special interest in thermal reactors are the fission products xenon-135 and samarium-149. Xenon-135 which originates from the decay of iodine-135 and directly from fission (see Fig. 3.17) is the most significant fission product poison in a thermal reactor, since it possesses a 2.65 megabarn absorption cross section for thermal energy neutrons. Because xenon-135 and its fission product precursors have short half-lives, the concentration of this isotope quickly reaches equilibrium or saturation values in a reactor.

Samarium-149, on the other hand, is a stable nuclide (see Fig. 3.18) that has a 41 kilobarn thermal neutron absorption cross section. Since it does not decay, samarium-149 disappears from the reactor core only as a result of neutron capture. Due to its relatively shortlived precursors, samarium-149 usually achieves equilibrium concentration in most reactors within a few days of operation. Unlike xenon-135, which essentially disappears through radioactive decay after the reactor is shut down, the ^{149}Sm concentration at shutdown remains in the core until the system is again made critical and samarium removal through neutron absorption can occur.

Space Reactor Control Techniques

Because of their generally compact size, reactivity control of space reactors is frequently accomplished through reflector manipulation, using drums in the radial reflector of the system [4, 20]. Rotation of these drums controls the amount of neutron leakage from the core. Reflector control configurations can take the form of full or partial cylinders, shutters, or windows. Cylindrical control drums, for example, sometimes incorporate a neutron poison, such as boron carbide B_4C, in a segment on one side of the drum to vary the amount of neutron absorption that occurs in the reflector. Shutters

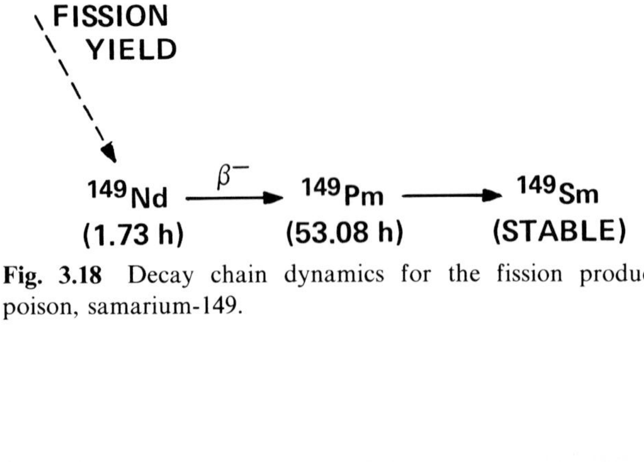

Fig. 3.18 Decay chain dynamics for the fission product poison, samarium-149.

and windows, on the other hand, simply create holes in the reflector, increasing neutron leakage to outer space.

In addition to reflector control, burnable poisons have also been considered for controlling the level of reactivity in a space reactor. These neutron poisons are depleted through neutron absorption, while the reactor operates. Their depletion increases core reactivity and compensates for fission product buildup and for the depletion of fissile material. Because of the small absorption cross section that most materials have for fast neutrons, the use of poisons is not particularly appropriate in fast reactors. With thermal reactors, however, such techniques can be used to compensate for the buildup of xenon-135, samarium-149, and other fission product poisons.

Finally, as the size of a space reactor increases, reflector control techniques may give way to the use of control rods, especially in the case of large, thermal spectrum systems.

Summary

In this chapter we have provided a preliminary view of the theory and practices governing the operation of nuclear reactors. Obviously, specific design analyses require much more sophisticated, very often computer-assisted treatment of the phenomena and principles just identified. Frequently, the main design objective for a space reactor, after aerospace nuclear safety considerations, of course, is to provide as much thermal energy as possible, at the highest core exit temperature possible, consistent with material compatibility limits. Parameters such as mass, volume, and reliability are also dominant in the selection of a candidate reactor design and its companion power conversion system. These topics will be treated in subsequent chapters.

References

1. John R. Lamarsh, *Introduction to Nuclear Engineering* (2nd Edition), Addison-Wesley Publishing Co., Reading, Mass., 1983.
2. S. Glasstone and M. C. Edlund, *The Elements of Nuclear Reactor Theory*, VanNostrand Co. Inc., Princeton, NJ, 1952.
3. John R. Lamarsh, *Introduction to Nuclear Reactor Theory*, Addison-Wesley Publishing Co., Reading, Mass., 1972.
4. D. Buden et al., "Selection of Power Plant Elements for Future Reactor Space Electric Power Supplies," LA-7858, Los Alamos Scientific Laboratory, September 1979.
5. Atomic Industrial Forum, "Guidebook for the Application of Space Nuclear Power Systems," January 1969.
6. H. M. Dieckamp, "Nuclear Space Power Systems," Atomics International, September 1967.
7. J. H. Van Osdol et al., "Design and Integration of Zirconium Hydride Reactor Power Systems," AI-AEC-13072, Atomics International, June 1973.
8. P. N. Stevens and H. C. Claiborne, *Weapons Radiation Shielding Handbook*, Chapter 2, DASA-1892-5, Defense Atomic Support Agency, June 1970.
9. A. M. Weinberg and E. P. Wigner, *The Physical Theory of Nuclear Chain Reactors*, University of Chicago Press, 1958.
10. Argonne National Laboratory, *Reactor Physics Constants*, ANL-5800 (2nd edition), July 1963.
11. L. Cranberg, N. Nereson, and L. Rosen, "Fission Spectrum of U^{235}," *Phys. Rev.* 103, 662 (1956).
12. Brookhaven National Laboratory, *Neutron Cross Sections*, BNL-325 (3rd edition), January 1976, Volume II.
13. J. P. Holman, *Thermodynamics* (2nd edition), McGraw-Hill Book Co., New York, 1974.
14. R. Meghreblian and D. Holmes, *Reactor Analysis*, McGraw-Hill Book Co., New York, 1960.
15. F. A. Valente, *A Manual of Experiments in Reactor Physics*, Macmillan Co., New York, 1963.
16. M. Abramowitz and I. Stegun, *Handbook of Mathematical Functions*, National Bureau of Standards, Washington, DC, 1964.
17. M. M. El-Wakil, *Nuclear Heat Transport*, International Textbook Co., Scranton, PA, 1971.
18. American Nuclear Society, Subcommittee ANS-5, 1968.
19. D. C. Kocher, "Radioactive Decay Data Tables," DOE/TIC-11026, 1981.
20. R. V. Anderson et al., "Space Reactor Electric Systems," ESG-DOE-13398, Rockwell International, 29 Mar 1983.

Chapter 4

Thermal Energy Transport

The primary heat transport subsystem in a space nuclear power plant carries thermal energy from the nuclear heat source to a point of application, such as the power conversion subsystem (PCS), where it is partially converted into electric energy. The heat transport subsystem may also interface with a second heat transfer loop which in turn goes to the PCS. Excess thermal energy, i.e., that not converted into electricity, must be rejected to space by means of a thermal radiator. Power conversion fundamentals and heat rejection techniques are treated in subsequent chapters. In this chapter we will examine some fundamental principles of thermal energy transfer. Heat transport systems can be divided into two general types: pumped and heat pipe. Pumped-loop heat transport systems include the all-liquid working fluid loop, the liquid-vapor working fluid loop, and the gaseous working fluid loop. Fig. 4.1 depicts a variety of heat transport configurations for a space nuclear reactor [1]. In the primary reactor loop, an all-liquid working fluid loop is simply a heat transport loop, while liquid-vapor and certain gaseous working fluid loops may contain the power conversion subsystem working fluid. Liquids may be pumped by traditional canned-rotor mechanical pumps or (for liquid metals) by the action of electromagnetic (EM) pumps. Gaseous working fluids can be circulated by fans in heat transport-only loops and by compressors, if the gas also serves as the working fluid in the PCS. To promote thermodynamic efficiency, two-phase fluids are pumped in the liquid phase. Heat pipes achieve the transport of thermal energy by the evaporation, condensation, and return capillary flow of a two-phase working fluid in a sealed container. The heat pipe is self-contained and requires no pumps, compressors, or external power to achieve large quantities of nearly isothermal heat transport. Since its working fluid does not communicate directly with the power conversion subsystem, the heat pipe is generally classified as a heat transport device.

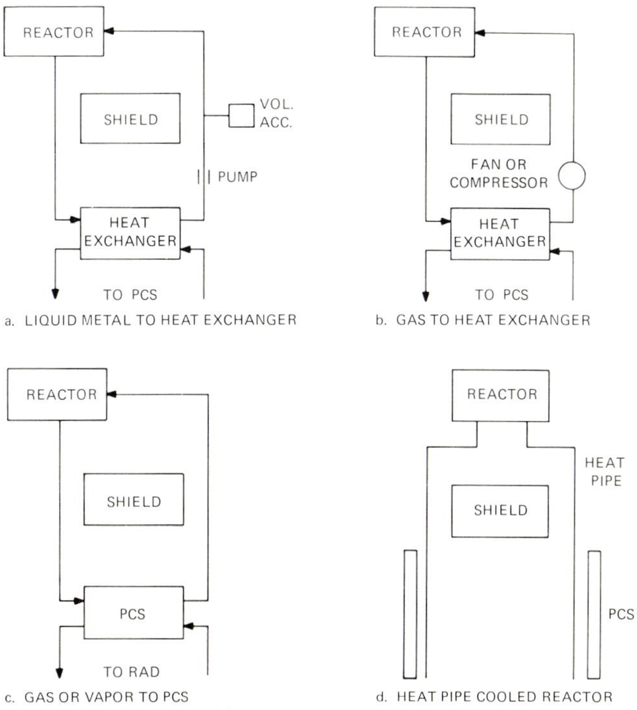

NOTE: PCS = POWER CONVERSION SYSTEM

Fig. 4.1 Various primary heat transport loop configurations for a space reactor power plant. *Courtesy of Department of Energy and Rockwell International.*

Thermodynamic Fundamentals

Thermal energy or heat (traditional symbols: Q or q) is a manifestation of disorganized (on a microscopic scale) energy flow from one region to another that occurs as a result of a temperature difference. Work (W), on the other hand, is defined as the energy expended by a force ($\mathbf{F}$) acting through a distance ($\mathbf{x}$). It represents a macroscopically organized flow of energy across the boundaries of a thermodynamic system [2, 3].

$$W \equiv \int_c \mathbf{F} \cdot d\mathbf{x}. \tag{4.1}$$

The line integral in eq. 4.1 indicates that only the product of force and displacement in the direction of the force is considered in the concept of work. It is important to realize that neither work nor heat are physically stored within a substance or thermodynamic system. Both are simply manifestations of energy transport across the boundaries of a thermodynamic system.

The concepts of heat and temperature are brought together by the Zeroth law of thermodynamics, a principle which states that if two objects are each in thermal equilibrium with a third body, they are also in thermal equilibrium with each other [2, 3]. Two objects or systems are considered to be in thermal equilibrium if energy transport as heat is possible but does not occur.

For the sake of continuity, the first and second laws of thermodynamics will also be introduced here. These fundamental principles describe the flow and transformation of energy in the physical universe and govern the performance of power conversion systems. The first law of thermodynamics is the conservation of energy principle. One statement of this law is that for an isolated system, the energy of the system remains constant. That is,

$$(dE)_{\text{iso}} = 0. \tag{4.2}$$

An isolated system is one that does not exchange energy or mass with its environment. In other words, the isolated system does not "communicate" with its surroundings.

The second law is a more complex concept linked to such physical observations as the fact that heat flows from a high temperature region to a low temperature region and that it is physically impossible to build a device which will operate continuously while receiving thermal energy from a single heat source and generating an equivalent amount of work [2, 3, 4]. One version of the second law is based on the increase of entropy principle. Entropy (S) is a thermophysical property defined in classical thermodynamics by the mathematical expression

$$dS \equiv \left(\frac{\delta Q}{T}\right)_{\text{rev}} \tag{4.3}$$

where dS is the macroscopic change in entropy (J/K)
δQ represents an infinitesimal amount of reversible energy transfer as heat (J) from a reservoir at absolute temperature in degrees Kelvin (K).

The subscript indicates a reversible process. On the basis of the entropy principle, a statement of the second law is that the entropy of an isolated system can never decrease. Mathematically, this can be written as

$$(dS)_{\text{iso}} \geqslant 0. \tag{4.4}$$

Entropy is an extensive thermodynamic property of matter. That is, its total value depends on the size or extent of the particular system. From statistical mechanics we can obtain another definition of entropy; namely, one based on the probability (p_j) of various allowed jth quantum states [2, 4]. That is,

$$S \equiv -k \sum_j p_j \ln p_j \ (\text{J/K}) \tag{4.5}$$

where $k = 1.3805 \times 10^{-23}$ J/K is the Boltzmann constant. This relationship, referred to as the Gibb's definition of entropy, represents entropy as a microscopic uncertainty measure [2].

Heat Transport Phenomena

There are essentially two basic mechanisms for thermal energy transport: conduction and radiation. We also frequently encounter another heat transfer mechanism called convection. It is simply the process in which a quantity of working fluid absorbs heat at one location in a system and transports this thermal energy by conduction and gross fluid motion to another location. If the working fluid moves under the action of a pump or fan, the process is called forced convection. If fluid motion is promoted by temperature-gradient-induced density differences, the phenomenon is called natural or free convection. In fact, this process involves the convection of internal energy by the working fluid, rather than the actual transport of heat. The total energy (E) of a working fluid may be divided into three subcategories: bulk kinetic energy (KE), bulk potential energy (PE), and internal energy (U). That is,

$$E = KE + PE + U. \tag{4.6}$$

The internal energy is thus an extensive thermodynamic property that macroscopically represents all the hidden microscopic energy modes of matter (e.g., molecular vibrational energies and molecular rotational energies) not accounted for by bulk mechanical energy terms. Specific enthalpy h is a thermodynamic property of convenience in flow systems defined as

$$h = u + pv \tag{4.7}$$

where h is the specific enthalphy (J/kg)
u is the specific internal energy (J/kg)
p is the pressure (N/m^2)
v is the specific volume (m^3/kg).

It should be noted that by convention total values are usually capitalized, while specific properties (total values per unit mass) are designated with the lower case letter symbol. For example, the total enthalpy of a system is H and has the units joules (J), while the specific enthalpy is h and has the units joules per kilogram (J/kg).

Conduction Heat Transfer

On a microscopic scale, conduction may be viewed as an atomic or molecular vibrational process. In solids, for example, higher temperature atoms will vibrate more vigorously than their lower temperature neighbors, achieving in time a net transfer of energy from the high temperature region to the low temperature region of the substance. Free electrons contribute to conduction heat transfer in metals, while molecular collisions support conduction in liquids and gases. Fourier's law is the fundamental rate equation of conduction heat transfer. In its simplest, one-dimensional form, this law states [2, 5, 6]

$$\dot{q} = -kA\frac{\partial T}{\partial x} \tag{4.8}$$

where $\dot{q}$ is the heat transfer rate (W or J/s)
A is the area (m^2)
k is the thermal conductivity of the medium (W/m-K)
$\partial T/\partial x$ is the temperature gradient (K/m).

The minus sign is a recognition of the consequence of the second law that heat flows from high to low temperature regions. For convenience, we often define a thermal energy flux or heat flux as

$$q'' \equiv \dot{q}/A \text{ (W/m}^2\text{)}. \tag{4.9}$$

The thermal diffusivity (α) of a substance is defined as

$$\alpha \equiv \frac{k}{\rho c} \tag{4.10}$$

where ρ is the density (kg/m^3)
c is the specific heat (J/kg-K).

The product ρc is often called the thermal capacity of a substance. Fourier's law can assume many forms. Its general three-dimensional form for planar geometries with temperature dependent thermal conductivity (k_T) and internal heat generation is [2, 5, 6]

$$\frac{\partial}{\partial x}\left(k_T\frac{\partial T}{\partial x}\right) + \frac{\partial}{\partial y}\left(k_T\frac{\partial T}{\partial y}\right) + \frac{\partial}{\partial z}\left(k_T\frac{\partial T}{\partial z}\right) + \dot{S} = \rho c\frac{\partial T}{\partial t} \tag{4.11}$$

where k_T is the temperature-dependent thermal conductivity (W/m-K)
$\dot{S}$ is the volumetric thermal source strength (W/m^3)
t is time (seconds).

In nuclear power supplies, the energy of the nucleus is tapped and ultimately appears as a rise in internal energy of the fuel material. Conduction is then the primary mechanism by which this energy is removed from the fuel for eventual transport by the primary heat transport subsystem to the power conversion subsystem. The governing conduction equations for one-dimensional planar, cylindrical, and spherical nuclear systems with steady-state volumetric heat sources are, respectively, [2, 5, 6]

$$\frac{d}{dx}\left(k_T\frac{dT}{dx}\right) + \dot{S} = 0 \quad \text{(planar)} \tag{4.12}$$

$$\frac{d}{dr}\left(k_T r\frac{dT}{dr}\right) + r\dot{S} = 0 \quad \text{(cylindrical)} \tag{4.13}$$

where r is the radius of the cylinder, and

$$\frac{d}{dr}\left(k_T r^2\frac{dT}{dr}\right) + r^2\dot{S} = 0 \quad \text{(spherical)} \tag{4.14}$$

where r is the radius of the sphere.

It is useful to examine the radial temperature distribution in a uniform solid sphere of radius (r_s), assuming a constant rate of internal heat generation ($\dot{S}$) and constant thermal conductivity ($\bar{k}$). This situation is analogous to nuclear energy generation involving a spherical uranium or plutonium fuel pellet in a steady-state reactor or in a spherical radioisotope fuel capsule consisting of a long half-life isotope, such as plutonium-238. Equation 4.14 describes the temperature field for radial heat flow. It can be shown [2] that the temperature distribution for this solid, uniform heat generating sphere is given by

$$T_s = T_0 - \frac{r_s^2\dot{S}}{6\bar{k}} \tag{4.15}$$

where T_s and T_0 are the absolute temperatures (K) at the surface and center, respectively, of the sphere.

Convection

If a working fluid is moved through the heat transport subsystem, energy and mass are transported by both simple conduction and the gross motion of the fluid itself. This complex process is called convection. In simple convection heat transfer, we define the convective conductance or the convective heat transfer coefficient h, which then appears in the convection rate equation

$$q'' = h\Delta T \tag{4.16}$$

where q'' is the surface heat flux (W/m^2)
h is the convective conductance (W/m^2-K)
ΔT is the temperature difference between the working fluid and solid surface (K).

Table 4.1 Dimensionless numbers encountered in convection.

Reynolds Number [Re]—defines the transition from laminar to turbulent flow

$$\mathrm{Re}_x = \frac{ux}{\nu} \; \frac{\text{inertial forces}}{\text{viscous forces}}$$

Prandtl Number [Pr]—relates relative thicknesses of hydrodynamic and thermal boundary layers

$$\mathrm{Pr} = \frac{c_p\mu}{k} = \frac{\nu}{\alpha} = \frac{\text{momentum transport}}{\text{heat transport}}$$

Nusselt Number [Nu]—measure of convective conductance

$$\mathrm{Nu}_x = \frac{h_x \cdot x}{k}$$

Stanton Number [St]—measure of convective conductance

$$\mathrm{St}_x = \frac{h_x}{\rho c_p u}$$

where: u is the free stream or bulk velocity (m/s)
x is some characteristic dimension or length (m)
μ is the dynamic viscosity (kg/s − m)
ρ is the density (kg/m^3)
ν is the kinematic viscosity (m^2/s), $\nu = \mu/\rho$
c_p is the specific heat at constant pressure (J/kg − K)
k is the thermal conductivity (W/m − k)
h is the convective conductance (W/m^2 − K)
α is the thermal diffusivity (m^2/s)

The trick, of course, is to obtain values for h and ΔT so that q'' can be calculated. In cases of boundary layer flow over a surface, the free stream temperature is frequently used in establishing a value for ΔT. The free stream temperature may be defined as the fluid temperature outside the boundary layer. The boundary layer is the layer of fluid in the immediate vicinity of a solid surface or boundary and represents the layer of fluid affected by its viscosity. If a fluid flows rapidly past a heated surface, the portion of the fluid actually heated by the wall (i.e., the portion experiencing conduction heat transfer) is restricted to this thin boundary layer. Conversely, in tube or pipe flow, since the boundary layer eventually fills the entire flow cross-sectional area, all of the flowing fluid eventually experiences wall heating effects. The computation of the convective conductance h for a particular system is a complex, often semiempirical, process. A variety of dimensionless numbers have been created to help understand and describe convective processes. These are summarized in Table 4.1 [2, 3, 6, 7].

Similarly, in cases of convective mass transport, we define a convective conductance or convective mass transfer coefficient (g) which relates the mass flux ($\dot{m}''$) to the thermodynamic driving force (D), such that [7]

$$\dot{m}'' = gD. \tag{4.17}$$

It should be noted that heat transfer problems involving convective mass tranport are generally nontrivial. When there are both heat and mass transfer processes occurring across boundaries of a thermodynamic system, as in ablation or transpiration cooling, the overall problem becomes extremely complex.

Radiation Heat Transfer

When an object is heated, it emits thermal radiation which can travel through a vacuum. This radiant energy can be viewed as either the propagation of electromagnetic waves or the transport of photons. When radiant energy is incident upon a material surface (see Fig. 4.2), a fraction of it is reflected (ρ), a fraction of it is absorbed (α), and the remaining fraction is transmitted (τ). This is summarized as

$$\alpha + \rho + \tau = 1 \tag{4.18}$$

where α is the absorptivity (fraction of incident radiation absorbed by the medium)
ρ is the reflectivity (fraction of incident radiation reflected by the medium)
τ is the transmissivity (fraction of incident radiation transmitted by the medium).

Reflection can be classified as either specular, in which the angle of incidence is equal to the angle of reflection, or as diffuse, in which the incident beam is uniformly distributed after reflection. Since the majority of solid objects do not transmit thermal radiation, we can rewrite eq. 4.18 for cases where $\tau = 0$ as

$$\alpha + \rho = 1 \qquad \text{(for opaque media).} \tag{4.19}$$

From the second law we find [8] that there is a maximum quantity of radiant energy that can be emitted by a body at a particular temperature T and wavelength λ. The emitter of this maximum amount of radiant energy is called a blackbody, which is an ideal system that also absorbs all incident radiation. In 1900, Max Planck discovered the formula for blackbody emissive power, $E_b(\lambda, T)$. Planck's radiation law is

$$E_b(\lambda, T) = \frac{2\pi hc^2}{\lambda^5(e^{(hc/\lambda kT)} - 1)} \tag{4.20}$$

where h is Planck's constant (6.626×10^{-34} J-s)
c is the speed of light (2.998×10^8 m/s)
λ is wavelength (m)
k is Boltzmann's constant (1.381×10^{-23} J/K)
T is absolute temperature (K).

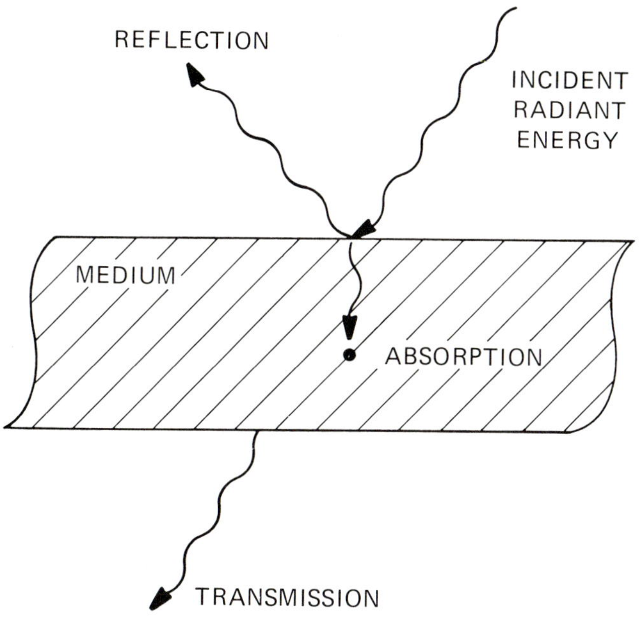

Fig. 4.2 Effects on incident electromagnetic radiation.

When this radiant energy distribution is integrated over all wavelengths, we find that the total radiant energy emitted by a blackbody (E_b) is proportional to the fourth power of absolute temperature. That is,

$$E_b(T) = \sigma T^4 \tag{4.21}$$

where $E_b(T)$ is the blackbody emissive power (W/m^2)
T is absolute temperature (K)
σ is the Stefan-Boltzmann constant—(5.669×10^{-8} W/m^2-K^4).

This relationship is called the Stefan-Boltzmann law. For a blackbody $\alpha = 1$ and $\rho = \tau = 0$. Since real objects do not behave as blackbodies but actually have an emissive power (E) from their surface less than that of a blackbody (E_b), we define the emissivity ϵ, as [2, 6, 8]

$$\epsilon \equiv E/E_b\,. \tag{4.22}$$

At thermal equilibrium, the emissive power of a body to the emissive power of a blackbody at the same temperature is equal to the absorptivity of the body [2, 6]. That is,

$$E/E_b = \alpha = \epsilon. \tag{4.23}$$

This relationship is known as Kirchhoff's law. The emissive power of a system at absolute temperature T is given by

$$E = \epsilon E_b = \epsilon\sigma T^4. \tag{4.24}$$

The emissivity of real materials varies with both temperature and wavelength. A graybody is defined as one for which the monochromatic emissivity ϵ_λ is considered independent of wavelength; that is,

$$\epsilon_\lambda = \frac{E_\lambda}{E_{b\lambda}} \tag{4.25}$$

where E_λ is the monochromatic emissive power of the body per unit wavelength
$E_{b\lambda}$ is the monochromatic emissive power per unit wavelength of a blackbody at the same temperature and wavelength.

In the graybody model, ϵ_λ is assumed constant over all wavelengths, so that

$$\epsilon = \epsilon_\lambda\,. \tag{4.26}$$

It should be noted that the emissivities of real materials vary considerably with surface condition, temperature, and wavelength. These factors must be taken into consideration in designing spacebased heat transport and rejection subsystems.

All surfaces emit thermal radiation, but only if there is a temperature difference between two objects will a net radiant energy exchange occur. The geometric configuration between two bodies at different temperatures must also be considered in evaluating a net radiant energy exchange. The overall problem then becomes a matter of establishing the amount of radiant energy leaving one surface that reaches the other. To help solve this problem, radiation shape factors or view factors (F_{ij}) have been created. By convention, F_{ij} is defined as the fraction of energy leaving surface i that reaches surface j. Thus, if we consider two blackbodies (1 and 2), each at a different temperature: $F_{12}A_1E_{b_1}$ is the radiant energy leaving surface 1 and arriving at surface 2; while $F_{21}A_2E_{b_2}$ is the radiant energy leaving surface 2 and arriving at surface 1. The reciprocity theorem for radiation shape factors states that [6, 8]

$$A_1F_{12} = A_2F_{21}. \qquad (4.27)$$

Since both objects are assumed to be blackbodies, all incident radiation will be absorbed. Then, the net radiant energy exchange $(Q_{1-2})_{rad}$ is given by

$$(Q_{1-2})_{rad} = F_{12}A_1E_{b_1} - F_{21}A_2E_{b_2}. \qquad (4.28)$$

Invoking the reciprocity theorem, eq. 4.28 becomes:

$$(Q_{1-2})_{rad} = F_{12}A_1[E_{b_1} - E_{b_2}]$$

or

$$(Q_{1-2})_{rad} = \sigma F_{12}A_1[T_1^4 - T_2^4]. \qquad (4.29)$$

Equation 4.29 represents the net radiant energy transfer rate equation between two blackbody surfaces at temperatures T_1 and T_2, respectively [2, 6, 8].

General Heat Transport Requirements

The primary heat transport subsystem must satisfy the following general requirements: (1) it must be hermetically sealed to prevent fluid leakage; (2) it must have a minimum mass and be mass-compatible with the overall power system mass budget; (3) it must have a satisfactory interface with the other components of the power system, e.g., nuclear heat source, shield, and PCS; (4) it must be capable of removing all the heat from the nuclear heat source and its components so that no component exceeds its design temperature; (5) it must be sufficiently reliable; (6) it must accommodate all thermally induced volumetric and dimensional changes; (7) it must minimize thermal energy losses; (8) it must be capable of startup with proper functioning after launch and after a normal or scheduled shutdown; and (9) for pumped-loop design systems, the fluid pumping power requirements must be consistent with the overall system mass and power requirements [1].

Working Fluids for Pumped-Loops

In pumped-loop heat transport subsystems, the fluids are generally metals in either the liquid or vapor phase or are gases. These working fluids are typically transported in thin-walled, metal alloy tubing. Pumped-loop arrangements are depicted in Fig. 4.1. The working fluid is circulated from the nuclear heat source (in this case, a reactor) to either a heat exchanger or the power conversion subsystem and then returned to the heat source.

For nuclear heat sources (especially reactors) with high thermal fluxes and high operating temperatures, liquid metals are of special interest as working fluids. These metals have superior thermophysical properties such as high thermal conductivity and low vapor pressure. In addition, they are stable at high temperatures and in intense radiation fields. Low atomic weight metals (e.g., lithium and sodium) also have relatively high specific heats and volumetric heat capacities. The main disadvantages in the use of liquid metals are: (1) their generally extensive chemical reactivity at high temperatures; (2) the necessity for using special containment materials to avoid corrosion; and (3) the fact that liquid metals must be protected from oxidation. Table 4.2 presents the physical properties of liquid metals most frequently considered as coolants for space nuclear power systems [1]. Of these, the sodium-potassium mixture (NaK) is often favored as the coolant because of its low melting point and, therefore, its simplified startup procedures. Sodium (Na) has limited application in very high temperature systems because of its low boiling point. In contrast, lithium (Li) is currently favored for very high temperature systems, mainly because of its high boiling point.

Frequently, these metals can be treated in their liquid phase as incompressible liquids. An incompressible liquid is one for which the specific volume (v) is a constant. For an incompressible liquid, the specific internal energy change is given by [2]

$$u_2 - u_1 = \int_1^2 c(T)\,dT \qquad (4.30)$$

where $c(T)$ is the temperature-dependent specific heat of the liquid (J/kg-K).

Similarly, the specific entropy change for idealized incompressible liquid is given by

$$s_2 - s_1 = \int_1^2 \frac{c(T)}{T}\,dT. \qquad (4.31)$$

Equations 4.30 and 4.31 can be further simplified if we assume that the specific heat is a constant ($\bar{c}$). Thus

$$u_2 - u_1 = \bar{c}[T_2 - T_1] \qquad (4.32)$$

and

$$s_2 - s_1 = \bar{c}\ln(T_2/T_1). \qquad (4.33)$$

Finally, the specific enthalpy change for an idealized incompressible fluid with constant specific heat is given by

Table 4.2 Physical properties of commonly used liquid metals [1].

General physical properties	Lithium (natural)	Mercury	Potassium	Sodium	Na (22%) K (78%)
Atomic number	3	80	19	11	
Atomic weight	6.94	200.61	39.100	22.997	
Density [kg/m^3 (lb/ft^3)]	479.0 (29.9)	13,231 (826)	714.4 (44.6)	823.3 (51.4)	741.7 (46.3)
Viscosity [MN · s/m^2 (lb/h · ft)]	339 (0.82)	1141 (2.76)	169 (0.41)	227 (0.55)	178 (0.43)
Surface tension [N/m (lb/ft)]	0.3472 (0.0238)	0.4482 (0.0307)	0.0834 (0.0057)	0.1471 (0.0101)	0.1138 (0.0078)
Electrical resistivity ($\mu\Omega$/cm)	36	108.3	48	29	71
Thermal properties					
Melting point [K (°F)]	454 (357)	234.5 (−37.97)	337 (146)	371.2 (208.1)	262 (12)
Boiling point [K (°F)]	1604 (2428)	631 (675)	1034 (1402)	1154 (1618)	1057 (1443)
Specific heat [J/g · K (Btu/lb · °F)]	1.286 (0.996)	0.0421 (0.0326)	0.235 (0.182)	0.389 (0.301)	0.270 (0.209)
Thermal conductivity [kJ/h · m · K (Btu/h · ft · °F)]	172 (27.6)	36.1 (5.8)	132 (21.2)	231 (37)	93.8 (15.05)
Heat of fusion [kJ/kg (Btu/lb)]	88.96 (185.9)	2.39 (5.0)	12.20 (25.5)	23.31 (48.7)	
Heat of vaporization [kJ/kg (Btu/lb)]	3995 (8349)	59.8 (125)	408.2 (853)	822.2 (1718)	
Prandtl number ($c_p\mu/k$)	0.0295	0.0154	0.0035	0.0044	0.0060
Change of volume on fusion (% solid volume)	1.5	3.6	2.41	2.5	2.5
Nuclear properties					
Fast activation cross section (mb)	0.03	89.0	0.59	0.87	0.64
Nonelastic scattering cross section at 2 MeV (barns)	< 1	2	4 (3.7 MeV)	< 1	NA
n, γ daughter half-life	None	48 days	12.4 h	15 h	See K, Na
n, γ daughter activity (MeV)	None	0.28	0.32, 1.51	2.775, 1.368	NA
ξ, average log energy decrement	0.2643	0.0100	0.0507	0.0852	NA
General properties					
Container material	Cb-Zr	Ferrous metals	304 SS	304 SS	304 SS
Reaction with uranium	Slight	Soluble	Slight (1)	Slight (1)	Slight
Reaction with plutonium	Slight	Soluble	Slight (1)	Slight (1)	Slight
Toxicity	Moderate	High	High	High	High
Fire or explosion hazard	High	None	High	High	High

$$(h_2 - h_1) = \bar{c}(T_2 - T_1) + (p_2 - p_1)v. \qquad (4.34)$$

The use of gases as a working fluid in a pumped-loop subsystem provides several distinct advantages: (1) thermal and radiation stability; (2) ease of handling; and (3) the absence of hazardous conditions. A major disadvantage would be an increased pumping power requirement. Hydrogen (H_2) and helium (He) are two often considered coolant gases (see Table 4.3). From both heat transfer and pumping power considerations, hydrogen appears to be the best gaseous coolant. However, its containment, especially under pressure, requires the use of materials which are not subject to attack or embrittlement. Hydrogen appears more suitable for use in short-duration power systems, where long-term material compatiblity concerns are not dominant.

Even though helium does not have as favorable heat transport properties as hydrogen, it has several distinctive advantages as a coolant, especially for graphite-moderated reactors operating above 875 K (1560 R). These include: (1) negligible neutron capture cross section; (2) thermal and radiation stability; and (3) the fact that helium is chemically inert and, therefore, neither hazardous nor corrosive to structural and containment materials. In space power systems using helium as the working fluid, high gas pressures are needed to decrease reactor core size and to minimize pumping power requirements.

Mixtures of helium and xenon have also been proposed. The advantages of this mixture arise from a trade-off between the higher heat transport rates obtainable with low molecular weight gases versus the greater overall cycle efficiency (due to increased pumping efficiencies) obtained with high molecular weight gases, like xenon.

Ideal Gas Model

Frequently, a gaseous working fluid can be treated using the ideal or perfect gas approximation. Many gases behave according to the following equation

$$pv = \frac{\mathcal{R}}{M} T = RT \qquad (4.35)$$

where p is the pressure (N/m^2)
v is the specific volume (m^3/kg)

Table 4.3 Thermal and nuclear properties of coolant gases [1].

General Physical Properties	Hydrogen	Helium
Molecular weight	2.016	4.0
Density [kg/m^3 (lb/ft^3)]	0.030 (0.00185)	0.060 (0.00375)
Viscosity [N · s/m^2 (lb/h · ft)]	1.74×10^{-5} (0.042)	3.8×10^{-5} (0.094)
Thermal properties		
Heat capacity [J/g · K (Btu/lb · °F)]	4.68 (3.625)	1.61 (1.245)
Thermal conductivity [kJ/h · m · K (Btu/h · ft · °F)]	1.40 (0.224)	0.99 (0.159)
Prandtl number	0.660	0.735
Nuclear properties		
σ_a, microscopic absorption cross section (barns)	0.332	0.0070
n, γ daughter half-life	None	None
n, γ daughter activity (MeV)	None	None
ξ, average log energy decrement	1.0	0.4281
General properties		
Container material	Carbon steel	Carbon steel
Reactions with U	Reacts	None
Reactions with Pu	Reacts	None
Toxicity	None	None
Fire or explosion hazard	High	None

$\mathscr{R}$ is the universal gas constant (8314.5 J/kg-mole K)
M is the molecular weight of the gas
R is the gas constant for a particular gas (J/kg-K).

The gas constant for a particular gas is given by

$$R = \mathscr{R}/M \text{ (J/kg-K)}. \tag{4.36}$$

Equation 4.35 is called the ideal gas equation of state. It has many useful forms and establishes relationships between the thermophysical properties needed to define the state of a thermodynamic system. A gas generally exhibits ideal behavior when the density is low, that is, at pressures low compared to the critical pressure p_{cr} or at temperatures high when compared to the critical temperature T_{cr}. The critical temperature and critical pressure are associated with the critical point, the thermodynamic state where the pure liquid phase has identical properties with the pure vapor phase at the same pressure and temperature. For helium, T_{cr} is 5.19 K and p_{cr} is 229.0 kilopascals (33.21 psi); while for hydrogen (H_2), T_{cr} is 33.24 K and p_{cr} is 1296.7 kilopascals (188.06 psi) [3].

A microscopic interpretation of the ideal gas model includes the following physical interpretations: (1) the gas atoms or molecules behave as point masses and are widely separated; (2) because of wide particle spacing, intermolecular forces exert very little influence on collisional processes; and finally (3) the physical volume occupied by the gas molecules is very small compared to the total volume occupied by the gas [2, 3]. For our ideal gas approximation the following thermodynamic relationships apply.

The constant pressure specific heat (c_p) is defined as

$$c_p \equiv \left(\frac{\partial h}{\partial T}\right)_p \tag{4.37}$$

where h is the specific enthalpy (J/kg)
T is the absolute temperature (K).

The constant volume specific heat (c_v) is defined as

$$c_v \equiv \left(\frac{\partial u}{\partial T}\right)_v \tag{4.38}$$

where u is the specific internal energy (J/kg). In accordance with Joule's law, the internal energy and enthalpy of an ideal gas are functions of temperature only; therefore,

$$u_2 - u_1 = \int_1^2 c_v(T)\,dT \tag{4.39}$$

and

$$h_2 - h_1 = \int_1^2 c_p(T)\,dT. \tag{4.40}$$

In addition, for an ideal gas, the gas constant (R) is related to the specific heats as follows:

$$R = c_p - c_v\,. \tag{4.41}$$

Similarly, the entropy change for an ideal gas going from state one to state two is given as [2, 3]

$$s_2 - s_1 = \int_1^2 \frac{c_v(T)}{T}\,dT + R\ln(v_2/v_1) \tag{4.42}$$

or

$$s_2 - s_1 = \int_1^2 \frac{c_p(T)}{T} dT - R \ln(p_2/p_1). \quad (4.43)$$

Equations 4.39 through 4.43 can be further simplified assuming that the specific heats $\bar{c}_p$ and $\bar{c}_v$ are constant. The resulting relationships for an ideal gas are summarized in Table 4.4.

Of course, the ideal gas model is just an approximation of real gas behavior. A compressibility factor Z has been developed to evaluate the non-ideal gas behavior of such real gases. That is,

$$Z = pv/RT. \quad (4.44)$$

As Z approaches unity, the real gas approaches ideal gas behavior. For the use of this compressibility factor and its associated generalized compressibility chart in examining real gas behavior, see references [2] and [3].

Table 4.4 Ideal gas relationships (assuming constant specific heats) [2, 3].

equation of state:	$pv = RT$
internal energy change:	$u_2 - u_1 = \bar{c}_v[T_2 - T_1]$
enthalpy change:	$h_2 - h_1 = \bar{c}_p[T_2 - T_1]$
entropy change:	$s_2 - s_1 = \bar{c}_p[\ln(T_2/T_1)] - R\ln(p_2/p_1)$
	$s_2 - s_1 = \bar{c}_v[\ln(T_2/T_1)] + R\ln(v_2/v_1)$
	$s_2 - s_1 = \bar{c}_p[\ln(v_2/v_1)] + \bar{c}_v\ln(p_2/p_1)$
gas constant:	$R = c_p - c_v$
for an isothermal process:	$p_1v_1 = p_2v_2$
for an isentropic process:	$p_1 \cdot v_1^\gamma = p_2 \cdot v_2^\gamma$
	where $\gamma \equiv c_p/c_v$

Liquid Metal Pumping Considerations

Pumped liquid metal heat transport loops have been constructed using thin-walled metal piping for many working fluids, including sodium, potassium, lithium, mercury, and the eutectic mixture of sodium and potassium (NaK). The piping and containment subsystems have been made of stainless steels for lower temperature (800 to 980 K) NaK systems. Higher temperature piping requires the use of super alloys or refractory alloys, depending to a large extent on the operating temperature. The primary heat transport subsystem is near the peak temperature of the space power system. The working fluid experiences a temperature change as it cycles through the loop. If the coolant undergoes a small temperature change, high flow rates and high pumping power are implied. Conversely, a large working fluid temperature change can impose undesirable thermal gradients on the nuclear heat source and other components in the loop.

A liquid metal loop's piping must have a method of accommodating the liquid metal inventory and thermal expansion and contraction of the loop itself. This is necessary to maintain a void-free system at appropriate operating pressures. Volume accumulators were developed for this purpose in the space nuclear power program. The accumulator consists of a variable volume container (bellows assembly) that is designed to keep a volume of liquid metal in compression by means of either a mechanical spring or a gas spring. The accumulator is generally located in a relatively low temperature region of the loop and connected to the loop by means of small diameter tubing. The volume accumulator shown in Fig. 4.3 uses gas to maintain system pressure. There are primary and secondary bellows in this concept and redundant gas volumes so that a single gas leak will not reduce the liquid metal pressure to an unacceptable level [1].

Liquid metals can be pumped by electromagnetic or mechanical devices. Two-phase fluids are pumped in the liquid phase. There is a considerable amount of engineering experience on the design of dc electromagnetic pumps for liquid metals. This experience goes back to the 1960s with the development of the SNAP-10A pump (see Fig. 4.4). This pump circulated heated NaK which exited the reactor core at 816 K and at a rate of 0.8 liter per second. (The SNAP-10A system is described elsewhere in this book.) This pump employed the Faraday principle in which a current from an external source is made to flow through liquid metal that is confined by a permanent magnetic field. Faraday's law of induction states that the electro-motive force (emf or $\mathscr{E}$) induced in a circuit is directly proportional to the time rate of change of the magnetic flux (Φ_M) of this circuit. That is, $\mathscr{E}$ (in volts) is given as

$$\mathscr{E} = -\frac{d\Phi_M}{dt}. \quad (4.45)$$

The magnetic flux threading the circuit is defined as

$$\Phi_M = \int \mathbf{B} \cdot d\mathbf{A} \quad \text{(webers)} \quad (4.46)$$

where **B** is the magnetic field strength (Wb/m^2); also called magnetic flux density
A is the conducting loop area (m^2)
$\mathscr{E}$ is the induced electromagnetic force (volts).

NOTE: The weber (Wb) is the SI unit of magnetic flux, 1 Wb = 1 volt-second. The tesla (T) is the SI unit of magnetic field strength,

$$1\ T = 1\ Wb/m^2 = \frac{1\ newton\ (N)}{ampere\ (A) \cdot meter}.$$

The CGS unit for magnetic field is the gauss (G), $1\ T = 10^4\ G$.

The Faraday principle can also be applied in a pump

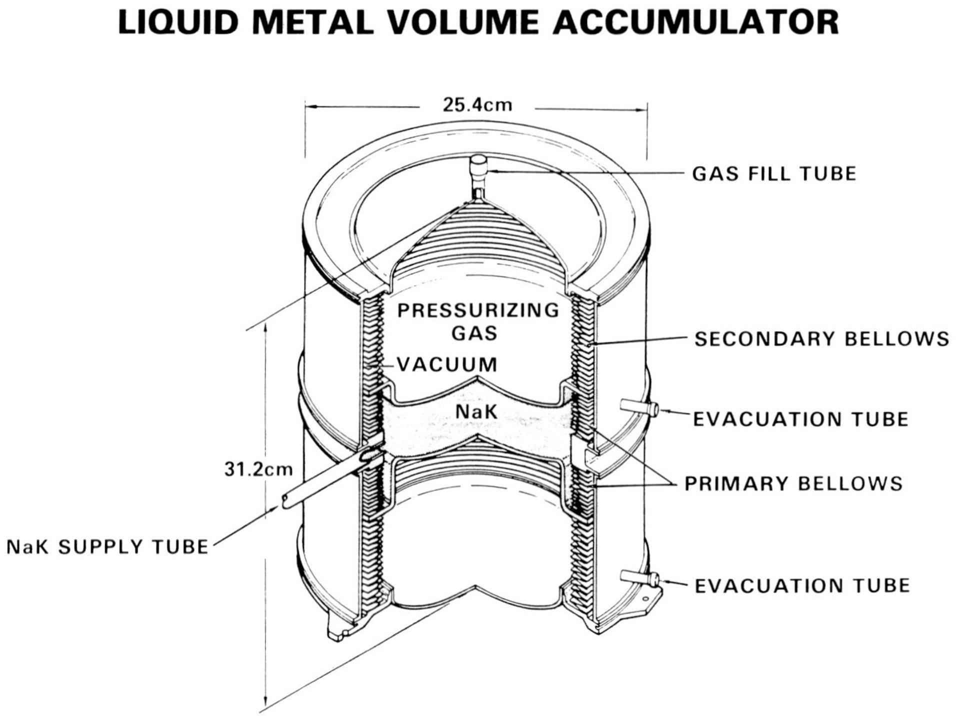

Fig. 4.3 Liquid metal volume accumulator [1].

that uses alternating current (ac). Here, the stator winding is connected to a source of polyphase power and a traveling magnetic field is produced which induces currents in the liquid metal working fluid. The electromagnetic interaction between the traveling magnetic field and induced currents causes the liquid metal to follow the traveling magnetic field [1].

Hermetically sealed centrifugal pumps, driven by canned electric motor assemblies, can also be used in liquid metal loops. The canned rotor-type pump gener-

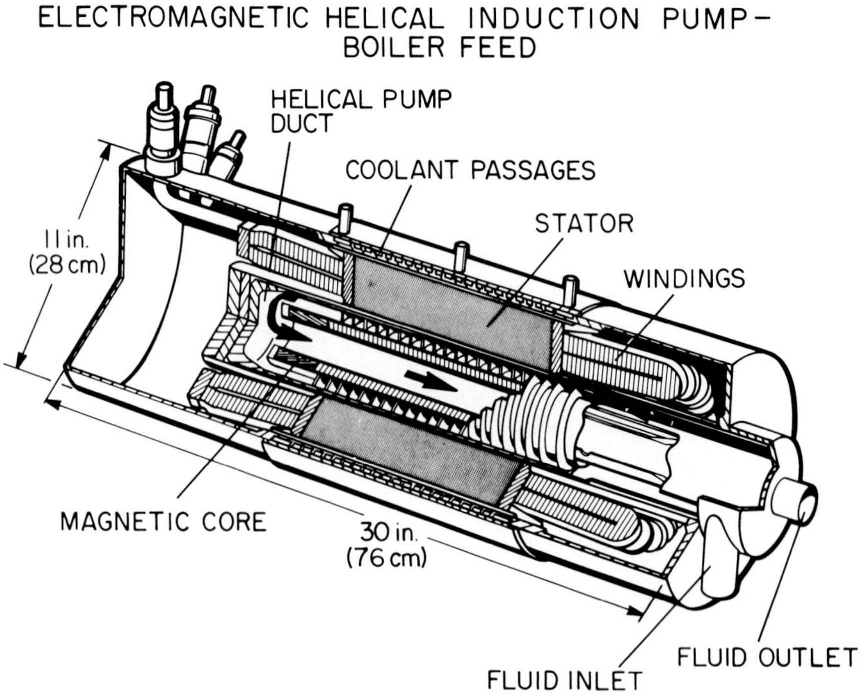

Fig. 4.4 SNAP-10A dc electromagnetic pump (powered by thermoelectrics) [1].

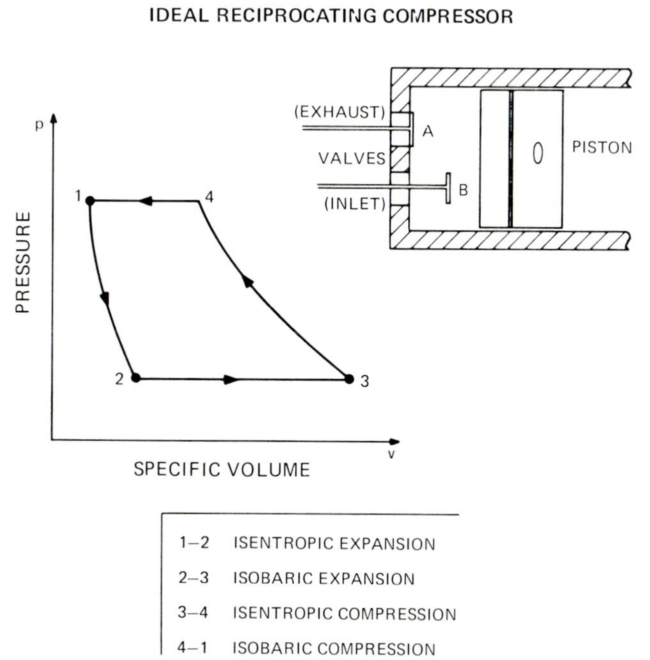

Fig. 4.5 Ideal reciprocating compressor.

ally has a higher efficiency than either the dc or ac electromagnetic pump.

Gas Pumping

Fans, blowers, and compressors can be used to circulate gases in the primary heat transfer loop. Fans are used to move gases against small pressure differentials. By convention, the designation "fan" is applied to devices that do not increase the density of the gas by more than 7 percent. At present, fans qualified for use in space power systems have not been developed [1].

Compressors can be divided into two general categories: displacement and rotodynamic machines. Displacement devices (e.g., the reciprocating compressor) are available with capacities ranging from 0.01 to 4.0 m^3/s. Fig. 4.5 depicts the pressure-volume diagram for an ideal reciprocating compressor. At point 1, called the top-dead-center (tdc) position, the piston has completed its compression stroke and delivery of gas in the cylinder and the exhaust valve (A) has just closed. The volume of gas remaining in the cylinder at this time is called the clearance volume. The piston then undergoes an isentropic (constant entropy) expansion from point 1 to point 2. At point 2, the intake valve (B) is opened and the gas to be compressed is pulled into the cylinder by the continuing reverse motion of the piston. In going from point 2 to point 3, the gas experiences an isobaric (constant pressure) expansion. At point 3, called the bottom-dead-center (bdc) piston position, the intake valve (B) is closed. From point 3 to point 4, the gas undergoes an isentropic compression. When the piston reaches point 4, the exhaust valve (A) is opened, enabling the piston to force gas out of the cylinder in an isobaric compression process from point 4 to point 1. Of course, actual reciprocating compressors do not follow this ideal cycle precisely. Intake and exhaust do not open instantaneously, and the gas itself experiences frictional effects and undesirable pressure losses as it goes through the expansion-compression cycle [3]. Reciprocating compressors have not yet been proposed for use in space nuclear power systems.

The rotodynamic-type compressor depends upon its high impeller velocity to impart sufficient momentum to the gas to achieve the necessary increase in pressure. The centrifugal compressor dominates this class of devices with capacities of up to 95 m^3/s available [1]. The major application of centrifugal compressors in space power systems is to provide for working fluid flow in gas turboelectric power conversion equipment, i.e., Brayton cycle systems, as described in Chapter 5. In general, centrifugal compressors share a common shaft with the gas turbine.

The most commonly used method for transporting coolant gases from one point to another in the heat transport subsystem is to force the fluid to flow through a piping configuration. Circular cross-section pipes are generally used because such geometry provides a greater flow area (cross-sectional) per unit wall surface area than other shapes and it also provides greater structural strength. Refractory materials may be needed for piping high temperature gases. In addition to pipes, gas loops may also contain other auxiliary elements such as orifices, valves, filters, flow straighteners, and flow splitting devices. The primary method of preventing gas leakage is to use all-welded piping and auxiliary equipment.

Pumped-Loop Design Constraints

The space nuclear reactor imposes several design constraints on the pumped-loop heat transport subsystem. First, the working fluid must adequately remove heat from the reactor core, while maintaining a reasonable nuclear fuel temperature and coolant loop pressure drop. The coolant (liquid or gas) must also have nuclear properties that are compatible with reactor criticality and control requirements. Similarly, the chemical properties of the working fluid must be compatible with piping and reactor fuel cladding materials.

The reactor's nuclear radiation shield also imposes constraints on the pumped-loop subsystem design. First, and perhaps foremost, pipes penetrating through the shield should not compromise the overall radiation attenuation effects of the shield. In addition, pipes that go outside the shield diameter should not significantly increase radiation scattering into the dose plane. Fur-

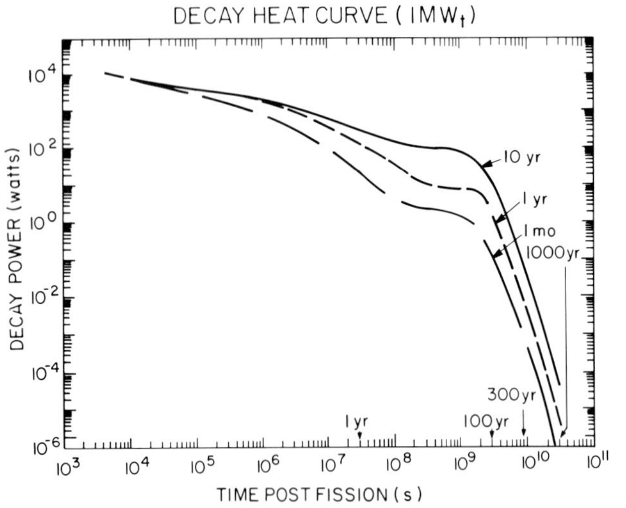

Fig. 4.6 Reactor decay heat phenomena [1].

thermore, thermal energy transfer from the high temperature piping must not heat the radiation shield materials beyond their specified limits. Finally, radioactivity induced in the coolant must not rise to levels which compromise the performance of the shield [1].

The pumped-loop heat transport subsystem must be designed to withstand launch loads (static and dynamic) and any loads caused by spacecraft on-orbit maneuvers. Since reactor startup may not be undertaken for some time following launch, the loop must be designed so that either the coolant does not freeze or startup can still be safely performed should coolant freezing happen to occur. This may require the use of loop heaters supplied by an auxiliary power supply.

For pumped-loop subsystems, auxiliary power supplies may also be required for pumps to circulate the working fluid during reactor startup and after reactor shutdown in order to remove decay heat. The amount of decay heat is a function of: (1) the power level of the operating reactor; (2) the length of time the reactor was operated at power; and (3) the time since reactor shutdown. The decay heat (expressed as a percent of full reactor power) is shown in Fig. 4.6 as a function of time after shutdown for two reactor operating times: 30 days and infinite [1].

Heat Pipes

Since the heat pipe first emerged in the 1960s as a by-product of the early space nuclear program, it has undergone an interesting transition from laboratory curiosity to a very useful thermal energy transfer device in a variety of applications [9, 10]. Heat pipes today are employed in a wide variety of special heat transfer applications including permafrost control for the Alaska pipeline; waste heat recovery; temperature control for space vehicles, for electronic equipment, and for irradiation experiments in nuclear reactors; and as thermal energy transfer devices in solar energy collection systems [11, 12]. Heat pipes are used where high thermal energy transfer rates are needed or where precise temperature control of a device or process is required.

The heat pipe is a self-contained device that attains very high thermal conductances by means of two-phase fluid flow with capillary circulation. In ordinary conduction heat transfer, thermal energy is transported by the motion of atoms and electrons. Metals, because of their special atomic structure, have exceptionally high thermal conductivities. Pure silver, for example, has the highest known thermal conductivity of all metals. Yet, the effective thermal conductivity of a heat pipe can be thousands of times greater than that of solid silver.

The high thermal conductance property of the heat pipe is achieved by making use of the latent heat of vaporization [L] (also called the enthalpy of evaporation) of a working fluid. As indicated in Fig. 4.7, an operating heat pipe is basically a container in which a working fluid circulates between heated and cooled regions. In the heat pipe, this fluid exists as two phases: liquid and vapor. In the heated or evaporator region, thermal energy transfer to the liquid causes it to experience phase change and become a vapor. The vaporized working fluid then flows from the evaporator region to the heat pipe's cooled or condenser region. There, the thermal energy transfer process is reversed. As the vapor cools, it becomes a liquid and gives up energy (i.e., it releases the enthalpy of evaporation). A wick structure within the heat pipe provides capillary action, permitting the return of the liquid to the evaporator region. The ability of a heat pipe to transfer thermal energy is proportional to the latent heat of vaporization of the working fluid. Some typical values are shown in Table 4.5.

In a properly designed heat pipe, working fluid evaporation and condensation occur at approximately the same temperature and thermal energy transfer between a source and sink occurs isothermally. Although the working fluid circulates within the heat pipe, there are no moving mechanical parts and it appears externally as a passive device.

The heat pipe principle is applicable over a great variety of shapes, sizes, temperatures, and materials. Most contemporary heat pipe applications involve low to moderate operating temperatures [11]. High temperature applications have been slow to evolve, mainly because above about 750 K liquid metals become the desired working fluid. Their use in high temperature heat pipes involves an increase in fabrication costs and the creation of a perception of increased hazard. The

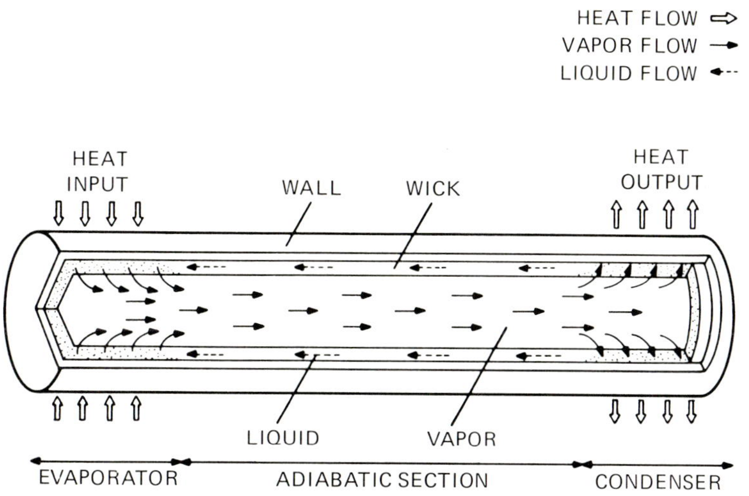

Fig. 4.7 Heat pipe operating principle. *Courtesy of Los Alamos National Laboratory.*

major contemporary stimulus for the application of high temperature (liquid metal) heat pipes is their potential role in space nuclear power systems, especially reactor systems [11]. In fact, the high temperature heat pipe was originally developed to solve thermal energy transport problems in the early space nuclear power systems, and it now appears that this same technology represents a key heat transfer component in advanced space nuclear systems.

Table 4.5 Typical heat pipe working fluids. *Courtesy of Los Alamos National Laboratory.*

Working Fluid	Temperature (K)	Latent Heat of Vaporization (kJ/kg)
Ammonia	273	1263
Freon II	273	190
Acetone	273	564
Methanol	273	1178
Water	273	2492
Mercury	773	291
Cesium	773	512
Potassium	773	2040
Sodium	773	4370
Lithium	1273	20525

Heat Pipe Design Considerations

Full advantage of the heat pipe is best realized when the devices are long and thin, as for example, forming long cylinders or extended thin planar structures [10]. Consider the right circular cylindrical heat pipe depicted in Fig. 4.8. This heat pipe is made up of an encapsulating tube of length (l) and has an outer radius (r_0). The annular capillary structure, bounded by an outside radius (r_w) and an inner or vapor space radius (r_v), is saturated with a working fluid which wets the wick.

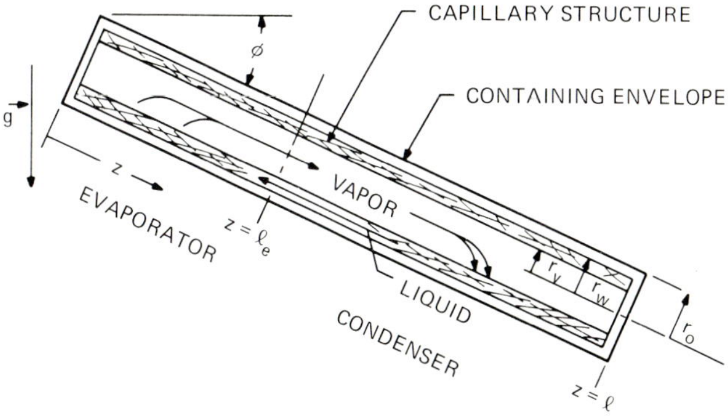

Fig. 4.8 Cylindrical heat pipe structure. *Courtesy of Los Alamos National Laboratory.*

NOTE: A liquid is said to wet a solid surface if the contact angle (θ) is less than 90°. The contact angle is defined as the angle that the tangent to the liquid surface makes with the solid. If the contact angle is greater than 90°, the liquid does not wet the solid. Consider both mercury and water in a glass container, as shown in Fig. 4.9. Here we say that the water wets the glass, while the mercury does not. The relative strengths of the cohesive forces between the molecules of the liquid in comparison to the adhesive forces between the molecules of the liquid and the molecules of the solid, determine whether a liquid wets a solid surface or not.

Since thermal energy is added to and removed from the heat pipe through the container wall by conduction heat transfer, this wall should be as thin as other conditions permit in order to keep radial temperature gradients to a minimum. The heat pipe container must be capable of sustaining the difference between the internal and environmental pressure, including, of course, the vacuum conditions of outer space. Heat pipe

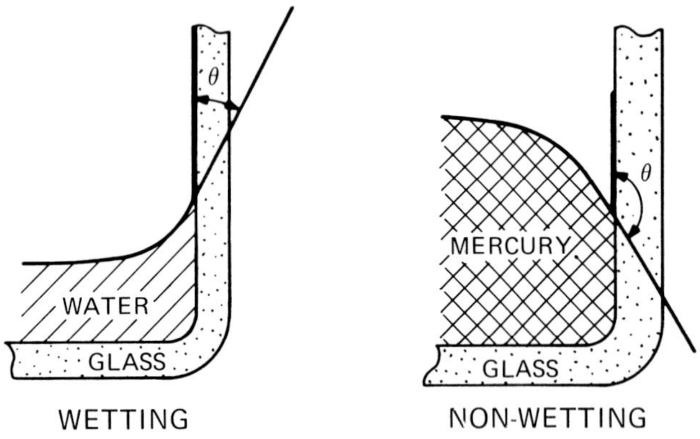

NOTE: θ = CONTACT ∡
DIMENSIONS EXAGGERATED

Fig. 4.9 Wetting of a solid surface. *Courtesy of Los Alamos National Laboratory.*

performance becomes effective at internal vapor pressures as low as 1 kilopascal (approximately one-hundredth of an atmosphere) and improves with increasing vapor pressure. The long-term compatibility of the working fluid with the container material is critical and influences container wall thickness. The heat pipe container itself can also serve as a structural member in the heat transport subsystem of a space power plant.

The choice of wick material and design is open to wide variation. The wick does not have to be placed against the inside surface of the heat pipe container, although this will generally be the best wick location. Since evaporation and condensation occur at the liquid-vapor interface, placing the wick along the inside wall permits radial heat transfer to occur under conditions of highest thermal conductance. Consequently, radial temperature differences are minimized. This location of the wick also makes the hydraulic diameter of the vapor space as large as possible, thereby minimizing axial pressure gradients in the flowing vapor.

The wick can be a woven cloth, a fine-meshed metal screen, or simply grooves or slots machined into the inner container wall. The capillary structure created by the wick is characterized by the following parameters: mean pore radius, liquid volume fraction, and permeability. It is also quite useful if the heat pipe is self-priming. A heat pipe is said to be self-priming if, when the required amount of working fluid (liquid) is placed anywhere in the heat pipe, it spontaneously saturates the entire wick structure [10]. This condition is achieved in practice by establishing a sufficiently small capillary pore size throughout the wick structure.

Finally, in a properly designed heat pipe, the working fluid must wet the wick structure. It is also desirable (though not essential) that the working fluid wet the container wall, since this improves thermal energy transport into and out of the pipe. While there is little penalty for a modest excess of working fluid over the amount of liquid necessary to saturate the wick, a deficiency of working fluid decreases the maximum amount thermal energy transport by reducing the effective wick volume in the evaporator region [10].

Heat Pipe Performance Factors

Consider now a heat pipe with constant heat addition along the evaporator and constant heat removal along the condenser. The axial heat flux distribution $Q_0(z)$ is then given by [10]

$$Q_0(z) = L\dot{m}_v(z) = \begin{cases} \left(\dfrac{z}{l_e}\right) Q_e & \text{for } 0 \leqslant z \leqslant l_e \\ \left[\dfrac{l-z}{l-l_e}\right] Q_e & \text{for } l_e \leqslant z \leqslant l \end{cases} \tag{4.47}$$

where $\dot{m}_v(z)$ is the vapor mass flow rate (kg/s)
L is the latent heat of vaporization (kJ/kg)
Q_e is the total heat input to the evaporator (W)
l_e is the length of the evaporator region (m)
l is the length of the heat pipe (m).

While the heat pipe is a device of very high thermal conductance, there are limitations on its total and local thermal energy fluxes. The total amount of heat transported axially may be increased only if the capillary force within the wick structure can sustain the necessary circulation of working fluid. The maximum difference in pressure between the vapor phase, $p_v(z)$, and the adjacent liquid phase, $p_l(z)$, that can be supported by the surface tension γ in the capillary structure is $2\gamma\cos\theta/r_c$, where θ is the contact angle and r_c is the average radius of the capillary pore. It is therefore a necessary condition that [10]

$$p_v(z) - p_l(z) = \frac{2\gamma\cos\theta}{r_c} \qquad \text{for } 0 \leqslant z \leqslant l. \tag{4.48}$$

Typically, a heat pipe will be required to transport the largest amount of thermal energy possible, subject to specific subsidiary constraints that could arise in a particular application. Considering our heat pipe with constant heat addition and removal, the maximum heat flux criterion can be written as

$$\Delta p_l - \Delta p_v - \frac{2\gamma\cos\theta}{r_c} = 0. \tag{4.49}$$

Considering the case of uniform heat addition and removal, p_l is given by

$$\Delta p_l = p_l(l) - p_l(0)$$

$$= \rho_l g l \sin\phi + \frac{b\eta_l Q_e l}{2\pi(r_w^2 - r_v^2)\rho_l e r_c^2 L} \quad (4.50)$$

where ρ_l is the density of the liquid (kg/m^3)
r_v is the radius of vapor space (m)
r_w is the outer radius of the wick structure (m)
ϕ is the angle to a gravitational field of acceleration (g)
η_l is the viscosity of the liquid
e is the fraction of wick volume occupied by the liquid
b is a dimensionless constant which is dependent on wick geometry.

If the wick consists of nonconnected parallel cylindrical pores, b has a value of approximately 8. If the wick structure consists of tortuous, interconnected pores, b varies between 10 and 20 [10].

The optimum choice of r_c is given by [10]

$$r_c = \frac{b\eta_l Q_e l}{4\pi(r_w^2 - r_v^2)\rho_l e L \gamma \cos\theta}. \quad (4.51)$$

The maximum heat transport is then given by

$$Q_e = \frac{\pi r_w^2 L \gamma \cos\theta}{3l}\left[\frac{e\rho_v\rho_l}{3b\eta_v\eta_l}\right]^{1/2} \quad \text{for } R_r \ll 1$$

$$= \frac{4\pi r_w^2 L}{3}\left[\frac{2\rho_v\rho_l e\gamma^2\cos^2\theta}{(\pi^2-4)bl\eta_l}\right]^{1/3} \quad \text{for } R_r \gg 1 \quad (4.52)$$

where ρ_v is the density of the vapor
η_v is the viscosity of the vapor
R_r is the Reynolds number based on radial flow velocity.

High Temperature Heat Pipe Applications

The high temperature heat pipe has a number of technical features that makes it especially attractive as a key component in advanced space nuclear reactor systems. First, and perhaps foremost, such heat pipes represent a very high thermal energy transport capability in a relatively simple, self-contained device capable of high temperature operation [11]. Reactor cores may thus be designed with many independent coolant loops, the failure of any one of which would not jeopardize the safe operation of the space nuclear power plant. The elimination of such single-point failure mechanisms in the primary reactor coolant system establishes a power plant design environment essentially devoid of catastrophic failure modes. This is an extremely important technical feature in space power systems which may be required to operate reliably for five or ten years without failure.

Table 4.6 Typical space reactor core heat pipe performance requirements [11].

Working Fluid	Lithium
Operating Temperature	1500 K
Axial Power Density*	
Normal Operation	8.0 kW/cm^2
Contingency Operation	10.5 kW/cm^2
Maximal Radial Power Density	
Evaporator	105 W/cm^2
Condenser	6.5 W/cm^2
Lifetime	
7 years in normal mode	0.98 reliability
7 years in contingency mode	0.96 reliability

*assumes: 17.4 kW per heat pipe for a 120 heat pipe, 1600 kilowatt-thermal reactor.

The series of independent, self-contained coolant loops in the heat pipe reactor eliminates the need for a bulky pressure vessel. This not only saves weight, but also enhances overall aerospace nuclear safety by permitting the design of a space reactor system that can break up more easily, if it should ever reenter the Earth's atmosphere. The residual radioactive contents of the core would then be widely dispersed at extremely high altitudes where they would represent a minimum threat to the biosphere far below.

The self-contained, self-operating nature of the heat pipe supports a primary coolant loop design that avoids potentially cumbersome valves, pumps, or compressors. The heat pipe's self-operating feature also guarantees the automatic removal of reactor core afterheat, once the reactor has been shut down intentionally or in response to a space system failure and all operating power is lost.

The ability of a heat pipe to operate a high temperature supports compact, high specific power reactor system designs. Furthermore, the ability of a heat pipe to start up directly from a frozen state (i.e., the working fluid has become a solid) eliminates any need for coolant preheating as is necessary with liquid metal cooling loops.

The nearly isothermal and high operational temperature characteristics of heat pipes permit direct energy conversion at maximum design efficiency. In heat pipe reactors, for example, thermal energy is delivered to the conversion device at what is essentially the reactor core exit temperature. In space nuclear power systems, where core thermal energy is ultimately transported to the conversion system by radiation heat transport, it is important that the radiating surface be isothermal and near the reactor exit temperature if maximum conversion efficiencies are to be achieved. This isothermal

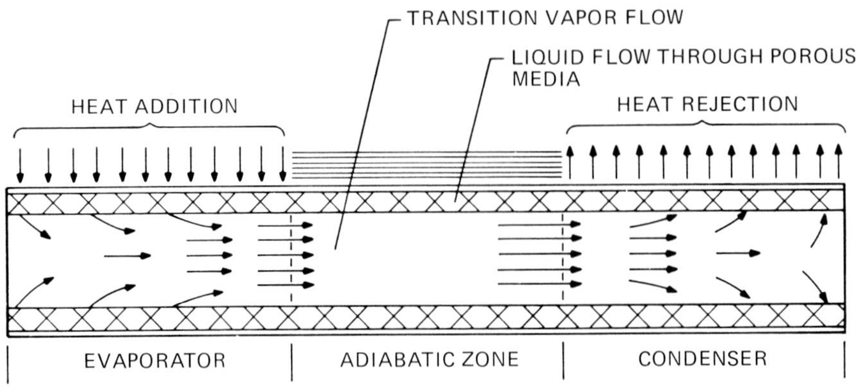

Fig. 4.10 Heat pipe flow regimes. *Courtesy of Los Alamos National Laboratory.*

radiating condition is not achievable with pumped loops because of inherent temperature drops associated with circulating single-phase fluids [11].

Table 4.6 lists the performance requirements that an advanced-design space nuclear reactor could place on its core heat pipes [11]. These are demanding, yet nonetheless, realistic heat pipe performance requirements. For the heat pipe flow regimes shown in Fig. 4.10, performance curves of the general shape shown in Fig. 4.11 may be anticipated. The sonic limit of a heat pipe occurs when the vapor velocity at the entrance to the condenser region reaches sonic velocity. Wicking limits are experienced when the forces required to drive the working fluid down the pipe as vapor and back through the wick as liquid exceed the available capillary force in the evaporator. For liquid metal heat pipes with compound wick structures, the liquid pressure drop limit is rarely a factor. Similarly, the liquid-dominated wicking limit is rarely a factor with lithium heat pipes. The boiling limit is established by the rate at which evaporation can take place without excessive temperature differentials occurring in the evaporator section. The boiling limit has been observed in high temperature sodium heat pipes, but is much less significant for lithium heat pipes operating in the 1,500 K temperature regime [11].

The heat transport performances of lithium and sodium working fluids are compared in Fig. 4.12 [11]. Represented are data for an annular wick heat pipe with a 14.1 mm inside diameter, a 33 micrometer capillary pore diameter, two 90° bends (0.2 m radius), a 0.30 m long evaporator section, a 1.03 m adiabatic section, and a 0.6 m long condenser region. For temperatures below 1,500 K, the low vapor pressure of lithium makes its performance less than that of sodium. However, increasing the operating temperature above 1,500 K permits very impressive quantities of thermal energy to be transported down a channel, which, for this case, is only 1.6 cm^2 in area [11].

The design of heat pipes for use with space nuclear reactors will ultimately be determined by materials and operating environment considerations. The selection of materials from which to manufacture high temperature reactor core heat pipes is severely limited by three requirements. The first involves chemical compatibility between the working fluid and the reactor fuel. Another is creep strength at 1,500 K. The third is ductility at launch and deployment temperatures, which could be as low as 200 K or less while the space system is transferred from low Earth orbit to some appropriate high Earth orbit operating location.

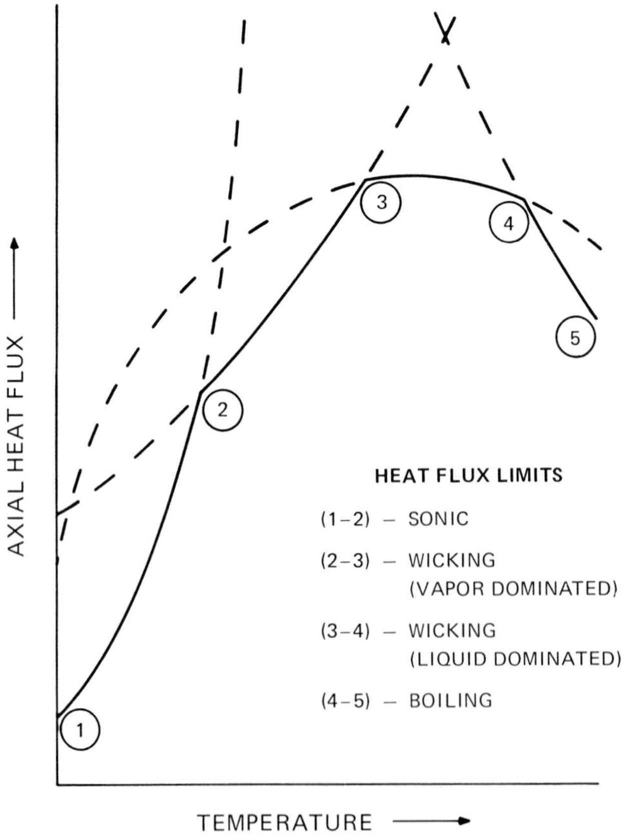

Fig. 4.11 Phenomena limiting heat pipe performance. *Courtesy of Los Alamos National Laboratory.*

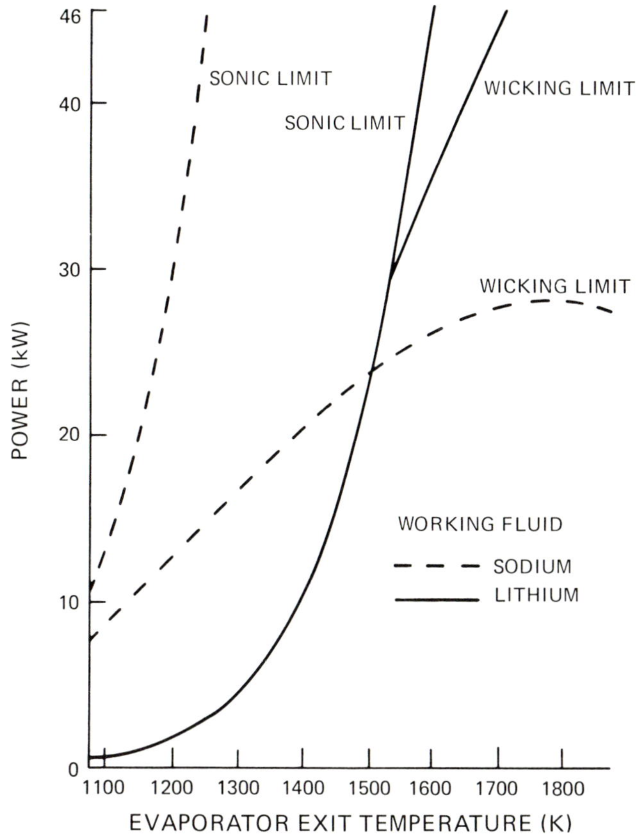

Fig. 4.12 A comparison of lithium and sodium heat pipe performance. *Courtesy of Los Alamos National Laboratory.*

Because the evaporator regions of reactor heat pipes form an integral part of the core, the materials compatibility question becomes considerably intensified by the reactor environment itself. For example, the high temperature, high thermal flux environment might support the diffusion of oxygen from the nuclear fuel (e.g., UO_2) through the heat pipe wall. This is an important consideration because lithium metal will take oxygen away from even stoichiometric UO_2 fuel [11]. In a stoichiometric mixture the molal proportions of the reactants are precisely given by the stoichiometric coefficients so that there is no excess of any constituent; the stoichiometric coefficients are the coefficients in a chemical equation.

Other effects of the reactor environment include stresses caused by fuel swelling, the effect of such stresses on grain growth, and the effect of grain growth on heat pipe wall integrity. Fast neutron flux effects on grain growth and creep strength could also be significant.

A small amount of hydrogen and helium generation will occur in the heat pipes, caused primarily by neutron capture in the lithium working fluid, especially by the isotope lithium-7. The generation of such gases might modify heat pipe behavior. Finally, fission products generated in the nuclear fuel could also interact with the heat pipe wall material [11]. These effects must all be considered in the final design of a heat pipe reactor system.

References

1. R. V. Anderson et al., "Space Reactor Electric Systems," ESG-DOE-13398, Rockwell International, 29 Mar 1983.
2. W. Reynolds and H. Perkins, *Engineering Thermodynamics*, McGraw-Hill Book Co., New York, 1970.
3. J. P. Holman, *Thermodynamics* (2nd Edition), McGraw-Hill Book Co., New York, 1974.
4. M. Zemansky and R. Dittman, *Heat and Thermodynamics*, McGraw-Hill Book Co., New York (6th Edition), 1981.
5. H. S. Carslaw and J. C. Jaeger, *Conduction of Heat In Solids*, 2nd edition, Oxford Press, 1959.
6. J. P. Holman, *Heat Transfer*, 4th Edition, McGraw-Hill Book Co., New York, 1976.
7. W. M. Kays, *Convective Heat and Mass Transfer*, McGraw-Hill Book Co., New York, 1966.
8. E. M. Sparrow and R. D. Cess, *Radiation Heat Transfer*, Brooks/Cole Publishing Co. Inc., Belmont, CA, 1966.
9. G. M. Grover, T. P. Cotter and G. F. Erickson, "Structures of Very High Thermal Conductance," *J. Applied Physics*, Vol. 35, No. 6, p. 1990, June 1964.
10. T. P. Cotter, "Theory of Heat Pipes," LA-3246-MS, Los Alamos Scientific Laboratory, 1965.
11. W. A. Ranken, "Status of High-Temperature Heat Pipe Technology," LA-UR-82-3340, Los Alamos National Laboratory, 1982.
12. F. C. Prenger, Jr., "Heat Pipes: Effective Devices for Heat Transport and Temperature Control," LASL-79-22, Los Alamos National Laboratory, March 1979.

Chapter 5

Energy Conversion Techniques

Introduction

In space nuclear power applications the thermal energy released by radioactive decay or nuclear fission can either be used directly in propulsion or environmental control devices or else converted into electric energy. There are a variety of possible energy conversion techniques (see Fig. 5.1) [1]. These techniques can be divided into two general classes: static and dynamic. A static energy conversion system, such as a thermoelectric (TE) or thermionic device, uses thermophysical principles that require no moving mechanical parts to convert heat into electricity. Such devices are commonly referred to as direct energy conversion (DEC) devices. A dynamic energy conversion system, on the other hand, uses the heat engine principle to provide the mechanical work necessary to generate electricity in a turboalternator assembly. Dynamic conversion systems can be further subdivided into rotating (i.e., Brayton and Rankine cycle) or reciprocating (i.e., Stirling engine) devices.

Energy conversion systems may also be classified as either steady-state or intermittent. In general, steady-state devices can be required to operate continuously for months or years in space, while intermittent devices may be pressed into service on short notice and operate for just a few hours or days at a time [2].

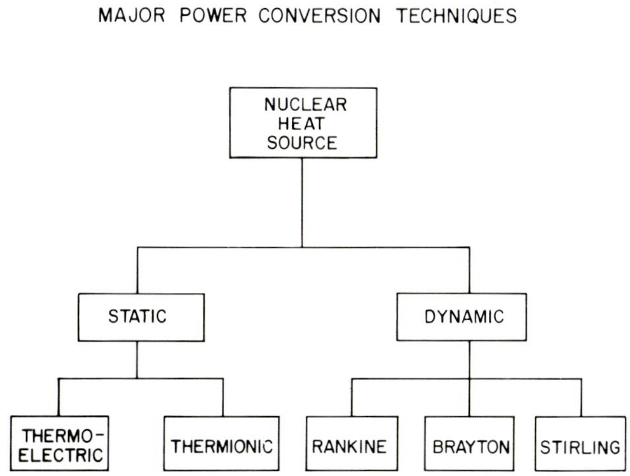

Fig. 5.1 Major power conversion techniques.

Another way of differentiating energy conversion systems is on the basis of whether alternating current (ac) or direct current (dc) is generated. Dynamic conversion systems use an alternator to produce ac power, while static or DEC devices convert thermal energy directly into dc power.

As a point of nomenclature, power is defined as the time rate of doing work [3, 4]. That is, power = work/time; hence, 1 watt equals 1 joule/second. We traditionally describe energy conversion systems by their power generation ability, using such terms as a 100 kilowatt-thermal (100 kW_{th}) system or a 10 megawatt-electric (10 MW_e) system. The former designation refers to a thermal energy source of 100 kilojoules per second, while the latter identifies the output of an energy conversion device that is providing 10 megajoules of electric energy per second. The term "power cycle" is often used to describe energy conversion devices that continuously convert thermal energy into mechanical work or electric energy in a repetitive process.

In this chapter we explore methods of converting thermal energy into useful spacecraft electric energy. Particular emphasis is given to heat engines and DEC devices.

The Heat Engine

A heat engine can be defined as a controlled-mass, thermodynamic device that receives thermal energy from a source, converts a portion of this energy into mechanical work (which is transferred across the system boundaries to the surroundings for application), and then rejects the unused amount of input thermal energy to a lower temperature sink. The second law of thermodynamics establishes the maximum amount of input thermal energy that can actually be converted into work. As we shall see shortly, this maximum thermal efficiency of a heat engine is called its Carnot efficiency

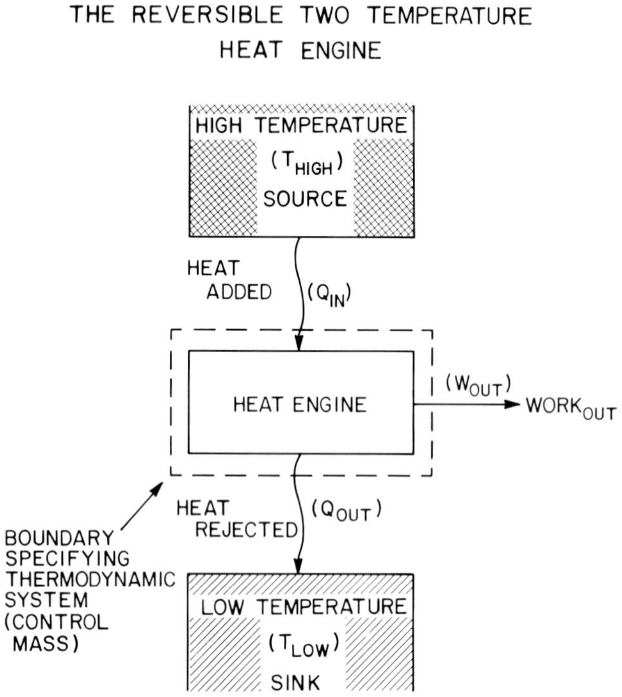

Fig. 5.2 The reversible two temperature heat engine.

(η_{Carnot} or η_{th}). Real heat engines fall short of this performance optimum because of irreversibilities caused by heat transfer with large temperature differentials and working fluid-induced frictional effects. The working fluid within the heat engine undergoes cyclic operation, so that after some period of time all the working fluid within the device returns to its initial state. Figure 5.2 shows a generic heat engine that operates between a high temperature heat source (T_H) and a low temperature heat sink (T_L) [3,4]. We call this type of engine a $2T$ heat engine, since thermal energy is transferred into the system at one temperature and rejected from the system at another (lower) temperature. Space nuclear system energy conversion efficiencies can range from 1.8 percent (i.e., the SNAP-10A thermoelectric converters) to over 40 percent with certain proposed dynamic converters [1,2].

Efficiency is introduced to express the effectiveness of using the heat available. Nevertheless efficiency is a very elusive and often misleading term. In a general engineering sense, the efficiency of a system is simply [3,4,5]

$$\eta \equiv \frac{\text{some useful effect}}{\text{energy that must be expended to achieve that effect}} \tag{5.1}$$

In a more rigorous thermodynamic sense, the thermal efficiency or Carnot efficiency is given by

$$\eta_{\text{th}} = 1 - T_L/T_H \tag{5.2}$$

where T_H is the higher absolute temperature (K) at which heat is added
T_L is the lower absolute temperature (K) at which heat is rejected.

From eq. 5.2, we can perceive two fundamental relationships. First, the thermal efficiency of the system is greater, the higher the heat addition temperature (T_H); second, the thermal efficiency is greater, the lower the heat rejection temperature (T_L).

Sometimes in treating a spacecraft's energy conversion system, we would like to separate thermal efficiencies (i.e., Carnot limitations) from other system losses, such as electrical or mechanical. In that case, we define a power conversion system (PCS) efficiency as

$$\eta_{\text{PCS}} = \eta_{\text{th}} \cdot \eta_{\text{D}} \tag{5.3}$$

where η_{th} is the thermal efficiency
η_{D} is the device efficiency.

The latter is a measure of the internal conversion efficiency of various energy conversion devices, excluding thermodynamic limitations. However, η_{D} should never be used as a stand-alone measure of PCS performance.

The Carnot Engine

The Carnot cycle is a fundamental theoretical concept

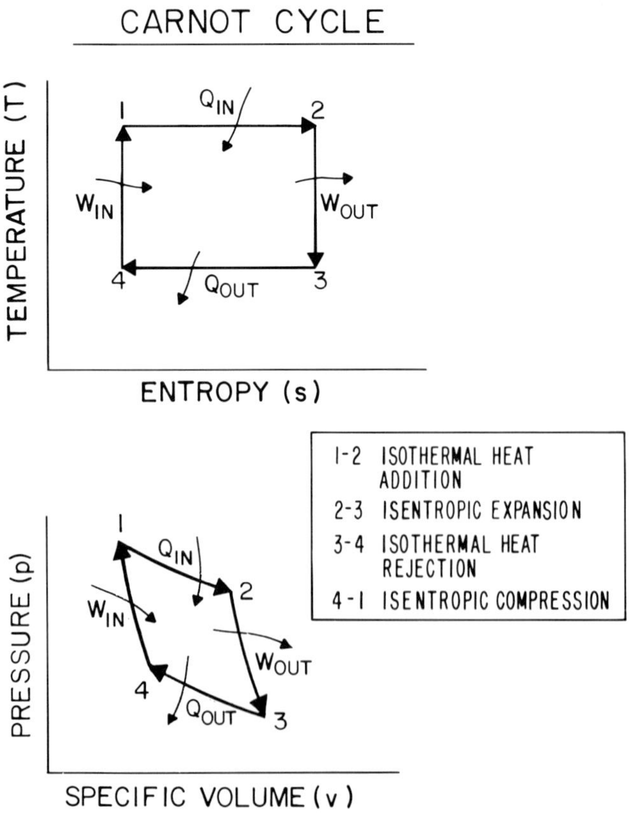

Fig. 5.3 Carnot cycle.

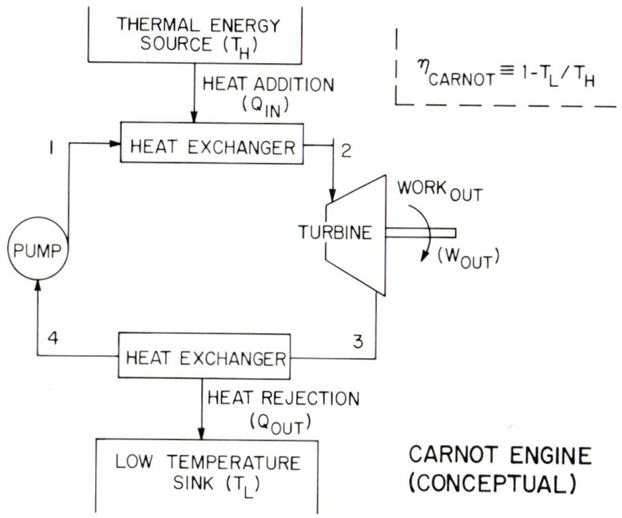

Fig. 5.4 Carnot engine (conceptual).

in dynamic energy conversion, because the thermal efficiency of this cycle is the maximum possible value for any heat engine operating between the same two temperature limits [3, 4, 5]. Carnot (1796–1832) was a French military engineer who performed pioneering examinations of heat engine efficiencies in the early 19th century. Although he incorrectly supported the caloric theory of heat which assumed that heat was a fluid, his very valid theortical work on heat engine efficiencies established the quantitative basis for Rudolf Clausius's postulation of the second law of thermodynamics in the 1850s [6]. Figure 5.3 shows the pressure-volume and temperature-entropy diagrams of the Carnot cycle. In this idealized cycle, a working fluid experiences: reversible, isothermal heat addition from state 1 to state 2; reversible, adiabatic (i.e., isentropic) expansion from 2 to 3 during which work is transferred from the system to its environment; reversible, isothermal heat rejection from state 3 to state 4; and finally, a reversible, adiabatic (i.e., isentropic) compression process from 4 to 1 during which work is transferred to the system from the surroundings. By definition, an isentropic or constant entropy process is a reversible, adiabatic process [3, 4, 5]. Figure 5.4 is a diagram of a conceptual steady flow Carnot engine. In theory, isothermal heat addition and rejection can be achieved by using suitable heat exchangers, while isentropic turbines and compressors or pumps may be used to transfer energy as work to or from the working fluid. The theoretical efficiency of this cycle is given by

$$\eta_{\text{CARNOT}} = \frac{\text{WORK}_{\text{out}}}{\text{HEAT}_{\text{IN}}} = \frac{T_H - T_L}{T_H} = 1 - T_L/T_H \tag{5.4}$$

where T_H is the higher absolute temperature (K) at which heat is added
T_L is the lower absolute temperature (K) at which heat is rejected.

As mentioned previously, the Carnot efficiency represents the maximum thermal efficiency (η_{th}) of any heat engine operating between the same source (T_H) and the sink (T_L) temperatures.

Of course, the Carnot cycle and engine just described are idealizations. After a brief discussion of power cycle components such as turbines, pumps, and compressors, we will investigate actual heat engine cycles, including Rankine, Brayton, and Stirling devices.

Idealized Turbines, Compressors, and Pumps

The one-dimensional, steady-flow energy balance equation for the control volume shown in Fig. 5.5 is [3–5]

$$\dot{W} + \dot{Q} + \dot{m}\left[(e_2 + p_2v_2) - (e_1 + p_1v_1)\right] = 0 \tag{5.5}$$

where $\dot{W}$ is the (mechanical) work delivered to the control volume by the surroundings per unit time ($-\dot{W}_{\text{in}}$ by convention here); or the work delivered by the control volume to the surroundings per unit time ($+\dot{W}_{\text{out}}$ by convention here). (J/s or W)
$\dot{Q}$ is the thermal energy delivered to the control volume per unit time ($-\dot{Q}_{\text{in}}$ by convention); or the heat rejected by the control volume to the surroundings per unit time ($+\dot{Q}_{\text{out}}$ by convention) (J/s or W)
$\dot{m}$ is the mass flow rate of the working fluid (kg/s)
e is the specific total energy content of the working fluid (J/kg)
p is the working fluid pressure (N/m^2 or Pa)
v is the specific volume of the working fluid (m^3/kg)
pv is the "flow work" (N-m/kg)

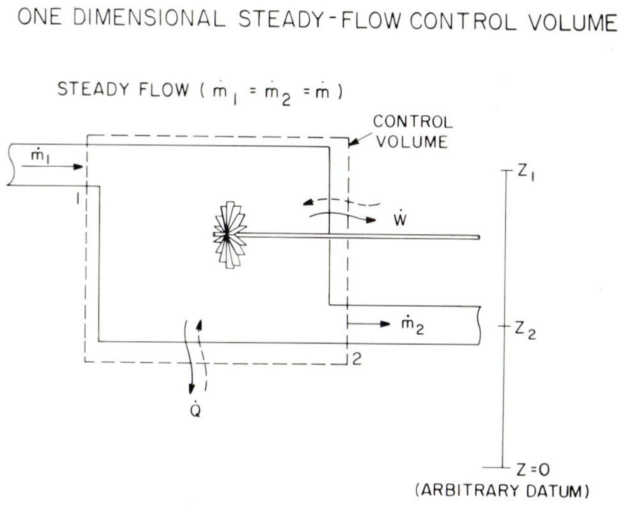

Fig. 5.5 One dimensional steady-flow control volume.

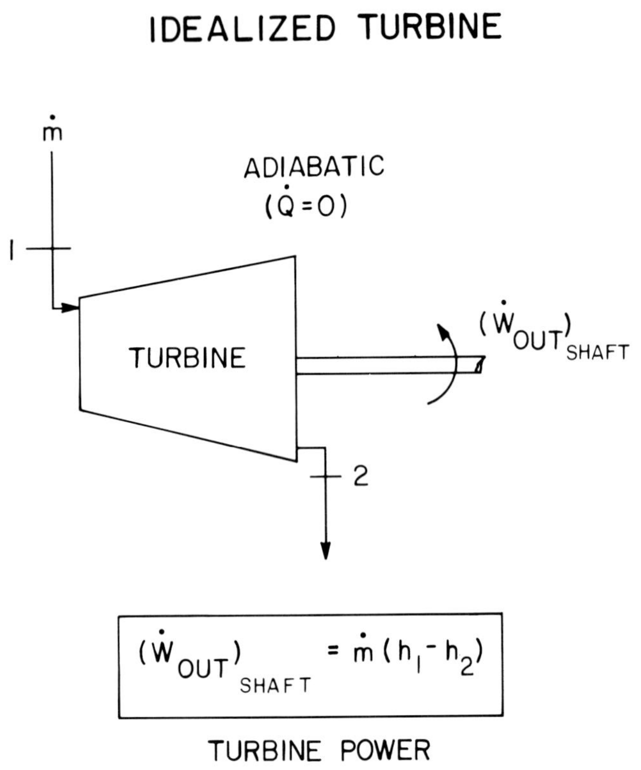

Fig. 5.6 Idealized turbine.

subscript "*1*" refers to inlet conditions
subscript "*2*" refers to exit conditions.

The term $\dot{W}$ is frequently called the "shaft work." It represents the amount of work transferred to the working fluid by a compressor or pump or extracted from the working fluid by a turbine.

The specific total energy term, e, includes the internal energy (u), bulk potential energy (gz), and bulk kinetic energy ($V^2/2$) of the working fluid. That is,

$$e_x = u_x + gz_x + V_x^2/2 \tag{5.6}$$

where u is the fluid specific internal energy (J/kg)
g is the acceleration due to gravity at sea level on Earth (9.8 m/s^2)
z is the height above some reference level z_0 (m)
V is the fluid velocity (m/s)
subscript "x" indicates a cross-sectional region of the control volume.

In the one-dimensional flow assumption, all fluid properties are considered equal across any given cross-sectional region, x.

A turbine is a device in which the working fluid (either gas or liquid) impinges upon a blade or fin configuration mounted on a shaft (see Fig. 5.6). The fluid imparts a portion of its energy to this shaft which then transmits this mechanical work out of the control volume where it can be used as rotary motion. Normally, the potential energy changes in the working fluid can be neglected so that the one-dimensional, steady-flow energy balance for a turbine is simply

$$\dot{Q}_{\text{out}} + \dot{W}_{\text{out}} = \dot{m}\left[(h_2 - h_1) + (V_2^2 - V_1^2)/2\right] \tag{5.7}$$

where h_1, h_2 are specific enthalpy values for the working fluid at the inlet and outlet, respectively (J/kg). It should be remembered that, by definition, $h \equiv u + pv$. In many cases, the idealized turbine can be represented as an adiabatic device ($\dot{Q}_{\text{out}} = 0$) in which the overall fluid kinetic energy change is actually small compared to the change in enthalpy. Thus, the shaft work out of the idealized turbine becomes

$$\dot{W}_{\text{shaft}_{\text{out}}} = \dot{m}(h_1 - h_2). \tag{5.8}$$

Similarly, in an idealized compressor (see Fig. 5.7) or an idealized pump (see Fig. 5.8) work is done on the working fluid in the control volume by the surroundings, resulting in a pressure increase for the fluid. Compressors are usually associated with flowing gases and pumps with liquids. Again, if we assume that our ideal-

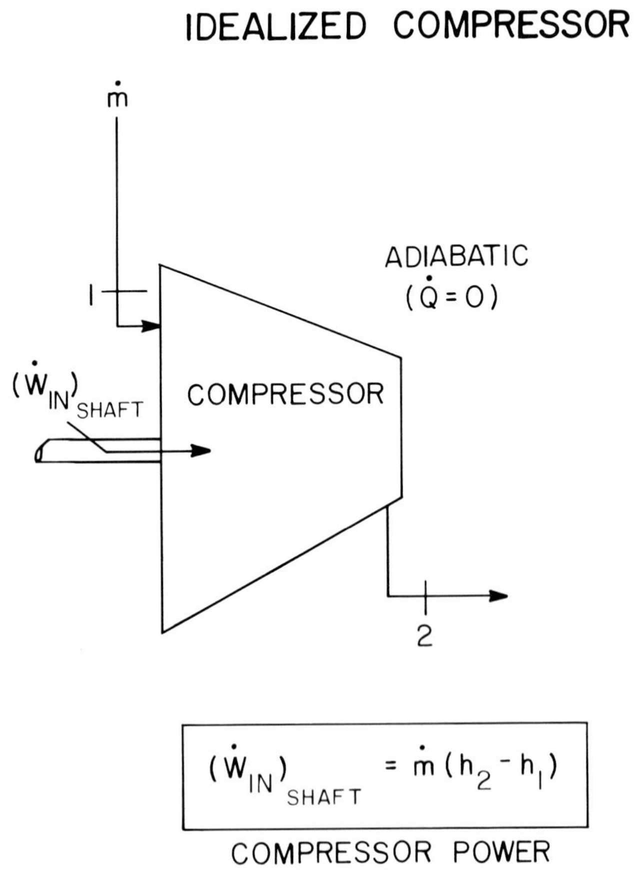

Fig. 5.7 Idealized compressor.

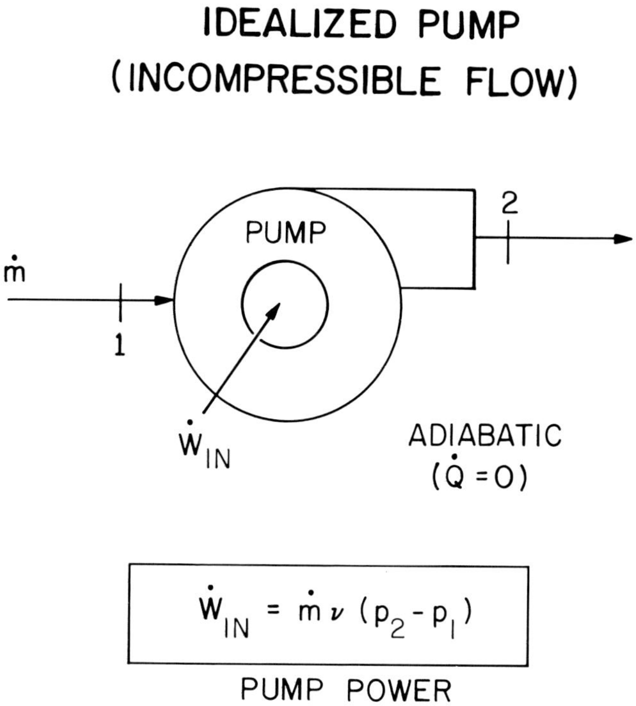

Fig. 5.8 Idealized pump (incompressible flow).

ized compressor experiences adiabatic operation in which the overall bulk kinetic energy changes are small compared to the change in working fluid enthalpy and neglecting changes in potential energy, we can write

$$\dot{W}_{\text{shaft}_{\text{in}}} = \dot{m}(h_2 - h_1). \tag{5.9}$$

Note that here we are doing work on the fluid in the control volume, while in the turbine we were extracting work from the fluid. Finally, consider the pumping of an incompressible fluid (i.e., v = constant), as shown in Fig. 5.8. Assuming an adiabatic process and again neglecting kinetic and potential energy changes of the liquid as it passes through the pump, the one dimensional energy equation for steady flow yields [3, 4]

$$\dot{W}_{\text{in}} = \dot{m}v(p_2 - p_1) \tag{5.10}$$

where $\dot{W}_{\text{in}}$ is the input pump power (W).

Ideal Rankine Cycle

Figure 5.9 illustrates the basic (ideal) Rankine cycle for a space nuclear power system. This thermodynamic cycle is named after the Scottish engineer, William Rankine (1820–1872), who made fundamental contributions to engineering thermodynamics especially with respect to the steam engine [6]. The basic Rankine cycle is a modification of the two-phase working fluid Carnot cycle, differing from the latter in that heat addition does not occur isothermally. In the basic Rankine cycle shown here, the working fluid undergoes an isentropic expansion in the turbine from state 1 to state 2. The "wet" liquid-vapor mixture then experiences a phase change to a saturated liquid as it undergoes isothermal heat rejection in the condensing radiator (state 2 to state 3). It next undergoes isentropic compression in the pump from state 3 to state 4. Here, the liquid is assumed to be incompressible and actually goes from the saturated liquid state (3) to the subcooled liquid state (4). Finally, the working fluid experiences constant pressure heat addition from state 4 to state 1. At state 4′ sensible heat addition gives way to phase change—so that, as more thermal energy is added, the working fluid gradually changes from a saturated liquid (point 4′) to a saturated vapor (point 1) of 100 percent quality. This phase change process occurs at both constant temperature and constant pressure [3–5].

The quality (x) of a liquid-vapor mixture is defined in thermodynamics as the fraction of mass of a working fluid in the vapor phase. Similarly, the wetness of a mixture $(1 - x)$ describes a liquid-vapor mixture with a quality less than 100 percent. By convention, the thermodynamic properties of a two-phase mixture are designated in terms of the following specific enthalpies:

$$h_{\text{mix}} = (1 - x)h_f + xh_g \tag{5.11}$$

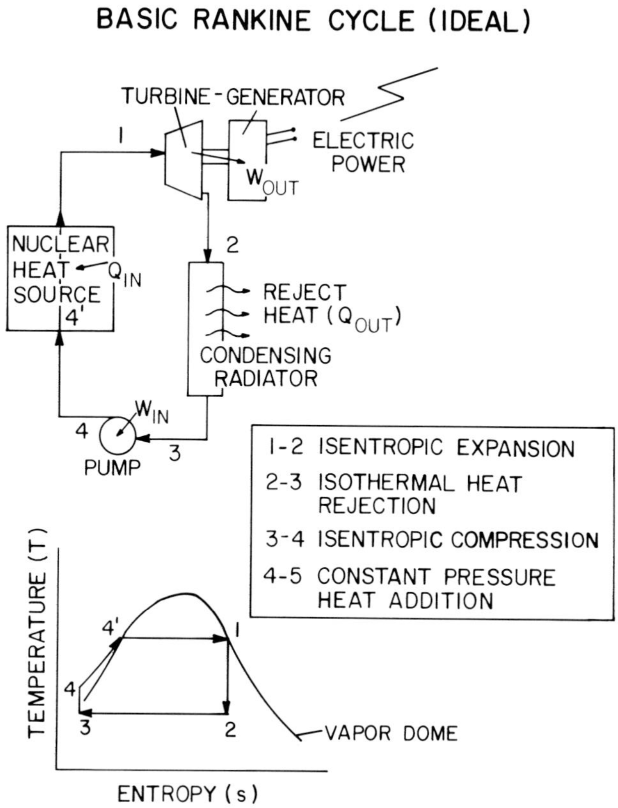

Fig. 5.9 Basic Rankine cycle (ideal).

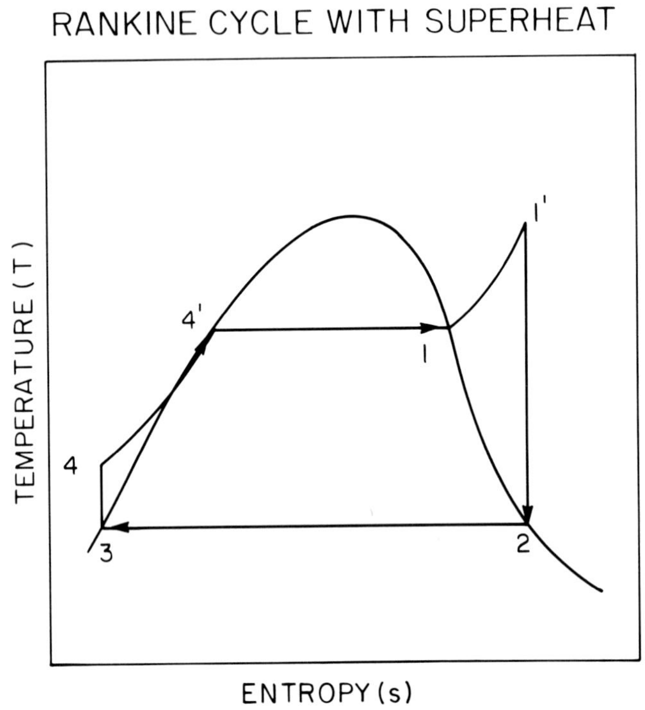

Fig. 5.10 Rankine cycle with superheat.

where h_{mix} is the enthalpy of the mixture (J/kg)
h_f is the enthalphy of the saturated liquid (subscript "f" for liquid phase) (J/kg)
h_g is the enthalphy of the saturated vapor (subscript "g" for vapor phase) (J/kg)
x is the quality of the mixture.

A subcooled or compressed liquid is one that exists at a temperature lower than the saturation temperature, corresponding to its pressure; while a superheated vapor is a vapor existing at a temperature greater than the saturation temperature, corresponding to its pressure. For a more extensive discussion of saturation states and the vapor dome region of a thermodynamic mixture the reader is referred to references [3–5].

If we again neglect changes in the potential and kinetic energy of the working fluid as it goes around the basic Rankine cycle, we can calculate the thermal efficiency from

$$\eta_{th} = \frac{(\text{work out})_{\text{turbine}} - (\text{work in})_{\text{pump}}}{\text{heat added}} . \quad (5.12)$$

Assuming ideal components throughout the system yields

$$\eta_{th} = \frac{(h_1 - h_2) - (h_4 - h_3)}{(h_1 - h_4)} . \quad (5.13)$$

One method of improving the thermal efficiency of the Rankine cycle is to superheat the working fluid as shown in Fig. 5.10. In this approach, the constant pressure heating of the working fluid is continued past the saturated vapor state (state 1) and into the superheated vapor region (state 1′).

Another approach used to improve the thermal efficiency of the basic Rankine cycle involves the reheating process shown in Fig. 5.11. In this technique, the working fluid is expanded in the first turbine stage (from point 1′ to 2) until it reaches pressure p_2 (on the saturated vapor line). Then, the working fluid is returned to the nuclear power source where it is reheated at constant pressure to temperature T_2 in the superheated vapor region (state 2′). It next undergoes expansion to state 3 in the seond stage of the turbine.

Superheat and reheat reduce the amount of moisture in the working fluid entering the turbine thus preventing liquid droplet erosion of turbine blades. They also raise the average temperature at which thermal energy is added to the working fluid. Therefore, both techniques improve overall Rankine cycle efficiency.

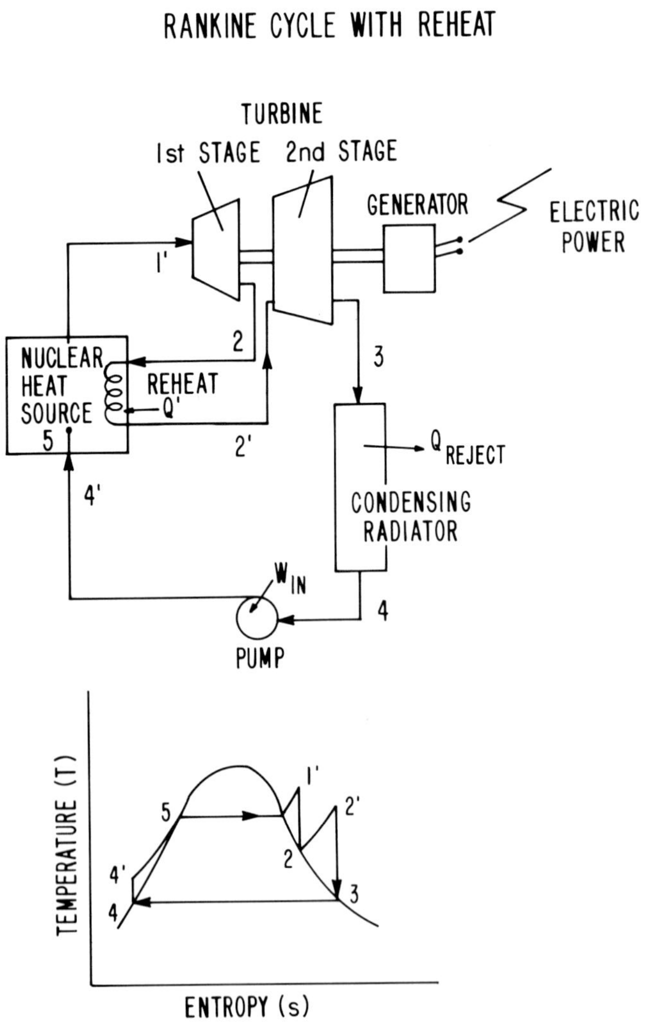

Fig. 5.11 Rankine cycle with reheat.

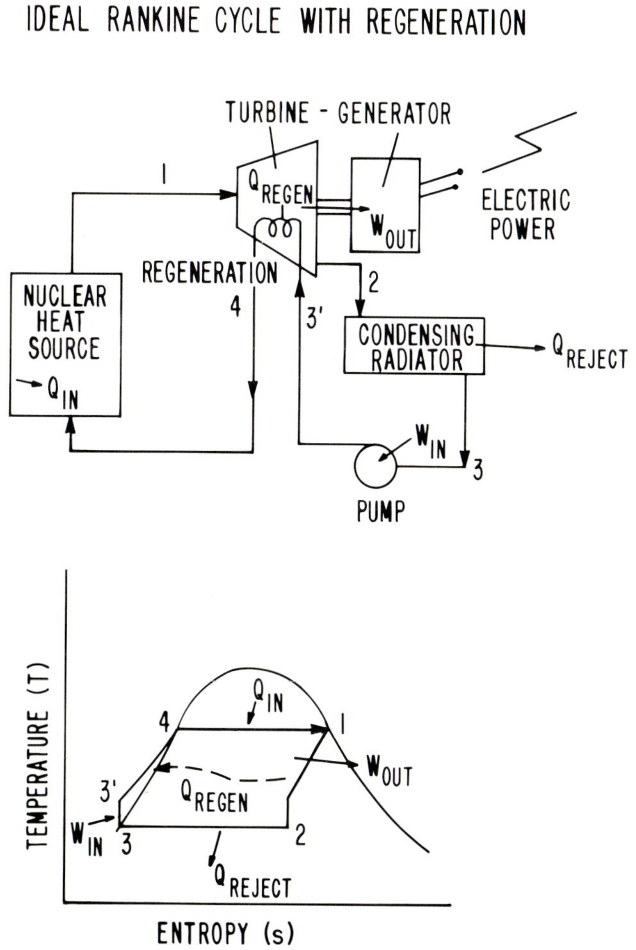

Fig. 5.12 Ideal Rankine cycle with regeneration.

We can also employ the concept of regeneration to improve the overall efficiency of the Rankine cycle. As shown in Fig. 5.12, when the working fluid expands in the turbine some of its thermal energy content is used to preheat the liquid-phase working fluid before it enters the nuclear heat source. This technique allows us to closely approach the ideal Carnot cycle efficiency, since thermal energy transfer in the overall cycle now occurs isothermally. Of course, an ideal turbine as a power producing device is not also an ideal heat exchanger. Many engineering trade-offs are made in practical Rankine cycle applications, including the use of superheat, reheat, and regeneration.

Rankine cycles using mercury, potassium, and organic working fluids have been studied for use in space nuclear power systems [1,2]. For example, the potassium Rankine cycle shown in Fig. 5.13 was considered for a high temperature (1400 K turbine inlet temperature) power system with very small radiator. The major components of this proposed space power system include a heat-pipe reactor linked to a potassium boiler, a multistage turbine with potassium reheat and condensate removal, an alternator, a potassium-to-liquid metal (NaK) condenser, a NaK thermal radiator (rejecting waste heat to space), and electro-magnetic pumps (EMP) for liquid potassium and liquid NaK circulation. Small secondary thermal energy transfer loops were considered for cooling the following: the potassium turbine bearing coolant, the potassium alternator bearing lubricant, and an organic coolant for electronic components [1].

In this proposed concept, the potassium working fluid is vaporized and superheated in a boiler, supplied with thermal energy by sodium heat pipes from the reactor core. The vaporized potassium is expanded through the turbine, driving the alternator assembly to produce electric power (a small portion of which output is used to run the electromagnetic pumps). The potassium working fluid is then condensed at low pressure. At the proposed turbine inlet temperature of 1420 K

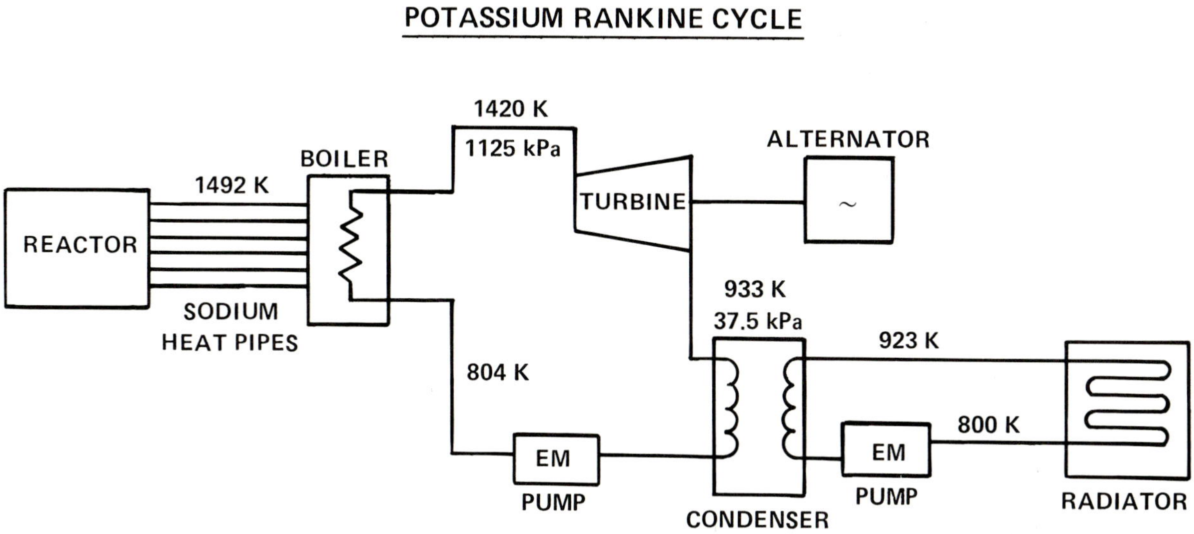

Fig. 5.13 Proposed potassium Rankine cycle [1].

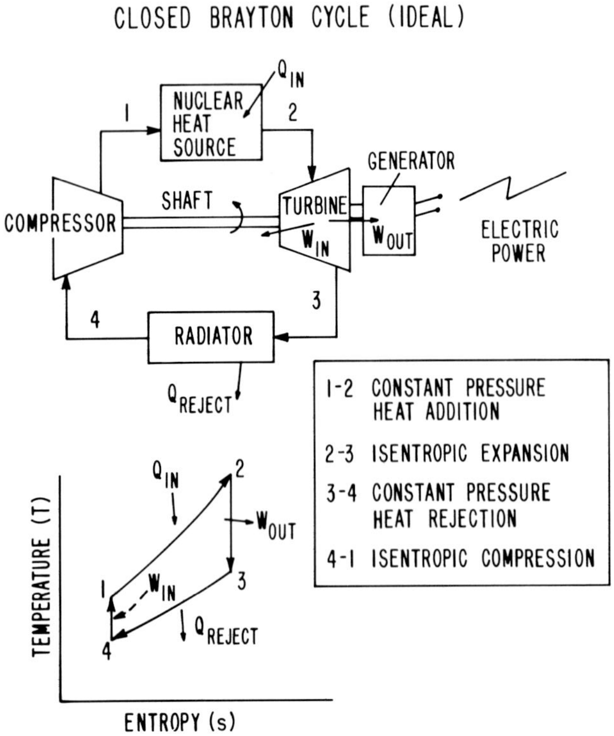

Fig. 5.14 Closed Brayton cycle (ideal).

and condenser inlet temperature of 933 K, the overall cycle efficiency is estimated to be 19 percent [1]. The thermal radiator is assumed to be liquid metal (NaK) cooled and operated over an 800 K to 923 K temperature range. Of course, the schematic Rankine cycle diagram shown in Fig. 5.13 is considerably simplified. It does not illustrate, for example, interstage condensate removal, interstage reheat loops, the auxiliary turbine alternator, or cooling loops for electronic components.

One problem with using potassium as the working fluid in a Rankine cycle is that because of material-caused temperature limits, the turbine operates in the condensing region of the potassium phase diagram. It is not practical to use a large enough superheat to avoid entry into the mixture region during expansion. Reheat helps, but it does not completely eliminate condensation. For example, for the conditions shown in Fig. 5.13, the turbine exit quality would be 80 percent, should the design not include condensate removal. If condensation is permitted in the turbine, it will promote turbine blade erosion, thereby shortening the turbine's life. Consequently, both reheat and condensate removal would be required with this version of the Rankine cycle.

Ideal Brayton Cycle

A closed cycle gas turbine power system, called the Brayton cycle, is illustrated in Fig. 5.14. In the idealized Brayton cycle, heat addition and rejection occur at constant pressure, while expansion and compression processes are assumed to be isentropic. The working fluid undergoes the following: constant pressure heat addition from state 1 to 2; isentropic expansion in the turbine from state 2 to 3 (a portion of the output work goes to compressing the working fluid); constant pressure heat rejection from state 3 to 4; an isentropic compression from state 4 to 1 [3, 4, 5]. Again, neglecting working fluid potential and kinetic energy changes, the thermal efficiency of the ideal closed Brayton cycle is given by

$$\eta_{th} = \frac{(W_{out})_{turbine} - (W_{in})_{compressor}}{Q_{in}} \tag{5.14}$$

or,

$$\eta_{th} = \frac{(h_2 - h_3) - (h_1 - h_4)}{(h_2 - h_1)} . \tag{5.15}$$

We can frequently treat the working fluid as an ideal gas, greatly simplifying the analysis. Invoking the ideal gas assumption, the thermal efficiency is given by [3, 4]

$$\eta_{th} = 1 - (p_1/p_4)^{(1-k)/k} \tag{5.16}$$

where $k = c_p/c_v$ is the ratio of the specific heats of the gas at constant pressure and volume respectively.

To improve the thermal efficiency of the basic closed

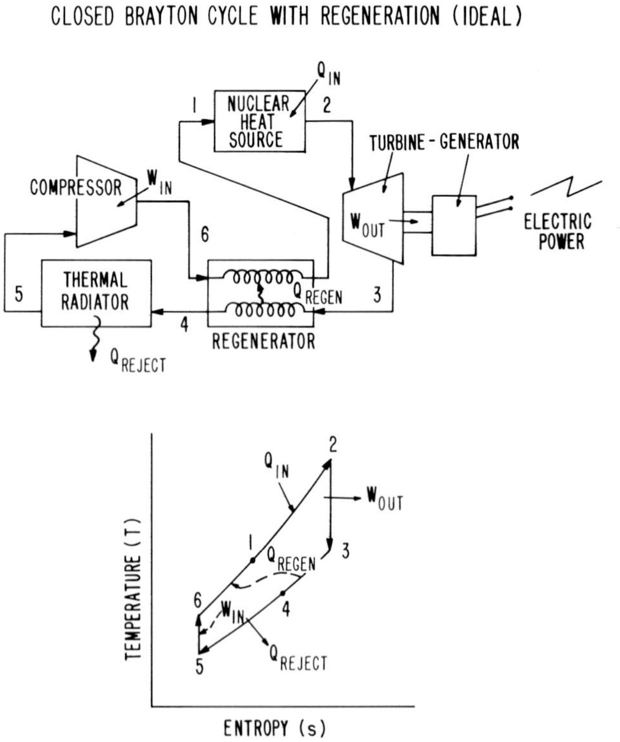

Fig. 5.15 Closed Brayton cycle with regeneration (ideal).

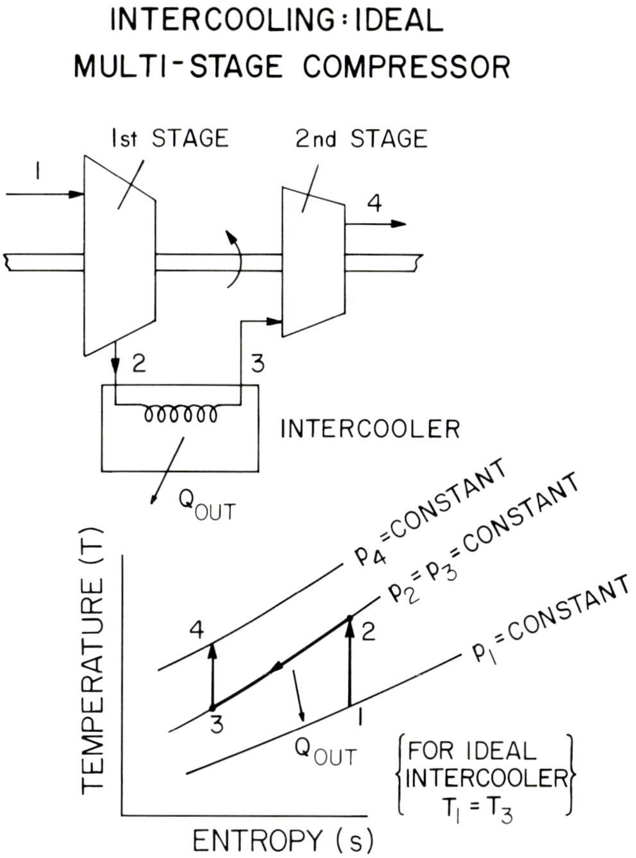

Fig. 5.16 Intercooling: ideal two-stage compressor.

Brayton cycle, a regenerator (or recuperator) is employed, as shown in Fig. 5.15. In the regenerator, the hot gases, which are exhausted from the turbine (i.e., state 3), preheat the working fluid as it exits the compressor (state 6) and returns to the nuclear power supply (state 1).

It is also possible to improve Brayton cycle efficiency through the use of multistage compression with intercooling and multistage expansion with reheat. In an ideal intercooler (see Fig. 51.6) the gas is cooled to the initial inlet temperature T_1 before entering the next state; that is, $T_1 = T_3$ and $p_2 = p_3$. Reheating was discussed in the treatment of the Rankine cycle.

From a practical engineering point of view, the Brayton cycle efficiency is improved most when a combination of reheating and intercooling is used in conjunction with regeneration [4, 5]. If regeneration is used along with a very large number of reheating stages and intercooling states, then, in principle, all thermal energy input eventually appears in the reheat-heat exchangers (when the gas is at its maximum temperature) and all heat rejection occurs in the intercoolers (when the gas is at its lowest temperature). In the limit, this condition approaches the Carnot efficiency of a reversible $2T$ heat engine [3] (see Fig. 5.17).

The helium-xenon working gas closed Brayton cycle has been under study for space power applications since the mid-1960s [1, 7]. Inert gases, such as a helium-xenon mixture, reduce the turbine blade erosion problem encountered in two-phase working fluid power cycles. Turbine studies have shown that as the molecular weight of the working gas increases, the required number of turbine blade rows (stages) decreases [5]. For example, if a particular turbine needs ten stages for helium (molecular weight = 4), it will require only one stage for argon as the working fluid (molecular weight = 40). However, the thermal energy transport properties of gases are much better at lower molecular weight [1]. Consequently, a mixture of helium and xenon (molecular weight $\approx$ 40) was chosen as an optimum trade-off between turbomachinery compactness and heat transport ability.

The principal components of this Brayton cycle configuration and representative operating parameters are shown in Fig. 5.18. These components include a turbine-compressor-alternator rotating unit; a high tem-

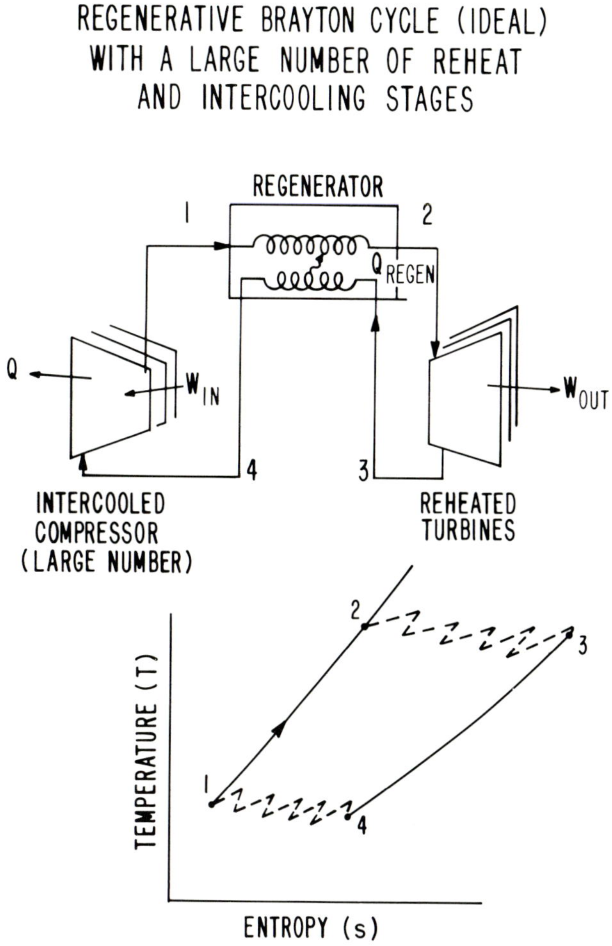

Fig. 5.17 Regenerative Brayton cycle (ideal) with a large number of reheating and intercooling stages.

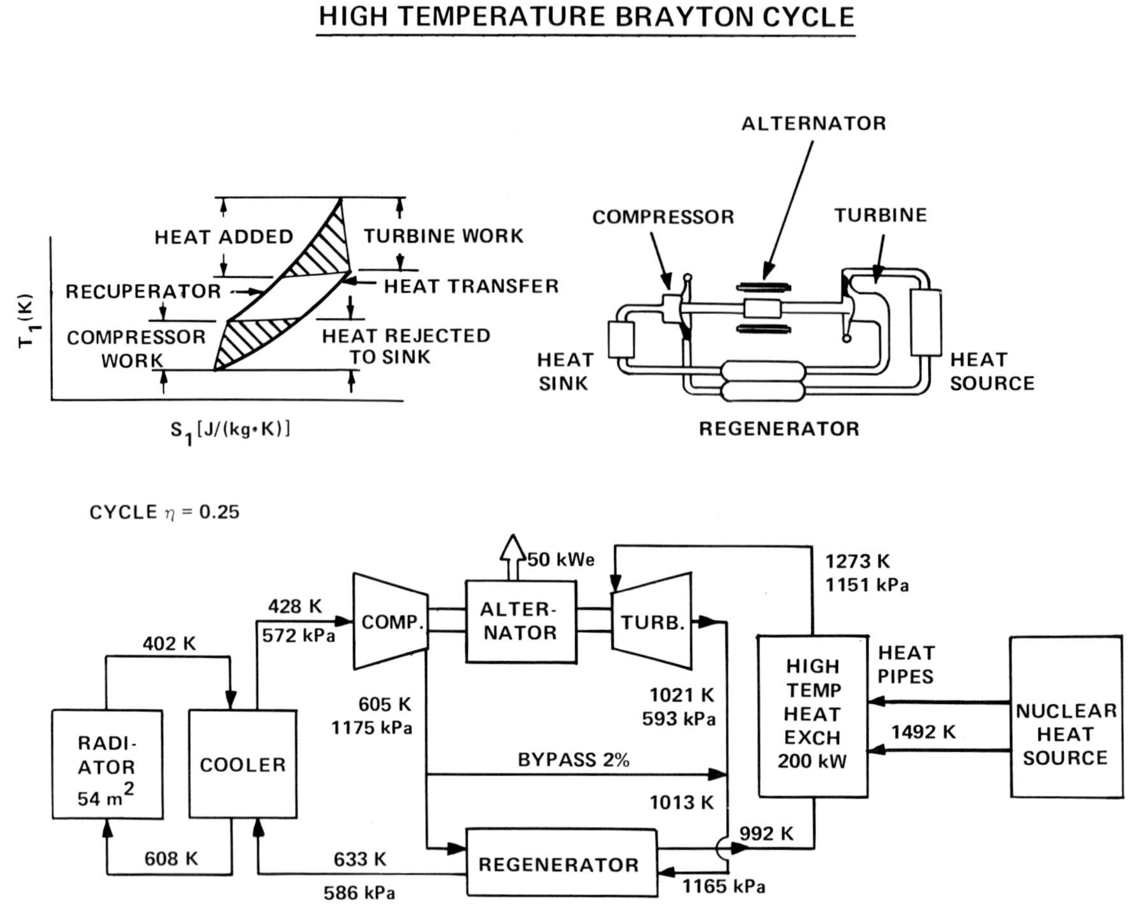

Fig. 5.18 Helium-xenon closed Brayton cycle configuration [1].

perature heat exchanger (nuclear heat source to working gas loop); a low temperature heat exchanger called the cooler (working gas loop to thermal radiator); a regenerator (also called a recuperator); and suitable gas inventory, ducting, and flow control equipment [1]. The rotating unit consists of a single-stage radial turbine, a single-stage radial compressor, and an alternator, all mounted on a common shaft. Foil-type gas bearings are used throughout, effectively eliminating all bearing wear. The foil (compliant) gas bearings are especially well-adapted to the relatively low-mass rotor designs needed in space power systems. They are also preferred over rigid geometry gas bearings because of their tolerance to thermal gradients, misalignments, vibration, and shock [2]. Alternator cooling can be provided through either a working gas bypass or a secondary coolant (organic or NaK) loop.

The helium-xenon mixture exits the high temperature heat exchanger at approximately 1270 K. As shown here, thermal energy is transported to the high temperature heat exchanger from the nuclear power source by sodium heat pipes, although a pumped liquid metal loop can also be considered for this heat source-to-heat exchanger interface. Work is extracted from the high temperature gas as it is adiabatically expanded in the turbine—a portion of this work is used to run the compressor. The ratio of compressor input work $(W_{comp})_{in}$ to turbine output work $(W_{turb})_{out}$ is called the back work ratio of the Brayton cycle. That is,

$$\text{Back Work Ratio} \equiv \frac{(\text{Compressor Work})_{in}}{(\text{Turbine Work})_{out}} . \quad (5.17)$$

For the ideal Brayton cycle shown in Fig. 5.14, this ratio becomes

$$\text{Back Work Ratio} = \frac{(h_1 - h_4)}{(h_2 - h_3)} . \quad (5.18)$$

In engineering practice, the back work ratio typically ranges from 0.4 to 0.8 [5]. When this ratio is unity, all the power extracted by the turbine is used to run the compressor—an obviously self-defeating situation!

The working fluid exits the turbine (here at 1021 K) and enters the high temperature pass of the regenerator, where it exchanges some of its thermal energy with the low temperature (602 K) working gas leaving the compressor, preheating it to 922 K prior to its entrance into

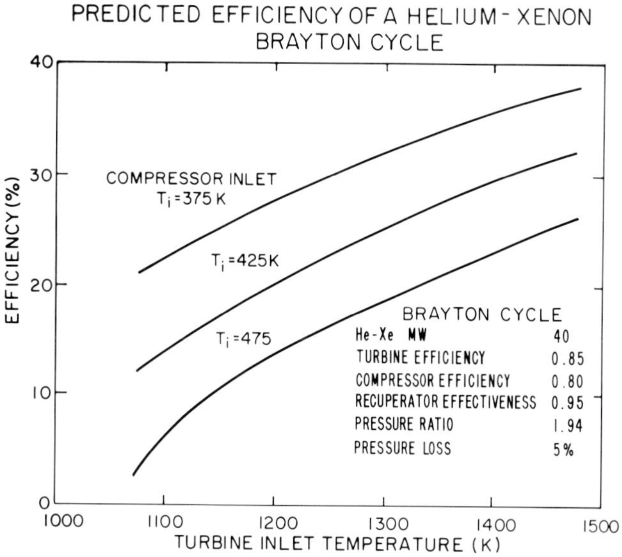

Fig. 5.19 Predicted efficiency of a helium-xenon Brayton cycle [1].

the high temperature heat exchanger. The expanded working fluid leaves the high temperature regenerator pass at 633 K and is further cooled in the heat sink heat exchanger (or cooler). Thermal energy is delivered to the radiator from the cooler by a secondary coolant loop. (If desired, the working gas can be cooled directly by flow through the radiator—but radiator mass and surface area design constraints may warrant the use of a more thermophysically efficient liquid working fluid.)

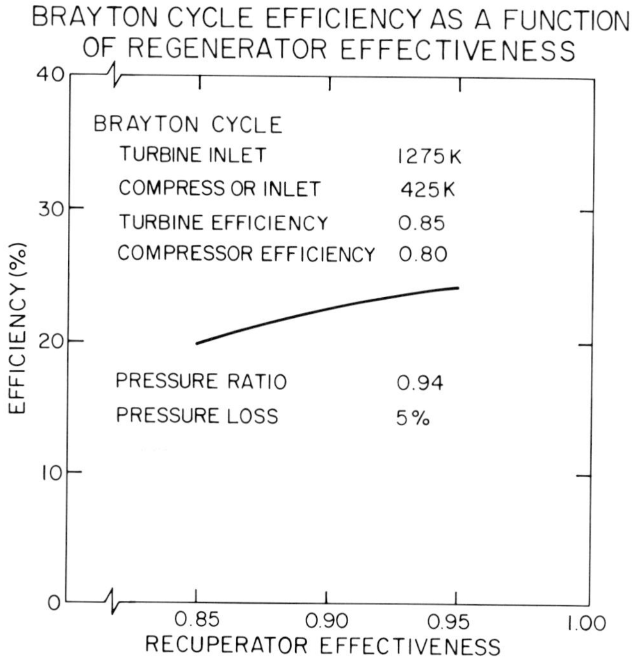

Fig. 5.20 Brayton cycle efficiency as a function of regenerator effectiveness [1].

Finally, at the low temperature region of the Brayton cycle (~428 K), the helium-xenon working gas is compressed and then continues on through the low temperature pass of the regenerator on its way to the high temperature heat exchanger to start the cycle again [1].

Turbine inlet temperature and regenerator effectiveness control the performance of a closed Brayton cycle. The anticipated performance of the helium-xenon Brayton cycle discussed here is shown as a function of inlet temperature in Fig. 5.19. At a turbine inlet temperature of 1275 K, the efficiency is estimated to be 25 percent with a compressor inlet temperature of 425 K. These performance estimates assume a turbine efficiency of 85 percent, a compressor efficiency of 80 percent, a regenerator effectiveness of 0.95, and a loop pressure drop of 0.05 of the turbine inlet pressure. Fig. 5.20 shows the impact of regenerator effectiveness on cycle efficiency; while Fig. 5.21 shows the effect of working fluid pressure loss on cycle efficiency.

Ideal Stirling Cycle

The requirement to achieve a totally reversible heat engine involves devising a means by which all thermal energy transfer to and from the system takes place both isothermally and reversibly. In the early 19th century a Scottish minister, Robert Stirling (1790–1878), designed such a cycle. In fact, the Stirling engine is the earliest example of a reversible heat engine. Working with his brother James, Robert Stirling employed the practice of

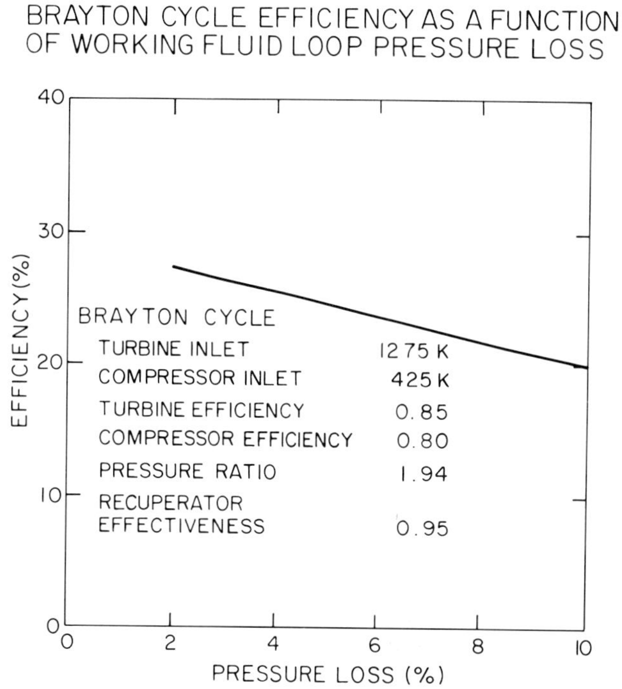

Fig. 5.21 Brayton cycle efficiency as a function of working fluid loop pressure loss [1].

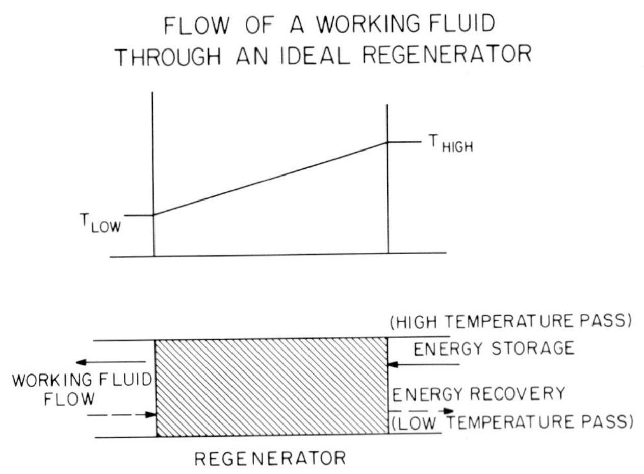

Fig. 5.22 Flow of a working fluid through an ideal regenerator.

regeneration in developing this engine. Stirling engines enjoyed considerable success in the last century as quiet pumping engines. However, with the development of more compact internal combustion engines at the turn of the century, Stirling engines soon fell into disuse. Space power applications are once again reviving interest in this thermodynamic cycle [8].

An ideal regenerator is the component of a heat engine into which the working fluid transfers heat on a temporary basis, while dropping in temperature from some upper value (T_{high}) to some lower value (T_{low}). Inherent in this ideal regenerator concept is the assumption that heat transfer occurs reversibly—that is, there is no temperature difference between the heat absorbing regenerator material (such as wire mesh or tiny thin-walled tubes) and the working fluid at any point where they are in thermal contact (see Fig. 5.22). Then, when the working fluid passes back through the regenerator, entering at T_{low}, it recovers the heat which was originally stored and leaves at T_{high}. Through the regenerator, heat is alternately stored and then recovered, reversibly. Since the regenerator is in exactly the same state after completion of a cycle as it was before, it is considered as a component of the heat engine and not as a part of some external heat source or sink. In the Stirling engine this reversible heat exchange in the regenerator occurs at constant volume.

The basic Stirling cycle with an ideal gas working fluid is shown in Fig. 5.23. From state 1 the gas, initially at the lower temperature limit, recovers the stored thermal energy from the regenerator in a reversible, constant volume process. This continues until the working fluid reaches the upper temperature limit, state 2. Then, from state 2 to state 3 the working substance experiences reversible heat addition from the external heat source (also at T_{high}) and expands isothermally to state 3. From state 3 to state 4, the working fluid again interacts with the regenerator in a constant volume process. This time, however, thermal energy is transferred from the working fluid to the regenerator. The ideal gas goes from T_{high} at state 3 to T_{low} at state 4. This is followed by an isothermal compression from state 4 to state 1. During this final portion of the Stirling cycle, the working fluid rejects heat to an external sink that is also at T_{low}. The thermal efficiency of the ideal Stirling cycle is [5]

$$\eta_{Stirling} = 1 - T_{low}/T_{high} \tag{5.19}$$

where T_{low}, T_{high} are absolute values of temperature (K). Equation 5.19 is the identical mathematical expression for the thermal efficiency of the Carnot cycle operating between the same temperature limits.

The free piston Stirling engine (FPSE) is a thermally driven mechanical oscillator, using the Stirling cycle and deriving its output power from heat flow between a source and a sink [2, 9, 10]. The FPSE displacer motion is produced by gas pressures rather than by mechanical linkages. This type of engine operates at the highest device efficiency of all known heat engines and is uniquely suited to drive direct-coupled reciprocating loads (such as linear generators) in a hermetically sealed configuration and without requiring high pressure shaft seals or contaminating lubricants.

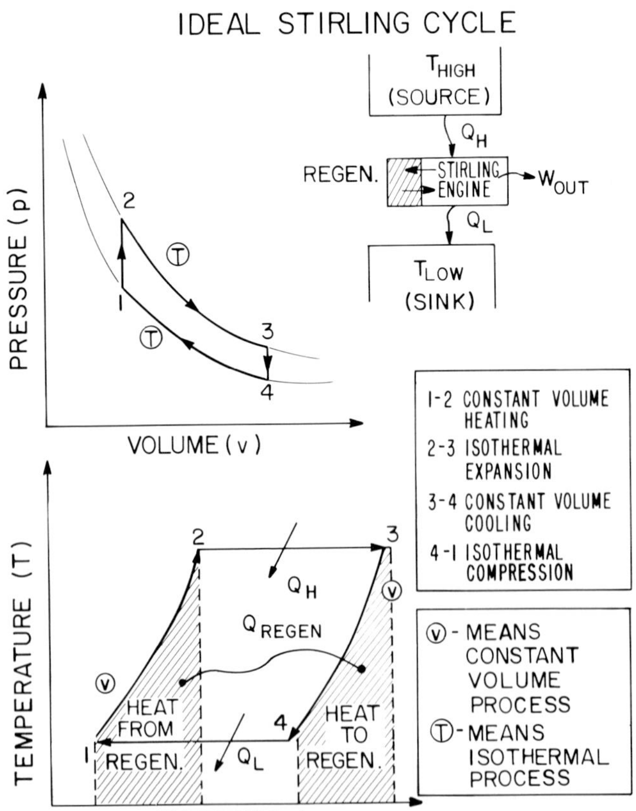

Fig. 5.23 Ideal Stirling cycle.

OPERATING PRINCIPLES OF FREE PISTON STIRLING ENGINE (FPSE)

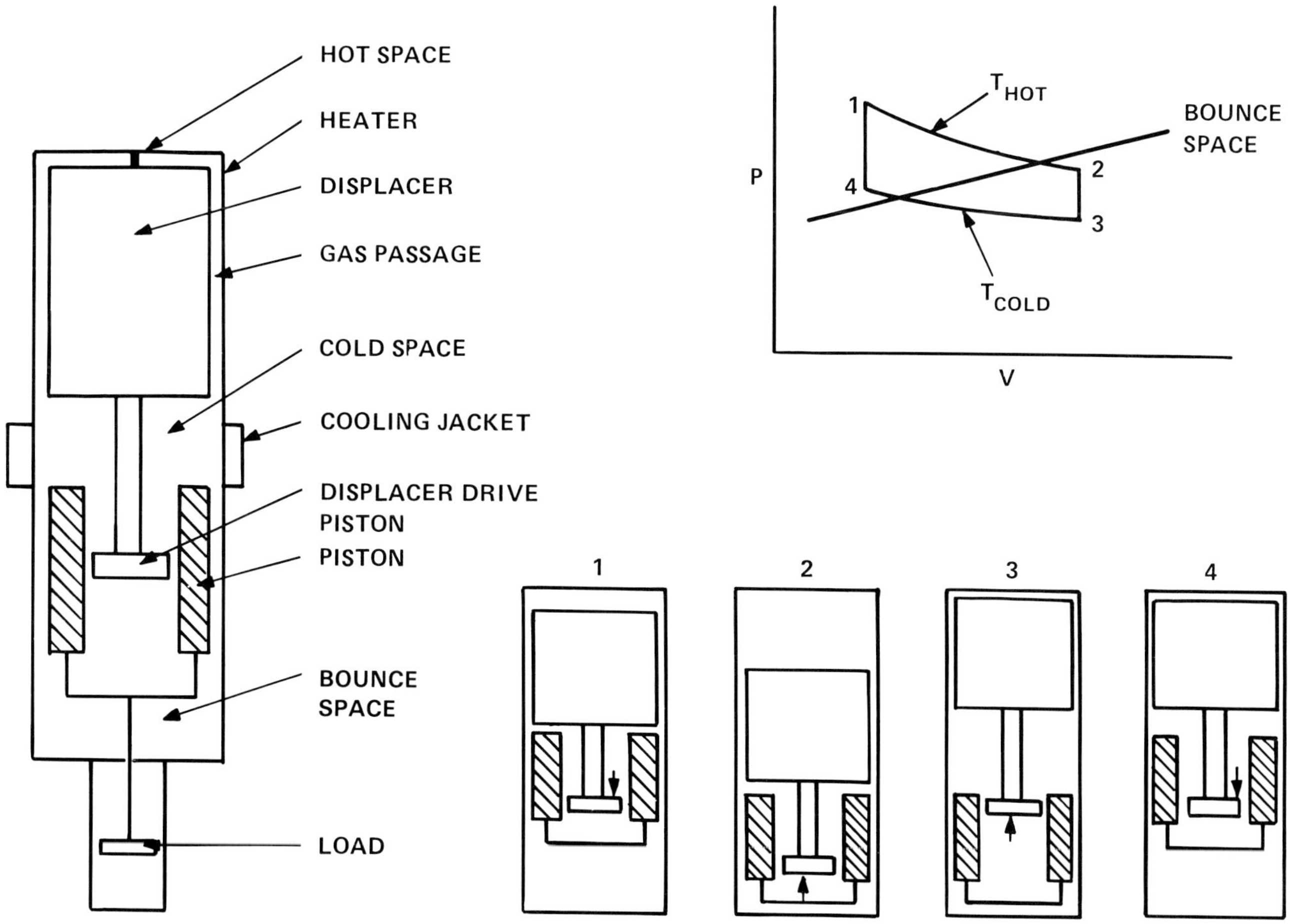

1-2 PISTON EXPANDS WORKING GAS; DISPLACER ON PISTON.

2-3 PRESSURE IN BOUNCE SPACE GREATER THAN PRESSURE IN WORKING SPACE, FORCING DISPLACER TOWARD HOT SPACE, WORKING GAS MOVED INTO COOLED SPACE, PRESSURE DROPS.

3-4 PISTON DRIVEN INTO WORKING SPACE BY HIGHER BOUNCE SPACE PRESSURE.

4-1 DISPLACER DRIVEN TOWARD COLD SPACE BY WORKING SPACE PRESSURE HIGHER THAN BOUNCE SPACE PRESSURE.

Fig. 5.24 Operational principles of free piston Stirling engine (FPSE) [2].

The FPSE consists of three basic components: a heavy piston (called the power piston), a low mass piston (called the displacer), and a sealed cylinder. As shown in Fig. 5.24, the displacer rod passes through the power piston and is in communication with the bounce space, which is essentially a constant pressure chamber [2]. The displacer and the power piston assembly are the only two moving parts in the FPSE engine-alternator unit. The power piston assembly is composed of the power piston directly connected to the alternator armature. This eliminates the need for converting reciprocating motion to rotary motion and also the need for lubricating such a Stirling device. The assembly is supported by two pressurized (hydrostatic) helium gas bearings. Long-term reliability is inherent in this design, because compressor inlet and discharge flow is controlled by means of fixed ports rather than valves.

A matrix-type regenerator is located between the heat source and the heat sink. It can consist of stacked layers

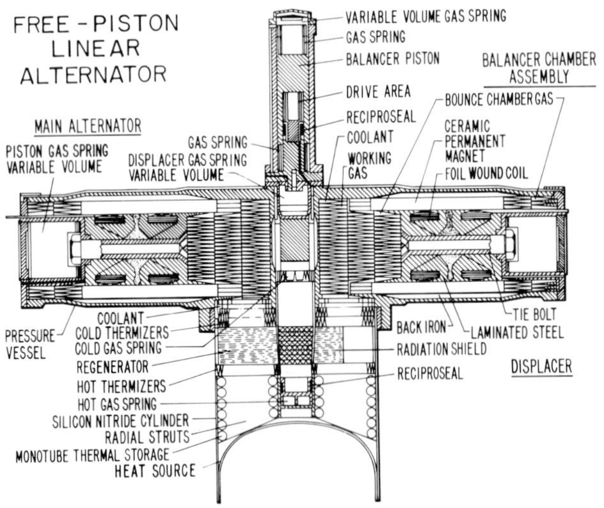

Fig. 5.25 Conceptual free piston Stirling engine—alternator configuration [2].

of wire screen or wrapped foil [2].

The electrodynamic linear alternator design is based upon the well-known principles of electromagnetic induction (see Fig. 5.25). A dc flux loop is created by the dc field coil. Then, as the armature moves back and forth in the stator, the dc flux links the ac coils and induces an ac voltage [2].

A reference 50 kW_e free piston Stirling engine-alternator configuration is shown in Fig. 5.26 [1]. The energy conversion system consists of a four-cylinder opposed configuration, four alternators, with a common bumper space between the opposed power pistons. Table 5.1 lists the performance characteristics for this reference design.

Direct Energy Conversion Devices

In a direct energy conversion (DEC) device different forms of energy (such as thermal, solar, or chemical) are converted directly into electricity without the use of a vapor power cycle (Rankine), gas power cycle (Brayton), or reciprocating machinery (Stirling engine). The absence of components to handle a working fluid, such as pumps, compressors, and turbines, offers an inherent reliability advantage. All U.S. space nuclear power systems flown to date, both radioisotope and reactor types, have used thermoelectric conversion devices to transform the energy of the atomic nucleus into useful spacecraft electric power. We will now discuss two especially important DEC systems: thermoelectric and thermionic.

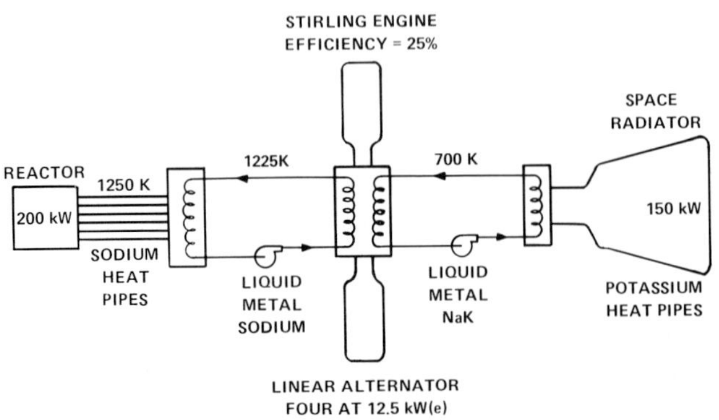

Fig. 5.26 Reference design 50 kW_e free piston Stirling engine with linear alternator [1].

Table 5.1 Performance characteristics for reference Stirling cycle power conversion system [1].

STIRLING CYCLE POWER-CONVERSION SYSTEM	
SYSTEM POWER (kW_e)	50
ENGINE GROSS POWER (kW_e)	58
NET ELECTRIC POWER (kW_e)	51
WORKING FLUID	HELIUM
HEAT SOURCE FLUID	NaK
TEMPERATURE HEATER FLUID (K)	1500
COOLANT TEMPERATURE (K)	700
EFFICIENCY (NET)(%)	30
WEIGHT (kg)	
ALTERNATOR	347
PRESSURE VESSEL	39
PISTON AND DISPLACER	5
HEATER	6
COOLER	3
TOTAL	400
SIZE (m)	
LENGTH	1.76
WIDTH	0.55
HEIGHT	0.27
MATERIALS	
ALL HOT COMPONENTS	316 ss
PRESSURE ENCLOSURE	TITANIUM
PISTON	ALUMINUM
DISPLACER	TITANIUM
CONFIGURATION	
FOUR CYLINDER OPPOSED	
FOUR ALTERNATORS	
COOLER – COUNTERFLOW TUBE	
HEATER – TUBE BUNDLE, PARALLEL FLOW	
REGENERATOR – SPIRAL COIL TYPE	

Thermoelectric Conversion

Direct conversion of thermal energy into electricity in a thermoelectric (TE) device is based upon the Seebeck effect. In 1821, the Russian-German physicist, Thomas Seebeck (1770–1831), observed that an electromotive force (emf) is generated when the junctions of two dissimilar metals are maintained at two different temperatures [6] (see Fig. 5.27). This thermally induced emf

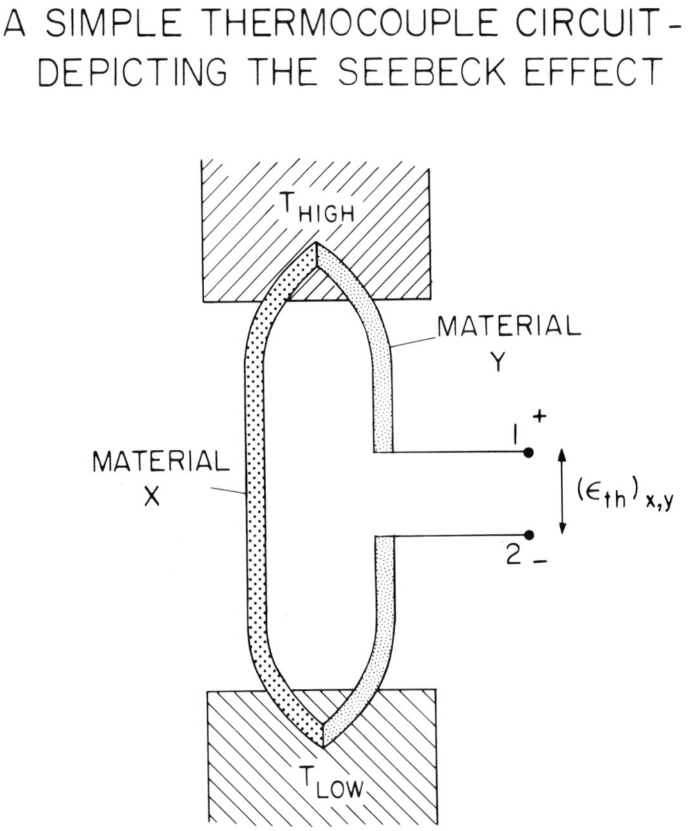

Fig. 5.27 A simple thermocouple circuit—depicting the Seebeck effect.

($\mathscr{E}_{\text{th}}$) forms the basis of the thermocouple, a widely used device for temperature measurements. The first derivative of this thermally induced emf with respect to temperature, $d\mathscr{E}_{\text{th}}/dT$, is related to the thermoelectric power (ϵ) of the thermocouple as follows [3, 4, 11]:

$$\epsilon \equiv -\left(\frac{d\mathscr{E}_{\text{th}}}{dT}\right)_{i=0}. \tag{5.20}$$

The subscript $i = 0$ expresses the fact that even if there is no current flow through the two dissimilar materials, an emf or electric potential will still exist, providing there is a thermal gradient present. The Seebeck coefficient for the two materials X and Y shown in Fig. 5.27 is defined as

$$\alpha_{x,y}(T) \equiv \frac{d\mathscr{E}_{x,y}}{dT}. \tag{5.21}$$

The Seebeck coefficient for these two dissimilar materials may also be expressed in terms of their thermoelectric powers. That is,

$$\frac{d\mathscr{E}_{x,y}}{dT} = \epsilon_x(T) - \epsilon_y(T) = \alpha_{x,y}(T) \tag{5.22}$$

assuming the output taps 1 and 2 are at the same temperature (i.e., $T_1 = T_2$) [3].

With the development of special semiconductor materials that combine a high Seebeck coefficient (α), relatively low electrical resistivity (ρ), and low thermal conductance (k), electric power generation with TE converters has become practical. Modern TE devices often use n-type and p-type semiconductor materials in order to create larger voltage outputs per degree of temperature difference. Figure 5.28 illustrates the basic operating principle of a thermoelectric conversion device. An individual TE converter consists of two semiconductor legs that are bonded to two heat transfer surfaces called the hot and cold shoes or junctions. One of the semiconductor materials is a p-type material in which the cold shoe region becomes positively charged by the migration of holes under the influence of a thermal gradient. The other semiconductor material is an n-type material in which a temperature difference causes electrons to diffuse to the cold shoe. One region of this TE couple is maintained by the bonded hot shoe at the high temperature, while thermal energy is removed from the other end of the couple at the cold shoe. The thermally driven flow of electrons and holes creates a voltage across both cold shoe plates. By connecting an external load across the two cold shoe plates, a current is made to flow through this external circuit. The power flowing through this external circuit is maximized when the load resistance is matched to the internal TE converter resistance. Such TE unicouples can then be connected in an external series-parallel circuit

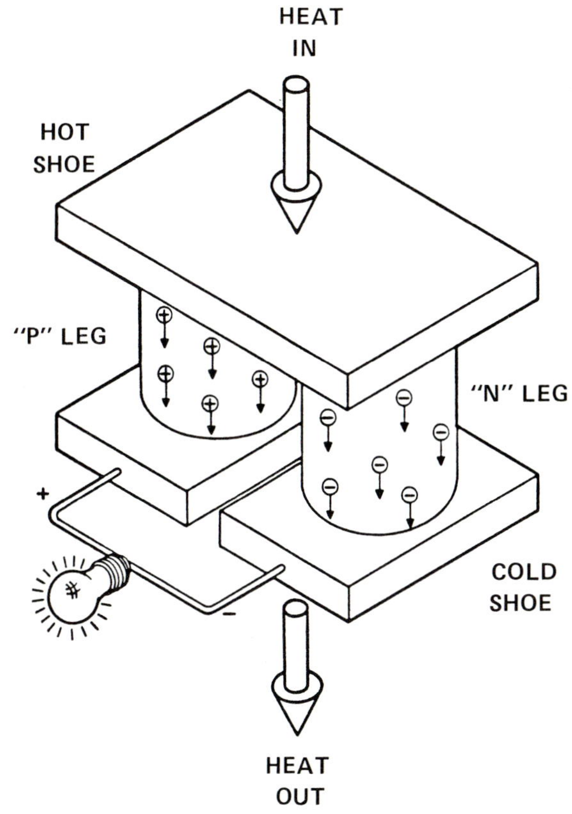

Fig. 5.28 Operating principle of the thermoelectric converter [2].

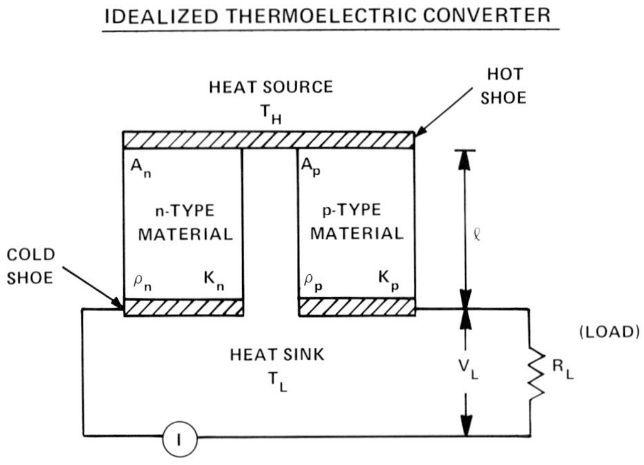

Fig. 5.29 Idealized thermoelectric converter [1].

to provide redundance against failures that create open circuits [1, 2].

The output of a TE converter cell depends on the operating temperatures, properties of the n-type and p-type semiconductor materials, and specific design details. The induced voltage in the cell is equal to the product of the overall Seebeck coefficient (n- and p-type material) and the temperature difference. The current flow is equal to the voltage divided by the sum of internal and external resistances. The net output power is equal to the current squared divided by the external load resistance. Again, maximum power occurs when this load resistance is equal to the internal resistance. Thermal energy transferred from the hot shoe to the cold shoe through either semiconductor material leg represents a loss. This conduction process is dependent upon the temperature gradient, the thermal conductivity of the TE materials, and the device design [1]. These relationships will become more apparent as we explore some idealized mathematical relationships that describe the design characteristics of a simple thermoelectric converter.

Consider the TE converter presented in Fig. 5.29. This idealized system consists of a heat source, a heat sink, and two semiconductor legs (one n- and one p-type) that are connected electrically in series with an external load. With the temperature differential ($T_H - T_L$) imposed between the hot and cold shoes, the induced voltage (in the absence of current flow) is given as

$$V_{I=0} = \alpha_{n,p}(T_H - T_L) \tag{5.23}$$

where

$$\alpha_{n,p} = |\alpha_n| + |\alpha_p| \tag{5.24}$$

and α_p is the p-type material Seebeck coefficient ($\mu V/K$)
α_n is the n-type material Seebeck coefficient ($\mu V/K$)
T_H is the heat source temperature (K)
T_L is the heat sink temperature (K).

When an external resistance (R_L) is connected across the two cold shoes, a current (I) will flow in this external circuit. The voltage across the load (V_L) is equal to the induced voltage ($V_{I=0}$) less internal voltage losses (i.e., IR voltage drops). Namely,

$$V_L = \alpha_{n,p}(T_H - T_L) - (R_n + R_p)I. \tag{5.25}$$

The n-leg total internal electric resistance is given by

$$R_n = \rho_n l_n / A_n \tag{5.26}$$

where ρ_n is the electrical resistivity of n-type semiconductor material ($m\Omega$-cm)
l_n is the length of n-type semiconductor (cm)
A_n is the cross-sectional area of n-type semiconductor material (cm^2).

Similarly, the p-leg total internal electric resistance is given by

$$R_p = \rho_p l_p / A_p . \tag{5.27}$$

It is assumed that the contact resistance and other external resistances are zero. The total power output ($\dot{W}_L = dW_L/dt$) is then

$$\dot{W}_L = R_L I^2. \tag{5.28}$$

From Kirchoff's law, the current can be found as the total induced voltage divided by the sum of the resistances in the circuit. Namely,

$$I = \frac{\alpha_{n,p}(T_H - T_L)}{(R_L + R_p + R_n)} . \tag{5.29}$$

Substituting eq. 5.29 for I in eq. 5.28, we obtain

$$\dot{W}_L = \left[\frac{R_L}{[R_L + R_p + R_n]^2} \right] (\alpha_{n,p}^2 (T_H - T_L)^2). \tag{5.30}$$

The efficiency of this thermoelectric unit can be defined as the electric power output divided by the thermal power input, or simply

$$\eta_{TE} \equiv \dot{W}_L / \dot{Q}_{in} . \tag{5.31}$$

Now, the thermal power input $\dot{Q}_{in}$, consists of three components: direct heat conduction across the semiconductor elements; plus the thermal energy required to compensate for Peltier cooling at the hot shoe; minus the Joule heating effect (I^2R) which represents the amount of electric energy dissipated as heat within the circuit.

A brief description of the Peltier effect can be given as follows. Consider the simple thermocouple shown in Fig. 5.27, but now assume that both of its junctions are at the same temperature (i.e., assume here that $T_{\text{high}} = T_{\text{low}}$). If an external current is placed through the thermocouple (using taps 1 and 2), the temperatures at both junctions will change. After compensating for the Joule effect, any additional change in the temperature of the junctions is due to the Peltier effect. This effect occurs whether the current flowing through the thermocouple is internally generated or provided by an external source [3, 11].

Assuming that half of the Joule heat appears at each junction, the heat input is

$$\dot{Q}_{\text{in}} = \left[\left(\frac{k_p A_p}{l_p}\right) + \left(\frac{k_n A_n}{l_n}\right)\right]\left[T_H - T_L\right] + \alpha_{n,p} I T_H - \frac{1}{2}\left[\left(\frac{\rho_p l_p}{A_p}\right) + \left(\frac{\rho_n l_n}{A_n}\right)\right] I^2. \tag{5.32}$$

If we define

$$\kappa_p \equiv k_p A_p / l_p$$

$$\kappa_n \equiv k_n A_n / l_n \tag{5.33}$$

and substitute R_n and R_p from eqs. 5.26–27, then, eq. 5.32 becomes

$$\dot{Q}_{\text{in}} = (\kappa_p + \kappa_n)(T_H - T_L) + \alpha_{n,p} I T_H - \frac{1}{2}(R_p + R_n) I^2. \tag{5.34}$$

We also define the ratio of load resistance to internal resistance as

$$m \equiv R_L / (R_p + R_n). \tag{5.35}$$

Equations 5.30, 5.31, 5.32, and 5.34 can be combined to form [1, 3]

$$\eta_{\text{TE}} = \left(\frac{T_H - T_L}{T_H}\right)\left[m \Big/ \left[(m+1) - \left(\frac{T_H - T_L}{2T_H}\right) + \frac{(m+1)^2 (R_p + R_n)(\kappa_p + \kappa_n)}{\alpha_{n,p}^2 T_H}\right]\right]. \tag{5.36}$$

The first term in eq. 5.36 is simply the Carnot efficiency (η_{th}) described by eq. 5.2. The remainder of the expression may be considered as a material efficiency (η_m) that is a function of thermoelectric material properties, geometries, the ratios of resistances and operating temperatures. Thus, for our TE converter

$$\eta_{\text{TE}} = \eta_{\text{th}} \eta_m \tag{5.37}$$

where the Carnot thermal efficiency from eq. 5.2 is

$$\eta_{\text{th}} = (T_H - T_L)/T_H \tag{5.38}$$

and the material efficiency is given as

$$\eta_m = \left[m \Big/ \left[(m+1) - \left(\frac{T_H - T_L}{2T_H}\right) + \frac{(m+1)^2 (R_p + R_n)(\kappa_p + \kappa_n)}{\alpha_{n,p}^2 T_H}\right]\right]. \tag{5.39}$$

To achieve a maximum material efficiency, it is necessary that the product $[(R_p + R_n)(\kappa_p + \kappa_n)]$ have a minimum value. Using previously defined terms, we can rewrite this product as

$$(R_p + R_n)(\kappa_p + \kappa_n) = \left[\left(\frac{\rho_p l_p}{A_p}\right) + \left(\frac{\rho_n l_n}{A_n}\right)\right] \times \left[\left(\frac{k_p A_p}{l_p}\right) + \left(\frac{k_n A_n}{l_n}\right)\right]. \tag{5.40}$$

Having chosen thermoelectric materials for the converter, the thermal conductivity (k) and the electrical resistivity (ρ) are established, and the product $[(R_p + R_n)(\kappa_p + \kappa_n)]$ is minimized with respect to the geometry, that is, minimized for any given area to length (A/l) ratios of both semiconductor material legs. Consequently, the condition for optimization becomes

$$\frac{d\left[(R_p + R_n)(\kappa_p + \kappa_n)\right]^2}{d\left[(l_n/A_n)/(l_p/A_p)\right]} = 0. \tag{5.41}$$

Applying the condition expressed in eq. 5.39 to eq. 5.40, we obtain [3]

$$\left(\frac{(l_n/A_n)}{(l_p/A_p)}\right)_{\text{opt}} = \left(\frac{(k_n \rho_p)}{(k_p \rho_n)}\right)^{1/2}. \tag{5.42}$$

Thus,

$$\left[(R_p + R_n)(\kappa_p + \kappa_n)\right]_{\text{opt}} = \left(\sqrt{\rho_n k_n} + \sqrt{\rho_p k_p}\right)^2. \tag{5.43}$$

If we differentiate eq. 5.39 with respect to m, the optimum resistance ratio (m_{opt}) becomes [1, 3]

$$m_{\text{opt}} = \sqrt{1 + Z\overline{T}} \tag{5.44}$$

Table 5.2 Typical calculated figure-of-merit values.

MATERIAL	TEMPERATURE	SEEBECK COEFFICIENT (α)	FIGURE OF MERIT (Z)
METALS	300 K	5 μV/K	3×10^{-6} K^{-1}
SEMICONDUCTORS	300 K	200 μV/K	2×10^{-3} K^{-1}
INSULATORS	300 K	1 mV/K	5×10^{-17} K^{-1}

where $\bar{T} = (T_H - T_L)/2$ and

$$Z = \frac{\alpha_{n,p}^2}{\left[(\kappa_n + \kappa_p)(R_n + R_p)\right]} . \tag{5.45}$$

Equation 5.45 describes Z, a thermoelectric converter property grouping called the figure of merit. Here Z is defined for the optimum resistance ratio (m_{opt}). Therefore, $(\eta_{\text{TE}})_{\text{opt}} = \eta_{\text{th}}(\eta_m)_{\text{opt}}$ or

$$(\eta_{\text{TE}})_{opt} = \left(\frac{(T_H - T_L)}{T_H}\right) \frac{(m_{\text{opt}} - 1)}{(m_{\text{opt}} + T_L/T_H)} \tag{5.46}$$

where m_{opt} is given by eq. 5.44.

If the n-type and p-type materials have equal thermal conductivities $(k)_{n,p}$ and equal electrical resistivities $(\rho)_{n,p}$, then the figure of merit for our idealized TE converter becomes

$$Z = \alpha^2/k\rho. \tag{5.47}$$

The figure of merit is usually considered the most significant parameter in the selection of materials for

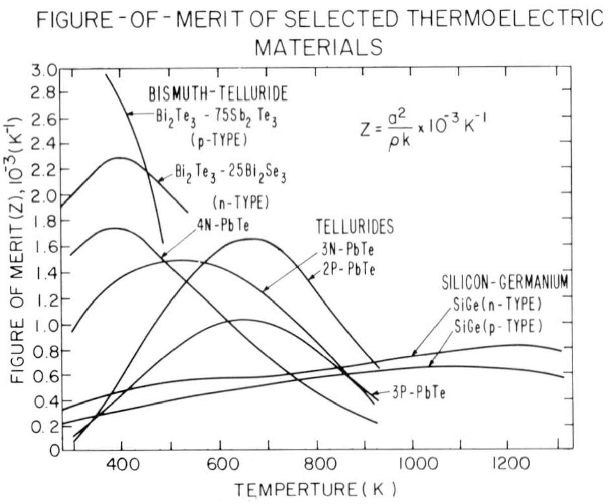

Fig. 5.30 Figure-of-merit of selected thermoelectric materials [12].

TE power generators [1,3]. Table 5.2 and Fig. 5.30 present figure of merit data for selected TE materials [12]. The larger the value of Z, the higher the overall efficiency of the TE converter; in the limit as Z becomes infinite, we have essentially negligible internal resistance and the overall thermoelectric converter efficiency (η_{TE}) approaches the Carnot efficiency (η_{th}).

Space nuclear power systems flown by the United States (see Table 5.3) have used thermoelectric heat-to-electric power conversion systems. The TE materials used included lead telluride (PbTe), silver-antimony-germanium telluride (called TAGS) and silicon germanium (SiGe). The telluride-based TE converters have been employed for many years and their performances have proven exceptional. The tellurides are limited, however, to operation at temperatures below 825 K because of material sublimation. For space nuclear power plants using TE converters and delivering 10 to 100 kilowatts-electric, it appears desirable to operate the system at temperatures of 1275 K or higher [1]. Silicon germanium TE converters can operate in this temperature regime. In addition, SiGe alloys have demonstrated reliable, long-term operation at high temperatures in the space environment. These materials can be metallurgically bonded even at their highest operating temperatures (1300–1400 K) and possess strength characteristics considerably greater than those of all other commonly used TE materials. In general, SiGe alloy thermoelectric converters can be operated at much higher hot shoe temperatures, permitting an elevated cold shoe operation without severely penalizing overall conversion efficiency [1]. As will be discussed in Chapter 6, an elevated cold shoe temperature supports a reduction in the size and mass of the radiator.

The addition of gallium-phosphorus (GaP) to silicon germanium alloys appears to enhance the overall figure of merit of these TE converter materials. However, a considerable effort will be required before these enhanced performance TE materials can be considered for application in operational space nuclear power systems [1].

An examination of the Seebeck coefficients and elec-

Table 5.3 TE materials used in space nuclear systems (1961–1983).

TE SEMICONDUCTOR MATERIAL	SPACE SYSTEM APPLICATION
TELLURIDES	• TRANSIT NAVIGATION SATELLITES (SNAP-3A; -9A)
	• NIMBUS III METEOROLOGICAL SATELLITE (SNAP-19B)
	• PIONEER 10/JUPITER FLYBY (SNAP-19)
	• VIKING MARS LANDER (SNAP-19)
	• APOLLO LUNAR LANDINGS (SNAP-27)
SILICON GERMANIUM	• LINCOLN EXPERIMENTAL SATELLITES (LES) 8/9 (MULTIHUNDRED WATT GENERATOR-MHW)
	• VOYAGER 1, 2 MISSIONS TO JUPITER, SATURN AND BEYOND (MHW)
	• SNAP-10A U-ZrH EXPERIMENTAL NUCLEAR REACTOR (EARTH ORBIT)

trical resistivities for commonly used TE materials reveals that these properties are quite similar for each material. This should not be particularly surprising, however, since all these materials are extrinsic semiconductors that obtain their electrical properties from doping. The differences in the figure of merit values for semiconductor TE materials arise from differences in thermal conductivities. The tellurides and selenides all have very low values of thermal conductivity, typically 0.005 to 0.015 W/K-cm; while the SiGe alloys have values generally in the range of 0.04 to 0.06 W/K-cm [1].

A variety of more exotic or advanced TE materials are also being investigated [1, 2, 12].

Thermionic Conversion

Thermionic energy conversion is another method of directly transforming heat into electricity. The thermionic (TI) converter is a static energy conversion device that essentially "boils electrons" from a hot emitter surface (typically 1800 K) across a small interelectrode gap ($\lesssim$0.5 mm) to a cooler ($\sim$1000 K) collector surface. Figure 5.31 depicts these operating principles and presents projected operating parameters and potential component materials [2, 3, 13, 14]. In effect, the TI converter is a heat engine that uses electrons as its working fluid. The Carnot efficiency (see eq. 5.6) is therefore appropriate as an upper performance limit. In a thermionic converter, the voltage potential between the emitter (cathode) and collector (anode) is used to drive a current through an externally connected electric load.

Several factors inhibit thermionic converters from achieving Carnot efficiency performance. These include: (1) radiation heat transfer between the cathode and the anode; (2) the space charge effect which impedes or prevents the flow of electrons from the emitter to the collector; and (3) thermal energy losses to the surroundings [1, 3].

It is possible to reduce the space charge effect in a thermionic converter by either moving the electrodes closer or by injecting positively charged ions into the interelectrode gap, thereby neutralizing the electron-produced negative space charge. A thermionic converter that has closely spaced electrodes and no vapor in the interelectrode gap is called a vacuum diode. When a vapor, like cesium, is introduced in the gap to reduce the space charge effect, the converter is called a plasma diode [3].

Using an ideal thermionic diode, we can develop a simple analytic model of thermionic converter performance [1, 3, 13, 14]. As depicted in Fig. 5.31, thermal energy is supplied to the emitter, keeping it at a sufficiently high temperature so that electrons can be emitted or boiled-off. These electrons then cross the interelectrode gap and are collected at a cooler electrode, called the collector or anode. Thermal energy is re-

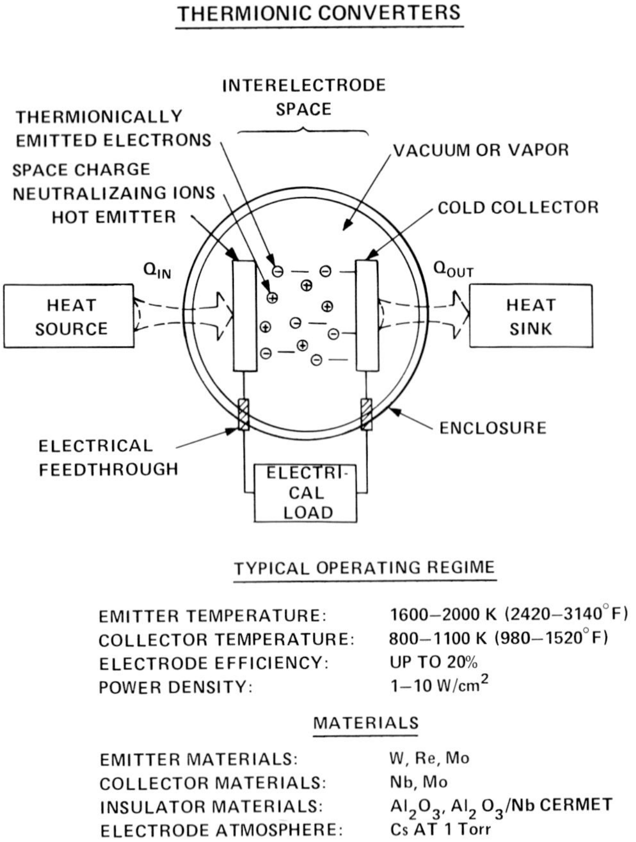

Fig. 5.31 Schematic of thermionic converter [*12*].

moved from the collector, keeping its temperature sufficiently low to prevent electron back emission. The electrons flow from the anode through an external load and return to the cathode.

Motive diagrams and converter current voltage characteristics for an ideal diode are shown in Fig. 5.32. In the motive diagram ϕ_e and ϕ_c represent the emitter and collector work functions, respectively. The work function is the potential barrier that must be overcome by electrons leaving either electrode. The Fermi–Dirac statistical model for electrons can be used to predict the number of electrons that will break through the (surface) potential barrier found in all metals [3]. The output voltage (V) of the thermionic converter is the difference between the emitter and collector Fermi levels.

The Fermi level (E_F) is the electron energy at which the Fermi–Dirac distribution function is 0.5. In solid-state physics, this distribution function represents the probability that an electron will occupy a particular quantum state at thermal equilibrium (see Fig. 5.32). The Fermi level may also be viewed as the maximum electron energy at $T = 0$ K. At temperatures above absolute zero, electrons acquire thermal energy and their distribution in particular energy levels is described by Fermi-Dirac statistics [3].

Assuming the absence of space charge and other such phenomena between the electrodes, the current density (J) from the emitter to the collector is determined by the emitter temperature (T_e) and the potential barrier (ϕ) the electrons must overcome. For an ideal diode, the current density is given by the Richardson–Dushman equation [1, 3, 13]

$$J\ (\text{amps/cm}^2) = AT_e^2 e^{-\phi/(kT_e)} \tag{5.48}$$

where A is a constant (A = 120 amp/cm^2 – K^2)
ϕ is the work function (eV)
k is the Boltzmann constant (8.6168×10^{-5} eV/molecule-K).

There are basically two distinct regions of operation of the TI diode. As long as the sum of the collector work function (ϕ_c) and the converter output voltage (V) is less than the emitter work function (ϕ_e), the barrier to electron flow is $\phi = \phi_e$. The current density (J) is independent of the output voltage and this region of the current-voltage (I-V) curve is called the saturation region. The saturation current density is expressed as

$$J_s = AT_e^2 e^{-\phi_e/(kT_e)} \tag{5.49}$$

where ϕ_e is the emitter work function (eV).

As soon as the output voltage is increased to the point where $\phi_c + V = \phi_e$, the potential barrier becomes $\phi = \phi_c + V$, and the converter current density falls exponentially with further increases in output voltage. The current density in this region of the I-V curve, called the retarding region, is given by

$$J = AT_e^2 e^{-[(\phi_c + V)/(kT_e)]}. \tag{5.50}$$

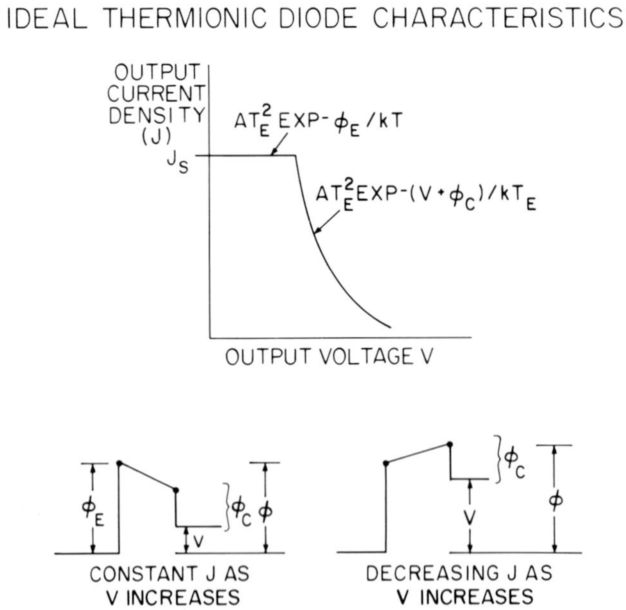

Fig. 5.32 Ideal thermionic diode characteristics [*1*].

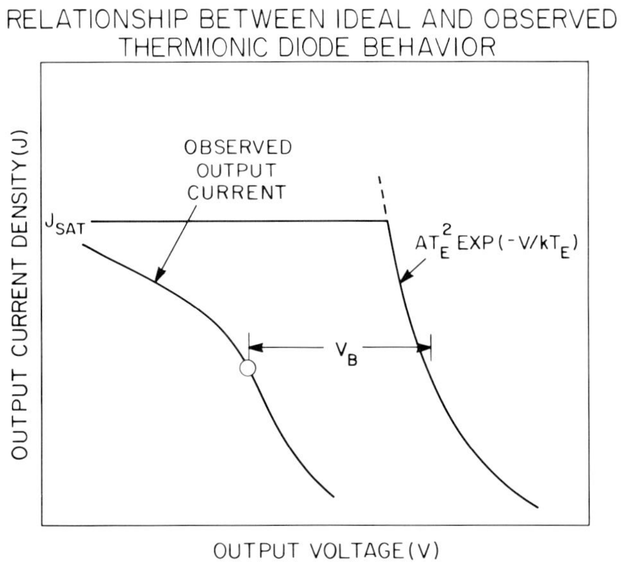

Fig. 5.33 Relationship between ideal and observed thermionic diode behavior [1].

Several factors complicate the operation of this ideal diode converter. First, the work functions of refractory metals are too high to permit their application at reasonable operating temperatures. For example, polycrystalline tungsten has a work function of approximately 4.6 eV. To obtain a current density of 1 amp/cm^2 with this material requires an operating temperature of 2600 K. Even tungsten has a significant evaporation rate at such temperatures. Second, electron space charge effects in a vacuum diode configuration with a design-practical interelectrode gap limit current density to less than one milliampere per centimeter squared. (Interelectrode gaps on the order of 10^{-2} mm are actually needed to achieve useful current densities in a vacuum diode [1, 13].)

Both these engineering difficulties are mitigated by adding a low pressure cesium vapor (typically 13 Pa to 1.33 kPa) in the interelectrode gap. This type of low pressure vapor can be established by including a reservoir of liquid cesium in the TI converter enclosure at a suitable temperature, generally betwen 500 K and 600 K. In the presence of this cesium vapor, a partial monolayer of cesium will form on both the emitter and collector, reducing their work functions to near optimum values. By varying the cesium temperature (and therefore its pressure), a range of work functions can be obtained at given emitter temperatures [1]. In order to maximize output voltage, the collector is operated at a low work function. The electron space charge effect to TI current flow can be neutralized by ionizing the cesium vapor, creating a plasma. Cesium is a particularly good choice because it has the lowest ionization potential of any element. Ions are generated in two ways in the thermionic converter. At higher emitter temperatures, a sufficient number of ions can be created thermally by means of contact ionization with the emitter. This process is called the unignited mode of converter operation. At lower temperatures, these ions are created electrically in a low voltage arc between the electrodes. In this situation, a portion of the electric power generated by the TI converter is dissipated internally to maintain the plasma. The plasma-sustaining voltage drop across the interelectrode gap is called the arc drop (V_d). Including the influence of electron scattering, its effective value is usually about 0.5 volt. This type of TI converter operation is called the ignited mode [13].

Since losses such as arc drop and electron scattering are incurred when cesium is added to the converter, its performance is no longer accurately represented by the ideal diode model. It has been observed, however, that a thermionic converter can be well-characterized by simply adding another voltage loss term in the ideal diode equation. As illustrated in Fig. 5.33, the sum of these losses, including the collector work function, is called the barrier index (V_B). The actual converter performance is then given by [1, 13, 14].

$$J = AT_e^2 e^{-[(V+V_B)/(kT_e)]}. \tag{5.51}$$

Typical converter values for V_B are ~1.9 eV to 1.5 eV [1].

While the actual power output of the converter is the product of current and voltage, only about 90 percent of this power is available at the external load due to voltage drops in the converter's electrical leads. Consequently, the useful output power density of a practical TI converter is given by [1]

$$P_L(\mathrm{W/cm^2}) = 0.9J\left[kT_e \ln(AT_e^2/J) - V_B\right]. \tag{5.52}$$

The collector temperature (T_c) is of crucial importance to the power conversion system. The highest possible collector temperature (and therefore heat rejection temperature) is desired in space power applications, since all waste heat must be radiated away to space and the size of the radiator is inversely proportional to T_c^4. However, if the collector has too high a temperature, electrons will be emitted back into the flow of electrons from the emitter. This collector back emission must be compensated for either by increasing the collector work function (thereby reducing back emission) or by increasing the number of electrons leaving the emitter. As a general consequence, increasing the collector temperature decreases the thermionic converter performance. So a design trade-off between heat rejection optimization and converter performance has to be made. In most TI converters the back emission is limited to 10 percent or less of the emitter current density [1].

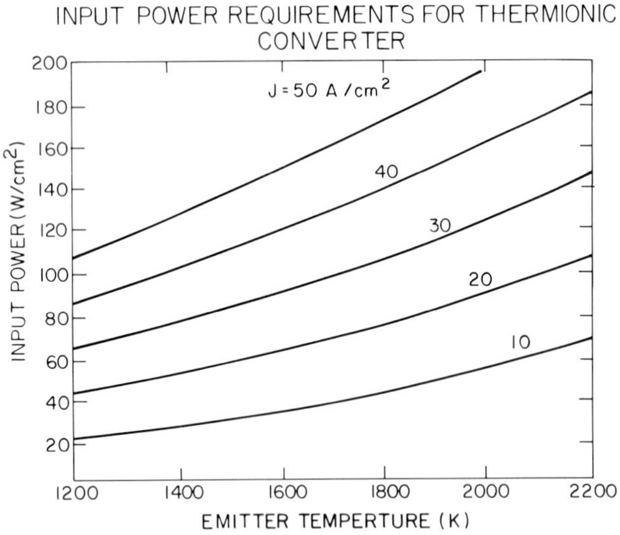

Fig. 5.34 Input power requirements for thermionic converter [1].

The emitter input power ($\dot{Q}_{in}$) is determined by emitter radiative heat transport, cesium gas thermal transport, conductive losses through the electric leads, and electron cooling of the emitter. Of these, electron cooling and radiative heat transfer are the dominant mechanisms. A good approximation to many converters in the region of interest is provided by the expression [1, 14]

$$\dot{Q}_{in}(\mathrm{W/cm^2}) = 1.8 \times 10^{-3} J T_e + 1.2 \times 10^{-12}(T_e^4 - T_c^4). \tag{5.53}$$

Fig. 5.34 shows the input power as a function of both emitter temperature and current density [1]. If we combine eq. 5.52 and eq. 5.53, the efficiency of a thermionic converter (η_{TI}) at its leads is approximated by

$$\eta_{TI} = \frac{0.9J\left[kT_e \ln(AT_e^2/J) - V_B\right]}{1.8 \times 10^{-3} J T_e + 1.2 \times 10^{-12}(T_e^4 - T_c^4)}. \tag{5.54}$$

As illustrated in Fig. 5.35, η_{TI} is relatively insensitive to the current density. Consequently, converter performance remains near its design level under partial load conditions. The output power density, however, is almost directly proportional to current density. Thus, the design current density must be carefully selected to optimally match the heat flux from the nuclear heat source, the electric load requirement, and space system size and mass constraints.

Thermionic converter technology has been under development in the United States since the 1960s at varying levels of support. First generation converters tested under the in-core (i.e., in the reactor core) thermionic program in the early 1970s produced power densities of 6 W/cm^2 at 1850 K [1, 2]. The design effort was directed toward the 120 kW$_e$ power level with probable lifetimes of 10,000 and 20,000 hours because of high temperature-induced materials and nuclear fuel problems. This development effort was cancelled in 1972, when the majority of the U.S. space nuclear power effort was terminated [1]. Continuing thermionic research has maintained this converter performance objective, but at lower emittance temperatures. This permits the converter system to become an out-of-core design—that is, the reactor core and the power conversion system are physically separated, resulting in a much more effective optimization of both the nuclear reactor and the thermionic power conversion system.

General Power Processing Considerations

Power processing with respect to space nuclear power systems has as its major function the maintenance of bulk electrical power within required voltage and (for ac power) frequency limits. This must be accomplished over the required range of payload power demands and be consistent with the range of heat source and power converter degradation. Regulation is usually accomplished with static conversion systems by means of shunt regulators (parasitic loads) across the full power converter, by tapping off a variable fraction of the power converter devices, or by a combination of the two. In dynamic (alternator) systems, voltage and frequency regulation is accomplished through a combination of variable parasitic loads and, where permitted by alternator design, a variation of the alternator's parameters.

Compensation for shorter-term payload power demand changes can usually be accommodated by the spacecraft power processing equipment. Longer-term power balancing may require compensation for radioisotope fuel decay or an adjustment of the reactor

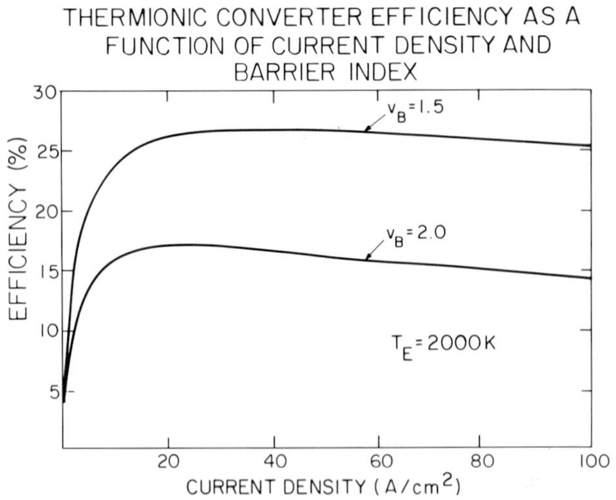

Fig. 5.35 Thermionic converter efficiency as a function of current density and barrier index [1].

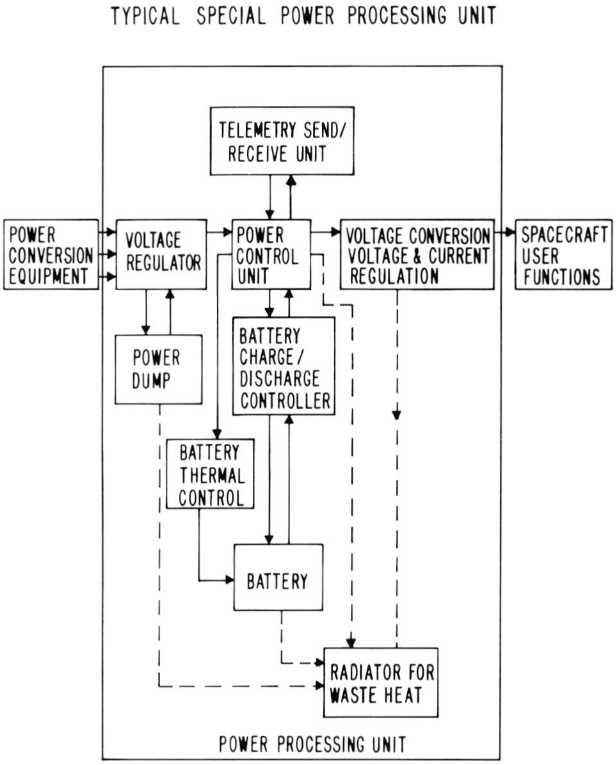

Fig. 5.36 Typical spacecraft power processing unit [2].

thermal power level. This is dependent on the type of power conversion system employed and long-term payload power demands. Power dissipated in parasitic loads must be radiated away to space, further complicating the thermal control design of the spacecraft and adding mass to the overall system.

The power processing subsystem may include batteries, battery charge and discharge controls, switching apparatus, voltage conversion equipment, cabling and auxiliary equipment (such as coolers, radiators, and controllers). These components depend on the spacecraft–nuclear power supply interface. The auxiliary equipment is that needed to keep components of the power processing unit functioning properly. Fig. 5.36 is a block diagram for a typical spacecraft power processing unit [2].

Selection Considerations for Power Conversion Systems

Many radioisotope thermoelectric generator (RTG) powered spacecraft have been used successfully in a variety of exciting U.S. space missions. These have all employed thermoelectric power conversion techniques, typically operated at peak temperatures of 800 to 1275 K and generally attaining lifetimes of 10 to 15 years. While dynamic conversion systems have not yet experienced such space-qualified performances, large amounts of engineering performance data have already been accumulated. Dynamic conversion techniques appear to offer higher conversion efficiencies, leading to power plants that are inherently smaller and lighter. At spacecraft power levels of approximately 1 kilowatt-electric, this difference of overall conversion efficiency is not as significant as with higher power level systems (i.e., > 300 kW_e), where the dynamic conversion techniques appear to offer distinctive advantages over static conversion approaches [1, 2].

There are both advantages and disadvantages with each power conversion approach, whether static or dynamic. The optimum power conversion system (PCS) will be driven by such basic mission requirements as power level, operating lifetime, mass, reliability, cost, and schedule. These parameters must be carefully weighed and assessed on a case by case basis. Table 5.4 lists some key factors in the selection and design of power conversion systems. Of these, efficiency, mass, and reliability are frequently considered to be the most important [2].

The reliability of a power conversion system is a function of such factors as the number of moving parts, the number of welds, cycle temperatures and pressures, tolerances, and wear modes. direct energy conversion devices, because of their lack of moving mechanical parts, have an inherent reliability advantage over dy-

Table 5.4 Key design factors for power conversion systems.

KEY DESIGN FACTORS FOR POWER CONVERSION SYSTEMS

- EFFICIENCY
- MASS
- RELIABILITY
- PEAK OPERATING TEMPERATURES
- HEAT REJECTION TEMPERATURES
- WORKING FLUID PROPERTIES (FOR DYNAMIC CONVERSION TECHNIQUES)
- LIFETIME
- VIBRATION AND TORQUE
- POWER RANGE
- MODULARITY
- STARTUP AND SHUTDOWN
- POWER PROCESSING
- RADIATION HARDENING

namic conversion systems. Some of this reliability advantage disappears, however, when the static devices are operated at very high temperatures or require very close component tolerances. Overall PCS reliability also depends on design trade-offs between redundancy and failure rate. As shown in Fig. 5.37, the use of redundant components and subsystems appears to provide a higher degree of reliability [2]. However, the use of redundancy in a spacecraft power system also introduces several potential drawbacks. These include: (1) higher mass, (2) additional design complexity, (3) higher cost, (4) the possibility of unknown interdependencies among redundant pieces of equipment, (5) common cause failures, and (6) the possibility of synergistic failures, where one component failure enhances another failure.

To obtain a higher thermal efficiency in a PCS that uses a working fluid in a thermodynamic cycle, temperatures and pressures are generally increased to a design point where any further increases would cause a significant decrease in reliability and operating lifetime. However, extracting improved performance through the use of extreme cycle temperatures has historically also been accompanied by many engineering problems. It may, therefore, be more reasonable to initially select design-conservative cycle temperatures for the PCS and then evolve to higher temperatures to support higher performance requirements in later design iterations.

The heat rejection temperature is generally chosen on the basis of power level, system mass, and radiator area. The lower the heat rejection temperature, the higher the thermal efficiency. However, as will be discussed in Chapter 6, the higher the rejection temperatures, the smaller and lighter the radiator. Consequently, some intermediate heat rejection temperature is chosen that reflects an optimum overall power plant design.

Working fluids are an integral part of dynamic power conversion systems. They can be liquid metals, such as mercury, sodium, potassium and a NaK mixture; organic liquids, such as Dowtherm A; or gases, such as helium. The thermophysical properties of some of these potential working fluids were presented in Chapter 4.

An extended PCS lifetime is desirable for nuclear powered spacecraft. For example, RTG systems currently have projected lifetimes of at least 15 years [2]. As the design of advanced space reactors matures in the 1990s, unattended lifetimes in excess of 10 to 15 years may become the engineering norm.

Vibrations in the PCS could interfere with spacecraft payloads, such as antennas, electro-optical sensors, or materials experiments and manufacturing in microgravity. This vibrational problem does not generally exist with static conversion systems but may be a consideration in rotating machinery (e.g., Rankine or Brayton cycle systems). On the other hand, vibrational interference problems could represent a very significant issue with reciprocating devices (e.g., Stirling engines) [2]. Torques can be generated in both rotating equipment and in circulating fluids. One design approach involves the use of multiple power conversion equipment in a symmetrical pattern so that every torque is neutralized by an opposing torque. This solution, however, places design restrictions on the overall PCS.

Power level will influence the voltage levels at which it is desirable to operate spacecraft equipment. For example, in the kilowatt-electric regime, previous spacecraft have generally employed 28 volts (dc). In the multikilowatt regime, the 100–400 volts (dc) range has been considered. The lower voltage leads to higher currents, larger cables, and potential switching problems; the higher voltage could lead to plasma losses and arcing. As spacecraft power demands increase, higher voltages must be considered to reduce mass, volume, cost, and parasitic loss per kilowatt [2].

In a truly modular PCS design two or more conversion units per power plant would be capable of operating independently of each other. One of the main advantages of PCS modularity is that at least partial power would be available, should one conversion system fail. Modular design also provides the engineer greater flexibility in component arrangement and power level selection.

PCS design must include an ability to be started in the space environment, as well as a positive shutdown mode. It is highly conceivable that a typical advanced nuclear system will not only be started and stopped several times during its mission, but that it will also be required to operate at different power levels. Consequently, the PCS must be capable of accommodating the electrical and thermal transients that accompany

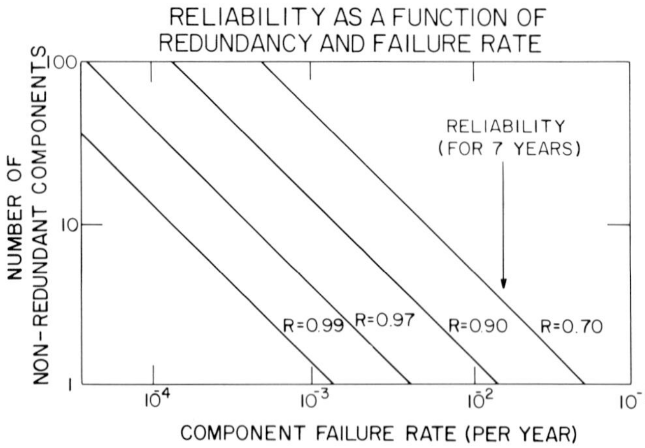

Fig. 5.37 Reliability as a function of redundancy and failure rate [2].

such operating modes. Another potential requirement is that the power plant remain quiescent for many months or even years before being brought up to rated power. Storage in the space environment can cause many working fluids to freeze, thereby creating problems during startup. This is especially significant for intermittent power plants.

There are several major design decisions that must be made regarding the power processing equipment. These include: (1) the generation and distribution of voltage levels; (2) the type of voltage and frequency regulation; and (3) the nature and capacity of storage devices. The merit or value of power processing equipment must be measured in terms of how well it achieves the following: (1) maintaining bulk output voltage and (if ac) frequency within required limits for both transient and long-term power level changes; (2) requiring mininum mass and volume, including auxiliary thermal control equipment; (3) providing voltages optimized for payload needs; (4) minimizing parasitic power losses in regulation and distribution; (5) providing power for on-orbit startup and restart; and (6) meeting established reliability requirements [2].

For static conversion techniques, the minimum PCS output voltage (dc) will be determined by the lowest voltage in standard use at the payload (currently, 28 Vdc nominal) and by the mass and power loss penalties that result from high current. The maximum voltage will be determined by the potential for arcing and plasma losses [2].

For dynamic conversion systems, similar considerations apply with standard minimum voltages being above 120 V (ac). The choice of frequency is based upon a trade-off among the competing goals of minimizing mass, volume, and parasitic power loss.

Finally, electronic components in the PCS, especially those found in the processing unit, must be designed to operate in a radiation environment for their entire mission lifetime. With the extensive use of semiconductor devices and integrated circuits, this represents a challenging design problem for nuclear power plants with multiple-year missions.

References

1. D. Buden et al., "Selection of Power Plant Elements for Future Reactor Space Electric Power Systems," LA-7858, Los Alamos National Laboratory, September 1979.
2. R. V. Anderson et al., "Space Reactor Electric Systems," ESG-DOE-13398, Rockwell International, 29 March 1983.
3. J. P. Holman, *Thermodynamics* (2nd Edition), McGraw-Hill Book Company, New York, 1974.
4. W. Reynolds and H. Perkins, *Engineering Thermodynamics*, McGraw-Hill Book Company, New York, 1974.
5. K. Wark, *Thermodynamics* (3rd Edition), McGraw-Hill Book Company, New York, 1977.
6. I. Asimov, *Biographical Encyclopedia of Science and Technology* (Revised Edition), Doubleday and Company, New York, 1972.
7. J. L. Klann and W. T. Wintucky, "Status of the 2 to 15 kWe Brayton Power System and Potential Gains from Component Improvements," Intersociety Energy Conversion Engineering Conference, Boston, 3–5 August 1971, pp. 195–200.
8. B. Goldwater, "Study of a Free Piston Stirling Engine-Linear Power Conversion System for a 10.5 kWe and 51.0 kWe Output Power," Mechanical Technology Inc., November 1975.
9. G. Walker, *Stirling Cycle Machines*, Oxford University Press, Ely House, 1973.
10. W. Beale et al., "Free Piston Stirling Engines—A Progress Report," SAE Paper 730647, June 1973.
11. M. Zemansky and R. Dittman, *Heat and Thermodynamics* (6th Edition) McGraw-Hill Book Company, New York, 1981.
12. C. Wood, "Optimization of Thermoelectric Materials," *Proceedings of 1982 Working Group on Thermoelectrics*, 7–9 Dec 1982, Jet Propulsion Laboratory, JPL-D-497, January 1983.
13. Rasor Associates Inc., "Direct Conversion Nuclear Reactor Space Power Systems, Volume I," NSR-17-1, Sunnyvale, CA, December 1981.

Chapter 6

Heat Rejection Techniques

Introduction

Space nuclear power plants must reject their waste heat by thermal radiation. Radiator design depends on both operating temperature and power level. The amount of waste heat that can be radiated to space by a given surface area is determined by the Stefan–Boltzmann law and is proportional to the fourth power of the radiating surface temperature. The surface area and mass of the thermal radiator are very sensitive to the heat rejection temperature. Higher heat rejection temperatures correspond to smaller radiator areas and, thus, lower radiator masses. Typical nuclear reactor power plants with thermoelectric converters have minimum weight when the rejection temperature ranges from 775 to 875 K. On the other hand, space nuclear power plants using the Brayton cycle for energy conversion generally have minimum weight at a radiator temperature of approximately 475 K [1]. For radioisotope thermoelectric generator power systems, radiator temperatures are usually about 575 K.

However, the efficiency of a thermodynamic cycle is also sensitive to the heat rejection temperature, demanding the lowest possible rejection temperature to yield the highest thermal (Carnot) efficiency. Frequently, a space power plant will be operated at much lower thermodynamic efficiencies than experienced in terrestrial plants operating on similar cycles. This occurs because a high premium is often placed in space power system design on minimizing radiator area and mass, resulting in the selection of higher heat rejection temperatures. Launch vehicle compatibility, such as the ability to fit inside the Space Shuttle's 18.3 m (60 ft) long and 4.6 m (15 ft) wide payload bay is another design constraint placed on the radiator system. Fig. 6.1 portrays a functional relationship between the radiator area, mass, and temperature; while Fig. 6.2 describes the temperature range of potentially useful radiator materials [1].

Radiator Configurations

A wide variety of radiator configurations have been considered for use with space nuclear power systems as shown in Fig. 6.3. These designs can generally be categorized as either fixed or deployable [2]. The cylindrical/conical configuration (either a complete conic or as a combination of cone and cylinder) supports simplified integration with the reactor, shield and spacecraft payload; provides for structural integrity; and promotes launch vehicle compatibility through design compactness. The flat plate and cruciform radiator designs have also been considered. Of course, compatibility with the available payload volume for a particular launch vehicle is a most important design constraint. When the full radiator size and configuration exceed the payload envelope, then deployable radiator concepts must be considered.

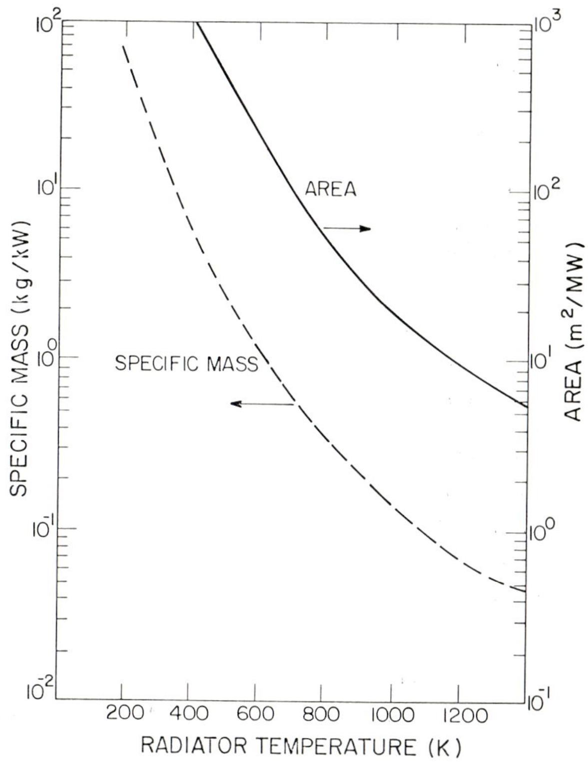

Fig. 6.1 Radiator area and mass as a function of temperature.

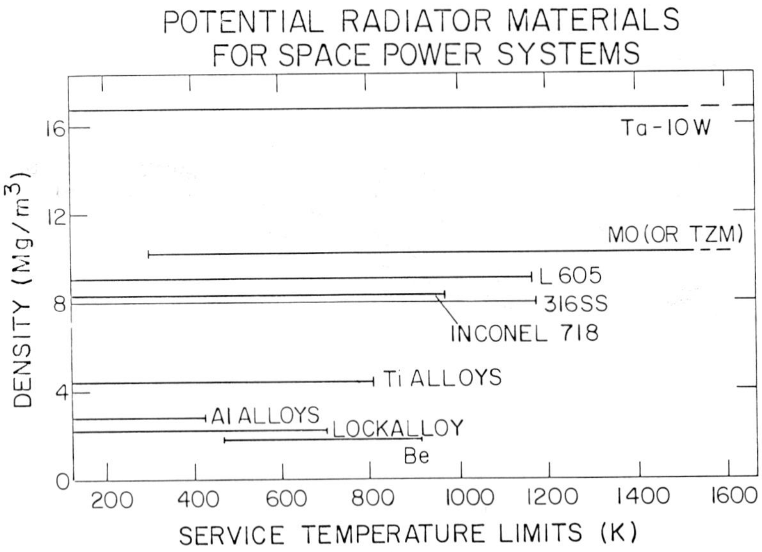

Fig. 6.2 Potential radiator materials for space power systems.

The heat pipe can also be used in a variety of configurations in a thermal radiator design. Fig. 6.4 shows a heat pipe–fin scheme in which thermal energy to be rejected is received at the evaporator region of the pipe, transported isothermally to the condenser region, and then radiated away to space after conduction through the wall to the flat fin [1, 2]. The heat pipe can either have a high emissivity coating or else be attached to a high-emissivity fin that serves as the radiator surface. Another heat pipe radiator concept is shown in Fig. 6.5. Here the radiator is configured as a frustum of a right circular cone, using two sets of heat pipes in series [1, 2]. The axial stringer heat pipes (arranged on the cone slant surface), uniformly transfer thermal energy to a radiating surface that is made up of many heat pipes placed side by side at right angles to the stringers around the radiator circumference. In operation, thermal energy is conducted from the stringer heat pipes to the cross heat pipes and is then rejected to space by radiation.

A generic liquid metal pumped-loop heat rejection arrangement is shown in Fig. 6.6. This configuration includes the following components: (1) a waste heat exchanger, (2) a radiating surface, (3) a volume accumulator, and (4) an electromagnetic (EM) pump [2]. If an organic liquid is chosen as the loop working fluid, a rotary pump would be used in place of the liquid metal EM pump. The volume accumulator serves two functions in such a closed loop. First, it compensates for thermal expansion of both the fluid and the container from startup to operating temperature. Second, it keeps the fluid at a preselected pressure. The volume accumulator is not employed when the working fluid is a gas. Unlike heat pipe radiators, the pumped-loop heat rejection systems suffer temperature losses around the loop. This non-isothermal condition forces the radiator to operate below the waste heat rejection temperature of the power conversion system with a subsequent penalty of larger radiator area [1, 2].

Radiator Thermal Design

The thermal environment for a spacecraft operating in Earth orbit includes solar radiation (1371 ± 5 W/m^2); Earth radiation ($\sim$240 W/m^2); and the sink temperature for outer space ($\sim$0 K) [1, 3, 4]. The thermal energy balance for a radiator must consider the follow-

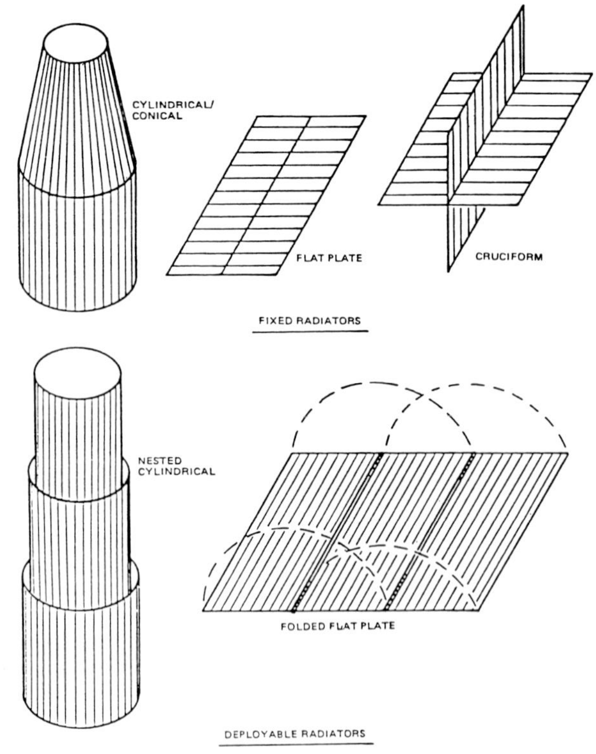

Fig. 6.3 Typical radiator configurations [2].

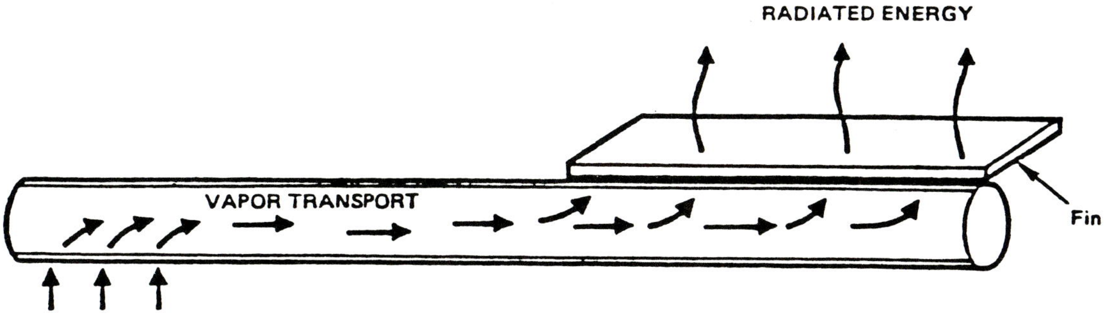

Fig. 6.4 Heat pipe–fin radiator configuration [2].

ing input terms: (1) direct solar radiation; (2) Earth-emitted radiation; (3) Earth-reflected solar radiation (2 and 3 are frequently called "earthshine"); and (4) any internally generated heat load from the spacecraft and its nuclear power plant. Earthshine consists of reflected sunlight and thermal radiation from the Earth and its atmosphere. All of these factors represent thermal energy that must ultimately be rejected to space by the radiator. From the first law of thermodynamics, we can construct an overall thermal energy balance equation for the radiator [1, 5].

$$\text{Energy}_{\text{in}} + \left[\begin{matrix}\text{Change in Energy}\\ \text{Stored}\end{matrix}\right] = \text{Energy}_{\text{out}}. \quad (6.1)$$

Assuming steady-state conditions, the change in energy stored is equal to zero, therefore

$$\text{Energy}_{\text{in}} = \text{Energy}_{\text{out}}. \quad (6.2)$$

$$\begin{aligned}&[\text{Direct Solar Radiation}]\\ &\quad + [\text{Earth-Reflected Solar Radiation}]\\ &\quad + [\text{Earth Direct Radiation}] \qquad (6.3)\\ &\quad + [\text{Internally Generated Heat Load}]\\ &= [\text{Energy Rejected by Radiator}].\end{aligned}$$

In symbols, eq. 6.3 becomes

$$\alpha_s F_s G_s + \alpha_r F_r A_p G_s + \alpha_e F_e E_e + P_i/A = \epsilon\sigma T^4 \quad (6.4)$$

where α_s is the solar absorptivity (0 to 1.0)
F_s is the cosine of the angle between the unit surface normal vector and the direction to the Sun (0 to 1.0)
G_s is the solar radiation incident on a plane normal to the Sun [At one astronomical unit (AU) the solar constant = 1371 ± 5 W/m^2] (4)
α_r is the absorptivity to solar radiation reflected from the Earth (0 to 1.0)
F_r is the view factor for solar radiation reflected by Earth (Typically, $F_r = 0.1$ for low-Earth-orbit and 0.02 for geosynchronous orbit)
A_p is the Earth's albedo; the fraction of incident solar radiation that is reflected by the Earth and its atmosphere to space (Typical value ≈ 0.3)
α_e is the absorptivity to radiation emitted by the Earth (0 to 1.0)
F_e is the view factor for radiation emitted by the Earth to the radiator surface
E_e is the earthshine radiation (typically ~240 W/m^2)
P_i is the internal waste heat load (W)
ϵ is the emissivity of the radiator surface
σ is the Stefan–Boltzmann constant (5.67×10^{-8} W/m^2 – K^4)
A is the area of the radiator (m^2)
T is the radiator operating temperature (K)

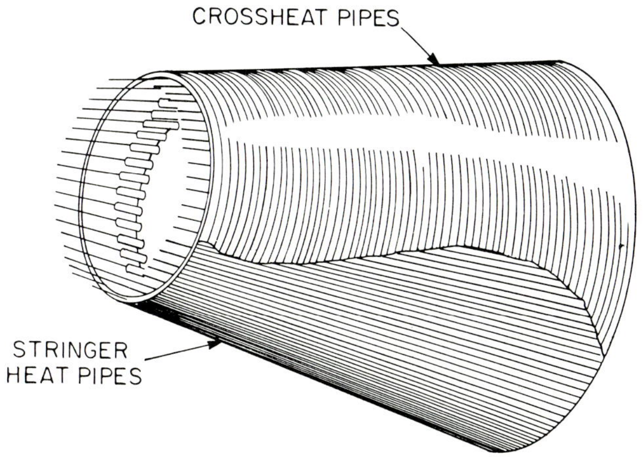

Fig. 6.5 Stringer and cross heat pipe radiator [1].

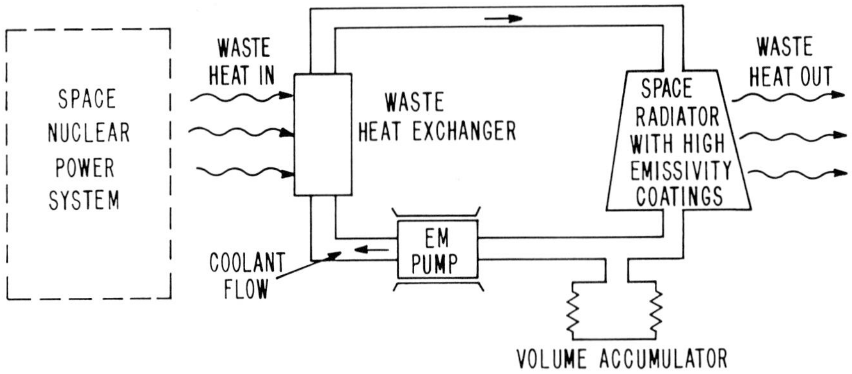

Fig. 6.6 Generic liquid-metal pumped-loop radiator configuration.

Reference High Temperature Radiator Design

A conceptual design of a heat pipe radiator has been developed for a thermoelectric or other high temperature reject heat conversion system for use with nuclear power plants [6]. The basic shape of this high temperature radiator system is a frustum of a right circular cone followed by a cylindrical section. The maximum diameter of this design is compatible with the cargo size limits of the Space Shuttle Orbiter's payload bay. The design includes heat pipes to carry thermal energy from the thermoelectric modules and to provide a radiation surface to space (see Fig. 6.7). The heat pipes are armored to resist small meteoroid punctures and have sufficient redundancy to compensate for losses from large meteoroids. The reference design involves the rejection of 1010 kilowatts-thermal at a radiator operating temperature of 775 K for a period of seven years with a 0.99 probability of survival for the radiator. Titanium was chosen as the heat pipe container material and potassium as the heat pipe working fluid. Table 6.1 presents typical radiator system data for this reference design. Maximum values of view factors (i.e., F_s, F_e, and $F_r = 1$) can always be assumed to create a worst-case solar and earthshine heat input.

The radiator itself is assumed to be a frustum of a right circular cone plus a cylindrical extension, (see Fig. 6.8) whose overall size is compatible with the Shuttle Orbiter's payload bay width dimension limits of approximately 4.57 m (diameter). For the frustum, the effective radiator area is taken as the outer slant surface of the cone plus in some cases the fraction of area visible through the cone base. The slant surface area is given by [7]

$$A = \pi(R_1 + R_2)S \tag{6.5}$$

where A is the area of the slant surface (m^2)
S is the slant height of the cone (m)
R_1 is the small radius (m)
R_2 is the large radius (m)

The cone slant height (S) can be calculated from

$$S = \frac{-2\pi R_1 + \left[(2\pi R_1)^2 - 4\pi A \sin\phi\right]^{1/2}}{2\pi \sin\phi} \tag{6.6}$$

where ϕ is the half angle cone.

For configurations that permit radiation to shine through the base of the cone

$$A_{\text{eff}} = A + F_{1-2}A = A\left[1 + F_{1-2}\right] \tag{6.7}$$

where F_{1-2} is the radiation view factor.

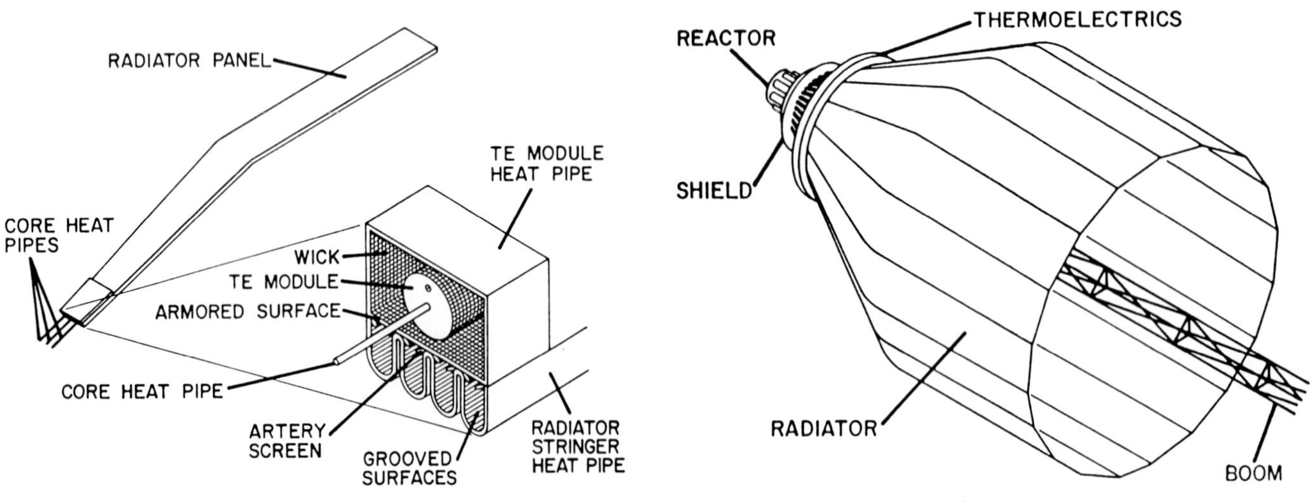

Fig. 6.7 Heat pipe panel radiator for thermoelectric space power plant [6].

Table 6.1 Typical thermal radiator design data.

TYPICAL THERMAL RADIATOR DESIGN DATA
(FOR REFERENCE HEAT PIPE RADIATOR SYSTEM)

α_s = 0.21		α_e = 0.30	
G_s = 1371 W/m^2		F_e = 0.30	
α_r = 0.21		E_e = 243 W/m^2	
F_r = 0.015 FOR GEOSYNCHRONOUS EARTH ORBIT; 0.10 FOR LOW EARTH ORBIT		ϵ = 0.9	
A_p = 0.30		T = 775 K	

To calculate S for cases with radiation shine through the base of the cone, use A_{eff} (eq. 6.7) for A in eq. 6.6.

Meteoroid Protection

Spacecraft will be subjected to meteoroid impacts during their time in orbit. Whenever a meteoroid-induced puncture would lessen or destroy the integrity of a space system component, such as a radiator, suitable precautions and protection from impact damage must be undertaken.

The solid objects encompassed by the term "meteoroid" range from micrometers to kilometers in diameter and from less than 10^{-12} g to more than 10^{+16} g in mass. By convention, objects less than one gram in mass are often called "micrometeoroids" [4].

The contemporary NASA near-Earth to lunar surface meteoroid environment model is given in Table 6.2 [3, 8]. This meteoroid model encompasses particles of cometary origin in the mass range between 10^{-12} g and one gram for sporadic meteoroids and 10^{-6} g and one gram for stream meteoroids [3]. Meteoroids are thought to originate primarily from comets and asteroids with perihelia near or inside the Earth's orbit around the Sun. Collisions have most probably broken down the original objects into a distribution of smaller bodies. Recently formed meteoroids tend to be concentrated near the orbital path of their parent body. These "stream meteoroids" produce the well-known terrestrial meteor showers that occur at certain dates. Meteoroids may be classified as follows by composition: stony, iron, and possibly icy [4]. They are attracted by the Earth's gravity field so that the flux from allowed directions in near-Earth orbit is increased by approximately 1.7 over the interplanetary values. The Earth also shields certain meteoroid arrival directions. To correct for the Earth's gravitational effect at a given distance above the Earth, the meteoroid mass flux (N_t) must be multiplied by the defocusing factor (G_e) that can be obtained from Fig. 6.9 [9]. To correct for planetary body shielding, the unshielded defocused mass flux is multiplied by the body shielding factor, e, which is determined as described in Fig. 6.10.

Table 6.2 NASA meteoroid environment model.

- AVERAGE TOTAL METEOROID ENVIRONMENT
  - PARTICLE DENSITY: 0.5 g/cm^3
  - PARTICLE VELOCITY: 20 km/s
- FLUX MASS MODELS*
  - FOR $10^{-6} \leqslant m \leqslant 1$

    $$\mathrm{LOG}_{10}\, N_t = -14.37 - 1.213\, \mathrm{LOG}_{10}\, m$$
  - FOR $10^{-12} \leqslant m \leqslant 10^{-6}$

    $$\mathrm{LOG}_{10}\, N_t = -14.339 - 1.584\, \mathrm{LOG}_{10}\, m - 0.063\, [\mathrm{LOG}_{10}\, m]^2$$
  - WHERE m IS THE METEOROID MASS (g)

    N_t IS THE NUMBER OF PARTICLES OF MASS (m) OR GREATER PER m^2–s

*DEFOCUSING FACTOR FOR PLANET EARTH (OR OTHER CELESTIAL BODY) AND SHIELDING FACTOR (IF APPLICABLE) MUST ALSO BE APPLIED

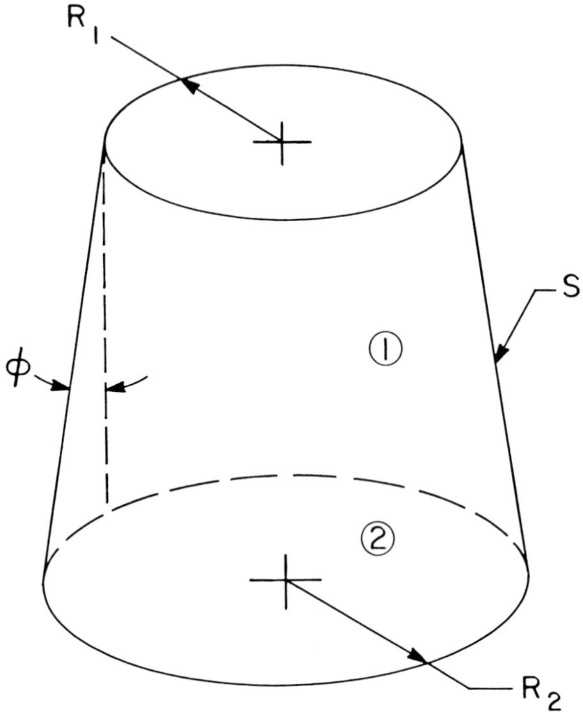

Fig. 6.8 Cone frustum [1].

There are three major mechanisms by which meteoroids can damage a space radiator. These are: (1) puncturing a fluid-carrying passage with the resultant loss of working fluid or coolant; (2) spalling, in which sizeable pieces of metal are expelled from the inside of the fluid-carrying passage and migrate to other components with possible catastrophic results; and (3) erosion of surface coatings, resulting in a change (loss) in emissivity value [2].

Two general approaches taken in meteoroid protection are armor and bumpers. The armor technique uses a solid material to stop an impacting meteoroid before it can inflict serious damage. In the bumper technique, a relatively thin shield of material, physically offset from the surface to be protected, is used to fragment an impacting meteoroid, spreading its energy over a greater, sacrificial area. A variety of finned-tube radiator configurations are possible (see Fig. 6.11). Some employ the armor technique against the meteoroid hazard, while other arrangements use the bumper concept in conjunction with armor. However, because a radiator heat pipe should be directly exposed to space, it is frequently not practical to protect it from meteoroid penetration with a bumper. A meteoroid puncture in an operating heat pipe container will result in a rapid loss of the working fluid to space. Meteoroid impacts that cause the heat pipe container wall to either dimple or spall without puncturing are not expected to significantly affect its operation. Therefore, when armoring the surfaces of a radiator heat pipe to protect it from meteoroids, emphasis should be given to determining some threshold penetration thickness of the container wall materials that is required to prevent penetration from a meteoroid of specified mass and velocity [9]. Several equations exist for predicting the infinite penetration thickness (p_∞) to defeat a specified mass particle traveling at hypervelocities. The armor equation given below was developed by Charters and Summers [10]. It can be used in radiator design.

$$p_\infty = \left(\frac{81}{8\pi} \frac{\rho_p}{\rho_t} m_p u_p^2 \right)^{1/3} \frac{1}{S_t^{1/3}} \tag{6.8}$$

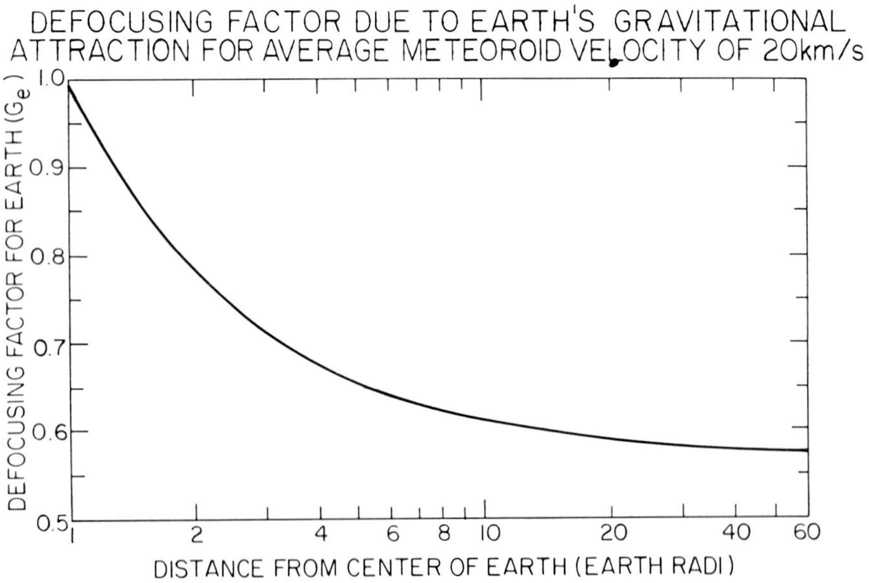

Fig. 6.9 Defocusing factor due to Earth's gravitational attraction for average meteoroid velocity of 20 km/s [1].

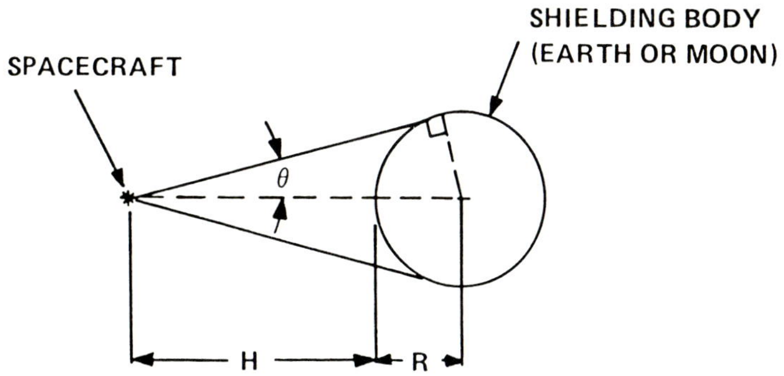

BODY SHIELDING FACTOR, e (DEFINED AS THE RATIO OF THE SHIELDED AND UNSHIELDED FLUX):

$$e = \frac{1 + \cos\theta}{2}$$

WHERE

$$\sin\theta = \frac{R}{R + H}$$

R = RADIUS OF SHIELDING BODY, AND
H = ALTITUDE ABOVE SURFACE.

Fig. 6.10 Method for determining planetary shielding factor for randomly oriented spacecraft [1].

where m_p is the meteoroid mass (kg)
- u_p is the meteoroid velocity (typically, u_p = 20,000 m/s)
- ρ_p is the meteoroid density (typically, $\rho_p = 0.5$ g/cm^3)
- ρ_t is the density of the target material (g/cm^3)
- $S_t^{1/3}$ is an experimentally determined target material property factor (J/mm^3)$^{1/3}$

The hypervelocity impact behavior of commercial pure, Grade 2 titanium and pure beryllium was studied at the University of Dayton Research Institute [9] by impacting targets made of these materials with millimeter-sized spheres of glass, copper, aluminum, and cadmium, propelled from a light-gas gun at velocities ranging from 4.5 to 7.6 km/s. Target temperatures ranged from 293 K to 775 K when impacted. Titanium was found to behave in a ductile manner when impacted, but beryllium was found to be brittle even at 775 K. The target material property factor ($S_t^{1/3}$) in eq. 6.8 was experimentally found to be 3.82 and 3.20 (J/mm^3)$^{1/3}$ for titanium at 293 K and 775 K, respectively. Since we are interested in the material thickness capable of preventing penetration by a specified mass particle, we can

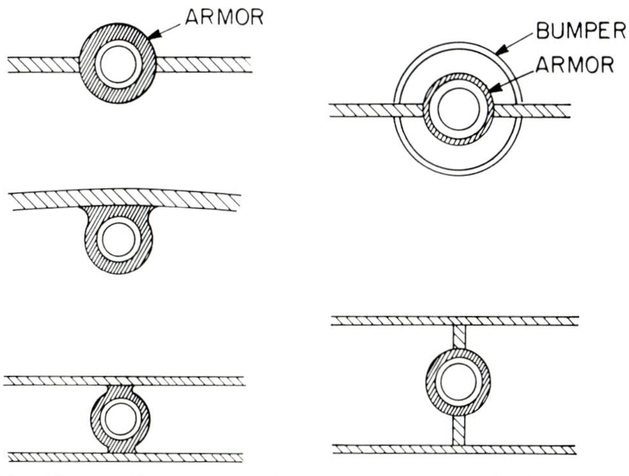

Fig. 6.11 Meteoroid armor configurations for finned-tube radiators [2].

Table 6.3 Titanium radiator parameters for 1010 kilowatt-thermal system [6]

TITANIUM RADIATOR SIZE AND MASS DATA FOR 100-kW$_e$ POWER LEVEL

Parameter		
Radiator power (kW)		1010
Total number of heat pipes		360
Number of heat pipes per TE module		4
Number of heat pipes surviving mission		321
Q_{max}/heat pipe (kW)		2.94
Critical meteoroid mass (g)		1.5×10^{-5}
Heat-pipe armor thickness (mm)		0.60
Cone half angle (deg)		30
Small radiator radius (m)		1.0
Large radiator radius (m)		2.25
Heat-pipe length (m)		5.15
Overall radiator length (m)		4.85
Armor (flat section) mass (kg)	170	
Container (curved section) mass (kg)	140	
Artery screen mass (kg)	9	
Distribution wick mass (kg)	35	
Fluid mass (kg)	30	
Total radiator mass (kg)		384

define a threshold penetration thickness (TPT) as that thickness of a plate that just provides protection against a particle impact. The plate may spall, but it must not be perforated. Empirical experience in ballistics and hypervelocity impact supports the approximation [6, 9]

$$\text{TPT} = 1.5p_{\infty} . \tag{6.9}$$

Equations 6.8 and 6.9 can be used along with empirically derived impact data to calculate the protective armor thickness for a radiator. Table 6.3 shows typical values for a reference 1010 kilowatt-thermal heat pipe radiator design [6].

Radiator Materials

The materials considered for the radiator depend upon the power conversion system selected, because it establishes the overall heat rejection temperature at which the radiator will function in space. As previously mentioned, radiators can be low reject temperature devices (typically 400 to 600 K for a Brayton cycle system) or high reject temperature devices (typically 750 to 950 K for a TE conversion system).

Radiator materials should exhibit the following general characteristics: (1) high strength-to-mass ratios; (2) high thermal conductivity; (3) high resistance to meteoroid damage; (4) compatibility with the working fluid or coolant; (5) high emissivity; and (6) good structural integrity.

A high strength-to-mass ratio enables the radiator structure to survive the launch environment while minimizing its mass. In configuring against the meteoroid hazard, beryllium and aluminum are frequently chosen as armor materials. There must be long-term compatibility between the heat transfer fluid and its containment material—whether the radiator uses a pumped-loop or heat pipe design. Finally, the material chosen should support reasonable fabrication operations.

Efficient radiators should have a high emissivity and low solar absorptivity to minimize the area and mass needed to accomplish the heat rejection job. High emissivity values can be achieved by applying a coating to the radiating surface or by choosing an appropriate base material that has an inherently high emissivity. Emissivity coatings, however, are subject to meteoroid erosion and to degradation with elapsed mission time. A carbon/carbon composite material with a high emissivity (~0.85) at 800 K appears to be a promising candidate for high rejection temperature heat pipe radiator systems [1, 2].

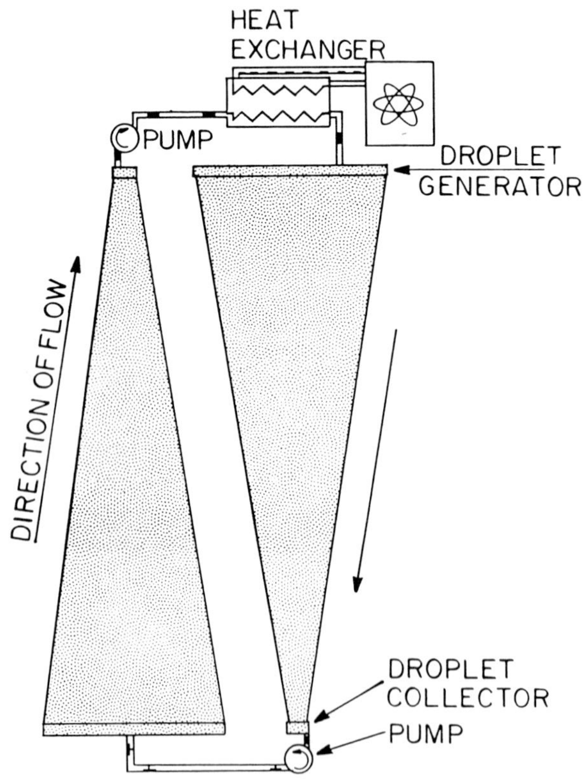

Fig. 6.12 Liquid droplet radiator (two-pass) concept [13].

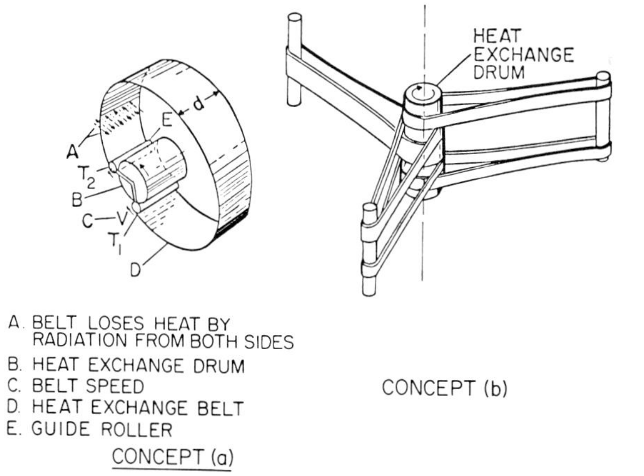

Fig. 6.13 Moving belt (direct-contact) radiator concept [*13*].

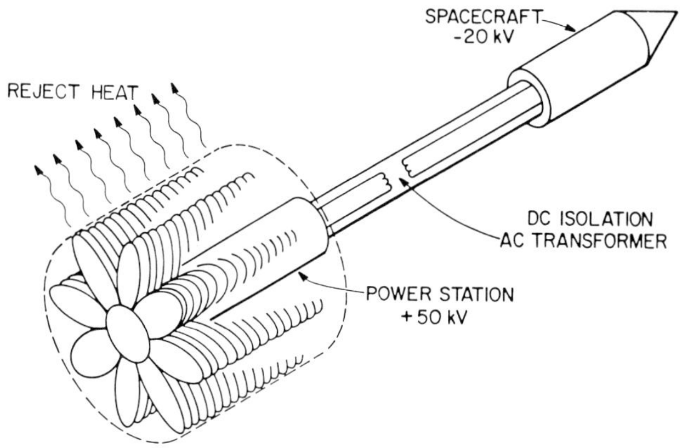

Fig. 6.14 Electrostatic thermal radiator concept [*13*].

Advanced Radiator Concepts

Up until now, we have focused our discussion on the more traditional engineering approaches to closed system heat rejection, namely, pumped-loop and heat pipe devices. As the power levels of tomorrow's space nuclear plants eventually reach the multimegawatt class, innovative radiator designs or open cycle heat rejection techniques will be required to keep the system within acceptable size and mass limits. Three such proposed advanced radiator concepts are the liquid droplet radiator, the moving belt radiator, and the electrostatic thermal radiator. While these concepts are all just in the preliminary idea stage, they are briefly discussed here as a technical stimulus, since advanced heat rejection techniques will be a major engineering design challenge for megawatt-class space nuclear power systems.

In the liquid droplet radiator concept (see Fig. 6.12), a liquid working fluid absorbs the thermal energy to be rejected and is itself projected through space as a thin converging sheet of droplets [2, 11]. During their passage through space, these droplets thermally radiate and are then gathered at the droplet collector for reuse. To avoid losing the working fluid inventory, the droplets must be a low vapor pressure liquid such as silicone oils (250 to 350 K), liquid metal eutectics (370 to 650 K), and liquid tin (550 to 1000 K) [2].

One version of the moving belt radiator concept has a section of the moving heat transfer surface in continuous contact with the waste heat exchanger (see Fig. 6.13) [2, 12]. Since the belt does not contain a working fluid, its penetration by meteoroid impact would not adversely affect its heat rejection operation.

Finally, the electrostatic thermal radiator concept (see Fig. 6.14) uses a liquid metal working fluid, such as tin, formed into tiny, electrically charged droplets. Typically, these droplets (some 50 μm in diameter and carrying a charge of 10^{-15} to 10^{-13} coulombs) are configured into a closely spaced train and ejected into space. An applied electric charge on the spacecraft (opposite to the charge on the droplets) causes the ejected droplets to circumscribe an elliptical orbit, while they reject heat to space. A collector on the spacecraft then gathers the droplets for reuse [2, 13].

References

1. D. Buden et al., "Selection of Power Plant Elements for Future Reactor Space Electric Power Systems," LA-7858, Los Alamos National Laboratory, New Mexico, September 1979.
2. R. V. Anderson et al., "Space Reactor Electric Systems," ESG-DOE-13398, Rockwell International, 29 March 1983.
3. NASA, "Space Shuttle Program–Space Shuttle System Payload Accommodations, Level II Program Definition and Requirements, Vol. XIV," JSC-07700, Rev H, 16 May 1983.
4. NASA, "Space and Planetary Environment Criteria Guidelines for Use in Space Vehicle Development," 1982 Revision (Volume 1), NASA TM-82478, January 1983.
5. F. Kreith, *Radiation Heat Transfer for Spacecraft and Solar Power Plant Design*, International Textbook Co., Scranton, 1962.
6. S. P. Girrens, "Design and Development of a Titanium Heat Pipe Space Radiator," Los Alamos National Laboratory Report LA-9251-MS, March 1982.
7. *CRC Standard Mathematical Tables*, CRC Press Inc., Cleveland, 1974.
8. B. G. Cour-Palais, "Meteoroid Environment Model-1969 (Near Earth Lunar Surface)," NASA-SP-8013, March 1969.
9. L. B. Lundberg, et al., "Hypervelocity Impact Studies on Titanium, Titanium Alloys, and Beryllium," Los Alamos National Laboratory, LA-9417-MS, August 1982.
10. E. Schneider, "Velocity Dependence of Some Impact Phenomena," presented at ESA Halley Comet Micrometeoroid Hazard Workshop, Nordwijk, April 1979.
11. A. T. Mattick and A. Hertzberg, "Liquid Droplet Radiators for Heat Rejection In Space," J. Energy, AIAA 80-9477R, Nov–Dec 1981, pp. 387–93.
12. R. C. Weatherston and W. E. Smith, "A Method for Heat Rejection from Space Power Plants," ARS Journal, March 1960, pp. 268–69.
13. F. C. Prenger and J. A. Sullivan, "Conceptual Designs for 150-MW Space Radiators," National Academy of Sciences Symposium on Advanced Concept Reactors, Washington, DC, 15–17 November 1982.

Chapter 7

Fundamentals of Radiation Protection and Shielding

Introduction

The use of nuclear power in space, reactor or radioisotope, will be accompanied by the presence of nuclear radiations. Since these radiations—alpha particles, beta particles, gamma rays, neutrons, etc.— are potentially harmful to the astronaut crew, delicate electronic components on the spacecraft, and scientific and applications payloads, space system planners must include radiation protection as a fundamental design factor.

Radiation protection may be defined as the protection of individuals, their offspring (progeny), and the global population and biosphere from potentially harmful effects of a radiation exposure, while still permitting the beneficial application of the nuclear energy source [1]. The harmful effects of radiation to human beings can be divided into two general classes: somatic and genetic (or hereditary). A somatic radiation exposure effect involves the individual, himself or herself, while a genetic effect involves descendants. Radiation exposure health effects are frequently divided into two other categories: stochastic and nonstochastic. Stochastic radiation exposure effects may be considered as those for which the probability of an exposure effect occurring (rather than the severity of the effect) is taken as a function of radiation exposure dose, without threshold. Nonstochastic radiation exposure effects, on the other hand, may be regarded as those effects for which the magnitude or the severity of the exposure effect varies with the dose, and for which, consequently, a threshold may exist [1]. Somatic exposure effects may be either stochastic, e.g., carcinogenesis, or nonstochastic, e.g., damage to the gonads that impairs fertility. Genetic effects are currently regarded as stochastic in nature for the dose range of interest in radiation protection [1].

Historically, radiation exposure limits have been recommended by the International Commission on Radiological Protection (ICRP) and the U. S. National Council on Radiation Protection and Measurements and their organizational predecessors [1–4]. These exposure limits have been derived to minimize the potential occurrence of the harmful effects that could result after continued exposure to nuclear radiation. According to the ICRP, the main objective of radiation protection efforts is to prevent harmful nonstochastic exposure effects and also to limit the possible occurrence of stochastic exposure effects—at least to risk levels that are considered acceptable. To prevent nonstochastic exposure effects from occurring, the radiation protection community sets dose limits at sufficiently low values. This procedure ensures that no known threshold exposure dose can be reached, even if the individual is exposed to that level for his or her entire period of employment (occupational career exposure limit) or a lifetime (general population exposure limit).

Some radiation exposure effects, however, are stochastic in nature; that is, some "damage" may occur at any level of exposure. But, because of the presence of natural radioactivity in the environment, it is very difficult for radiation protection experts (sometimes called health physicists) to unambiguously establish that some changes or perhaps harmful consequences will not actually occur in individuals or general populations exposed to low levels of ionizing radiation.

In fact, carcinogenesis is taken as the main somatic risk of individuals and the general population exposed to low levels of radiation. Just how much low level exposure should be accepted by occupational radiation workers or by the general public in developing the particular benefits of a nuclear power application is a matter of heated, often very emotional, contemporary debate. Health physicists recommend that all "justifiable" exposures should be kept "as low as reasonably achievable" (the ALARA premise). This ALARA approach to exposure or even the risk of exposure by large segments of the population would limit stochastic exposure effects. Of course, the established occupational and general public exposure limits would always serve as an upper limit or exposure level boundary. Thus, while there is general recognition that some small, frequently imperceptible, damage might occur from low level radiation exposures, there is also a general consensus in the radiation protection community that the limitations on low level exposure should be based upon a careful balance of the potential risk to an individual or population group and the gains from the overall activity involving a nuclear energy source [1].

One of the most fundamental principles of radiation protection is that exposure should always be kept to a minimum in achieving a particular goal or in performing a specific task; and exposure to an established limit should only be permitted when absolutely necessary.

Consequently, the ICRP has recommended a system of radiation exposure dose limitation with the following major features: [1]

1. no practice shall be adopted unless its introduction produces a positive net benefit;
2. all exposures shall be kept as low as reasonably achievable (ALARA), economic and social factors being taken into account; and
3. the dose equivalent to individuals shall not exceed the limits recommended for the appropriate circumstances by the Commission (ICRP).

Inherent in this radiation protection philosophy is the assumption that if human beings are being adequately protected, then other living things in the terrestrial biosphere would also be sufficiently protected. There are, of course, non-health related deleterious effects of radiation exposure. In space nuclear power applications, these include: radiation damage to payloads, spacecraft equipment, and instrumentation; and the possible radiological contamination of portions of the Earth in the event of a launch vehicle abort or nuclear powered spacecraft reentry.

Radiation Protection Units

As discussed in Chapter 2, when nuclear radiation passes through matter it gives up its energy by producing ionizations and excitations in the atoms and molecules of the medium. The concept of radiation dose has been based on the amount of energy absorbed by the medium. Traditionally, three fundamental radiation protection units have been used: the roentgen (R), the rad, and the rem [1–8]. The roentgen is a measure of radiation exposure; the rad a measure of absorbed dose; and the rem a measure of biological dose equivalent.

The roentgen is used to describe exposure to X-ray and gamma ray radiation and is a measure of the amount of ionization produced in a given volume of air. Originally, the roentgen was defined as the amount of X or gamma radiation that produces one electrostatic unit of charge in a cubic centimeter of dry air at standard conditions, i.e., 273 K and 101.32 kPa. This is equivalent to the production of 2.58×10^{-4} coulomb per kilogram of air [1–8]. The roentgen is considered a unit of radiation exposure because it does not express the direct interaction between the X-ray or gamma ray radiation and the material irradiated. As a rule of thumb, an exposure of 1 roentgen of X or gamma radiation in the 0.1 to 3.0 MeV energy range produces a 0.87 rad dose in air and approximately a 0.98 rad dose in soft tissue [3, 6, 7, 8].

Absorbed dose (D) is traditionally defined in terms of the rad (*r*adiation *a*bsorbed *d*ose). One rad is the amount of radiation required to cause the absorption of 100 ergs per gram of material, i.e., 62.4×10^6 MeV per gram. The SI unit of absorbed dose is called the gray (Gy) and is defined as the amount of radiation required to cause the absorption of 1 joule per kilogram. Comparing the two unit systems, we find that 1 gray = 100 rads or 1 rad = 10 milligray (10 mGy). To facilitate understanding, we will use both unit systems where appropriate in this book.

Experimental data clearly indicate that different types of radiation produce different biological effects. The biological effect of absorbing ionizing radiation depends not only on the amount of energy absorbed, i.e., the number of rads or grays, but also upon the rate or spatial distribution at which the incident radiation loses energy in living tissue. This energy loss rate is frequently expressed in terms of linear energy transfer (LET) factor. In general, the harmful effects of radiation on biological systems increase with increasing values of the LET. For example, radiation that is more easily stopped, such as alpha particles or protons, produces more damage in cells and tissue when it reaches sensitive organs than does less heavily ionizing radiation, such as X-rays, gamma rays, or beta particles (see Chapter 2).

The concept of dose equivalent (H) has been developed to indicate the severity or probability of harmful effects from radiation on living things. The dose equivalent at a point in living tissue is defined as [1]

$$H \equiv D \cdot \mathrm{QF} \cdot N \tag{7.1}$$

where D is the absorbed dose

Table 7.1 Linear energy transfer factor versus quality factor.

Quality Factor (QF)	Average LET in Water (LET*) (keV/μm)
1	$\leqslant 3.5$
2	7
5	23
10	53
20	175

Table 7.2 Recommended quality factor values.

QF	Type of Radiation
1	X-rays, gamma rays, electrons, positrons.
10	Neutrons (fast), protons, and singly charged particles of rest mass greater than 1 amu of unknown energy.
20	Alpha particles, multiply charged particles, and particles of unknown charge and/or unknown energy.
[WHEN NEUTRON ENERGY SPECTRUM IS KNOWN]	
2	thermal neutrons
2.5	0.01 MeV neutrons
7.5	0.1 MeV neutrons
11	0.5 MeV neutrons
11	1.0 MeV neutrons
6.5	10 MeV neutrons

QF is the quality factor
N represents any other modifying factors necessary to describe the biological effects of a particular irradiation situation. (N is currently assigned a value of 1 by the ICRP [1]).

Traditionally, the dose equivalent H has been expressed in terms of the rem (roentgen equivalent man), such that

$$\underset{\text{(rem)}}{\text{dose equivalent}} = \underset{\text{(rad)}}{\text{absorbed dose}} \times \underset{\text{(QF)}}{\text{quality factor}} \times \underset{(N)}{\text{modifying factor}}\,. \tag{7.2}$$

Taking $N = 1$, eq. 7.2 becomes [1–8]

$$\text{rem} = (\text{rad}) \times \text{QF}. \tag{7.3}$$

Similarly, a special dose equivalent unit, called the sievert (Sv), has also been defined in the SI unit system, such that

$$\underset{\text{(sievert)}}{\text{dose equivalent}} = \underset{\text{(gray)}}{\text{absorbed dose}} \times \underset{\text{(QF)}}{\text{quality factor}} \times \underset{(N)}{\text{modifying factor}}\,. \tag{7.4}$$

Again taking $N = 1$, eq. 7.4 for dose equivalent becomes

$$\text{sievert} = (\text{gray}) \times \text{QF}. \tag{7.5}$$

The traditional and SI units for dose exposure are related as follows

$$1 \text{ rem} = 10 \text{ millisieverts } (10 \text{ mSv})$$

or

$$1 \text{ sievert (Sv)} = 100 \text{ rem}. \tag{7.6}$$

The quality factor (QF) correlates the effect of a particular type of radiation with the way it microscopically gives up its energy. It is defined in terms of the collision stopping power in water (LET*). Values of the quality factor as a function of LET* are presented in Table 7.1 [1, 5–8]. Approximate values of the quality factor are used, when the distribution of radiation in terms of LET* values is not precisely known at all locations in the irradiated volume of tissue of interest. The QF values shown in Table 7.2 are currently recommended for both internal and external radiations [1, 5–8]. The values of QF presented in Tables 7.1 and 7.2 have been chosen by the radiation protection community on the basis of appropriate relative biological effectiveness (RBE) values. The RBE has been used in radiobiology to specify the differences in the degrees of effectiveness of various types of radiations in producing harm in biological systems. It is defined as [7, 8]

$$\text{RBE} \equiv \frac{\text{dose of 250 keV X-rays producing a certain biological effect}}{\text{dose of a given radiation required to produce the same biological effect}} \tag{7.7}$$

In contemporary radiation protection activities, however, the use of the quality factor in estimating dose equivalent is preferred [1, 5–8]. For example, the quality factor values take into account the fact that the established dose equivalent limits (e.g., 5 rem [5 mSv] per year wholebody irradiation for occupational workers) are, in fact, based upon extrapolations from higher absorbed doses at which harmful biological effects can be directly observed. Table 7.3 summarizes the major radiation protection units and terms described up to this point [1–8].

Table 7.3 Fundamental radiation protection terms.

EXPOSURE
roentgen (R) unit of radiation exposure
1 roentgen is defined as the amount of X- or gamma radiation that creates 2.58×10^{-4} coulomb/kg of air.
1 roentgen (X- or gamma ray in the 0.1–3.0 MeV range) $\approx$0.87 rad (8.7 mGy) in air $\approx$0.98 rad (9.8 mGy) in soft tissue
ABSORBED DOSE (D)
1 rad $\equiv$ 100 ergs/g ($= 62.4 \times 10^6$ MeV/g)
1 gray (Gy) $\equiv$ 1 joule/kg
1 rad = 10 mGy
DOSE EQUIVALENT (H)
dose equivalent (rem) = absorbed dose (rads) $\times$ quality factor
$H = D \cdot \text{QF}$
dose equivalent (sieverts) = absorbed dose (grays) $\times$ quality factor
sievert (Sv) = Gy $\cdot$ QF
1 rem = 10 mSv

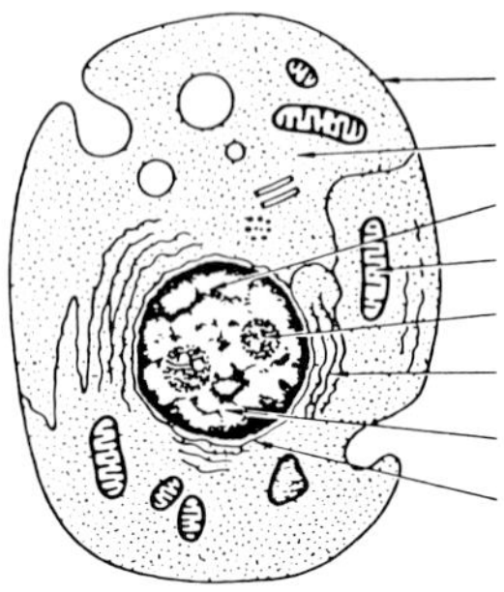

Fig. 7.1 Generalized diagram of a typical somatic cell; significant organelles are identified. *Courtesy of U.S. Department of Energy.*

The ICRP has also developed several other concepts and terms to describe the harmful effects of radiation exposure [1]. Two of these are the detriment to health (G) and collective dose equivalent (S).

The concept of the detriment (G) is an attempt to quantify all the harmful effects of radiation exposure, including stochastic and nonstochastic health effects involving exposed individuals, stochastic effects involving exposed populations, and even nonhealth effects such as denied use of areas or restricted use of certain material or products. According to the ICRP, the detriment to health (G) in a group of persons (P) is given by

$$G \equiv P \sum_i p_i f_i \tag{7.8}$$

where p_i is the probability of suffering (health) effect i
f_i is a weighting factor that expresses the severity of the ith effect.

The detriment takes into account the population exposed, the probability of each harmful effect, and also the severity of each effect. The collective dose equivalent S in a population P is defined as

$$S \equiv \sum_i H_i P_i \tag{7.9}$$

where H_i is the dose equivalent per person in terms of wholebody or specified organ exposure of P_i members of the ith subgroup of the exposed population (P).

These terms are often useful in evaluating the potential risk of large population exposures.

Biological Effects of Ionizing Radiation

Each organ of the body is composed of tissues that consist of a variety of living cells. In fact, an adult human body contains about 40×10^{12} cells. There are two broad classes of living cells: somatic and germ. The majority of cells in the body are somatic cells; while germ cells, or gametes, function in reproduction. A typical somatic cell is illustrated in Fig. 7.1 [9, 10]. This somatic cell consists of a dense inner structure called the nucleus that is surrounded by a less dense mass of cytoplasm. The nucleus is separated from the cytoplasm by a perforated double envelope called the nuclear membrane. The cytoplasm has a network of membranes that form the boundaries of numerous canals and pouches (or vesicles) and it is laden with small bodies called ribosomes. This membranous network is called the endoplasmic reticulum and is distinct from the mitochondria, which are membranous organelles structurally independent of other components of the cytoplasm. The cell membrane or plasma membrane is the outer coat and forms the cell boundary.

The nucleus is of special significance, since it contains a number of threadlike bodies called chromosomes that are the carriers of the cell's heredity-controlling system. These chromosomes contain granules of chromatin, a material rich in deoxyribonucleic acid (DNA). The chromosomes generally are not readily observed in the cellular nucleus, except when the cell, along with its nucleus, is dividing. When the nucleus is not dividing, a spherical body called the nucleolus can be seen. However, when the nucleus is dividing, the nucleolus disappears.

Not every cell possesses all the structures indicated in Fig. 7.1. Red blood cells, for example, do not have a nucleus; while in other type cells the endoplasmic reticulum may be at a minimum. Nonetheless, the generalized somatic cell depicted in Fig. 7.1 is representative of a great majority of the cells of higher organisms [9, 10].

The major classes of cell chemical constituents are: the carbohydrates (sugars), the lipids (fats), the proteins, and the nucleic acids. A cell also contains protoplasm (primarily water) and several other organic and inorganic compounds such as minerals and vitamins.

Carbohydrates serve mainly as foodstuff within the cell and can be stored in several related forms. The carbohydrates can also serve a number of functions outside the cell, i.e., as structural units.

Lipids occur in the cell as alcohols, steroids, aldehydes, and phospholipids and can be found in all fractions of the cell. Their most significant functions appear to be the formation of membranes and the establishment of a specific level of permeability within these membranes. Lipids also store chemical energy, mostly in the form of neutral fats.

Proteins are found in many cell structures and occur in many varieties. For example, enzymes are proteins that serve as catalysts for the cell's metabolic processes. The nucleic acids, DNA and RNA (ribonucleic acid), function together to produce the cell's proteins. Both

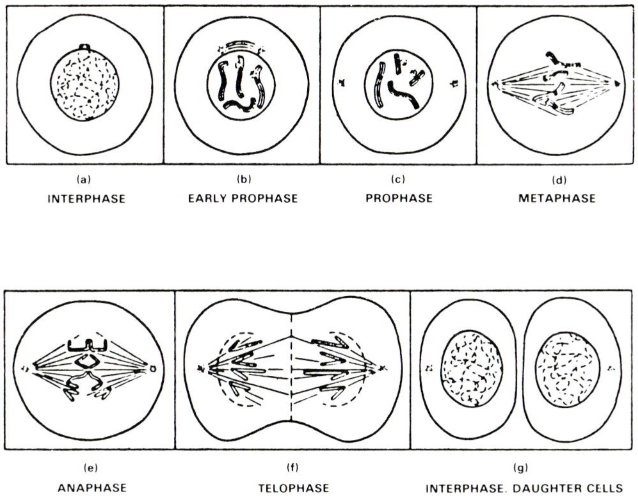

Fig. 7.2 Stages of mitosis (mitotic cycle) for a hypothetical somatic cell with four chromosomes. *Courtesy of U.S. Department of Energy.*

DNA and RNA are essential for cell vitality. Essentially all the DNA of most cells is concentrated in the nucleus, while RNA can be found throughout the cell. Some RNA is located in the nucleus, but the majority of it is associated with minute particles in the cytoplasm called microsomes. Some of these minute particles are particularly rich in RNA and they are consequently referred to as ribosomes. These microsomes are much smaller particles than the mitochondria.

Perhaps one of the most remarkable features of living cells is their ability to grow and divide—in essence, replicating themselves. When a cell reaches a certain stage in its development, it divides into two parts. Then, after another period of development, these two cells can in turn divide. In this remarkable way, plants, animals, and human beings develop to their normal size and injured tissues can be repaired. The process of somatic cell division is called mitosis (see Fig. 7.2) [9, 10]. Cell division occurs when essential elements of the cell have been doubled by replication and the division of the cell results in two essentially equal new parts, called the daughter cells. Mitosis is a continuous process. During interphase, the cell metabolizes, synthesizing new cellular materials, and prepares for self-duplication by synthesizing new chromosomes. In prophase, the chromosomes (each now made up of two identical strands called chromatids) shorten by coiling and the nucleolus and nuclear membrane disappear. During metaphase, the chromosomes line up in one plane near the cell equator. At anaphase, the sister chromatids of each chromosome separate and each part travels toward the poles or ends of the cell. During telophase, the chromosomes now uncoil and become essentially invisible once again; a new nucleus, nucleolus, and nuclear membrane are reconstituted at each pole of the cell; and the division of the cell body occurs between the newly formed nuclei, thereby creating two new cells. In this process, each daughter cell receives a full set of chromosomes and, since the genes are in the chromosomes, each daughter cell has the same genetic complement [9, 10].

All life processes consume energy and consequently require fuel. The mitochondria (small structures in the cytoplasm) are important in controlling the metabolism of a cell and play a central role in the processes by which the energy contained in sugars is made available for cell activity. The gene action system, involving nucleic acid and protein processes, is the series of biochemical events that regulate and direct all life processes by transcribing the genetic information contained in the molecules of DNA.

Germ cells are quite different from somatic cells. They are found in two varieties: ova or egg cells which are produced in the ovaries of the female of the species and sperm cells, produced in the testes of the male of the species. The union of a sperm cell and ovum (each containing 23 chromosomes) during the act of reproduction creates the zygote, the first living cell (containing 46 chromosomes) of a new offspring. The splitting of the zygote and its cellular progeny ultimately creates a full size individual.

This brief exploration of cellular biology yields one very essential fact—living cells are extremely complicated entities. The passage of ionizing radiation through a living cell ionizes and excites its atoms and molecules, breaking molecular bonds, forming new molecules and radicals (such as the oxy (O) and hydroxyl (OH) radicals), and causing chemical reactions to occur. In a very general sense, radiation alters cellular chemistry and the biological effects of radiation may be thought of as the biological effects of various chemical imbalances. Such radiation-induced chemical changes in any part of the complex structure of a living cell can result in a variety of deleterious effects. These include impairment of cell functioning, inhibition of mitosis, alteration of the chromosome structure of the reproductive cells that could ultimately result in genetic changes, and in the extreme, death of the cell. If enough cells are either impaired or destroyed, the organ or tissue involved is also impaired or destroyed. Taking this chain one step further, if the organ is a vital organ or if enough tissue is impaired or destroyed, the living creature irradiated will become physically impaired up to and including the point of death.

There are essentially two classes of radiation effects involving living cells: direct and indirect. The direct

Table 7.4 Summary of estimates of annual average wholebody dose rates in the United States (1970).

Radiation Source	Average Dose Rate mrem/yr	mSv/yr
ENVIRONMENTAL		
(1) Natural		
a. External		
Cosmic Rays	28	0.28
Terrestrial	50	0.50
b. Internal		
Radioisotopes In Body (natural)	22	0.22
(2) Global Fallout	4	0.04
(3) Nuclear Power Industry	0.003	3×10^{-5}
Environmental Subtotal	(~104)	(~1.04)
MEDICAL AND THERAPEUTIC		
(1) Diagnostic	72	0.72
(2) Radiopharmaceuticals	1	0.01
Medical Subtotal	(~73)	(~0.73)
OCCUPATIONAL RADIATION WORKER	0.8	0.008
MISCELLANEOUS	2.0	0.02
TOTAL	~180	~1.8

radiation effect involves primary ionization and excitation of atoms in the cell, a process independent of the surrounding medium. The indirect effect involves the interaction of radiation-created chemical products. These chemical products include H^+, OH^-, and H_2O_2, and other highly reactive chemical species. As these chemicals diffuse through the cell's protoplasm, they react with other cellular molecules. Of course, these direct and indirect radiation effects are not mutually exclusive and can even be considered in many instances as complementary. In both effects, a chain of chemical reactions is started that leads to complex, potentially significant biological consequences. For example, if molecules of DNA are involved, it could be lethal to the cell.

Somatic effects relate to cell injuries that involve the maintenance of bodily functions, as for example, the formation of blood or bone marrow. Genetic effects relate to cell injuries in the reproductive organs (gonads). Germ cells are responsible for the propagation of genetic characteristics to subsequent generations. Radiation in these cells can create mutations that appear in later generations. These mutations are, in general, harmful and permanent.

Exposure to radiation, however, is part of our natural environment. For example, the average adult in the United States receives about 100 mrem per year (1 mSv/yr) dose equivalent from the cosmic radiation penetrating the Earth's atmosphere and from naturally occurring radioactive materials in the lithosphere and in the human body itself. Table 7.4 lists the total average annual wholebody dose rates experienced in the United States in 1970 [2, 6]. A wholebody dose is a dose of radiation absorbed more or less uniformly over the entire body.

Acute Effects of Ionizing Radiation Exposure

In considering the acute effects of ionizing radiation exposure, many scientists contend that there are presently no clinically observable effects apparent from radiation doses within occupational exposure limits. Actually, the first such physical effects occur at about 50 rem (0.5 Sv), a level that is ten times the 5 rem (0.05 Sv) annual exposure limit for radiation workers.

Wholebody Exposure to Gamma and X-Rays

There are four major sources of information on acute effects due to radiation exposures to human beings:

1. Exposure of the Japanese at Hiroshima and Nagasaki
2. Exposure of fishermen and the Marshallese Islanders to fallout at Rongelap in 1954
3. Reactor and accelerator radiation accidents in the United States, Soviet Union, and Yugoslavia
4. Exposure of patients to therapeutic radiation

Acute radiation effects occur within one or two months after exposure and are only observed when relatively large doses of radiation, i.e., greater than about 50 rem (500 mSv), are absorbed in a significant portion of the

Table 7.5 Acute radiation effects from wholebody exposure to gamma radiation.

Acute Irradiation Level (rem)	(sievert)	Acute Somatic Effect
15–25	0.15–0.25	Subtle white blood cell reductions; not generally apparent from exposure to one individual unless a blood sample was taken before as well as after exposure.
50	0.5	Reduction in white blood cell count (WBC) after exposure; the count returns to normal in a few weeks.
75	0.75	10% chance of nausea.
100	1.0	10% chance of temporary loss of hair.
200	2.0	90% chance of radiation sickness; moderate depression of WBC fractions.
400–500	4–5	50% chance of death within 30 days without extensive medical treatment.
> 600	> 6	Lethal to most people in 3 to 30 days; even with extensive medical treatment, death is likely within a few months from infection and hemorrhage.
> 10,000	> 100	Lethal with 24 hours from central nervous system damage.

body in a short period of time. The symptoms are collectively called radiation sickness. Table 7.5 shows the clinical effects of large doses of penetrating gamma radiation to the entire body [6].

The onset, duration, and severity of the radiation sickness symptoms depend on the radiation dose and the dose rate. Three organ systems in the human body are of primary concern in cases of acute radiation syndrome. These critical organs are the blood forming organs; the gastrointestinal tract, primarily the small intestine; and the central nervous system.

In the 15–100 rem (0.15–1.0 Sv) range, most individuals do not have any apparent clinical symptoms, although some depression in circulating white blood cells may be seen above 50 rem (0.5 Sv). Nausea and vomiting are noticeable within a few hours after 100–200 rem (1–2 Sv). Blood studies indicate a definite decrease in the number of circulating white blood cells. Complete recovery usually occurs within a few weeks.

In the 200–600 rem (2–6 Sv) range, the white blood cells are severely depressed, internal bleeding leads to anemia, and systemic bacterial infection resulting in death may occur, if untreated. Hair loss may be extensive above 300 rem (3 Sv). Finally, it may be at least a year before complete recovery of the circulating blood cells is observed.

Beyond 600 rem (6 Sv), death occurs more frequently from extreme damage to the blood-forming organs and to the small intestine. In approximately the 600–1000 rem (6–10 Sv) range, death occurs in 60 to 100 percent of the cases from massive hemorrhage and infection. In the 1000–5000 rem (10–50 Sv) exposure dose range, death is usually caused by circulatory collapse or as a result of generalized infection and electrolyte imbalance. Above 5000 rem (50 Sv), the central nervous system is severely damaged leading to death within a few days.

Localized Exposure to Gamma and X-Rays

Acute local radiation exposure to small portions of the body causes different symptoms and effects than wholebody irradiation. Table 7.6 lists such localized radiation exposure effects [6]. An X-ray dose of greater than 300 rem (3 Sv) to the skin will cause a sunburn-type of reaction with possible hair loss and regrowth. Doses higher than 1500 rem (15 Sv) may cause permanent hair loss and ulceration.

Temporary sterility may result from doses of 50 rem (0.5 Sv) or more to the testes or 250 rem (2.5 Sv) or more to the ovaries. Permanent sterility may result from

Table 7.6 Effects to the body from localized exposure to X and gamma radiation.

Organ of Body	Acute Irradiation Level (rem)	(sievert)	Acute and Delayed Biological Effects
SKIN	300	3.0	Erythema or "sunburn" effect noticeable.
	1500	15.0	Raw, moist skin surface where irradiated.
	5000–7000	50–70	Ulceration, slow healing, possible skin cancer
GONADS	50	0.5	Brief functional sterility in males only.
	250	2.5	Sterility for 1 to 2 years in both male and female.
	600	6.0	Permanent sterility.
EYE	200	2.0	Change in optic lens opacity.
	600	6.0	Clinically significant cataract.
FETUS	10–20	0.1–0.2	Significant probability of malformation, if irradiation occurs in first 3 months of pregnancy.

doses in excess of 600 rem (6 Sv) to the gonads (ovaries or testes).

Delayed Effects of Radiation Exposure to the Individual

Radiation damage and effects that are visible many months or years after the exposure are collectively called the delayed effects of radiation exposure. Many types of delayed effects are difficult to investigate and quantify because of the long time interval between exposure and final biological effect. We will discuss three such delayed effects here: cataract formation; effect on the fetus; and cancer induction from radiation.

An acute exposure of at least 50 rads (0.5 Gy) of neutron or several hundred rads of gamma radiation to the lens of the eye may cause cataract formation many months or years after the irradiation. Changes in lens opacity are clinically apparent after 200 rem (2 Sv). Cataract formation is highly likely from exposures greater than 600 rem (6 Sv).

The effect of radiation on the fetus has been studied extensively from the Hiroshima and Nagasaki experience. High doses, probably greater than 50 rem (0.5 Sv), delivered in a short period of time during the first three months of pregnancy cause reduction in the number of brain forming cells with secondary reduction in skull growth and head size. Small head size, mental retardation, and decreased growth have been the major congenital anomalies documented. Because of the radiation sensitivity of the unborn child, radiation exposure to females of childbearing age should be limited to 500 mrem (5 mSv) during the nine month gestation period, instead of the normal 5 rem per year (50 mSv/y) occupational worker limit.

Cancer is another possible radiation effect that may become apparent many years after the initial exposure. Cancer induction is very difficult to study in human beings because cancer may also be caused by many other agents such as carcinogenic chemicals and viruses. The actual development and physical manifestation of these cancers may take 5 to 25 years. At present, radiation-induced cancers cannot be distinguished from cancers caused by other agents.

Most of the information concerning radiation-induced cancer in humans comes from several exposed groups: atomic bomb survivors, accident victims of the 1954 Bikini nuclear explosion test, pioneer radiologists, children who were exposed intentionally to X-rays for thymic enlargement, children who were exposed before birth during radiological examinations of the mother, radium dial painters, and uranium miners. Data from experimental animals, as well as the available data concerning humans exposed to radiation, have been used to estimate the probability of cancer induction.

Leukemia is the most common type of radiation-induced cancer in human beings. The incidence of leukemia cases increases with increased dose for accumulated doses above 50 rem (0.5 Sv) and high dose rates above 1 rem per minute (10 mSv/min). Other types of cancers can result from high radiation exposures. Bone cancer can be caused by high concentrations of bone-seeking radioisotopes deposited in the skeleton. This was the case with the radium dial painters. Thyroid cancer can be caused by high levels of radioactive iodine deposited in the thyroid. Uranium miners have a higher incidence of lung cancer due to long term retention of insoluble radon daughters in the lungs. High level radiation exposure of the skin may cause skin cancer after many years.

Below about 10 rem (0.1 Sv), neither human nor experimental animal exposure data currently available clearly define the real cancer risks from radiation. Con-

Table 7.7 Currently permissible occupational worker exposure levels.

Type of Radiation Exposure	Exposure Period	Maximum Possible Dose Equivalent (rem)	(mSv)
Wholebody, head and trunk, gonads, lens of the eye, red bone marrow, active blood forming organs	Year	5	50
	Calendar quarter (13 weeks)	3	30
Unlimited areas of the skin, except hands and forearms; other organs, tissues, and organ systems, except bone	Year	15	150
	Calendar quarter	5	50
Bone and forearms	Year	30	300
	Calendar quarter	10	100
Hands and feet	Year	75	750
	Calendar quarter	25	250

sequently, radiation standards are founded upon conservative extrapolations from higher doses. For exposure doses ranging from natural background levels to about 10 rads (0.1 Gy) cancer is a very rare phenomenon; and it is impossible to separate radiation-induced cancers from cancers caused by other carcinogenic agents. No clearly identifiable radiation-induced cancers have been caused in mammals at the 5 rem per year (50 mSv/y) level—the level corresponding to the radiation exposure limit presently recommended by the ICRP for radiation workers.

Genetic Effects of Radiation

Up to this point, we have discussed the effects of radiation on the individual (somatic effects). Genetic effects become significant when large populations, i.e., many hundreds of thousands of people, are exposed to a mutagen with sufficient strength to change the gene pool. We will now briefly discuss the effect of radiation on succeeding generations from a large population of irradiated individuals.

Although definite birth defects were observed among the survivors of Hiroshima and Nagasaki, genetic defects have not been statistically shown [6]. The normal incidence of all types of genetic birth defects in human beings is about 12,000 per million live births or 1.2 percent. Estimates have been made to determine the radiation dose that produces as many mutations as occur naturally. This doubling dose is somewhere between 20–200 rem (0.2–2.0 Sv) delivered to a large population (greater than one million) in the first 30 years of life. This doubling dose represents some 7 to 70 times the natural background exposure dose received by an average person in the United States. It would take about five generations to equilibrate dominant genetic defects and hundreds of generations to equilibrate recessive genetic defects from a doubling of the mutation rate through the exposure of a large population to some 20 to 200 rem (0.2–2.0 Sv) during their 30-year reproductive lifetime. Currently, the sum of the average environmental, medical, miscellaneous, and occupational dose to the American population is about 180 mrem per year (1.8 mSv/y), a dose corresponding to approximately 5.4 rem (54 mSv) per 30 year reproductive lifetime. This dose is still one or two orders of magnitude below the best available estimate of the doubling dose.

Radiation Protection Standards

Space nuclear power applications must keep the radiation exposure of astronauts, occupational workers, e.g., ground support personnel, and members of the general public as low as reasonably achievable (the ALARA premise) during all mission phases, normal and abnormal. This section introduces contemporary radiation protection standards that are based on guidelines and/or recommendations from the International Com-

Table 7.8 Currently permissible exposure standards for the general public.

Type of Radiation Exposure	Annual Maximum Permissible Dose Equivalent			
	Based on Dose to Individuals At Points Of Maximum Probable Exposure (rem)	(mSv)	Based on Average Dose To a Suitable Sample Of the Exposed Population (rem)	(mSv)
Wholebody, gonads, or bone marrow	0.5	5	0.17	1.7
Other organs	1.5	15	0.50	5.0

Table 7.9 Radiation exposure limits recommended for spaceflight crewmembers.

Constraint	Primary Reference Risk* rem [Sv]	Bone Marrow** (Dose Equivalent at 5 cm) rem [Sv]	Ancillary Reference Risks: Skin (Dose Equivalent at 0.1 mm) rem [Sv]	Ancillary Reference Risks: Ocular Lens (Dose Equivalent at 3 mm) rem [Sv]	Ancillary Reference Risks: Testes (Dose Equivalent at 3 cm) rem [Sv]
1 year average daily dose		0.2 [0.002]	0.6 [0.006]	0.3 [0.003]	0.1 [0.001]
30 day maximum		25 [0.25]	75 [0.75]	37 [0.37]	13 [0.13]
Quarterly maximum***		34 [0.34]	105 [1.05]	52 [0.52]	18 [0.18]
Yearly maximum		75 [0.75]	225 [2.25]	112 [1.12]	38 [0.38]
Career limit	400 [4.0]	400 [4.0]	1200 [12.0]	600 [6.0]	200 [2.0]

*Dose equivalent at 5 cm depth in tissue
**Wholebody exposure
***May be permitted for two consecutive quarters followed by 6 months of restriction from further exposure to maintain yearly limit and includes all occupational exposures

mission on Radiological Protection (ICRP), U.S. Environmental Protection Agency (EPA), U.S. Nuclear Regulatory Commission (NRC), U.S. Department of Energy (DOE), and U.S. National Council on Radiation Protection and Measurements (NCRP) [1, 4, 11, 12]. According to these recommended standards, the maximum permissible doses (MPD) that may be received by persons working with ionizing radiation on Earth are listed in Tables 7.7 and 7.8 [1, 4, 11, 12].

The natural radiation environment in outer space provides an additional source of exposure concern for the astronaut crews. In 1970, the Radiological Advisory Panel, Committee on Space Medicine of the National Academy of Sciences issued a set of recommended radiation exposure limits for use by NASA in planning and developing radiation protection strategies for future manned space missions, including those missions using nuclear power supplies [13, 14]. These recommended limits are presented in Table 7.9 [13, 14]. An individual mission exposure limit was set at 25 rem (0.25 Sv), because radiation exposure data indicated that this level would have a very low probability of causing radiation sickness, e.g., vomiting, diarrhea, and its subsequent adverse impact on the conduct of the space mission. An astronaut career exposure limit of 400 rem (4 Sv) was established on the basis of the doubling dose considered a risk for leukemia [13, 14].

As can be seen in Tables 7.5 through 7.9, all body organs and systems are not affected equally by a particular radiation exposure. Table 7.10 groups major body organs and elements according to their relative radiosensitivity [14]. The lymphoid system is considered the most radiosensitive body component from the viewpoint of somatic effects. This system is closely followed in radiosensitivity by bone marrow and the gastrointestinal epithelium. Body elements appearing in the high radiosensitivity group are characterized by rapidly growing tissues. Growing tissue, in turn, has rapidly dividing cells that are consequently the most sensitive to any ionizing radiation. On the other hand, tissues that exhibit moderate radiosensitivity have slow growth characteristics. The lens of the eye is particularly sensitive to neutron radiation, which can produce cataracts and lenticular opacities at lower exposure levels than would be anticipated. Finally, body elements such as muscle, cartilage, and the central nervous system reproduce quite slowly, are taken as very stable, and exhibit low radiosensitivity [14]. Body organs that exhibit the highest degree of radiosensitivity, such as bone marrow, are sometimes referred to as critical organs.

Inhalation, ingestion, and absorption are all pathways by which radioactive materials may enter the body, causing internal radiation exposure. The dose to the critical organ from these pathways is dependent on the type of radiation, the effective biological half-life, and the amount of material entering the body. The biological half-life is defined as the time required for a biological system, e.g., man or animal, to eliminate by natural processes half the amount of a radioactive material that has entered the system. The radiation limits for critical organs range from 5 to 30 rem per year (0.05–0.3 Sv/y), depending on the relative radiation sensitivity of the tissues.

External radiation exposure to the skin of the entire body is limited to 5 rem/calendar quarter (0.05 Sv/quarter) (except hands and forearms), not to exceed 15 rem per year (0.15 Sv/y).

The internal dose from radionuclides for which the entire body is the critical organ is combined with the

Table 7.10 Differential radiosensitivity of human tissue.

HIGH RADIOSENSITIVITY ORGANS
- Lymphoid tissue
- Bone marrow elements
- Gastrointestinal epithelium
- Gonads (testis and ovary)
- Embryonic tissues

MODERATE RADIOSENSITIVITY ORGANS
- Skin
- Vascular endothelium
- Kidney
- Liver
- Lung
- Eye (lens)

LOW RADIOSENSITIVITY ORGANS
- Endocrine glands (except gonads)
- Muscle
- Central nervous system
- Connective tissue
- Bone and cartilage

external wholebody dose. For example, tritium is a radionuclide in which the entire body is considered the critical organ. Therefore, if an individual received a tritium exposure and an external penetrating radiation exposure, the two exposures would be combined to determine an accumulated radiation dose. As previously mentioned, wholebody exposure is limited to 5 rem/y (0.05 Sv/y) from any occupational exposure situation on Earth.

For the general public, all radiation exposures should be kept to the lowest levels technically and economically practicable within the limits recommended. Table 7.8 lists the currently recommended exposure dose standards for the general public. These are equivalent to one tenth (1/10) of the current occupational exposure standard.

An individual should be issued some form of personal radiation monitoring equipment, i.e., film badge or dosimeter, if the potential exists for receiving a dose in any calendar quarter in excess of 10 percent of the quarterly dose standards stated in Table 7.7. Radiation workers, especially astronauts and ground crew personnel, should be aware of any radiation levels they are working in from a nuclear-powered payload and make every effort to minimize their overall exposure. The philosophy of keeping radiation exposures as low as reasonably practical (ALARA) should be central to the radiation protection policy of any space nuclear power program. Finally, the environment should be monitored to assure that radiation doses to the public are maintained at the lowest practical level, consistent with the dose standards appearing in Table 7.8. This is especially important in the event of a launch abort or reentry of a space nuclear power system.

A sometimes used mathematical relationship for establishing the accumulated wholebody dose equivalent permitted for a radiation worker is

$$H_{\max} = 5(N - 18) \text{ rem} \tag{7.10}$$

or

$$H_{\max} = 50(N - 18) \text{ mSv}$$

where $H_{\max}$ is the individual's accumulated dose (rem or mSv) to the age N

N is the individual's age in years.

Equation 7.10 ensures that the individual does not exceed the currently recommended 5 rem (0.05 Sv) per year maximum permissible wholebody dose. This equation also indicates the practice that occupational radiation exposure is to be prevented before the individual is 18 years old. It should be noted that eq. 7.10 is not presently recommended for use by the ICRP.

Example 7.1. What is the total maximum permissible accumulated dose equivalent for a 48 year old radiation worker who has been involved in the nuclear industry since age 18?

Solution. Current Age(N) = 48

Assuming the individual received the maximum annual permissible dose equivalent (5 rem or 50 mSv) each year, we can use eq. 7.10

$$H_{\max} = 5(48 - 18) \text{ rem} = 150 \text{ rem (1.5 Sv)}$$

This represents a total (30 year) maximum career dose limit of 150 rem (1.5 Sv). A 50 year career dose ($H_{\max_{50}}$) has the value of 250 rem (2.5 Sv). In light of the great uncertainty about the effect of chronic exposure to relatively low levels of radiation and the contemporary uncertainty about the actual physical significance of a 250 rem (2.5 Sv) dose equivalent delivered to an individual over a career lifetime (here 50 years), there exists serious consideration within the radiation protection community to not only limit annual occupational exposures to a maximum of 5 rem (0.05 Sv), but also to adjust the formula expressed in eq. 7.10 to reduce the maximum potential 50 year career dose equivalent ($H_{\max_{50}}$) from its current value of 250 rem (2.5 Sv) to 100 rem (1.0 Sv).

Shielding

Fundamental Considerations

The use of a nuclear power source in space is frequently accompanied by the need for a radiation shield to protect the astronaut crew, payload, and radiation-sensitive spacecraft equipment from the damaging ef-

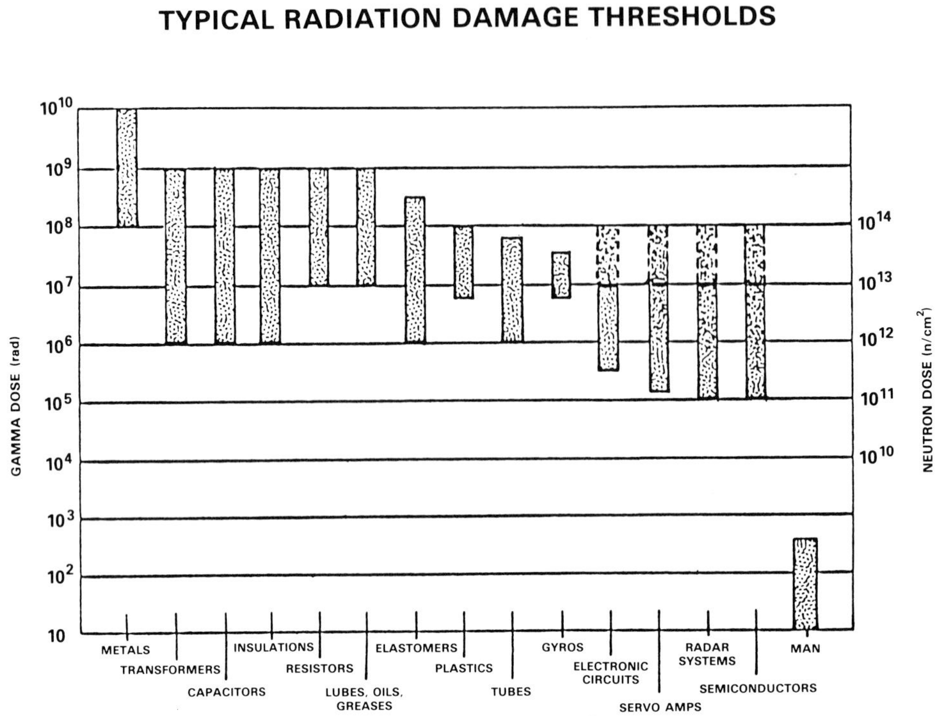

Fig. 7.3 Typical radiation damage thresholds. *Courtesy of U.S. Department of Energy and Rockwell International.*

fects of radiation. This is especially true for nuclear reactor systems and the copious quantities of neutrons and gamma rays that accompany the fission process. The establishment of permissible radiation levels to protect personnel and/or sensitive equipment is one of the major steps in the design of a radiation shield. Radiation dose limits for astronauts, ground support personnel (including radiation workers), and the general public were described in previous sections of this chapter.

Figure 7.3 presents a composite description of typical radiation damage thresholds for certain materials, components, and systems that might be used in a nuclear-powered spacecraft [15]. When such materials and components are exposed to nuclear radiations, their properties are frequently changed in a variety of ways that cannot be readily anticipated from nonradiation environment operating experience. Two general effects are especially significant in establishing the overall lifetimes of materials and components in a nuclear radiation environment: rate effects and cumulative effects.

The rate effect usually involves the delivery of an intense radiation dose in a short period of time, leading to ionization-induced damage. For example, sensitive electronic circuits will experience noise, electronic upset, or even burnout under influence of a rate effect. A cumulative radiation effect typically involves the long-term buildup of radiation-induced lattice defects in the material that then produce significant changes in its thermophysical properties. All materials and components eventually exhibit some degree of cumulative or integral damage effects after prolonged radiation exposures.

Many factors influence the geometry, composition, and mass of the radiation shield. Some of these factors include [15, 16]

- The size and nature of the nuclear power source, e.g., reactor or radioisotope
- The type of radiation—alpha, beta, gamma, neutron, etc.

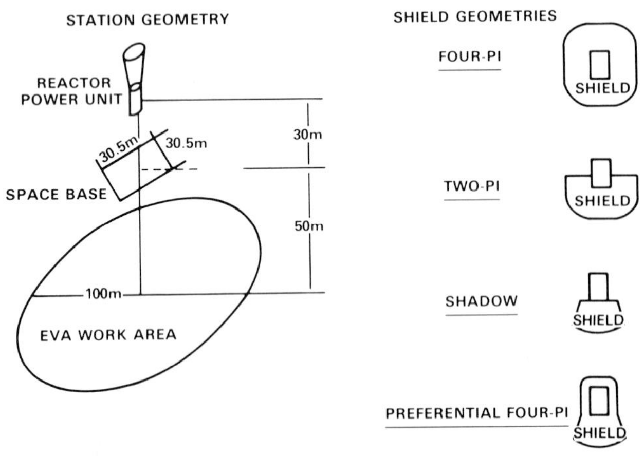

Fig. 7.4 Various shield design concepts. *Courtesy of NASA.*

- The configuration of the spacecraft or platform and its payload, including crew compartment, if appropriate
- Mission-generic operational procedures and requirments, such as the need for rendezvous and docking or on-orbit maintenance
- The length of the mission
- The total radiation dose levels that will be permitted

Shields can be configured to fit a variety of spacecraft or space platform requirements. Figure 7.4 shows a reactor-powered space base and the variety of shield configurations that might be selected. For example, a shadow shield will only partially surround the nuclear power source. The area within the shadow of the shield is protected, while radiation escapes to space from the reactor's non-shielded surfaces. If a shadow shield geometry is chosen for applications involving human beings or very sensitive equipment, then special consideration must be given to the regions of space around the vehicle that might be used for rendezvous and docking operations, on-orbit maintenance, or extravehicular activity (EVA) by the astronaut crew [16, 17]. The upper limit of a shadow shield is the four-pi (4π) shield—a shield design that provides approximately the same amount of shielding material in all directions around the radiation source. Numerous variations of these two basic shielding configurations can also be created, for example the 2π shield and the preferential 4π shield.

As introduced in Chapter 2, the passage of nuclear radiation through matter is a complicated process that often does not lend itself to simple mathematical description. In fact, shield design is a fairly rigorous and complicated task, the fine points of which are beyond the scope of this book. We will, however, illustrate some of the fundamental considerations involved in shielding calculations by introducing two basic examples: a monoenergetic beam of gamma rays (photons) incident upon a uniform, homogeneous slab shield; and an isotropic point source of monoenergetic gamma rays surrounded by a spherical shield of uniform thickness and composition.

We first consider the case of a narrow beam of monoenergetic gamma rays of energy E_0 passing through a homogeneous slab of shielding material of thickness L (see Fig. 7.5). If every time a gamma ray interacted with matter, it truly "disappeared," then it would be a very simple matter to calculate the "uncollided" gamma ray flux (ϕ_u) at dose point B from [18–21]

$$\phi_u = \phi_0 e^{-\mu L} \tag{7.11}$$

where ϕ_0 is the incident gamma ray flux (photons/cm^2-sec)

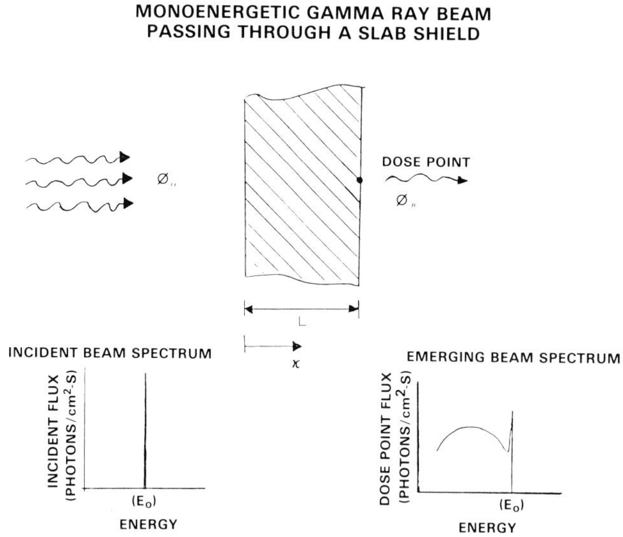

Fig. 7.5 Monoenergetic gamma ray beam passing through a slab shield.

ϕ_u is the uncollided gamma ray flux (photons/cm^2-sec) after passing through a shield thickness L

L is the shield thickness (cm)

μ is the total linear attenuation coefficient (cm^{-1}).

As discussed in Chapter 2, eq. 7.11 applies only to the "thin geometry" shield.

However, gamma rays do not actually disappear every time they interact with matter. In Compton scattering, for example, the photons are merely scattered, giving up a portion of their energy and continuing through the shield material in some other direction. The emission of X-rays will usually accompany the photoelectric effect and, even though the original high energy gamma ray disappears in pair production, the appearance of annihilation gamma radiation also accompanies this interaction. Consequently, a monoenergetic beam of gamma rays passing through a slab shield will actually emerge with a spiked, continuous spectrum as shown in Fig. 7.5 [18–21]. The sharp peak at energy E_0 in the emerging gamma ray spectrum represents the surviving, uncollided portion (ϕ_u) of the originally incident gamma ray flux (ϕ_0). The smooth, continuous portion of the spectrum displays the effects of Compton scattering and includes minor contributions from pair production and photoelectric effect processes.

A buildup factor must be used to compensate for the underestimation of the scattered radiation reaching a dose point beyond the shield. For slab geometry, we can define this buildup factor as [18–21]

$$B(\mu x) \equiv \frac{\text{true flux (buildup flux)}}{\text{"thin geometry" flux}} = \frac{\phi_b}{\phi_0 e^{-\mu x}} \tag{7.12}$$

or

Table 7.11 Total gamma attenuation coefficient μ (cm^{-1}).

Material	Density (g/cm^3)	E_γ (MeV)	0.1	0.5	1.0	2.0	4.0	6.0	10.0
beryllium (Be)	1.85		0.244	0.143	0.1045	0.0729	0.0492	0.0390	0.0298
aluminum (Al)	2.70		0.435	0.227	0.1658	0.1166	0.0837	0.0713	0.0618
iron (Fe)	7.86		2.704	0.6508	0.4677	0.3333	0.2594	0.2389	0.2311
lead (Pb)	11.34		55.99	1.644	0.7757	0.5182	0.4763	0.4944	0.5545
tungsten (W)	19.3		81.25	2.413	1.235	0.8434	0.7759	0.8067	0.8975
uranium (U)	18.7		198.2	3.291	1.416	0.9051	0.8228	0.8509	0.9556

(Column group header over the energy columns: Total Attenuation Coefficient μ (cm^{-1}))

Table 7.12 Total gamma ray mass attenuation coefficient μ/ρ (cm^2/g).

Material	Density $(g/cm)^3$	E_γ (MeV)	0.1	0.5	1.0	2.0	4.0	6.0	10.0
beryllium (Be)	1.85		0.132	0.0773	0.0565	0.0394	0.0266	0.0211	0.0161
aluminum (Al)	2.70		0.161	0.0840	0.0614	0.0432	0.0310	0.0264	0.0229
iron (Fe)	7.86		0.344	0.0828	0.0595	0.0424	0.0330	0.0304	0.0294
lead (Pb)	11.34		5.29	0.145	0.0684	0.0457	0.0420	0.0436	0.0489
tungsten (W)	19.3		4.21	0.125	0.0640	0.0437	0.0402	0.0418	0.0465
uranium (U)	18.7		10.60	0.176	0.0757	0.0484	0.0440	0.0455	0.0511

(Column group header over the energy columns: Total Mass Attenuation Coefficient μ/ρ (cm^2/g))

$$\phi_b = \phi_0 B(\mu x) e^{-\mu x} \tag{7.13}$$

where ϕ_b is called the buildup flux (photons/cm^2-sec)
$B(\mu x)$ is the slab geometry dose buildup factor for a monodirectional beam of gamma rays.

Table 7.11 provides the total gamma ray attenuation coefficient μ and Table 7.12 the mass attenuation coefficient (μ/ρ) for selected materials at various gamma ray energies [18–21]. Table 7.13 describes the slab geometry monodirectional beam buildup factor $B(\mu x)$ for selected materials and gamma ray energies [18–21]. It should be noted that the shield thickness (μx) identified in Table 7.13 is presented in mean free path lengths, computed at the energy of the incident gamma ray.

We now turn our attention to the case of an isotropic point source S (photons/sec) that is surrounded by a uniform spherical shield of radius R (cm) (see Fig. 7.6). The buildup flux ϕ_b at a dose point on the surface of this spherical shield is given by [18–21]

$$\phi_b = \frac{S}{4\pi R^2} B_{pt}(\mu R) e^{-\mu R} \tag{7.14}$$

where ϕ_b is the buildup flux (photons/cm^2-sec)

Table 7.13 Slab geometry monodirectional beam buildup factor.

Shield Material	Shield Thickness (μx)	E_γ (MeV)	0.5	1.0	2.0	4.0	6.0	10.0
iron (Fe)	1		2.07	1.92	1.69	1.48	1.35	1.22
	2		2.94	2.74	2.35	1.90	1.71	1.44
	4		4.87	4.57	3.76	2.95	2.48	1.95
	10		12.4	11.6	8.78	6.46	5.35	4.07
lead (Pb)	1		1.24	1.38	1.40	1.28	1.19	1.11
	2		1.39	1.68	1.76	1.56	1.40	1.24
	4		1.63	2.18	2.41	2.18	1.87	1.54
	10		2.08	3.40	4.35	4.69	4.69	3.54
uranium (U)	1		1.17	1.30	1.33	1.25	1.18	1.10
	2		1.28	1.53	1.62	1.49	1.37	1.21
	4		1.45	1.90	2.15	2.02	1.82	1.48
	10		1.73	2.70	3.56	4.06	4.12	3.21

(Column group header over the energy columns: Buildup Factor $B(\mu x)$)

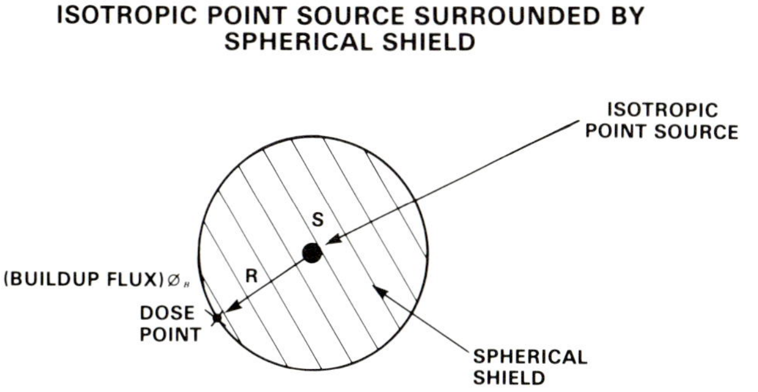

Fig. 7.6 Isotropic point source surrounded by spherical shield.

S is the source strength (photons/sec)

$B_{pt}(\mu R)$ is the dose buildup factor for an isotropic point source

μ is the total gamma ray attenuation coefficient (cm^{-1})

R is the shield thickness (cm)

μR is the shield thickness (in multiples of mean free paths).

Table 7.14 presents a set of dose buildup factors $B_{pt}(\mu R)$ for an isotropic point source of gamma radiation as a function of shield thickness in mean free paths (μR), materials, and gamma ray energy [18–21]. The data presented in Table 7.14 are appropriate for shields that have at least one mean free path thickness (that is, $\mu R = 1$).

For shielding calculations involving an isotropic point source, it is often useful to obtain the dose buildup factor $B_{pt}(\mu R)$ from a mathematical expression known as the Taylor form of the buildup factor [18–21].

$$B_{pt}(\mu R) = A_1 e^{-a_1 \mu R} + A_2 e^{-a_2 \mu R} \tag{7.15}$$

where

$$A_2 = 1 - A_1 \tag{7.16}$$

and A_1, a_1, and a_2 are coefficients found in Table 7.15 [20]. Equation 7.15 often provides sufficiently accurate dose buildup factor data for many initial shielding calculations.

Example 7.2. An isotropic point source of 2 MeV gamma rays of strength 10^8 photons/sec is surrounded by a uniform spherical shield of tungsten, 10 cm thick. What is the buildup flux ϕ_b on the outer surface of the shield?

Solution. For tungsten, assuming density (ρ) = 19.3 g/cm^3 and $E_\gamma = 2$ MeV, we find $\mu = 0.8434\ cm^{-1}$ from Table 7.11. Consequently,

$$\mu R = (0.8434\ cm^{-1})(10\ cm) = 8.434.$$

Table 7.14 Dose buildup factor $B_{pt}(\mu R)$ for a point isotropic gamma ray source.

Shield Material	Shield Thickness (μR)	Buildup Factor $B_{pt}(\mu R)$ E_γ (MeV): 0.5	1.0	2.0	4.0	6.0	10.0
aluminum (Al)	1	2.37	2.02	1.75	1.53	1.42	1.28
	2	4.24	3.31	2.61	2.08	1.85	1.55
	4	9.47	6.57	4.62	3.22	2.70	2.12
	7	21.5	13.1	8.05	5.01	4.06	3.01
	10	38.9	21.2	11.9	6.88	5.49	3.96
iron (Fe)	1	1.98	1.87	1.76	1.45	1.34	1.20
	2	3.09	2.89	2.43	1.94	1.72	1.42
	4	5.98	5.39	4.13	3.03	2.58	1.95
	7	11.7	10.2	7.25	4.91	4.14	2.99
	10	19.2	16.2	10.9	7.11	6.02	4.35
tungsten (W)	1	1.28	1.44	1.42	1.29	1.20	1.11
	2	1.50	1.83	1.85	1.62	1.43	1.25
	4	1.84	2.57	2.72	2.41	2.07	1.64
	7	2.24	3.62	4.09	4.03	3.60	2.62
	10	2.61	4.64	5.27	6.27	6.29	4.65
uranium (U)	1	1.17	1.31	1.33	1.24	1.16	1.09
	2	1.30	1.56	1.64	1.50	1.36	1.20
	4	1.48	1.98	2.23	2.09	1.85	1.51
	7	1.67	2.50	3.09	3.21	2.96	2.26
	10	1.85	2.97	3.95	4.66	4.80	3.78

Table 7.15 Coefficients for dose buildup factor equation.

Material	Coefficients	E_γ (MeV) 1.0	2.0	4.0	6.0	10.0
aluminum (Al)	A_1	8.0	5.5	3.8	3.1	2.25
	$-a_1$	0.11	0.082	0.066	0.064	0.060
	$+a_2$	0.044	0.093	0.130	0.152	0.128
iron (Fe)	A_1	8.0	5.5	3.75	2.9	2.0
	$-a_1$	0.090	0.079	0.075	0.083	0.095
	$+a_2$	0.04	0.07	0.082	0.075	0.0116
lead (Pb)	A_1	2.45	2.60	1.65	0.96	0.50
	$-a_1$	0.045	0.071	0.123	0.175	0.214
	$+a_2$	0.178	0.103	0.064	0.059	0.080
tungsten (W)	A_1	3.3	2.9	2.05	1.2	0.6
	$-a_1$	0.043	0.069	0.118	0.171	0.212
	$+a_2$	0.148	0.188	0.070	0.000	0.144

We next interpolate between values in Table 7.14 to find that for $E_\gamma = 2$ MeV, $B(\mu R) \approx 4.564$. Then, from eq. 7.14,

$$\phi_b = \frac{S}{4\pi R^2} B_{pt}(\mu R) e^{-\mu R}$$

$$\phi_b = \frac{(10^8 \text{ photons/sec})}{4\pi \ (10 \text{ cm})^2} [4.564] \ e^{-8.434}$$

$$\phi_b = 78.9 \text{ photons/cm}^2\text{-sec.}$$

If we had ignored the buildup factor in the shield, the residual uncollided flux (ϕ_u) would be estimated as only

$$\phi_u = \frac{S}{4\pi R^2} e^{-\mu R} = \left[\frac{10^8 \text{ photons/sec}}{4\pi \ (10 \text{ cm})^2}\right] e^{-8.434}$$

$$\phi_u = 17.3 \text{ photons/cm}^2\text{-sec.}$$

This calculation would represent a significant underestimation of the radiation field beyond the shield.

Shielding Considerations for Radioisotope Power Supplies

Radiation from radioisotope power supplies might involve alpha particles, beta particles, gamma rays, X-rays (from Bremsstrahlung), or neutrons (from (α, n) reactions and spontaneous fission). The nuclear radiations from a radioisotope source are typically well defined in comparison to those from a nuclear reactor. As will be discussed in Chapter 8, alpha emitters (especially plutonium-238) are the preferred radioisotope fuel for space power applications. Alpha particles are readily stopped within very thin pieces of shielding material (the thickness of this page for example) and generally present no significant shielding problems (see Chapter 2). However, reactions between the alpha particle and impurities in the heat source plutonium fuel or with the fuel capsule wall materials may result in secondary radiations which could require additional shielding.

Of particular concern are neutrons originating from (α, n) reactions and from spontaneous fission reactions within the radioisotope fuel itself. These neutrons are fast neutrons and will also produce secondary gamma rays as they slow down and experience inelastic scattering and radiative capture reactions. There will also be some gamma radiation directly associated with spontaneous fission. In fact, these secondary radiations often control the shielding requirements for a radioisotope power supply, fueled by plutonium-238 or other alpha emitter.

Table 7.16 describes typical isotopic weight percents (i.e., grams of Pu isotope/100 grams of Pu mixture) for heat source plutonium-238. Table 7.17 provides associated radiological properties for selected plutonium isotopes and americium-241 [22, 23].

Table 7.16 Approximate isotopic weight percents of heat source plutonium-238.

ISOTOPE	WEIGHT PERCENT (g Pu isotope/100 g Pu mixture)
238	90.0
239	9.1
240	0.6
241	0.03
242	< 0.01

Table 7.17 Radiological properties of selected plutonium isotopes and americium-241.

Plutonium-238 [Half-life = 87.75 years]
main decay mode: alpha decay feeding ^{234}U.
% spontaneous fission (SF) = 1.84×10^{-7}
major radiations:

type	energy	intensity (%)
$alpha_1$	5.358 MeV	0.1
$alpha_2$	5.457 MeV	28.3
$alpha_3$	5.499 MeV	71.6
X-ray	13.6 keV	11.6
average values for weak gammas	55.3 keV	0.05

* * *

alpha activity (α/min-μg)	$= 3.9 \times 10^7$
SF neutrons (n/s-g) [metal]	= 3420
(α, n) reaction neutrons (n/s-g)	$[PuO_2] = 1.4 \times 10^8$
	$[PuF_4] = 2.1 \times 10^6$

Plutonium-239 [Half-life = 24,131 years]
main decay mode: alpha decay feeding ^{235}U.
% spontaneous fission = 4.4×10^{-10}
major radiations:

type	energy	intensity (%)
$alpha_1$	5.105 MeV	11.50
$alpha_2$	5.143 MeV	15.10
$alpha_3$	5.155 MeV	73.30
X-ray	13.6 keV	4.4
average values for weak gammas	112.9 keV	0.05

* * *

alpha activity (α/min-μg)	$= 1.4 \times 10^5$
SF neutrons (n/s-g) [metal]	= 0.03
(α, n) reaction neutrons (n/s-g)	$[PuO_2] = 45$
	$[PuF_4] = 4300$

Plutonium-240 [Half-life = 6569 years]
main decay mode: alpha decay feeding ^{236}U.
% spontaneous fission = 4.95×10^{-6}
major radiations:

type	energy	intensity (%)
$alpha_1$	5.123 MeV	26.4
$alpha_2$	5.168 MeV	73.5
X-ray	13.6 keV	11.0
average values for weak gammas	54.3 keV	0.05

* * *

alpha activity (α/min-μg)	$= 5.0 \times 10^5$
SF neutrons (n/s-g) [metal]	= 1020
(α, n) reaction neutrons (n/s-g)	$[PuO_2] = 170$
	$[PuF_4] = 1.6 \times 10^4$

Plutonium-241 [Half-life = 14.4 years]
main decay mode: beta decay (99.998%) feeding ^{241}Am
alpha decay (0.00245%)
main radiations:

type	energy	intensity (%)
beta minus (max)	20.81 keV	—
beta minus (ave)	5.23 keV	99.9975
alpha	4.80–5.06 MeV	0.0025

* * *

alpha activity (α/min-μg)	= 5700
SF neutrons (n/s-g) [metal]	$\ll 1$
(α, n) reaction neutrons (n/s-g)	$[PuO_2] = 1.8$
	$[PuF_4] = 170$

Plutonium-242 [Half-life = 375,800 years]
main decay mode: alpha decay feeding ^{238}U.
% spontaneous fission = 5.5×10^{-4}
major radiations:

type	energy	intensity (%)
$alpha_1$	4.856 MeV	22.0
$alpha_2$	4.900 MeV	78.0
X-ray	13.6 keV	9.1
average values for weak gammas	56.5 keV	0.04

* * *

alpha activity (α/min-μg)	= 8700
SF neutrons (n/s-g) [metal]	= 1700
(α, n) reaction neutrons (n/s-g)	$[PuO_2] = 2.7$
	$[PuF_4] = 170$

Americium-241 [Half-life = 432.2 years]
main decay mode: alpha decay feeding ^{237}Np.
% spontaneous fission = 3.77×10^{-10}
major radiations:

	energy	intensity (%)
$alpha_1$	5.388 MeV	1.40
$alpha_2$	5.443 MeV	12.80
$alpha_3$	5.486 MeV	85.20
$alpha_4$	5.551 MeV	0.20
$alpha_5$	5.554 MeV	0.34
X-ray	13.9 keV	43.
$gamma_1$	26.345 keV	2.4
$gamma_2$	33.205 keV	0.106
$gamma_3$	59.537 keV	35.9
average values for weak gammas	69.2 keV	0.18

* * *

alpha activity (α/min-μg)	$= 7.6 \times 10^6$
SF neutrons (n/s-g) [metal]	= 1.6
(α, n) reaction neutrons (n/s-g)	$[PuO_2] = 2720$
	$[PuF_4] = 4.0 \times 10^5$

Shielding is used to reduce the level of secondary neutron and gamma radiations associated with an alpha-emitting radioisotope source. The radioisotope power supply can also be physically extended away

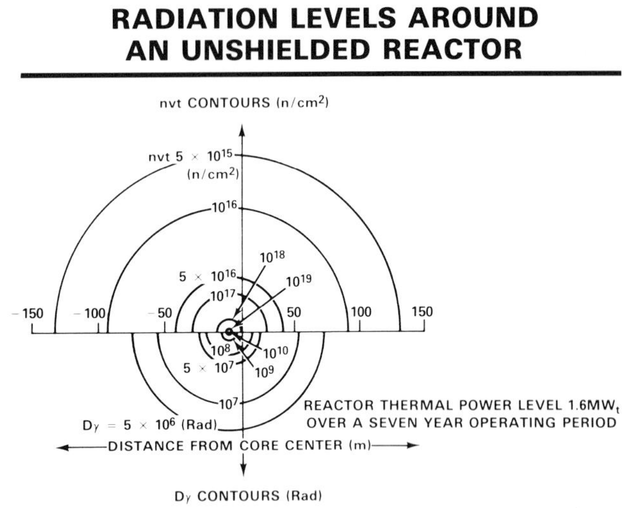

Fig. 7.7 Radiation levels around an unshielded reactor. *Courtesy of NASA.*

from the spacecraft using a boom structure. In this case, an engineering trade-off is needed to compare the advantages of a boomed deployment of the radioisotope source which takes advantage of the inverse square law in reducing radiation source intensity and the additional shielding mass needed in the vicinity of the radioisotope power supply to achieve the same reduction in radiation intensity.

Shielding Considerations for Space Nuclear Reactor Systems

Because they represent the most significant shielding problem in space nuclear power applications, we will focus the remainder of this chapter on reactor shielding issues. Figure 7.7 shows the total radiation doses (neutron and gamma) in the vicinity of an unshielded 1.6 megawatt-thermal space reactor that has operated for over seven years [16]. Because such reactors are operating in the vacuum of space, these radiation levels decrease as the square of the distance from the core. Some problems experienced in designing shields for space reactors are essentially independent of the application for which the reactor is intended, while other shielding issues are very much application-specific or mission-specific.

Some of the conditions encountered in a space mission that influence reactor shielding design include [15]

1. The absence of a surrounding gaseous medium or atmosphere which would scatter neutrons back toward the system
2. The need to reject waste thermal energy to space by radiation heat transfer
3. The fact that unmanned missions will usually have less restrictive radiation dose criteria than missions involving human crews
4. The shield mass-efficient design practice of locating the reactor and its payload at opposite ends of the spacecraft or space platform in some elongated geometric configuration

The vacuum of space makes it possible to employ a shadow shield and, consequently, leave a major portion of the reactor completely unshielded. For certain missions, usually unmanned, this represents a major savings in shield mass. A typical shadow shield configuration is shown in Fig. 7.8 [24]. The end of the reflector has been rounded to reduce both shield mass and the possibility of secondary scattering. Neutron attenuation is accomplished using lithium hydride (LiH) fabricated in the shape of a frustum, while a heavy metal (typically tungsten) gamma shield is added at the reactor end of the shield configuration.

By locating the reactor at one end of the spacecraft, invoking an elongated spacecraft design geometry, and placing the payload and sensitive spacecraft components in the dose plane or shadow of the shield, designers take advantage of the inverse square law for radiation attenuation as well as using the minimum shield mass needed to achieve a desired level of radiation protection [24].

Space reactor systems are often operated at fairly high temperatures to reduce the size and mass of thermal radiators (see Chapter 6). This design condition also can create an elevated (750–900 K) temperature environment for the radiation shield.

For a nuclear reactor used on a manned spacecraft or space station, a 4π shield is generally required to avoid exposure during docking and rendezvous operations and to prevent prohibitive constraints on crew activities in space around the facility. Figure 7.9 identifies typical

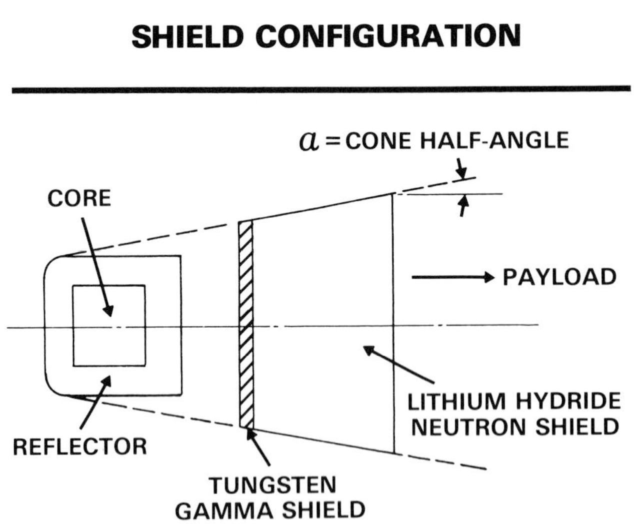

Fig. 7.8 Typical shadow shield configuration. *Courtesy of Los Alamos National Laboratory.*

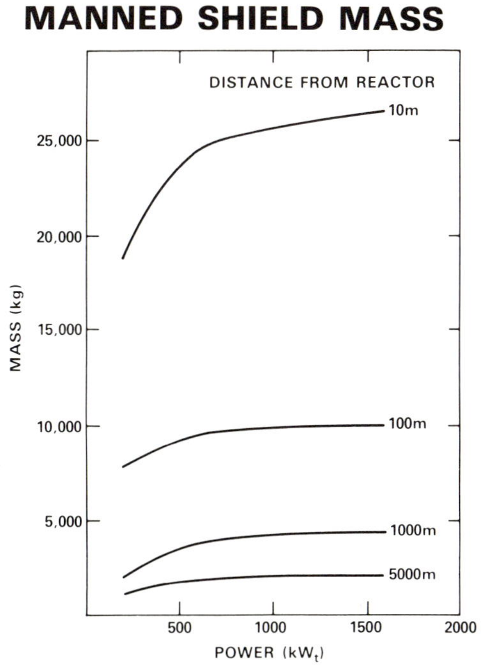

Fig. 7.9 Typical 4π shield mass needed for a manned space mission. *Courtesy of NASA.*

4π shield masses needed for a manned space system, as a function of separation distance from the reactor [16].

As shown in Fig. 7.10, a variety of design features, including booms and tethers, can be used to complement the 4π geometry shield mass in achieving acceptable radiation dose levels [16, 24].

A five-step iterative design procedure can be used in space reactor shield design activities [15, 24]. First, the designer must establish and quantify to the greatest extent possible the overall guidelines for the shield. In particular, the shield designer must know what portions of the spacecraft or platform are to be protected; what is the overall mission, including rendezvous and docking requirements if appropriate; and what vehicle configurations can actually be considered in light of launch vehicle and on-orbit assembly constraints.

Second, the shield designer must identify and establish acceptable radiation exposure levels for the payload, spacecraft, and if appropriate, crew. Third, the designer must quantify the primary and secondary sources of radiation that can contribute to the particular shielding problem. With shadow shields, particular care should be taken to ensure that no radiation scattering sources exist. A scattering source is an object that projects beyond the conical shadow shield and reflects or scatters radiation into the protected zone (dose plane) beyond the shadow shield. A rendezvousing spacecraft or a derelict space object can become an unintentional and undesirable scattering source. Neu-

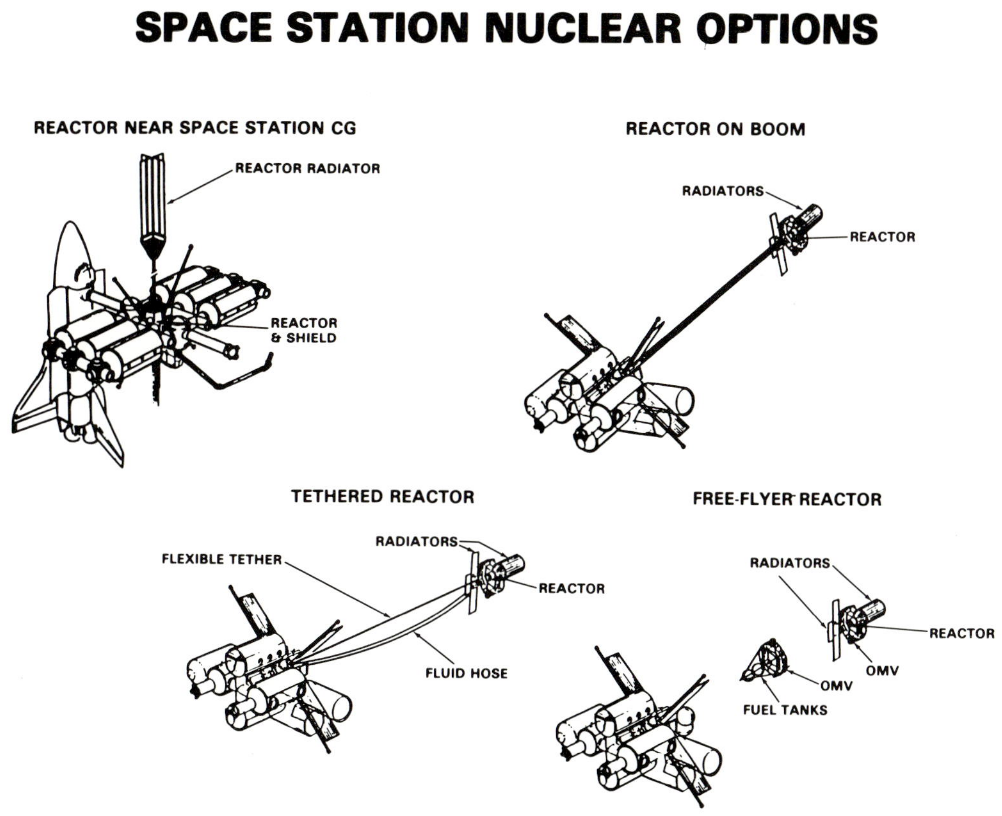

Fig. 7.10 Various reactor deployment options for a cluster-concept space station. *Courtesy of NASA.*

Table 7.18 Representative properties of lithium hydride (LiH).

Density (g/cm^3)	0.775
Molecular weight	5.95
Hydrogen content N_H (atoms H/cm^3)	5.85×10^{22}
Hydrogen content (weight %)	12.68
Melting point (K)	960
Volume change on melting (%)	$+25 \pm 2$
Crystal structure	face centered cube (fcc)
Heat of fusion (kJ/mole)	21.77 ± 1.3
Compressive strength, cold pressed (MPa)	96.5 – 165.5
Molar volume (cm^3/mole)	10.254

trons and gamma rays streaming through penetrations in shields also represent major shielding concerns that must be identified and quantified. These shield penetrations can include the primary coolant loop or core heat pipes, reactor control actuator drives, shield cooling pipes, and electrical cable ducts and conduits.

Fourth, the shield designer must select materials on the basis of their ability to attenuate the radiation sources and to withstand the rigors of the space environment. Minimum mass is another key design parameter that is invoked in selecting shield materials. Finally, the designer must perform detailed, often computer-based, shielding calculations in an effort to arrive at an optimum shield design. Of course, a thorough shield design effort involves numerous iterations of these steps, with each iteration characterized by many design trade-offs and compromises.

The primary function of the shield in a space reactor system is to reduce the neutron and gamma ray dose rates to levels acceptable for the payload, spacecraft components, and, if appropriate, the human crew. Shielding materials are often chosen on the basis of the following factors [15, 24]:

1. Ability to attenuate radiation
2. Minimum mass
3. Resistance to radiation-induced thermophysical damage
4. Stability at elevated operating temperatures, i.e., 700–900 K regime
5. Ease of fabrication
6. Availability
7. Cost

As discussed in Chapter 2, materials containing large amounts of hydrogen make the best neutron attenuating materials. One of the most important parameters for rating shielding materials for space applications is the number of hydrogen atoms (N_H) the material contains per unit volume. Water, with an N_H of approximately 6.7×10^{22} atoms of hydrogen/cm^3, is commonly used to moderate and shield low temperature terrestrial reactors, but it is not suitable as a shield above its critical point. Because of their high temperature stabilities and hydrogen content values, metal hydrides are the prime candidates for neutron shielding applications in space [24]. Of these materials, lithium hydride (LiH) is the favored choice because of its high hydrogen density ($N_H = 5.9 \times 10^{22}$ hydrogen atoms/cm^3), low mass density (0.775 g/cm^3), and moderately high melting point of ~960 K. Lithium hydride also produces a minimum amount of secondary radiation. Some LiH thermophysical properties are presented in Table 7.18, while Fig. 7.11 depicts typical LiH neutron attenuation. The thermal environment of a space reactor shield is a critical design consideration. For example, when LiH melts, it expands some 25 percent. Consequently, the shield must be kept below the LiH melting point temperature to avoid undesirable thermomechanical stresses.

Two principal mechanisms account for the heating of a lithium hydride shield: primary coolant loops or core heat pipes passing through the shield; nuclear radiation-induced heating by neutrons and gamma rays. Thermal insulation placed around those portions of the heat transport system that pass through the shield can prevent localized melting within the shield. Tight containment of the lithium hydride represents a shield design complication in high neutron flux regions, if natural lithium is used as the shield material. In natural lithium, approximately 7.5 percent of the lithium has a

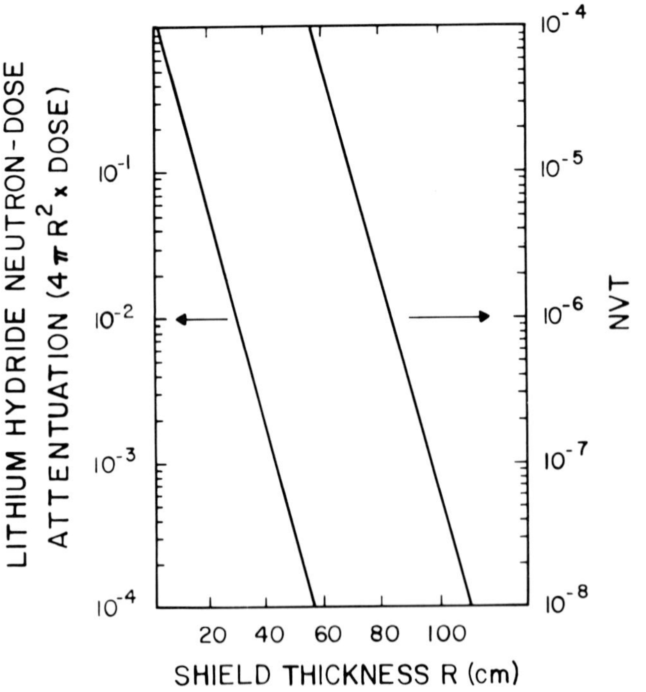

Fig. 7.11 Typical lithium hydride neutron attenuation. *Courtesy of Los Alamos National Laboratory.*

Table 7.19 Thermophysical properties of candidate gamma ray shielding materials.

Property	Uranium-8 wt% Molybdenum (U-8Mo)	Tantalum-10 wt% Tungsten (Ta-10W)
Density (g/cm^3) (at room temperature)	17.40	16.83
Thermal expansion (10^{-6} cm/cm-K)	14.4 at 673 K	6.3 at 373 K 6.9 at 773 K
Thermal conductivity (J/s-cm-K)	1.34 at 298 K 2.89 at 773 K	0.762 at 298 K 0.695 at 773 K
Specific heat (kJ/kg-K)	0.138 at 298 K 0.176 at 773 K	0.150 at 298 K 0.147 at 773 K
Melting point (K)	1403	—

very high thermal neutron cross section—that is, 945 barns at 2200 m/s for the ^{6}Li $(n, \alpha)^3$H reaction [15, 24].

$$^6_3\text{Li} + ^1_0\text{n} \rightarrow ^3_1\text{H} + ^4_2\text{He}. \qquad (7.17)$$

The helium generated by this reaction would build up significant pressures within the LiH shield containers during the lifetime of the space power system. Consequently, to avoid this problem, LiH isotopically enriched in $^7_3$Li, generally 99.99 weight percent $^7_3$Li, can be used in shield regions where the thermal neutron fluence is expected to be high [24].

Lithium hydride must be loaded, processed, and when at room temperature, maintained in a very dry atmosphere to prevent reaction with water or water vapor. At elevated temperature, a hydrogen overpressure must be maintained to prevent hydrogen loss [24].

Low power reactor systems, i.e., less than 100 kilowatts-thermal, generally require only a neutron shield for unmanned space missions. For higher power levels or for missions requiring more restricted radiation dose levels, e.g., less than 10^7 rad (10^5Sv) gamma radiation and less than 10^{13} nvt (neutrons/cm^2), a gamma ray shield will usually also be required. High density materials are used for such gamma ray shields to minimize overall shield volume and system mass. Table 7.19 lists properties for some of the high density, high melting point materials considered as suitable or potential gamma ray shielding materials for space power systems [15, 24].

Summary

Shield design for a space reactor depends to a large extent on whether the mission will be manned or unmanned. Design alternatives are also driven by the desire to minimize mass, using wherever appropriate a shadow shield configuration. Since the reactor system is generally smaller in size than the payload/spacecraft, the shield mass becomes a direct function of the solid angle subtended by the shadow shield. Consequently, minimum mass shadow shields require that the space reactor itself possess a minimum projected area and that all structural components of the spacecraft be contained within the conical region established by the dose plane and the extremities of the radiation source. Of course, in shadow shield geometries there will also be an engineering trade-off between boom/structure mass and additional shield mass needed to achieve the same dose at the payload.

Shield design for reactors used in manned missions is driven by much lower permissible radiation dose levels and the reactor shield must include both neutron and gamma ray attenuation components. In general, this type shield will be a laminated one with layers of heavy metal gamma ray shielding followed by tightly packaged containers of neutron attenuating lithium hydride. Secondary gamma rays from neutron radiative capture reactions can also make a significant contribution to the radiation level appearing in the dose plane and must be compensated for.

Optimum shield design for a space reactor depends on many factors, some mission-unique and others essentially mission-independent. These factors and considerations must be carefully blended and evaluated in creating a shield design that satisfies radiation protection requirements at minimum mass.

References

1. International Commission on Radiological Protection (ICRP), Publication 26, "Recommendations of the International Commission on Radiological Protection," Pergamon Press, Oxford, 1977.
2. John W. Healy, *Los Alamos Handbook of Radiation Monitoring*, Los Alamos National Laboratory, New Mexico, 1970.
3. Jacob Shapiro, *Radiation Protection* (2nd Edition), Harvard University Press, Cambridge, Mass., 1981.
4. U.S. Environmental Protection Agency, "Proposed Federal Radiation Protection Guidance for Occupational Exposure," (Background Report) EPA 520/4-81-003, 16 January 1981.
5. International Atomic Energy Agency, *Radiation Protection Procedures*, Safety Series No. 38, 1973.
6. Health, Safety, and Environmental Manual, "Biological Effects of Ionizing Radiation," Technical Bulletin 503, Los Alamos National Laboratory Manual LR-7940-M, October 1979.
7. International Commission on Radiation Units and Measurements, "Radiation Quantities and Units," *ICRU Report 19*, Washington, D.C., 1971.
8. International Commission on Radiation Units and Measurements, "Dose Equivalent," *Supplement to ICRU Report 19*, Washington, D.C., 1973.
9. W. E. Kisieleski and R. Baserga, "Radiation and Life Processes," U.S. Department of Energy (formerly Atomic Energy Commission), 1967.
10. I. Asimov and T. Dobzhansky, "The Genetic Effects of Radiation," U.S. Department of Energy (formerly Atomic Energy Commission), 1966.
11. *Standards for Protection Against Radiation*, Title 10, Code of Federal Regulations, Part 20, Washington, D.C. (continuously updated).
12. Los Alamos National Laboratory, Health, Safety and Environmental Manual, *Radiation Protection Standards*, Technical Bulletin 502, LANL Manual LA-7940-M, 1979.
13. *Radiation Protection Guides and Constraints for Space Mission and Vehicle Design-Studies Involving Nuclear Systems*, Radiological Advisory Panel, Committee on Space Medicine, National Academy of Sciences, Washington, D.C., 1970.
14. A. E. Nicogossian and J. F. Parker, Jr., *Space Physiology and Medicine*, NASA, SP-447, 1982.
15. R. V. Anderson et al., "Space Reactor Electric Systems," ESG-DOE-13398, Rockwell International, 29 March 1983.
16. J. Angelo, Jr. and D. Buden, "Shielding Considerations for Advanced Space Nuclear Reactor Systems," 1982 IEEE Conference on Nuclear and Space Radiation Effects, Las Vegas, Nev., 20–21 July 1982.
17. Atomic Industrial Forum, *Guidebook for the Application of Space Nuclear Power Systems*, January 1969.
18. S. Glasstone and A. Sesonske, *Nuclear Reactor Engineering* (3rd Edition), Van Nostrand Co., New York, 1981.
19. H. Goldstein, *Fundamental Aspects of Reactor Shielding*, Addison-Wesley, Reading, Mass., 1959.
20. Argonne National Laboratory, *Reactor Physics Constants* (2nd Edition), ANL-5800, 1963.
21. J. R. Lamarsh, *Introduction to Nuclear Engineering* (2nd Edition), Addison-Wesley Publishing Co., Inc, Reading, Mass., 1983.
22. Los Alamos National Laboratory, Health, Safety and Environmental Manual, *Plutonium*, Technical Bulletin 505, LANL Manual LA-7940-M, 1979.
23. D. C. Kocher, *Radioactive Decay Data Tables*, DOE/TIC-11026, U.S. Department of Energy, Technical Information Center, 1981.
24. D. Buden et al., "Selection of Power Plant Elements for Future Reactor Space Electric Power Systems," LA-7858, Los Alamos National Laboratory, September 1979.

Chapter 8

Radioisotope Space Power Generators

Introduction

Radioisotope power systems have been used by the United States since 1961 to support its space program. Starting with Transit, a navigation satellite, radioisotopes have powered weather and communication satellites, scientific stations on the Moon, robot explorer spacecraft on Mars, and highly sophisticated deep space interplanetary missions to Jupiter, Saturn, and beyond. Table 8.1 presents a detailed summary of the radioisotope generators developed by the United States for space power applications. These radioisotope systems have demonstrated the principles of safe and reliable operation, even though several missions were aborted by launch vehicle failures. In fact, radioisotope thermoelectric generators (RTGs) have proven in actual missions their long life, high reliability, and ability to operate both in hostile environments and independent of the solar flux. This outstanding performance record, summarized in Table 8.2, includes operations in such diverse and severe environments as the Van Allen radiation belts, the lunar surface with its temperature extremes, Martian dust storms, the asteroid belts, and in the vicinity of the frigid outer planets [1].

Figure 8.1 illustrates the steady increase in RTG power level. In 1961, SNAP-3B (the first RTG used in space) had a mass of 2.1 kg and delivered 2.7 watts of electric power to the Transit navigation satellite. The Voyager 1 and 2 missions to Jupiter, Saturn, and beyond, on the other hand, had power supplies that delivered 150 watts-electric per generator with a mass of only 37.6 kg. Although other radioisotopes have been considered, all RTGs flown by the U.S. to date have used the isotope plutonium-238. All these missions have also employed thermoelectric (TE) power conversion techniques with telluride materials in the Nimbus, Pioneer, Viking, Transit, and Apollo RTGs; and silicon germanium (SiGe) converters in the Lincoln Experimental Satellites (LES) and Voyager generators. Table 8.3 summarizes the major performance characteristics of various RTG systems [1–7].

By 1963, the power level for the Transit navigation satellites had increased to 25 watts-electric in the SNAP-9A generator units. Then, the first of a planned new generation of Transit spacecraft was launched in 1972, using a 36 watt-electric RTG. Waste heat from this radioisotope power supply was used to maintain spacecraft instrumentation at a temperature near 293 K. The SNAP-19 system supplied power to the Nimbus III weather satellite launched on 14 April 1969. The production of images of the Earth's cloud cover and expanded investigation of the atmosphere over a wide range of the electromagnetic spectrum depended on electric power from two 56 watt generators.

The Apollo Moon landings not only demonstrated

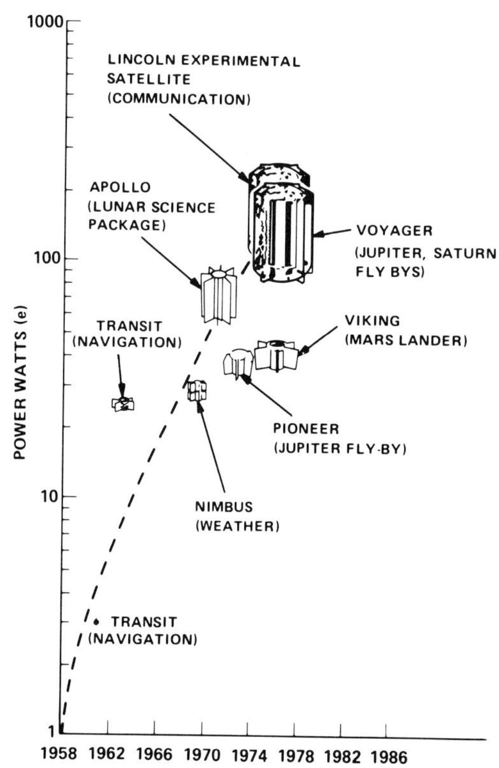

Fig. 8.1 Progress in RTG development. *Courtesy of U.S. Department of Energy.*

Table 8.1 Radioisotope generators developed for space electric power by the United States.

Desig-nation	Application	Electric Power (W)	Weight (kg)	Design Life	Fuel	Status
SNAP-1	Satellite (Air Force)	500	—	60 days	144_{Ce}	Replaced by longer-lived SNAP-1A in 1959
SNAP-1A	Satellite (Air Force)	125	91	1 year	144_{Ce}	Program canceled in 1959
SNAP-3	Thermoelectric demon-stration device (AEC)	2.5	2	90 days	*210_{Po}	Shown to President Eisenhower in 1960
SNAP-3B	Navigation satellite (Navy)(Transit 4A and 4B)	2.7	2	> 1 year	*238_{Pu}	Units *launched* in June and Nov. 1961; June unit oper-ational in 1976; Nov. unit operated to 1971
SNAP-9A	Navigation satellite (Navy/Transit 5BN)	25	12	> 1 year	*238_{Pu}	Units launched in Sept. and Dec. 1963 and operated some 20,000 hr before sharp de-crease in power; third unit aborted during launch in April 1964
SNAP-11	Surveyor lunar lander (NASA)	25	14	90 days	*242_{Cm}	Surveyor requirement canceled 1965; electrically heated unit delivered to NASA and life tested over 5 yrs; fueled demon-stration at ORNL in July 1966
SNAP-13	Thermionic demonstra-tion device (AEC)	12.5	2	90 days	*242_{Cm}	Fueled demonstration at ORNL in Nov. 1965; program com-pleted in 1966
SNAP-17A and 17B	Communication satel-lite (Air Force)	30	14	> 1 year	90_{Sr}	Design and component test phase completed Nov. 1965
SNAP-19A	Imp satellite (NASA)	20	8	> 1 year	238_{Pu}	Design and component tests com-pleted in 1963; not used on Imp because of radiation interfer-ence with payload instrumentation
	Various satellites (AEC)	250	—	> 1 year	90_{Sr}	Six different design studies completed in 1964
	Surveyor lunar roving vehicle (NASA)	40	10	1 year	238_{Pu}	Design and integration study completed in 1964
	Extended Apollo missions (AEC)	1500	—	30–90 days	210_{Po}	Design and feasibility studies completed in 1964
SNAP-19B1 2, 3	Meteorological satellite (NASA) (Nimbus III)	30	14	> 1 year	*238_{Pu}	First Nimbus launch aborted in 1968 and fuel recovered from off shore waters; replacement unit launched in April 1969, opera-ting continuously at gradually decreasing power level
SNAP-25	Various satellites (AEC)	75	16	> 1 year	238_{Pu}	Program canceled
SNAP-27	Apollo Lunar Surface Experiment Packages (NASA)	63.5	31 (excludes 11 kg cask)	1 year	*238_{Pu}	First unit placed on lunar surface by Apollo 12 astro-naut in Nov. 1969 and continu-ously powered ALSEP well be-yond design life; second unit landed in deep Pacific Ocean on Apollo 13 abort. Additional units used on Apollos 14 to 17
SNAP-29	Various satellite and lunar missions (DOD and NASA)	400–500	180–225	90 days	210_{Po}	Program canceled in 1969 with completion of component tests
	Various missions (DOD and NASA)	250	—	5 years	238_{Pu} 90_{Sr}	Three design studies completed in 1967; controlled intact re-entry and ground handling included

Table 8.1 Radioisotope generators developed for space electric power by the United States (continued).

Designation	Application	Electric Power (W)	Weight (kg)	Design Life	Fuel	Status
	Thermionic module development (AEC)	100	9	—	^{244}Cm	Program suspended after completion of preliminary design in 1967 and component tests in 1970
	Isotope Brayton ground test (AEC and NASA)	3000–15,000	—	> 1 year	^{238}Pu	Preliminary heat-source designs completed in 1966; fuel capsule development 1967–1970; NASA completed 2,500 hr life test on electrically heated system in 1970.
Transit RTG	Navigation satellite (Navy) improved Transit	30	14	5 years	*^{238}Pu	Launched in 1972
SNAP-19	Pioneer F and G (Jupiter flyby) (NASA)	30	14	3 years	*^{238}Pu	Launched in 1973; operated over 10 yrs
SNAP-19	Viking Mars lander (NASA)	35	14	> 2 years	*^{238}Pu	Launched in Aug. and Sept. 1975; operated well beyond design life
Multi-Hundred-Watt RTG	Various missions (DOD and NASA)	100–200	—	5–10 years	*^{238}Pu	Used in LES 8/9 in 1976 and Voyager 1 and 2 in 1977. Voyager still operating 7 yrs later beyond Saturn
Dynamic Radioisotope Power Systems (DIPS)	Various missions (DOD and NASA)	500–2000	215	7 years	^{238}Pu	Program 1975 to 1980 demonstrated 11,000 hr Rankine cycle including 2,000 hr endurance test
GPHS-RTG	Various missions (DOD and NASA)	290	54	40,000 hr	*^{238}Pu	Under development for Galileo mission

* Units fueled

the utility of nuclear power supplies in extraterrestrial exploration but also clearly indicated that such systems could be safely and effectively integrated into manned missions. For example, two 15 watt-thermal, plutonium-238 fueled heater units provided thermal energy for the early Apollo scientific experiment package (EASEP) which was left on the lunar surface by the Apollo 11 astronauts. These heaters successfully kept the scientific instruments warm during the frigid lunar night (approximately 14 Earth days in duration). Subsequent Apollo missions employed the SNAP-27 system to generate some 63.5 watts of electric power in support of the Apollo lunar scientific experiment package (ALSEP). These radioisotope units continued to function successfully far beyond their one year design life and returned valuable scientific data long after the last Apollo astronaut had walked on the lunar surface.

Nuclear powered Pioneer spacecraft took the first close-up imagery and performed valuable scientific measurements of the giant outer planets, Jupiter and Saturn. In these truly pathfinder missions of exploration, the Pioneer spacecraft carried four modified SNAP-19 units, each generating over 40 watts-electric. The three year mission to Jupiter included successful passage through the asteroid belt and operation in the highly intensive radiation belts surrounding Jupiter. In 1983, the Pioneer 10 spacecraft still functioning with its nuclear power supply became the first manmade object to leave the Solar System.

The Red Planet, Mars, has been the object of human curiosity since prehistoric times. In 1975, two Viking Mars Landers carrying SNAP-19 systems were launched as part of another historic planetary exploration mission. These robot landers used nuclear power to support the search for extraterrestrial life and to provide data about the Martian surface and atmosphere for many years after their planned two-year design life.

A new generation of radioisotope power systems, called multihundred watt (MHW), was initially developed for the experimental communication satellites, LES 8 and LES 9. These radioisotope units were also used on the spectacular Voyager 1 and 2 flyby missions to Jupiter, Saturn, and beyond. When the LES 8 and 9 spacecraft were launched in 1976, each carried two MHW units, providing some 300 watts of electric power. The Voyager 1 and 2 spacecraft, launched in 1977, each contained three MHW units. Plans are now underway to use even more advanced radioisotope sys-

Table 8.2 Radioisotope power systems launched by the United States (1961–1983) [1].

Power Source	Spacecraft	Mission Type	Launch Date	Status
SNAP-3B	Transit 4A	Navigational	29 June 1961	Successfully achieved orbit
SNAP-3B	Transit 4B	Navigational	15 Nov. 1961	Successfully achieved orbit
SNAP-9A	Transit-5BN-1	Navigational	28 Sept. 1963	Successfully achieved orbit
SNAP-9A	Transit-5BN-2	Navigational	5 Dec. 1963	Successfully achieved orbit
SNAP-9A	Transit-5BN-3	Navigational	21 April 1964	Mission aborted: burned up on reentry
SNAP-19B2	Nimbus-B-1	Meteorological	18 May 1968	Mission aborted: heat source retrieved
SNAP-19B3	Nimbus III	Meteorological	14 April 1969	Successfully achieved orbit
SNAP-27	Apollo 12	Lunar	14 Nov. 1969	Successfully placed on lunar surface
SNAP 27	Apollo 13	Lunar	11 April 1970	Mission aborted on way to moon. Heat source returned to South Pacific Ocean.
SNAP-27	Apollo 14	Lunar	31 Jan. 1971	Successfully placed on lunar surface
SNAP-27	Apollo 15	Lunar	26 July 1971	Successfully placed on lunar surface
SNAP-19	Pioneer 10	Planetary	2 March 1972	Successfully operated to Jupiter and beyond; first manmade object to leave Solar System
SNAP-27	Apollo 16	Lunar	16 April 1972	Successfully placed on lunar surface
Transit-RTG	"Transit" (TRIAD-01-1X)	Navigational	2 Sept. 1972	Successfully achieved orbit
SNAP-27	Apollo 17	Lunar	7 Dec. 1972	Successfully placed on lunar surface
SNAP-19	Pioneer 11	Planetary	5 April 1973	Successfully operated to Jupiter and Saturn and beyond
SNAP-19	Viking 1	Mars	20 Aug. 1975	Successfully landed on Mars
SNAP-19	Viking 2	Mars	9 Sept. 1975	Successfully landed on Mars
MHW	LES/ 8/9	Communications	14 March 1976	Successfully achieved orbit
MHW	Voyager 2	Planetary	20 Aug. 1977	Successfully operated to Jupiter and Saturn and beyond
MHW	Voyager 1	Planetary	5 Sept. 1977	Successfully operated to Jupiter and Saturn and beyond

tems to power an intensified exploration of Jupiter and its moons in a project called Galileo.

Radioisotope Generators

Radioisotope generators are composed of a nuclear heat source and appropriate power conversion equipment. Nuclear heat source designs have been dominated by aerospace nuclear safety philosophy, which has undergone revision since the start of the space nuclear power program. For example, the SNAP-3B and -9A systems were designed for nuclear fuel burnup and high altitude dispersal in the event of an atmospheric reentry of the nuclear-powered spacecraft; the SNAP-19 system was designed for fuel containment in event of reentry; and the SNAP-27 for containment during reentry and after impact. A key design feature, past and present, is the immobilization of the plutonium-238 fuel during all phases of the mission, both normal and potentially abnormal, including launch abort, reentry into the Earth's atmosphere, and post-reentry (possibly random) impact.

Super alloys (Haynes-25) were used in the early "burnup and disperse" designs, because of the low operating temperatures of the RTG systems, necessitated by the choice of lead telluride (PbTe) thermoelectric converters. However, with the advent of higher operating temperature SiGe TE converters and the requirement for intact reentry designs for the RTG systems, the super alloys were replaced by higher melting point noble metals and refractory alloys. Present technology includes iridium alloys that can withstand plutonium oxide (PuO_2) fuel operating temperatures up to 1675 K.

Nuclear fuel capsules have been designed in various

shapes, such as right circular cylinders, hemispherical domed cylinders, and spheres. The fuel capsule configuration is established by the overall nuclear heat source design approach, the nuclear fuel form, the converter design, and the properties of the structural materials. Coatings have been used effectively to retard adverse material interaction by diffusion and to control emissivities, thereby reducing temperature drops among the various heat source components. However, with systems involving plutonium fuel and iridium alloys, such protective coatings are not generally required.

A variety of nuclear fuel selection factors should be considered in designing a radioisotope generator. These include

1. The power density (in W/cm^3) of the radioisotope fuel
2. Its half-life
3. The availability of the fuel and its cost
4. Any external nuclear radiations associated with the decay of the radioisotope (e.g., neutrons and gamma rays)
5. Contemporary developments in materials science and fuel form technology that directly influence any potential radiological and biological hazard of accidentally released fuel
6. Potential radiological consequences of the particular fuel should it inadvertently enter the terrestrial biosphere

Because of their relatively low shielding requirements and inherently high power densities, alpha-emitting radioisotopes are generally considered the most attractive nuclear fuels for space applications. The main unattractive feature associated with the use of alpha-emitters is helium gas generation which has the potential for creating undesirable high pressure conditions within the fuel capsule—a condition frequently alleviated by venting. In all radioisotope power generators flown in space to date by the United States, the alpha-emitting isotope, plutonium-238, has been the fuel of choice. Though costly, it is readily available and has a long half-life (87.7 years) and a favorable power density.

As mentioned previously, metallic fuel capsules that contain long-lived alpha emitters, such as plutonium-238, must contain a vent for the helium gas generated by the radioactive decay of the nuclear fuel—unless, perhaps, adequate space is provided within the system for such venting to occur, as was incorporated in the SNAP-27 design. These vents can be either selective or nonselective. Selective vents pass helium, but retain any larger gas molecules and solid particulates from the fuel. Nonselective vents also retain solid particulates from the fuel, but pass helium and other gaseous effluents, including uncondensed fuel, impurity vapors and possibly other fuel decay products such as radon. Several types of nonselective vents have been flight-qualified and flown, as for example with the SNAP-19 and MHW systems.

Various types of protective layers can be used around the radioisotope fuel capsule(s). For example, an impact shell can be employed to absorb the kinetic energy of explosion fragments or post-reentry abort impact on Earth. A wound-graphite impact shell is presently used in the MHW heat source. Graphite composites are also being investigated to determine whether they can provide significant impact protection in addition to ablation protection during possible launch pad fires, flight aborts, or atmospheric reentry. The graphite ablators used to date have been generally fine-grained ones that offer the best reentry environment ablation resistance,

Table 8.3 Radioisotope generator characteristics [*3, 4, 5, 6, 7*].

	SNAP-3B	SNAP-9A	SNAP-19	SNAP-27	Transit-RTG	MHW	GPHS-RTG
Mission	Transit	Transit	Nimbus Pioneer Viking	Apollo	Transit	LES 8/9 Voyager	Galileo
Fuel form	Pu metal	Pu metal	PuO_2-Mo cermet	PuO_2 microspheres	PuO_2-Mo cermet	Pressed PuO_2	Pressed PuO_2
Thermoelectric material	PbTe	PbTe	PbTe-TAGS	PbSnTe	PbTe	SiGe	SiGe
BOL[1] output power (W_e)	2.7	26.8	28–43	63.5	36.8	150	290
Mass (kg)	2.1	12.2	13.6	30.8[2]	13.5	38.5	54.4
Specific power (W_e/kg)	1.3	2.2	2.1–3.0	3.2[3]	2.6	4.2	5.2
Conversion efficiency (%)	5.1	5.1	4.5–6.2	5.0	4.2	6.6	6.6
BOL fuel inventory (W_{th})	52	565	645	1480	850	2400	~4400
Fuel quantity (curies)	1800	17,000	34,400 – 80,000	44,500	25,500	7.7×10^4	1.3×10^5

[1] BOL = beginning of life
[2] without cask
[3] includes 11.1-kg cask

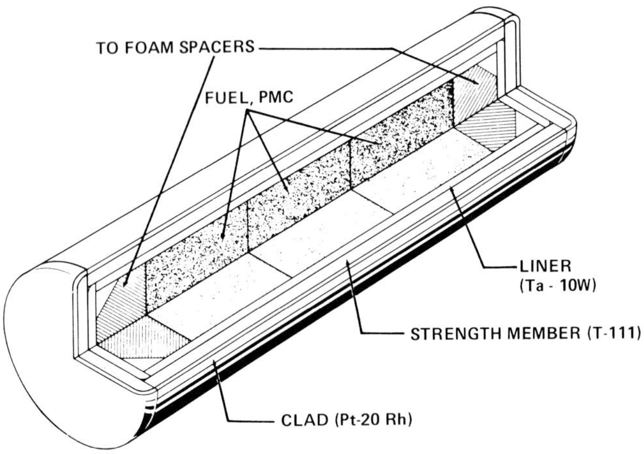

Fig. 8.2 Pioneer radioisotope heater unit. *Courtesy of Mound Laboratory, Monsanto Research Corporation.*

but possess only minimal strength with respect to thermal stresses of impact loads. When the ablator cannot provide enough thermal transport during a postulated reentry heating environment, a thermal insulator, such as pyrolytic graphite, is used to protect the fuel capsule. Such insulating materials are put around the fuel capsule to support aerospace nuclear safety requirements, even though their use also creates large, undesirable temperature gradients in the heat source during normal operations. Frequently, cladding is also placed around the hot graphite components to protect them from physical damage or from oxidation that might occur during prelaunch handling.

Several different approaches have been used for mechanically assembling the arrays of thermocouples between the heat source and the heat sink. Most telluride TE systems have been assembled with springs. This approach provides good thermal and electrical contacts at the junctions and mechanical forces for withstanding operational or launch payloads. It also accommodates thermal expansion changes that might occur during thermal cycling of the converter system. Two designs have been created without springs. These are the lightweight, unsealed, bonded panels (launched on the Transit spacecraft) and the hermetically sealed, close-packed tubular module. In SiGe converter systems, such as the MHW unit, the thermocouples are cantilevered from the cold side, and the heat input to silicon molybdenum hot shoes is accomplished by radiative transfer from the heat source.

Thermal insulation reduces heat losses, e.g., from the ends of the generator or between the thermocouples. The choice of insulation materials depends on the operating temperature, the geometry of the system, and the chemical compatibility of its various components. Good thermal insulation can ensure that 85 to 95 percent of the thermal energy from the heat source passes through the power conversion thermocouples.

The generator housing provides the structure for supporting the internal parts, attaching the heat rejection system (radiator), and mating with the spacecraft. This housing is also a major design consideration for maintaining the inert gas or vacuum operating environment within the generator. Generator housing penetrations must be made for the power leads and for any internal monitors. Gas management within the RTG system involves the heat source decay helium gas (and possibly other gases and vapors) and outgassing contaminants from the insulation, etc.

The heat rejection systems for previous, flight-qualified generators have all been static conduction and radiation heat transfer fin designs that were attached to the generator housing. The fins and the generator housing have been covered with coating materials that have a high emissivity-to-absorptivity ratio at normal operating temperatures. Radiator temperatures were selected to optimize the system size, power, and mass under the heat sink conditions anticipated during the mission.

Dynamic power conversion units can also be coupled with radioisotope heat sources to extend power generation range beyond that practically obtainable with TE converter systems and to achieve lower unit costs through higher power conversion efficiencies. However, such dynamic power conversion systems have moving parts that require long-lived bearings for years of maintenance-free operation in space. The most reliable

Table 8.4 Characteristics of Pioneer radioisotopic heater unit.

	Pioneer Radioisotopic Heater Unit
Fuel Data	
Fuel Form	PMC (Plutonia Molybdenum Cermet)
Fuel Required	
Weight Pu-238/ Source (Grams)	1.8
Weight Fuel/ Source (Grams)	3.02
Heat Sources	
Number Fabricated	50
Watts (th)/Source	1
Operating Temperature (K)	310
Materials of Construction Liner	Ta-10W
Strength Member	T-111
Clad	Pt-20 Rh
Geometry (cm)	Cylindrical (2.9 × 1.0)
Weight/Source (Grams)	23

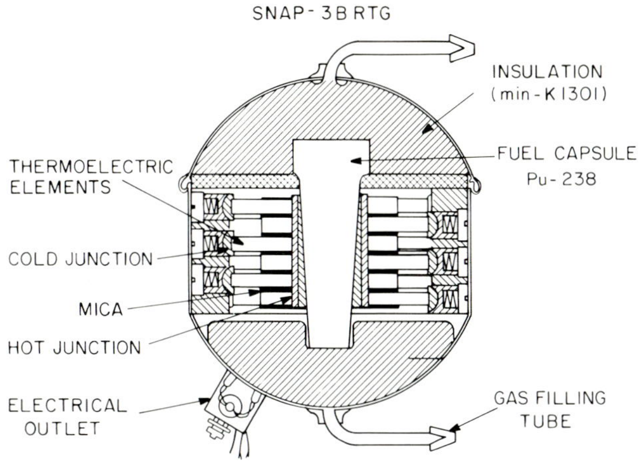

Fig. 8.3 SNAP-3B generator. *Courtesy of Mound Laboratory, Monsanto Research Corporation.*

bearings are those that support the moving parts on a film of the working fluid during operation. In addition, dynamic systems have inherent rotational torques and vibrational forces that must be considered in spacecraft integration. Dynamic systems can start up on the ground or in orbit. In the event that the working fluid is lost, an emergency cooling system must be provided to prevent the radioisotope heat source from reaching excessively high temperatures. This emergency cooling system can often be combined in the design of the dynamic power system with the auxiliary cooling systems needed for ground handling and launch operations. Multiple radioisotope heat sources can also be coupled to a common power conversion system or vice versa.

Radioisotope heat source technology developed for electric power generation is also applicable to spacecraft thermal control requirements. Two types of "heat-only" applications involve: (1) the thermal control of isolated spacecraft components by small (typically one watt-thermal) distributed nuclear sources; and (2) providing thermal power directly to cryogenic cooler units for active payload temperature control (see Fig. 8.2 and Table 8.4). For isolated component thermal control, the sources used are small, operate at low temperatures, and pose few engineering problems. On the other hand, the radioisotope heat sources used to power cryogenic coolers are relatively large (typically kilowatt-thermal class) and operate at temperatures comparable to PbTe RTGs. These units require the same kind of handling precautions, emergency cooling, auxiliary cooling, insulation, etc., as found in large radioisotope electric power generation systems. Even with these constraints, however, the overall size and weight of such heat sources are extremely favorable when compared to other spacecraft thermal control design options.

Because all nuclear power systems first generate heat that is then partially converted into electricity, there are substantial advantages in integrating the power system and the thermal control system in the overall design of the spacecraft. For example, if the power system radiator can be used as the external structure of the spacecraft, mass and volume will be saved. If the heat rejection system is properly designed, external sources of radiant thermal energy (such as the Sun or a nearby planet) will have little effect on the spacecraft's temperature and thermal cycles will be very modest over the entire mission profile. The thermal design of a nuclear-powered spacecraft can also be relatively independent of its orbital altitude or orientation.

SNAP-3B

The first use of nuclear power in space occurred in 1961 with the successful launches of the Transit 4A (June) and Transit 4B (November) navigational satellites. The SNAP-3B RTGs on board these spacecraft supplemented solar cell arrays and operationally demonstrated the use of nuclear systems for space power applications. The spacecraft were placed into 1,100 km orbits around the Earth, this altitude providing greater than 1,000 year orbital lifetimes. As part of the aerospace nuclear safety philosophy prevalent during their development, these generators were designed for burnup and high altitude fuel dispersal (to background radioactivity levels below those resulting from atmospheric nuclear weapons testing) in the event that a mission abort resulted in atmospheric reentry of the spacecraft and its nuclear power supply.

The SNAP-3B heat source (see Fig. 8.3) contained 92.7 grams of plutonium-238 in the form of plutonium metal. When used as a radioisotope heat source, plutonium metal contains approximately the following isotopic mixture: ^{238}Pu (80 wt %), ^{239}Pu (16 wt %), ^{240}Pu (3 wt %), and ^{241}Pu (1 wt %). The total heat source mass was 359 grams, generated 52.5 watts-thermal, and had an operating temperature of 810 K [8]. Tantalum was used as the liner material, complemented by a strength

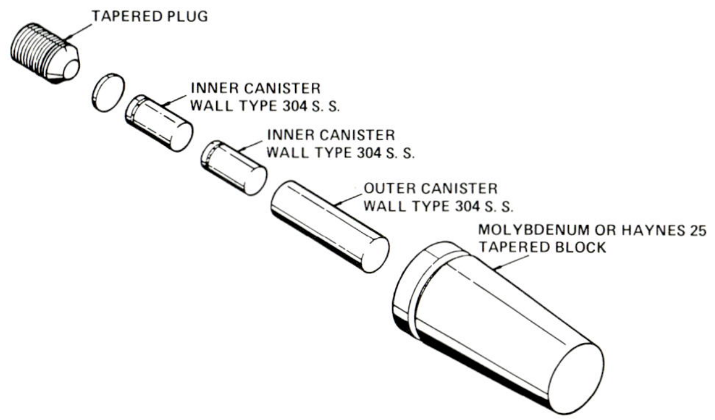

Fig. 8.4 SNAP-3B fuel capsule. *Courtesy of Mound Laboratory, Monsanto Research Corporation.*

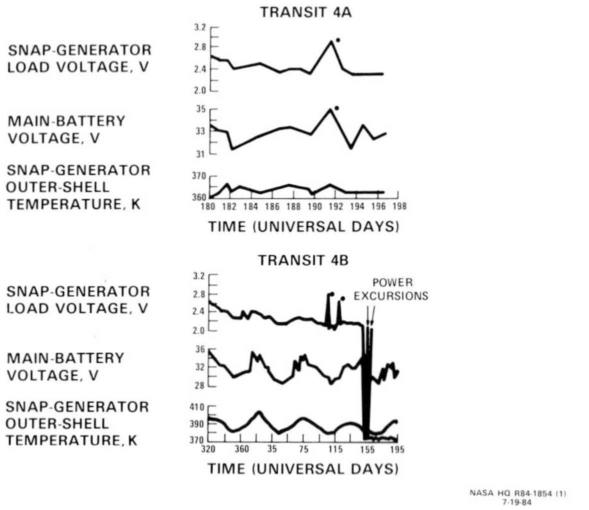

Fig. 8.5 SNAP-3B generator performance data as telemetered from the Transit 4A and Transit 4B spacecraft [7].

member consisting of a Haynes Alloy No. 25 tapered cylinder (see Fig. 8.4). The SNAP-3B generator used lead telluride (PbTe) TE converter materials. The system had an overall conversion efficiency of 5 percent. Thermal contact between the nuclear heat source and the TE converter was accomplished by means of springs. Fig. 8.5 presents SNAP-3B generator performance data as telemetered to Earth from the Transit 4A and 4B spacecraft [7]. The Transit 4A generator was still operational in 1976, 15 years after its launch, and the last reported signal from Transit 4B was in April 1971 [7].

SNAP-9A

The SNAP-9A system was an expanded version of the SNAP-3B and produced 25 watts-electric (see Fig. 8.6). The generator's nuclear heat source consisted of six plutonium-238 fueled capsules arranged symmetrically about the vertical center line of the generator. The SNAP-9A heat source was 14.6 cm long and 2.5 cm in diameter [8] (see Fig. 8.7). The outer capsule was constructed of Haynes Alloy No. 25 (Haynes 25, for short) and the inner liner of tantalum. The fuel capsules weighed 458 grams each.

After fueling, each capsule was placed on its side and was heated, causing the fuel to melt. The plutonium fuel form used melted at 890 K. Upon solidification, the fuel bonded itself to the capsule liner. This established a very good heat transfer environment. A segmented graphite block positioned the capsules and conducted the alpha decay heat to the thermoelectric converters. Consistent with aerospace nuclear safety philosophy of the day, the fuel capsules were designed for intact impact under launch abort conditions and for high altitude burnup and dispersal in the event that a mission abort caused the spacecraft to reenter the Earth's atmosphere. The segmented fuel block design permitted separation of the capsules for exposure to aerodynamic heating during a reentry abort. The liner was composed of super alloy Haynes 25 with a melting point of about 1650 K. This material accommodated atmospheric burnup, yet resisted oxidation and marine environment corrosion. A tantalum liner was used as a compatibility barrier between the plutonium-238 metal and the Haynes 25 structural member.

Table 8.5 provides the operating parameters for the SNAP-9A system, while Fig. 8.8 illustrates the SNAP-9A power performance [7]. The top plot in Fig. 8.8 presents smoothed data from the Transit 5BN-1 spacecraft; the bottom plot from the Transit 5BN-2 spacecraft. The gap in the Transit 5BN-2 data corresponds to the period of time when these data were not taken.

SNAP-19

The SNAP-19 systems employed plutonium-238 as the heat source and TE materials for power conversion. These RTGs used a new 645 watts-thermal heat source

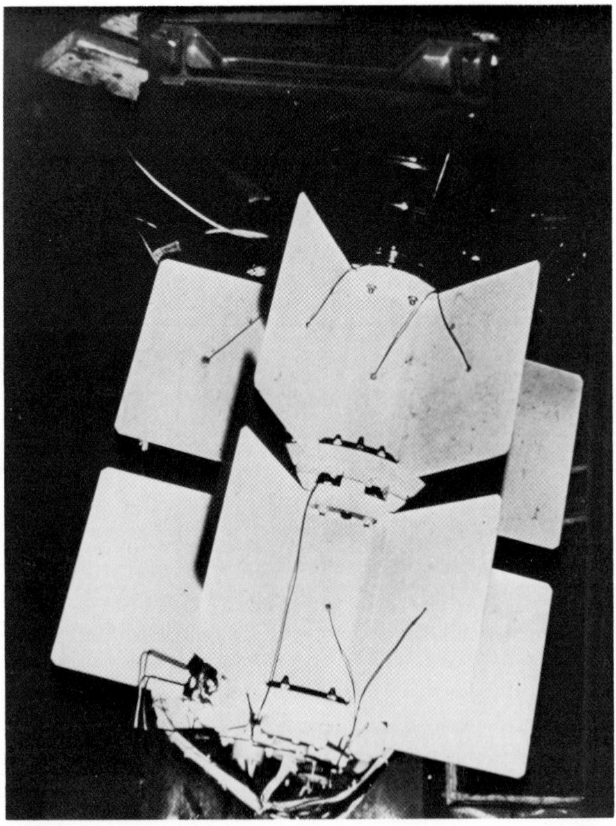

Fig. 8.6 SNAP-9A generator. *Courtesy of Mound Laboratory, Monsanto Research Corporation.*

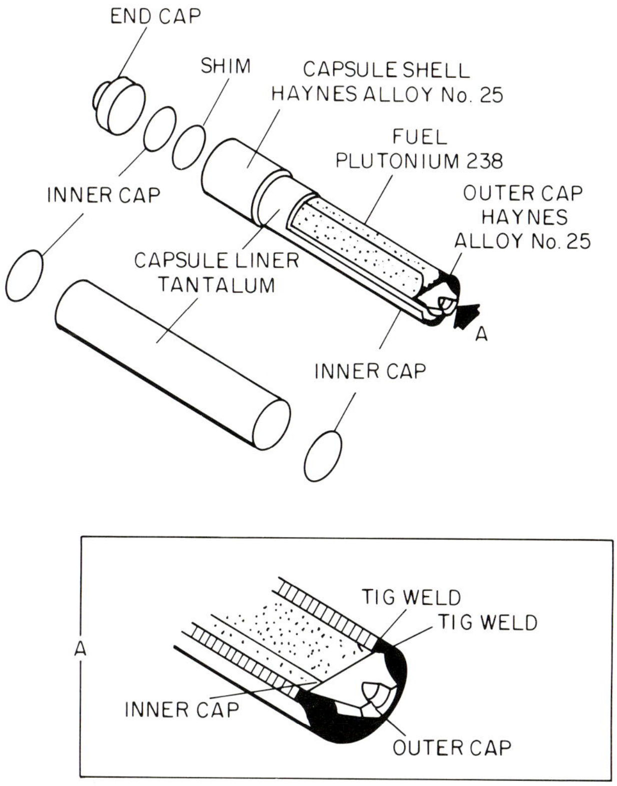

Fig. 8.7 SNAP-9A fuel capsule. *Courtesy of Mound Laboratory, Monsanto Research Corporation.*

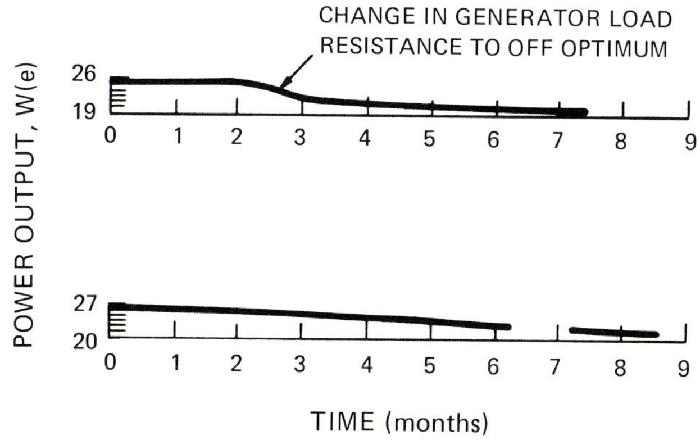

Fig. 8.8 SNAP-9A power performance data: Transit 5BN-1 (upper plot) and Transit 5BN-2 (lower plot) [7].

called the Intact Impact Heat Source, in conjunction with an array of 90 lead telluride (PbTe) and silver antimony germanium telluride (PbTe-TAGS) TE converter materials. It was designed to contain the radioisotope fuel under normal operational conditions and to limit the probability of contamination of the biosphere under abnormal launch and mission abort conditions. Safety design requirements included survival upon reentry and impact and containment or immobilization of the fuel [9–13).

The SNAP-19 generator was first successfully launched on a NASA Nimbus III meteorological satellite in April 1969. This RTG system operated properly for approximately 20,000 hours (over twice its design life) and then experienced a sharp degradation in performance [7, 9] (see Fig. 8.9). This decline was attributed to the sublimation of the thermoelectric material and loss of the hot junction bond due to internal cover gas depletion. The SNAP-19 generator design was then modified from the Nimbus III spacecraft configuration to provide up to a six year design life in support of the Pioneer missions to Jupiter and Saturn and the Viking missions to Mars. The design differences between the SNAP-19 generators used for the Pioneer spacecraft and for the Viking landers are shown in Table 8.6 [3, 10]. The unique mission requirements for each of these exciting space exploration programs created the need for such design variations. For example, the SNAP-19 generators used in the Viking landers had to withstand high temperature sterilization procedures in support of the planetary quarantine protocol, storage during the flight to Mars, and then the severe temperature extremes of the Martian surface.

Among all the deep space travelers launched by the United States, perhaps Pioneers 10 and 11 are the most appropriately named—for they truly performed pioneering exploration of the outer regions of our Solar System. Launched in 1972 and 1973, respectively, they

Table 8.5 SNAP-9A operating parameters.

Parameter	Beginning of Life	End of Life in Space (6 years)
Skin Temperature (K)	430	425
Cold Junction Temperature (K)	450	444
Hot Junction Temperature (K)	789	767
Fuel Centerline Temperature (K)	907	839
Heat Input (watts)	525	500

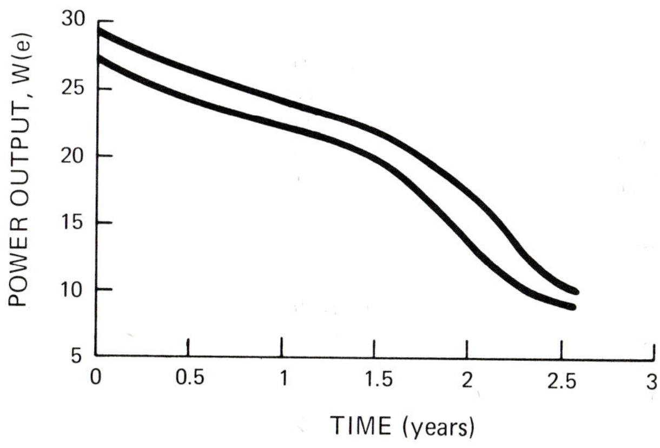

Fig. 8.9 SNAP-19 power output for the Nimbus III spacecraft [7].

Table 8.6 SNAP-19 comparative design data [3, 10].

	Pioneer	Viking/Lander
Design life (yr)	6	2
Overall Size (dia × lg)[1] (cm)	51 × 28	59 × 28(40)
Generator Weight (kg)	13.6	15.2
BOL Power Output (watts-electric)	41.2	42.5
Lead Voltage (volts)	4.0	4.4
BOL Fuel Inventory (watts-thermal)	648	683
BOL Pu-238 Inventory (curies)	19,500	20,600
Heat Shield Length (cm)	16.5	17.3
Capsule Length (cm)	12.2	12.7
Capsule Inner Liner Thickness (mm)	0.13	0.22
Capsule Weight (kg)	3.4	3.6
T/E Converter Initial Gas Fill (%)	75 He/25 Ar	90 He/10 Ar
Dome Reservoir Initial Gas Fill (%)	Not applicable	5 He/95 Ar
Capsule Cover Gas	Capsule exposed to T/E	Converter fill gas
Heat Shield Getter Recess	No	Yes
End Cover Attachment	Bolted/Seal-Weld	Lock Ring/Seal-Weld
Electrical Receptacle Seal	Single Viton O-Ring	Dual Viton O-Rings
Dome Reservoir-to-T/E Converter Seal	Not Applicable	Single Viton O-Ring
Hot Junction Temperature (K)	785	819
Number of Fins	6	6

[1] Diameter is across fin tips; length is mounting-flange-to-mounting-flange except value in parenthesis for Viking is overall length including dome reservoir and pressure transducer boss on lower cover.

were the first spacecraft to fly past Jupiter (in 1973 and 1974). At Jupiter, Pioneer 11's trajectory was carefully targeted to swing it (now renamed Pioneer Saturn) toward Saturn for an encounter in September 1979. Table 8.7 summarizes some of the discoveries made by these automated, nuclear-powered explorers [14]. The SNAP-19 generators continue to operate successfully well beyond their six year design life [11] (see Fig. 8.10).

Table 8.7 Selected scientific highlights of the Pioneer missions to Jupiter and Saturn [14].

Jupiter Highlights

- First spacecraft flybys of Jupiter (Pioneer 10-1973; Pioneer 11-1974).
- Confirmed Jupiter emits twice as much energy as it receives from Sun.
- Detailed studies of general banded structure of Jovian atmosphere, including the Red Spot.
- Detected huge Jovian magnetic field.
- Discovered Jovian radiation belt (energetic protons).
- Closeup observations of Galilean satellites (Europa, Io, Ganymede, and Callisto).

Saturn Highlights

- First spacecraft flyby (Pioneer 11/"Pioneer Saturn"—1979).
- Confirmed that (like Jupiter) Saturn radiates about twice as much energy as it receives from Sun.
- Discovered magnetic field around Saturn.
- Discovered Saturnian radiation belts.
- Observed that rings consisted of particles several centimeters in diameter (possibly made of frozen water and other ices).
- Two new rings discovered (called F and G Rings).
- Several new moons discovered.

On 13 June 1983 Pioneer 10 became the first manmade object ever to leave the Solar System.

Project Viking was the culmination of a series of U.S. missions to explore the planet Mars. These missions began in 1964 with Mariner 4 and continued with the Mariner 6 and 7 flybys in 1969 and the Mariner 9 orbital mission in 1971–1972. Viking was designed to orbit Mars and to land a robot laboratory on its surface. Two identical spacecraft were launched from Cape Canaveral: Viking 1 on 20 August 1975 and Viking 2 on 9 September 1975. The landers were sterilized before launch to prevent contamination of Mars by terrestrial microorganisms. Viking 1 reached Mars orbit on 19 June 1976 and the first landing on Mars occurred on 20 July 1976 on the western slope of Chryse Planitia. Viking 2 began orbiting Mars on 7 August 1976 and its lander touched down on the Red Planet on 3 September 1976 at Utopia Planitia. Among their many outstanding scientific accomplishments, these nuclear-powered robot laboratory landers discovered the following [14]:

a. The highly oxidized Martian soil produced unique chemical reactions in the life detection equipment.

b. The reddish color of the soil is due to oxidized iron.

c. The winds at the surface of Mars are light, approximately 24 km/hr, while the surface temperatures range from about −84° C at night to −29° C in the afternoon.

Over 4,500 high quality images of the Martian land-

Table 8.8 Viking/SNAP-19 heat source characteristics [3].

	Heat Source	Capsule
Configuration	Hexagonal prism	Cylinder with hemispheric end caps spheric end caps
Weight (kg)	5.2	3.6
Ballistic Coefficient[1]		
Side-Stable (N)	235	63
Dimensions (cm)		
Length	17.3	12.7
Diameter	8.9[2]	6.3[3]
Operating Temperature at launch (K)[4]	785[5]	985[6]

[1] Nominal, hypersonic continuum value
[2] Across flats
[3] Includes nominal thickness of emissive coating
[4] Assumed fin root is 395 K maximum allowable
[5] Heat shield surface
[6] Clad temperature

scape were returned by the two Viking landers. Since no evidence of the presence of living microorganisms was discovered in the soil near the landers—the question of life on Mars remains open, to be resolved by future expeditions, robotic and manned.

The Viking/SNAP-19 RTG is shown in Fig. 8.11. The heat source, plutonium-238 dioxide–molybdenum cermet hot-pressed into discs, was arranged in a three-layer tantalum alloy cylindrical fuel capsule contained within a hexagonal graphite shield (see Fig. 8.12). This design provided the thermal energy and proper temperature distribution for the TE converters and also satisfied aerospace nuclear safety criteria. It used a four-layer, vent capsule configuration enclosed in a right hexagonal prism graphite body. The design supported intact reentry through aerothermal heating and nuclear fuel retention in the event of impact. The nuclear fuel charge produced a nominal 682 watts-thermal at the beginning of life (BOL) with each capsule containing 20,600 curies. The fuel capsule contained 18 or 19 plutonium discs and the initial power density of each disc was 3.2 W/cm^3. Table 8.8 provides other characteristic data concerning the Viking/SNAP-19 heat source.

An oxygen barrier consisting of molybdenum-46% rhenium was incorporated between the fuel and other components. A liner provided the primary fuel encapsulant and facilitated easy decontamination during fuel assembly. It was a 0.5 mm layer of tantalum-10% tungsten. A pressure relief valve was included to vent decay gases. Impact resistance was provided by a shell of tantalum alloy T-111. The final capsule shell, composed of platinum-20% rhodium, was used as an external oxygen barrier. A high emissivity coating of platinized alumina was also placed on the capsule to facilitate heat transfer from the fuel capsule to the graphite heat shield. Graphite felt sleeves were used to protect the ends from vibrations during launch; while a zirconia support ring was used to relieve thermal stresses during reentry. A graphite heat shield surrounded the capsule and provided the ablation material needed during reentry. This graphite shield also functioned as a heat sink during normal operation.

The TE converters consisted of TAGS thermoelectric material with a thin layer of SnTe at the hot side for the p-type material leg and PbTe for the n-type material leg. Each TE module contained 15 couples with six

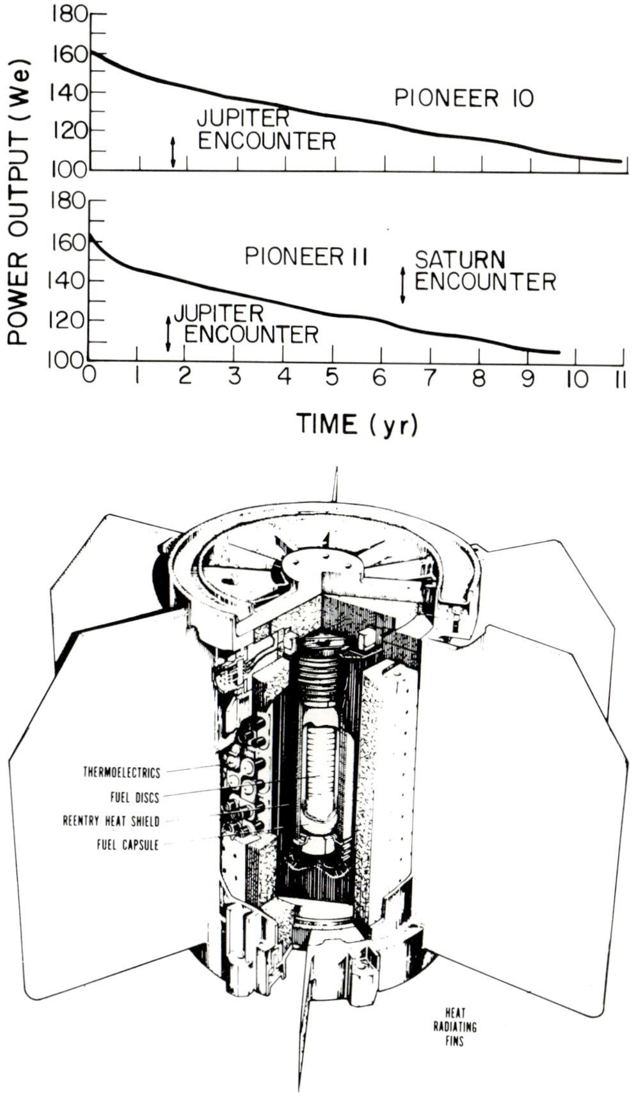

SNAP 19/PIONEER RADIOISOTOPE THERMOELECTRIC GENERATOR

Fig. 8.10 The SNAP-19 generator in the Pioneer program. (a) SNAP-19 power output history for the Pioneer 10 and 11 spacecraft (as of 1983) [7]. (b) SNAP-19/Pioneer RTG. *Courtesy of U.S. Department of Energy.*

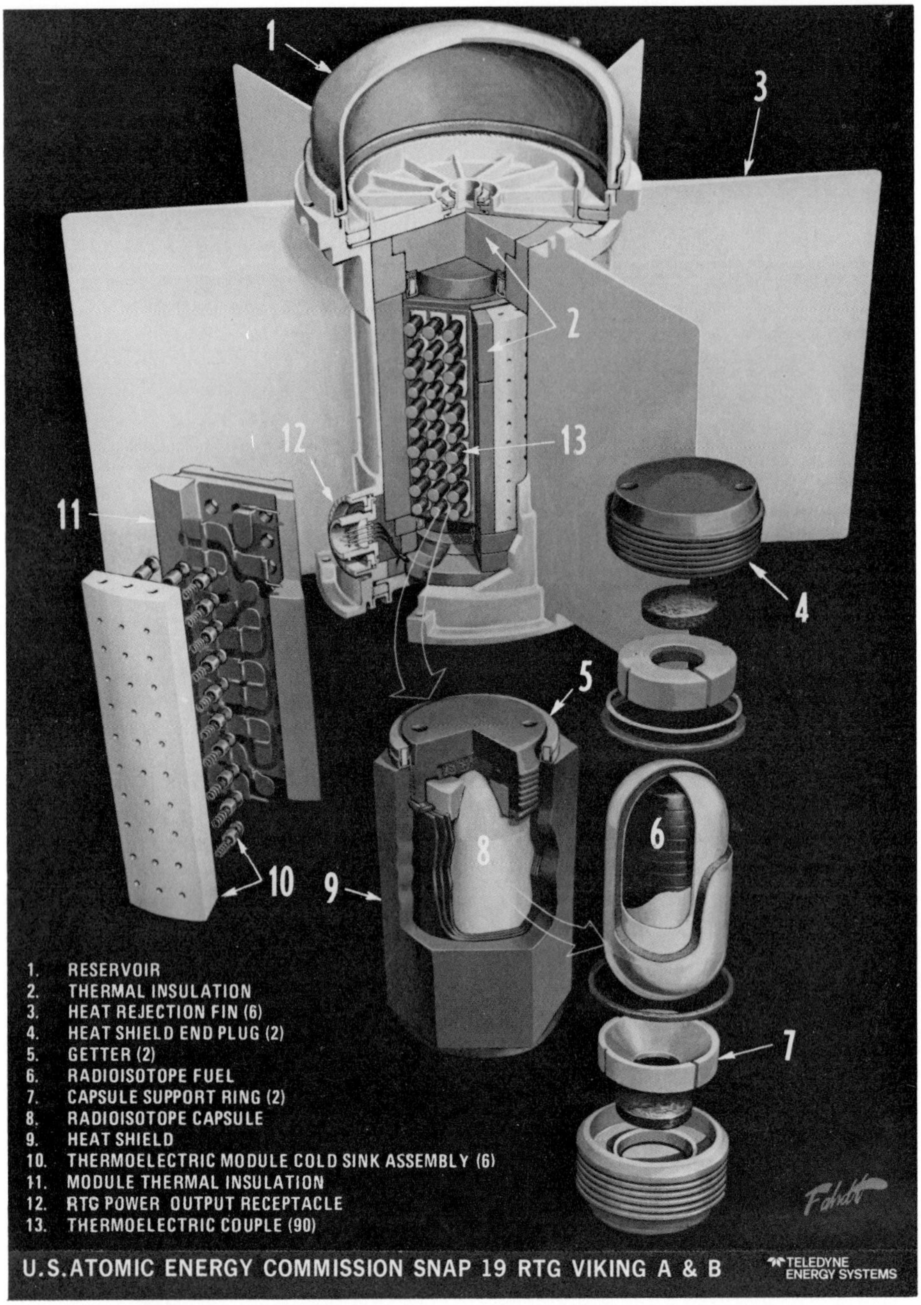

Fig. 8.11 Viking/SNAP-19 RTG. *Courtesy of Teledyne Energy Systems.*

modules in the overall assembly. The couples were electrically connected in a parallel-series configuration (see Fig. 8.13) [4]. The six modules surrounded the hexagonally shaped heat source with a module resting on each of its six flats. Electrical insulation at the hot side between the individual modules and the heat source was provided by a sheet of mica. Each module also had cold end hardware, including springs, pistons, alignment buttons, and heat sink bar. Thermal insulation was used in the TE converter to minimize thermal shunt losses.

The housing-radiator assembly consisted of a six-finned cylindrical housing, two ribbed end covers, and two seal-weld rings. These components were made of magnesium-thorium alloy. A dome reservoir, containing argon as a cover gas to retard sublimation of TE materials, was also provided. Thermoelectric converter and dome reservoir volumes were separated in a design configuration that permitted a controlled interchange of gases. The argon and the helium-rich TE cavity atmosphere were interchanged with time, promoting increased thermal efficiency and enhanced electrical performance. The housing, fins, and dome radiator surfaces were coated with a zirconium oxide/sodium sili-

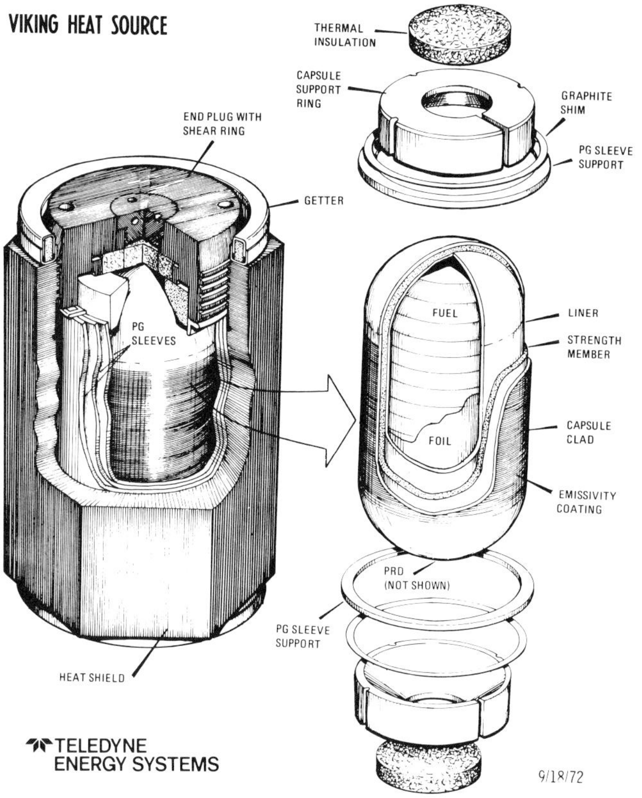

Fig. 8.12 Viking/SNAP-19 heat source. *Courtesy of Teledyne Energy Systems.*

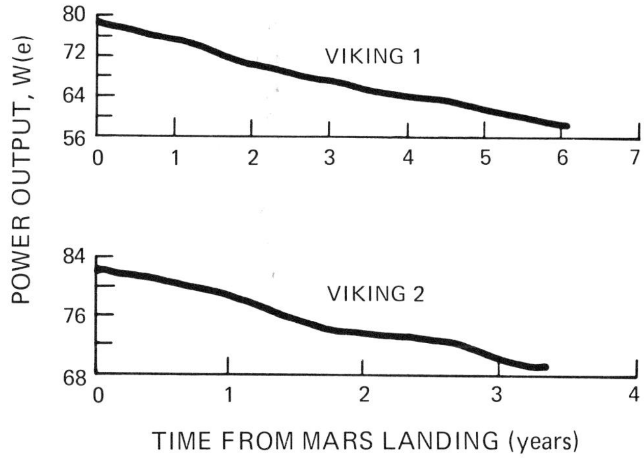

Fig. 8.14 Power history of Viking/SNAP-19 RTGs (smoothed data) [*7*].

cate emissivity coating. This coating material was attractive because it provided an emissivity greater than 0.83, accompanied by a solar absorptivity of only 0.2. The overall housing diameter across the fin tips was 58 cm and the overall height 29 cm. The power history of the Viking/SNAP-19 RTGs is given in Fig. 8.14 [7].

SNAP-27

The SNAP-27 system was designed for use during the Apollo Program to power instrument and experiment

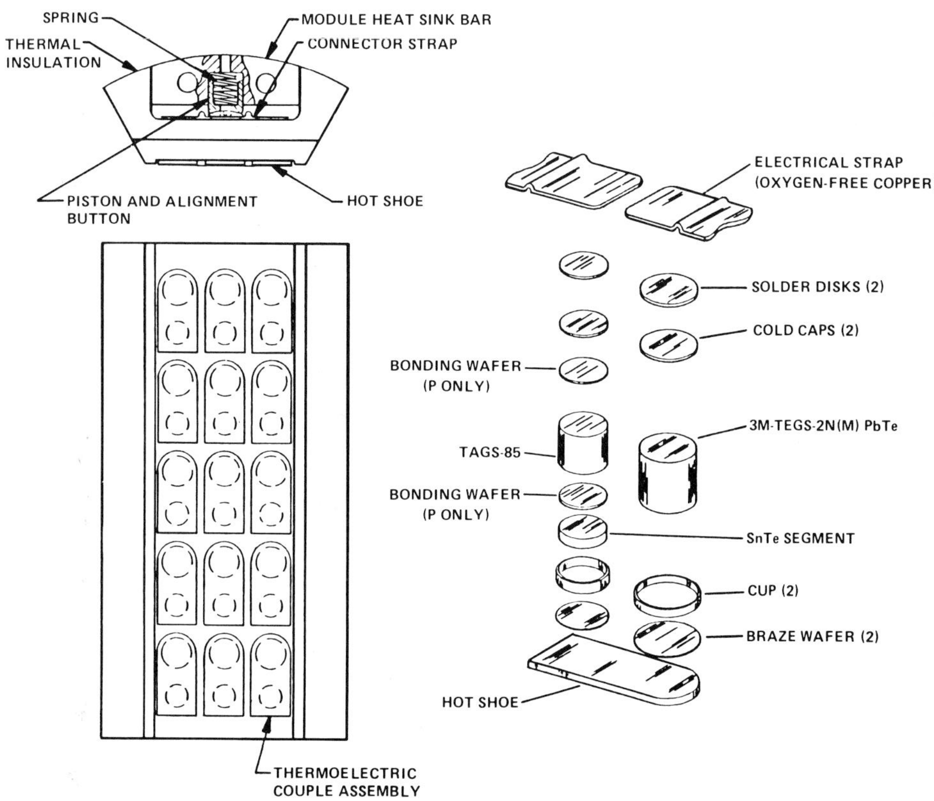

Fig. 8.13 Viking/SNAP-19 thermoelectric unit arrangement [*4*].

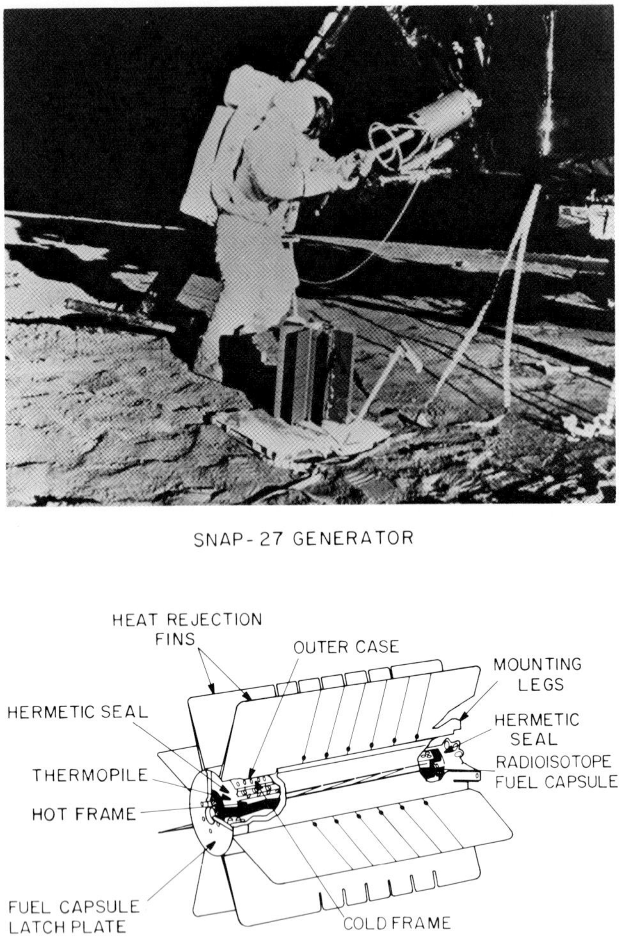

Fig. 8.15 The SNAP-27 system. (a) The SNAP-27 heat source being removed from the Lunar Excursion Module (LEM) by Astronaut Gordon Bean during the Apollo 12 mission to the Moon in 1969. *Courtesy of NASA*. (b) SNAP-27 generator.

packages deployed on the lunar surface [15, 16]. Unique design features were added to this radioisotope-powered system because astronauts were physically involved in its deployment (see Fig. 8.15). In fact, separate shipment packages were used for the radioisotope heat source and the TE generator. The major SNAP-27 hardware included: (1) a generator with a thermopile, structure, and heat rejection system; (2) a fuel capsule that included the plutonium fuel in a separate hermetically sealed structure which prevented fuel release; and (3) a graphite lunar module fuel cask (GLFC) (see Figures 8.16 and 8.17). The GLFC provided support for the fuel module during the journey to the Moon and also provided thermal and blast protection in the event of a launch pad or mission abort. The general aerospace nuclear safety criterion in effect during the Apollo Program was that, should there be an aborted mission, the probability of radiological exposure of the general public would be minimized.

The SNAP-27 fuel capsule and cask were, therefore, designed to guarantee survival should a hazardous environment occur. This included launch pad explosion, an atmospheric ascent abort, and reentry into the Earth's atmosphere and ground impact. The capsule consisted of two fuel compartments integrated into a single high strength super alloy outer structure. Each fuel compartment contained a nominal 740 watts-thermal of $^{238}PuO_2$ fuel in the form of 50 to 250 micrometer-size "microspheres." The plutonium fuel was held in an annular configuration with void volume that accommodated long-term storage by confining the helium decay gas. The welded structure was 6.3 cm in diameter and approximately 41.9 cm long. A backplate helped secure the fuel capsule and the assembly weighed less than 6.8 kg. An iron titanate high emissivity coating (greater than 0.85) was applied to the outer surface. The fuel capsule was qualified to the following thermal conditions: (1) a steady state air temperature of 725 K; (2) ground storage at 530 K; (3) Earth-to-Moon transit in the GLFC at 1035 K; and (4) generator operation at 1005 K.

The GLFC kept the fuel capsule together during the mission from launch to deployment on the lunar surface. It also served as an aerospace nuclear safety feature, enhancing containment during a reentry abort. The GLFC consisted of a cylindrical container with a primary and secondary thermal shield and a support structure for the fuel capsule. The primary heat shield, consisting of graphite, formed the outer cylinder and hemispherical end caps. It protected the fuel capsule during reentry through heat sink, reradiation, and ablation processes. The secondary thermal shield was a beryllium cylinder that was coated with successive layers of silver, rhodium, and a high emissivity coating (radifrax) to promote radiant heat transfer and to provide oxidation resistance. During normal operation, heat from the fuel capsule was radiated to this beryl-

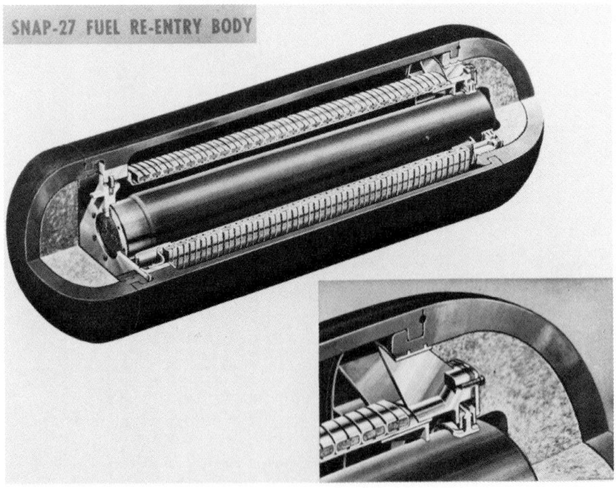

Fig. 8.16 SNAP-27 fuel capsule assembly [*15*].

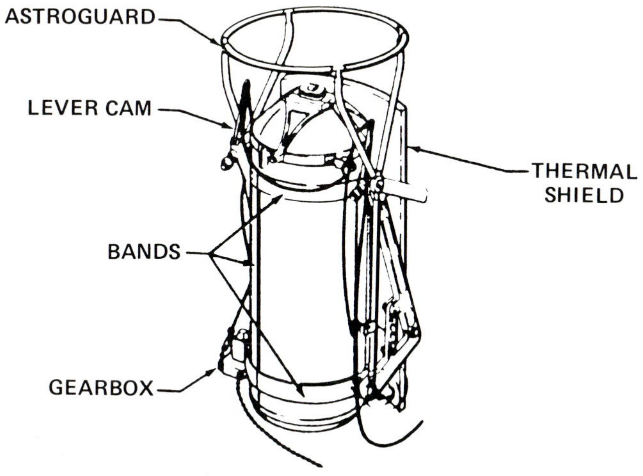

Fig. 8.17 SNAP-27 lunar fuel cask assembly [*16*].

lium cylinder, conducted through it, and then reradiated to the graphite cylinder. If reentry occurred, the beryllium cylinder would serve as a heat sink, protecting the fuel capsule. Thus, a combination of rings, backplate, secondary thermal shield, and forward capsule support provided an independent internal structure to maintain fuel capsule integrity during postulated mission abort conditions.

Thermal energy from the fuel capsule was transferred to the generator hot frame by radiative coupling. To promote radiation heat transfer, both surfaces were covered with a high emissivity iron titanate coating with the heat source surface having an emissivity of 0.85 and the hot frame surface an emissivity of 0.80. When deployed on the lunar surface, the fuel capsule operated at 1005 K, while the Inconel 102 alloy hot frame was 880 K. Thermal energy received by the hot frame was

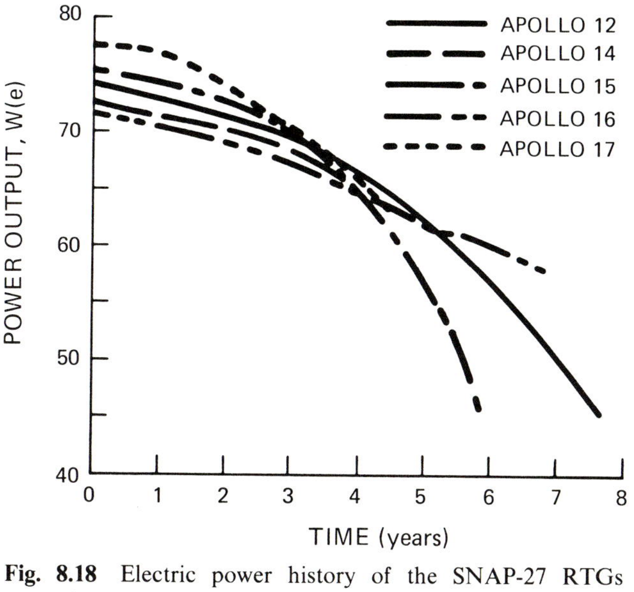

Fig. 8.18 Electric power history of the SNAP-27 RTGs (smoothed data) [*7*].

transported to the hot side of the TE converter materials by first passing through the electrical insulator, then through an electrode, and finally to a hot button. The hot junction temperature ranged between 855 K and 865 K, reflecting an overall temperature drop of 15 to 25 K. On the Moon's surface, where temperatures can vary from 350 K during the lunar day to a frigid 100 K during the lunar night, the generator's cold side temperature operated at 545 K. Insulation was provided around the radiative coupler to minimize heat losses. Lead telluride (PbTe) served as the TE material and the couples were assembled in a series-parallel electrical arrangement to prevent string loss, should a couple have failed open. The output voltage was between 14 and 16 volts DC. Each element was preloaded into its hot button by individual springs, sized and shimmed to establish a bearing pressure of 1.03 MPa. Thermal energy from these couples was then transferred through beryllium oxide followers that interfaced with the cold cap. To help smooth out temperature gradients, the heat was allowed to flow into a massive beryllium frame. This outer case also functioned as the prime structural member of the generator. The electric power histories of the SNAP-27 RTG units deployed on the lunar surface are presented in Fig. 8.18 [7].

Transit-RTG

The Transit-RTG, designed for navigational satellites, was the first radioisotope system developed for a five year life [17–19]. The unit was designed to have a BOL power output of 34.2 watts-electric, weigh less than 17 kg, and produce a minimum of 5 volts. Its TE converter employed PbTe materials and did not use a cover gas to inhibit sublimation. This system was also the first to use radiative coupling between the heat source and the TE converters, although this was accomplished at some loss in efficiency.

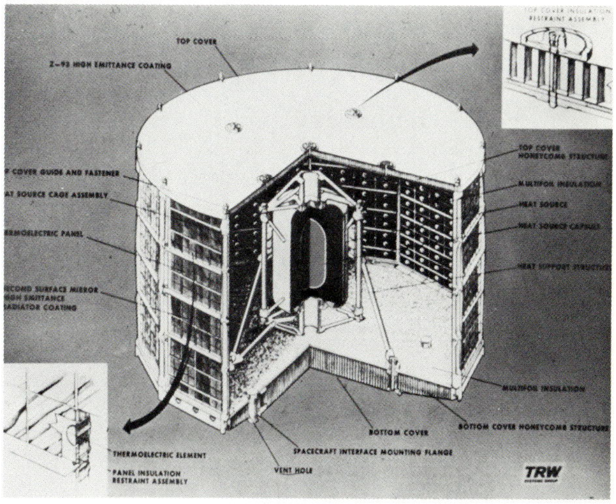

Fig. 8.19 Transit generator. *Courtesy of TRW.*

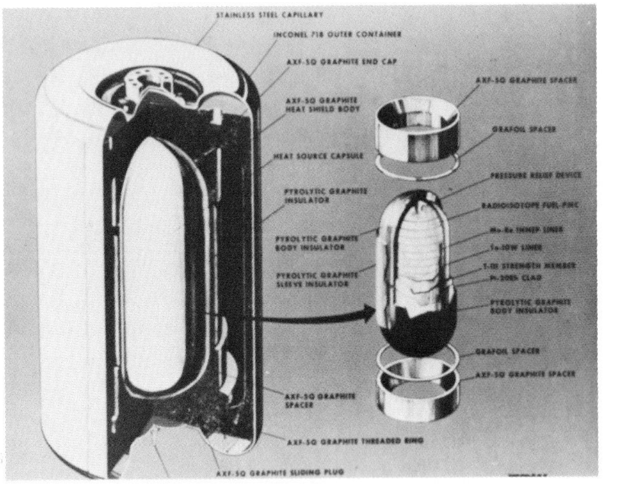

Fig. 8.20 Transit heat source design. *Courtesy of TRW.*

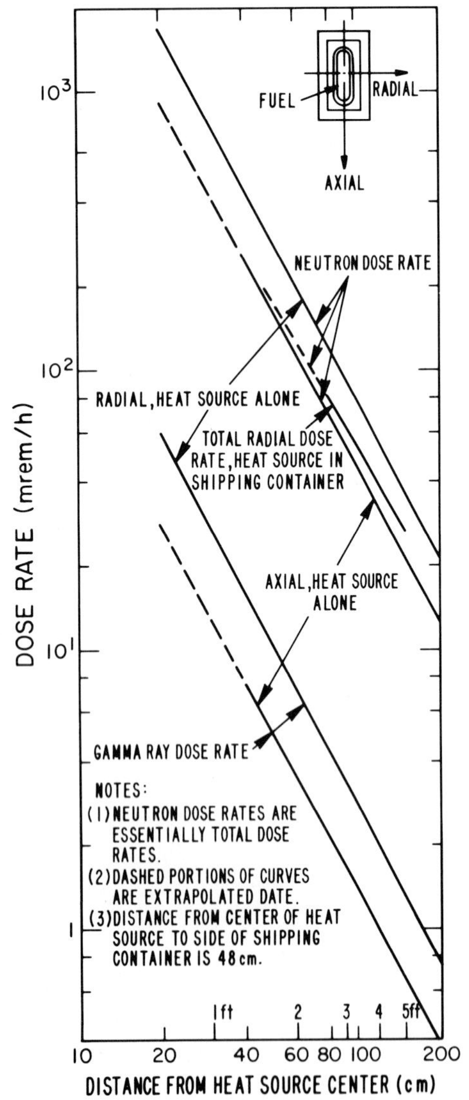

Fig. 8.21 Radiation levels as a function of distance from the Transit heat source. *Courtesy of TRW.*

As shown in Fig. 8.19, this generator had two major parts: the heat source and the converter. The 850 watts-thermal heat source had a power density of 3.2 W/cm^3 (see Fig. 8.20). Its fuel was a plutonium dioxide molybdenum cermet with 82.5 wt % PuO_2 and 17.5 wt % Mo. The fuel was pressed into discs 5.4 cm in diameter and 0.54 cm thick. No overcoat was used on these discs. Twenty-two such fuel discs were placed in the 850 W heat source, which had an overall length of 18.1 cm and an overall diameter of 12.7 cm. Approximately 2.64 kg of radioisotope fuel (some 25,500 curies) were used and Fig. 8.21 shows radiation levels in the vicinity of the heat source [17].

The fuel capsule had three subassemblies: (1) a three-layer refractory metal capsule that contained the fuel; (2) a two-layer graphite heat shield that resisted reentry environments and reduced the impact velocity of the heat source; and (3) an outer can that provided oxidation resistance during storage and and prelaunch operations for up to 1000 hours. A pressure relief device was vented through a 4.6 m long capillary tube. The fuel capsule was supported by the hemispherical ends. A preload was applied to the two graphite plugs to prevent vibrations during launch. A typical temperature distribution for the fuel capsule is shown in Fig. 8.22.

The converter was divided into twelve panels arranged into a regular polygon (see Fig. 8.23). The top closure plate and the bottom support plate did not contain TE converters and were insulated to minimize heat losses. The twelve TE panels formed a prism, 61.0 cm across the flats. The generator was attached to the spacecraft by means of a 34.3 cm diameter ring. Each of the twelve TE panels consisted of 36 PbTe thermocouples interconnected in an electrical series-parallel network. Each TE element was surrounded with molybdenum opacified quartz washers to reduce sublimation of the TE material and to prevent the TE elements from short circuiting with the thermal insulation materials. The space between the heat collecting surfaces and the cold junction radiator was filled with a close-fitted

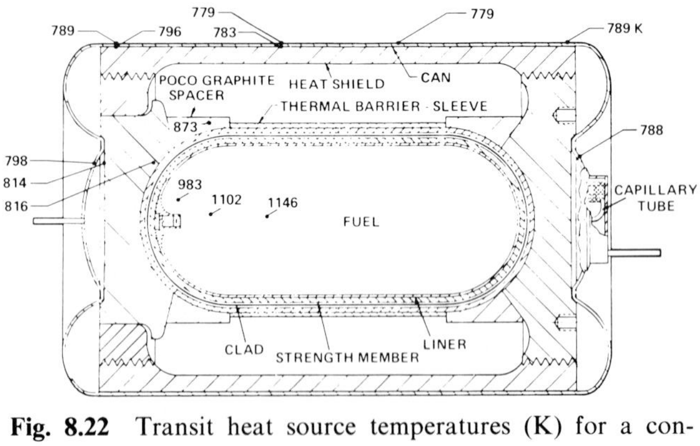

Fig. 8.22 Transit heat source temperatures (K) for a converter temperature of 673 K [*18*].

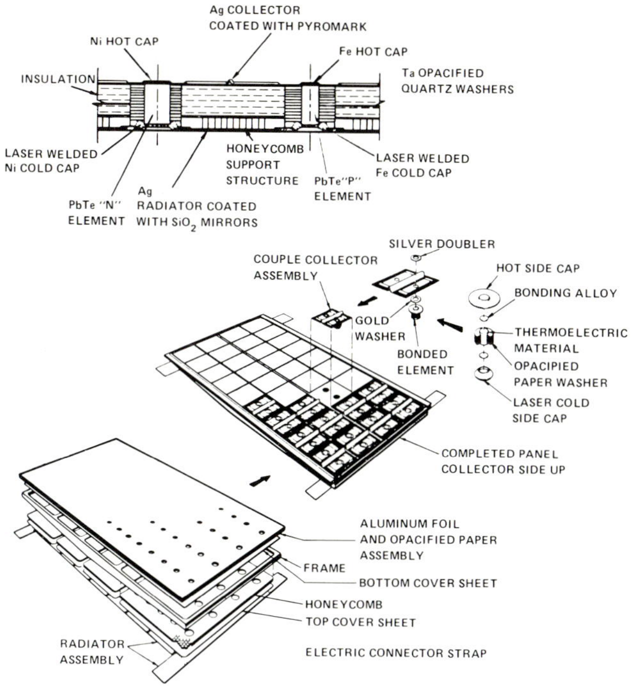

Fig. 8.23 Schematic of Transit thermoelectric converters. *Courtesy of TRW.*

multilayer insulation of alternate layers of aluminum foil and opacified quartz paper.

The Transit power plant weighed 12.8 kg with the converter weighing 5.8 kg, the heat source 6.3 kg, the support structure and cage 0.7 kg, and other miscellaneous components 0.11 kg.

Multihundred Watt (MHW) Generator

The Multihundred Watt (MHW) generators, with their higher temperature heat source and use of silicon-germanium (SiGe) as the thermoelectric material, represent a major step forward in RTG design [20–25]. Individual MHW generators produced 150 watts-electric with an efficiency of 6.7 percent. Two MHW generators were used on each of the LES 8/9 communications satellites and three MHW generators on each Voyager spacecraft. The Voyager missions performed spectacular encounters with the giant outer planets Jupiter and Saturn. The encounters with Saturn occurred over four years after launch with all RTG units working as anticipated. The Voyager 2 spacecraft is now on a trajectory that will encounter Uranus in 1986 and Neptune in 1989.

The MHW heat source design is governed by aerospace nuclear safety considerations with immobilization of the plutonium fuel as the primary objective. Its design philosophy supports the following objectives: (1) prohibit the release of nuclear fuel, especially in the form of biologically significant respirable particles; (2) minimize environmental (biospheric) contamination, particularly over populated land masses; and (3) maximize long-term immobilization following potential heat source accidents.

The MHW generator shown in Fig. 8.24 consists of the radioisotope heat source and the converter. Table 8.9 presents the MHW design performance data [20, 22]. The heat source (see Fig. 8.25) is in the shape of a right circular cylinder that has 24 fuel containers of 100 watts-thermal each. Each fuel container also has individual impact protection arranged around a spherical ball of plutonium dioxide (PuO_2). These fuel spheres have a metallic (iridium) shell that contains the radioisotope fuel and a graphite impact shell which provides the primary resistance to mechanical impact loads. A graphite aeroshell serves as the primary structural element and as a secondary ablator for shallow angle and orbital decay reentries. A graphite ablation sleeve provides the primary heat protection to the isotope heat source during reentry.

The individual fuel spheres are 3.7 cm in diameter with a power density of 3.6 W/cm^3. The 24 spheres contain a total of 6.05 kg of PuO_2. This fuel form has an activity level of 12.8 Ci/g which corresponds to

Fig. 8.24 MHW radioisotope thermoelectric generator. *Courtesy of General Electric Co.*

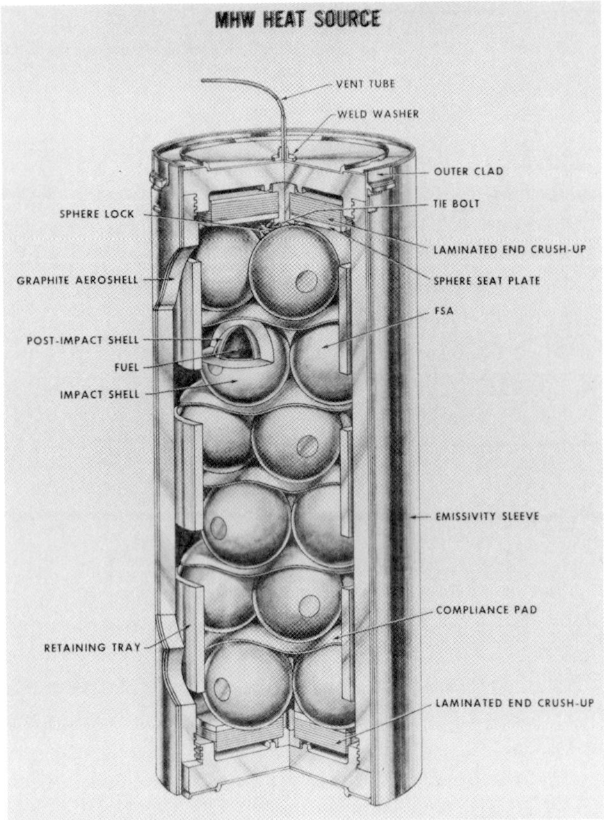

Fig. 8.25 MHW heat source. *Courtesy of General Electric Co.*

77,300 curies per heat source. The fuel spheres are arranged in six planes of four spheres each. The total weight of the heat source is 19.6 kg and it has a surface temperature of 1330 K. The thermal power level of this heat source is reduced approximately 1 percent a year, since the plutonium-238 has a half-life of 87.75 years.

As shown in Fig. 8.24, the converter design contains a beryllium outer core and pressure domes with the TE

Table 8.9 MHW design performance [*20, 22*].

Power output (watts)	
Beginning of mission	150
Five years	135
Weight (kg)	36
Output voltage (volts)	30 ± 5
Envelope (cm)	
Diameter	38
Length	61
Average hot junction temp (K)	1000
Thermoelectric material	SiGe
Magnetic field	$< 15 \times 10^{-9}$ tesla (15×10^{-5} gauss) at 1 meter
On pad	30 day unattended capability
Ground storage life (yr)	at least 2

unicouples attached directly to the outer shell. The converter has an overall diameter of 39.7 cm and an overall length of 58.3 cm. The heat source is supported as a column within the converter by compression pads located at each end. Loads from the heat source are

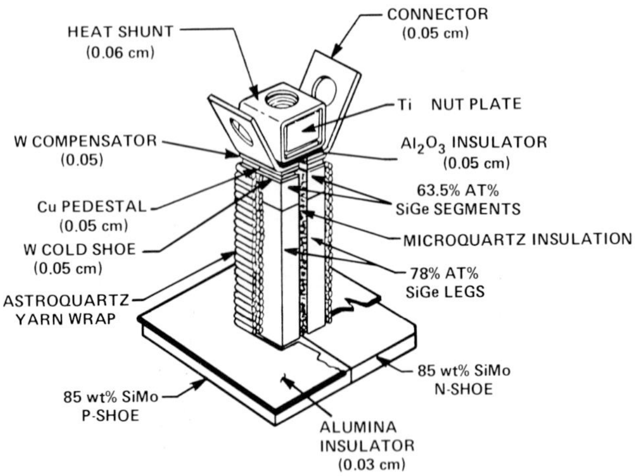

Fig. 8.26 SiGe unicouple for MHW generator. *Courtesy of General Electric Co.*

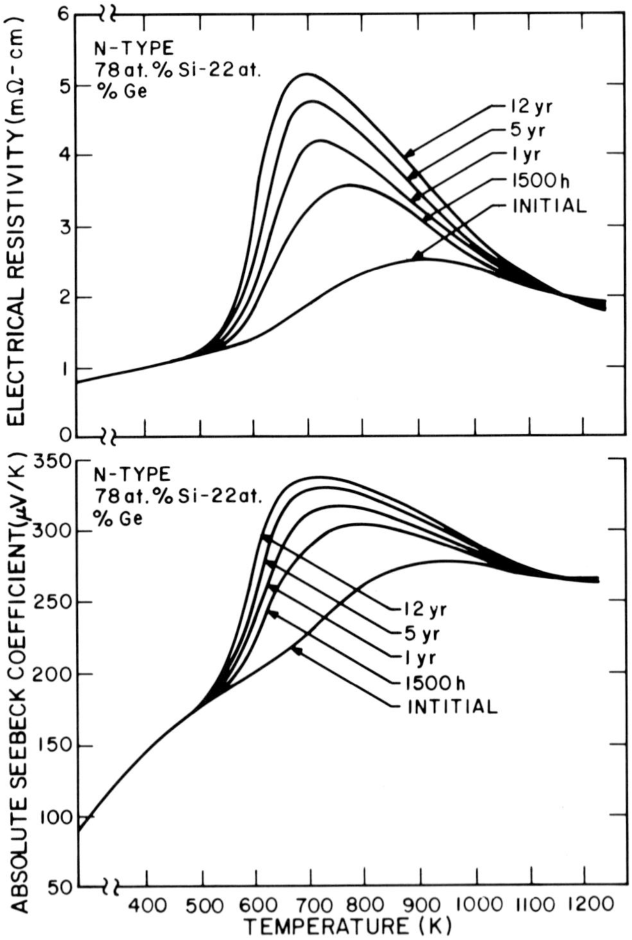

Fig. 8.27 Changes in the electrical resistivity and Seebeck coefficient of *n*-type SiGe thermoelectric material with time [*24*].

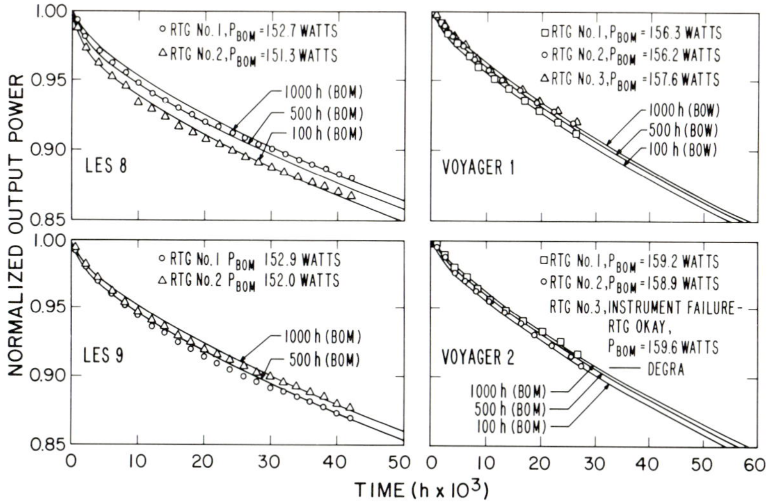

Fig. 8.28 MHW generator performance [25].

transmitted to the outer case through titanium end closure fittings that hold the compression pads and are attached internally to the outer case. The outer shell is coated with iron-titanate, with an emissivity of 0.85 to provide a heat rejection surface for the TE converters.

The power converter contains 312 SiGe unicouples arranged in 24 circumferential rows with each row containing 13 couples. These couples are made in unicouple assemblies (see Fig. 8.26) that are individually bolted to the outer case. Thermal energy from the radioisotope heat source is radiated to the unicouple hot shoe. This hot shoe, made of SiMo (85 wt % Si) is bonded to the SiGe TE material. Two compositions of SiGe are used in the TE converter legs: 78 atom percent silicon for most of the length; and 63.5 atom percent silicon for a short transition segment at the cold end of the couple to compensate for thermal expansion. For *n*-type TE material, the SiGe is doped with phosphorous; while for *p*-type material, boron is used as the dopant. The silicon alloy is bonded to a cold stack assembly of tungsten, copper, and alumina parts that separate the electrical and thermal currents. The unicouple legs are coated with a one micrometer layer of silicon nitride (Si_3N_4) to retard silicon sublimation. Heat losses are greatly reduced between the hot and cold shoes through the use of multifoil insulation composed of 0.008 mm-thick molybdenum foil and SiO_2 cloth (Astroquartz) separators. Astroquartz yarn is also wound around the unicouple legs to electrically insulate the unicouples from the multifoil insulation. The converter weighs 18.3 kg with the SiGe material making up 2.9 kg of this amount. The hot shoe temperature is 1310 K and the cold strap is 575 K.

Converter performance has been extensively evaluated in both ground and flight tests. Early testing identified: (1) silicon sublimation as a problem, leading to the addition of silicon nitride coating; and (2) the top layer of the molybdenum foil breaking and shorting to the SiMo hot shoe, leading to modifications in the design geometry and to the use of an alumina spacer to separate the hot shoe from the multifoil insulation. The major source of degradation is due to dopant precipitation [23].

Long-term degradation mechanisms have also been identified for the MHW converters [24]. Dopant precipitation changes the TE properties of SiGe as a function of time. These changes are attributed to a diffusion limited dopant precipitation process. Fig. 8.27 shows the changes in resistivity and Seebeck coefficient of the *n*-type TE material. Similar changes also occur in the *p*-type material, but in the 875 K to 1275 K temperature range.

Another performance degradation phenomenon is material sublimation and reactions that can induce changes in thermal conductance and electrical shorting. At temperatures around 1275 K, some of the silicon reacts with the SiO_2 insulation and produces SiO vapor. The SiO vapor then diffuses to the cooler zones (875 K–1075 K) of the thermopile and condenses. The condensing SiO tends to release free silicon and form SiO_2. Silicon and germanium also sublime and condense in the cooler zones. The condensate increases thermal conductance of the leg wrap material and also increases the emissivity of the molybdenum foils, resulting in higher heat losses and a decrease in power output. Partial shorts also developed between the TE legs and foils before the Si_3N_4 coatings were applied.

A change in thermal conductance has been observed, due to an inhomogeneity in the fabricated TE materials. During operation, germanium-rich regions diffuse, resulting in lower thermal conductivity. The power loss for uncoated elements is typically 10 watts for a five year mission. This loss, however, is reduced to just 3

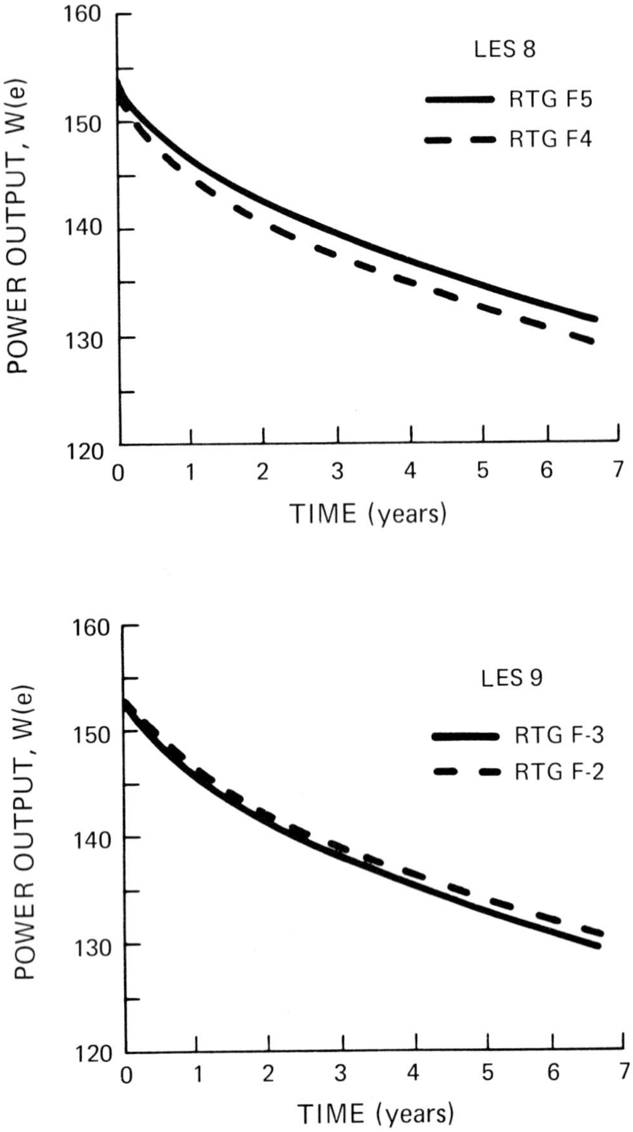

Fig. 8.29 Power history for LES 8 MHW RTGs (top plot) and LES 9 MHW RTGs (bottom plot) (smoothed data) [7].

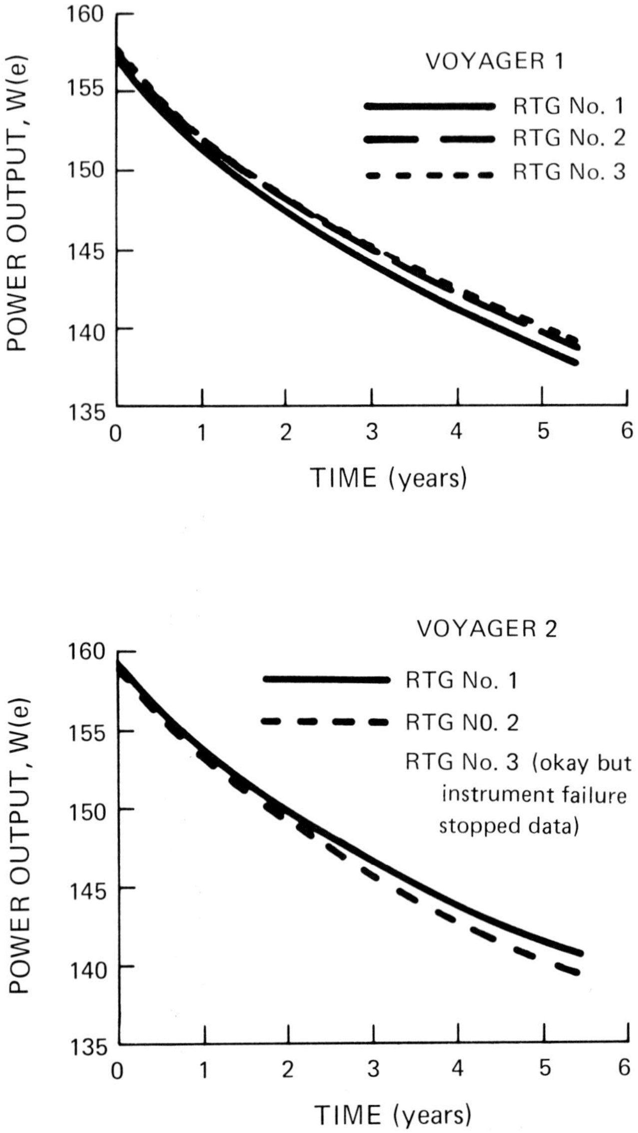

Fig. 8.30 Power history for Voyager 1 MHW RTGs (top) and Voyager 2 MHW RTGs (bottom) (smoothed data) [7].

watts in five years for coated elements. Other long-term losses include: the thermal degradation of insulation, which contributes a power loss of about 3 watts in five years; and changes in foil electrical resistance of greater than 500 ohms, which produce a power loss of 0.15 to 0.6 watt.

Studies have been made concerning the effect of storage on these converters [25]. Two years of storage, followed by extensive testing of up to 32,500 hours, did not reveal any significant degradation. Such testing indicates that a projected life of 600,000 hours at nominal RTG temperatures can be anticipated, alleviating concern about the possible failure of SiGe TE generators due to structural degradation. The very successful performance of the LES 8/9 and Voyager 1 and 2 MHW flight generators is presented in Figures 8.28 through 8.30 [7, 25].

General Purpose Heat Source–Radioisotope Thermoelectric Generator (GPHS-RTG)

Traditionally, radioisotope power systems have been designed for specific space missions or classes of missions. To meet an increased demand for power in a new mission, one would have to add integral numbers of complete conversion systems. This represents a design approach that often introduces weight and volume penalties [26]. A modular power unit approach, on the other hand, provides design flexibility in meeting rising demands for space power. One such modular heat source is called the General Purpose Heat Source (GPHS) [26–28]. It is being developed to replace the Multihundred Watt (MHW) generator system and will be used in future space missions, such as the Galileo

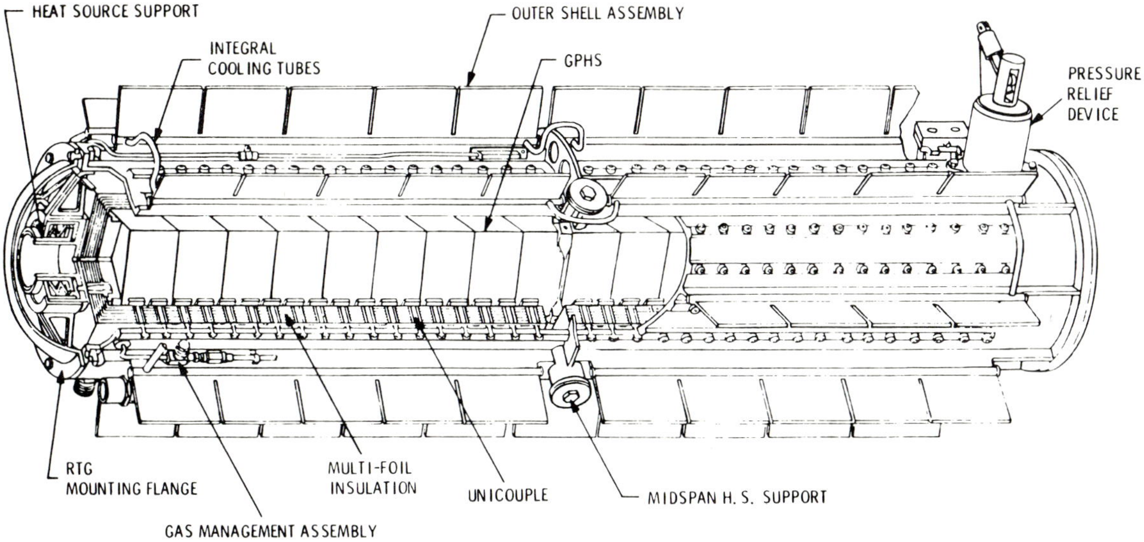

Fig. 8.31 General purpose heat source-RTG. *Courtesy of U.S. Department of Energy and General Electric Co.*

Project. As a successor to the Voyager Program, the Galileo Project will send a spacecraft to Jupiter to orbit the planet, making extensive studies of its satellites, and to launch a probe into the Jovian atmosphere.

The GPHS incorporates a modular design, making it adaptable to a wide range of future space power needs. Each 250 watt-thermal module is completely autonomous with its own passive safety provisions. Aerospace nuclear safety requirements for the GPHS center around immobilization of the plutonium fuel to the maximum extent possible during all phases of a mission, including ground transportation and handling, launch operations, launch, ascent and orbital insertion, on-orbit operations, and reentry, impact, and post-impact environmental behavior.

The GPHS-RTG is shown in Fig. 8.31. Eighteen individual modules are stacked in a column to provide an initial thermal source strength of 4500 watts. Each module has an aeroshell that contains two graphite impact shells (GISs) (see Fig. 8.32). The graphite aeroshell serves as the structural element and an ablator. A vent hole is provided to permit helium gas release from the radioisotope fuel pellet, but prevents a release of particulates. A fueled clad (as shown in Fig. 8.32) consists of a plutonium-238 fuel pellet (PuO_2) encased in an iridium shell that contains the fuel. Two of these fueled clads are combined in a graphite impact shell and are separated by a floating graphite membrane that provides resistance to mechanical impact loads. The GPHS has a specific power of 172 watts/kg; contains 10.7 kg of PuO_2; and has an activity level of 1.3×10^5 curies. The modules are constrained by locking members that minimize any relative lateral motion by individual modules. They are also packaged in a support system that provides axial compression to prevent separation of the modules.

The TE converter, the same as used in the MHW generator, surrounds the heat source, and 576 TE couples convert the heat of radioactive decay directly into electricity. These SiGe unicouples operate at a hot side temperature of 1275 K and a cold side reject temperature of 575 K. An outer case provides the main support for the TE elements and the heat source. The heat source support system provides preload to enable the GPHS-RTG unit to withstand the mechanical stress environments of launch and ascent, and to avoid separation of the heat source modules during mission operation. The converter also provides axial and mid-span heat source supports, a multifoil insulation packet and an internal frame, and the gas management system. A gas management valve in the pressure-tight outer case maintains the desired internal environment. An inert gas environment is used to permit partial power opera-

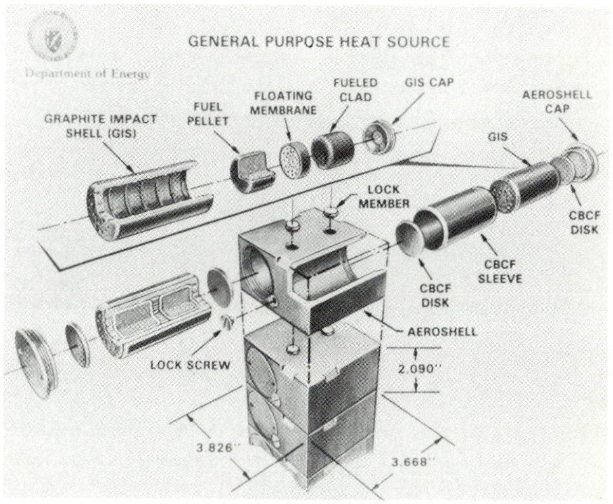

Fig. 8.32 GPHS aeroshell module. *Courtesy of U.S. Department of Energy and General Electric Co.*

Table 8.10 GPHS-RTG performance data [28].

RTG Power Output (watts)	290 (BOM) 250 (minimum at EOM)
Operation Life (hr)	40,000 after launch
Weight (kg)	54.1
Output Voltage (volts)	28($\pm$0.5)
Envelope (cm)	$<$ 46 dia. $\times$ $<$ 115 long
Hot Junction Temperature (K)	1275
Fuel	PPO (83.5 $\pm$ 1% $^{238}PuO_2$)
Thermoelectric Material	SiGe
Magnetic Field	$< 10 \times 10^{-9}$ tesla at one meter
On Pad	30 day unattended capability
Storage Life	1 year ground storage
Auxiliary Cooling	375 K avg. outer shell temp.

BOM = beginning of mission; EOM = end of mission.

tion on the launch pad. This inert gas also protects the molybdenum foil and refractory materials during storage and ground operations. Once in space, the gas is vented by a pressure relief valve. The outer case is actively cooled up to the time the spacecraft is deployed.

Each GPHS generator has an overall length of 114 cm and an overall diameter of 42.2 cm. The converter efficiency is 6.8 percent. Other GPHS-RTG performance data appear in Table 8.10 [28].

Dynamic Radioisotope Power System (DIPS)

For power demands above approximately 1 kilowatt-electric, radioisotope heat sources can be combined with dynamic thermal-to-electrical energy conversion devices to achieve efficiencies about three times higher than those obtainable with thermoelectric converters [29–31]. Since fuel costs are a major factor in the overall costs of radioisotope power supplies, this represents a potentially significant advantage. Extensive engineering investigations have been performed on both the Brayton (gas working fluid) and Rankine (two-phase working fluid) cycle radioisotope-powered systems. For example, the final ground demonstration configuration of an organic fluid Rankine cycle radioisotope system has completed a 2000 hour endurance test with no component failures [29].

The Dynamic Radioisotope Power System (DIPS) is an organic working fluid Rankine cycle system designed to provide 1 to 2 kilowatts-electric power. This power level can be expanded up to 5 kilowatts-electric, if needed. DIPS development reached the stage where a ground demonstration unit was completed, but the program itself has been inactive since December 1980.

Figure 8.33 is a schematic diagram of the DIPS power plant and Fig. 8.34 illustrates typical operating conditions. The organic working fluid is Dowtherm A, a eutectic mixture of biphenyl and biphenyl ether. The heart of the DIPS system is the turbine-alternator-pump power conversion assembly which is mounted on a common shaft (see Fig. 8.35). This single unit is supported by tilt-pad thrust and journal bearings, lubricated by the Dowtherm A working fluid. During system operation, there would be no contact between the rotating shaft and the bearings, except for a few seconds at startup.

As shown in Fig. 8.34, the DIPS has two major flow paths. Starting at the pump discharge, the Dowtherm A organic fluid is a liquid. Approximately 10 percent of the working fluid moves through the inside of the tubes in the regenerator, extracting heat from and cooling the turbine discharge vapor. The working fluid then passes through the spiral tubes surrounding the isotope heat source, where the Dowtherm A is vaporized and heated to its peak temperature of 630 K. This vapor is next expanded through the turbine, where energy is removed and converted to shaft power to drive the alternator

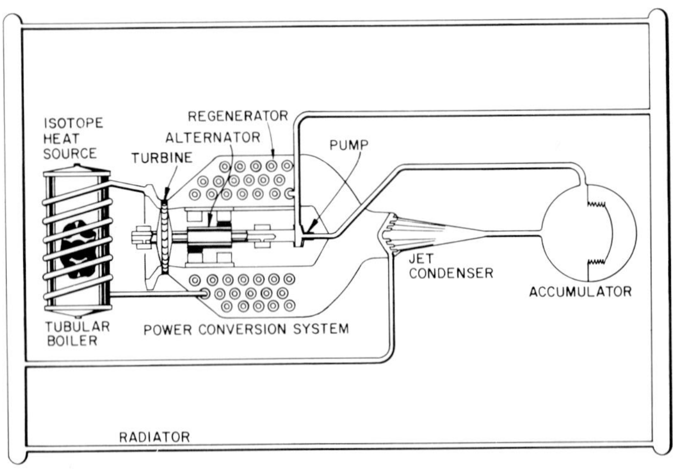

Fig. 8.33 Schematic of Dynamic Isotope Power System (DIPS) powerplant [1].

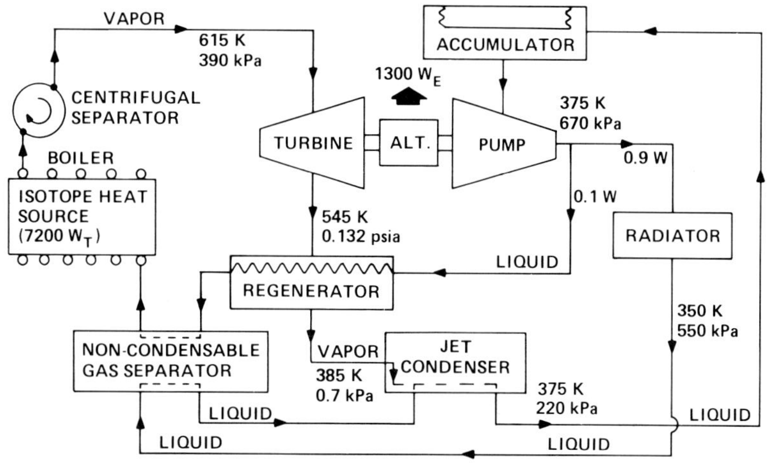

Fig. 8.34 Typical operating conditions for radioisotope organic Rankine cycle generator system; the working fluid is Dowtherm A [*31*].

and pump. After leaving the turbine, the spent vapor passes over the external surfaces of the regenerator tubes where it preheats the pump discharge flow, enhancing system efficiency. The vaporized working fluid is then condensed on the subcooled radiator flow in the jet condenser and proceeds to the accumulator, from which it is returned to the pump, completing the Rankine power conversion cycle.

In the second major working fluid path described in Fig. 8.34, approximately 90 percent of the organic working fluid flow passes through the radiator as a single phase fluid (liquid)—that is, there is no condensation or two-phase flow in this radiator. Instead, the working fluid is cooled in the radiator as a liquid, rejecting the waste heat extracted from the vaporized working fluid in the jet condenser. The liquid Dowtherm A then exits from the radiator to a shower-head type fixture in the jet condenser. The multiple liquid streams generated in the shower head fixture provide the surface area and heat capacity needed to condense the vapor as it exits from the power conversion system regenerator. The jet condenser design for the DIPS is fully effective in the microgravity environment of outer space and will also operate at acceleration levels up to 30 g's, as may be encountered during launch and orbital insertion maneuvers. Ancillary temperature regula-

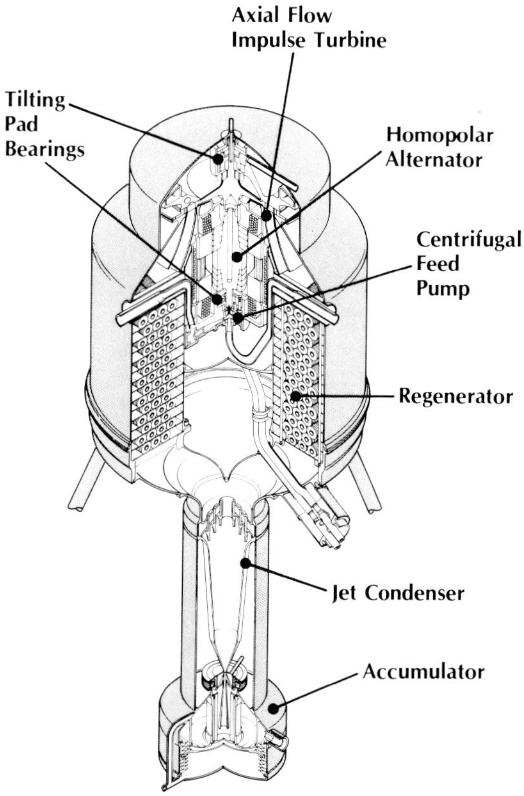

Fig. 8.35 DIPS power conversion system [*31*].

Table 8.11 Dynamic Isotope Power System (DIPS) flight development system characteristics.

Rated Output Power, EOM (W_e)	1300
Optional Ratings (W_e)	500 to 2000
Input Power @ 1300 W_e	7200
Overall Thermal Efficiency[1]	18.1% rectified to 28 VDC
Peak Working Fluid Temperature	630 K
Total Weight[2] (kg)	215
Envelope Dimensions for 1300 W_e	
BOM[3] (cm)	
Diameter	132
Length	24
Design Point Output Voltage (VDC)	24 ± 2
Response—Variable Output Power—0 to 100%	Milliseconds
Partial Output Power Capability	On Pad Launch to Parking Orbit, Orbital Transfer to Final Orbit
Output Power Capability—0 to 100%	Spin Stabilized Spacecraft
Lifetime (years)	7
Capable of Stable Operation with Unbalanced Solar Input	

Resistant to Natural and Induced Radiation

[1] With RTG topping system efficiency can be increased to 22.1% or higher.

[2] Based on updated flight system design utilizing conventional materials, weight reductions such as alternate radiator configuration and converter rotating unit housing machining, possible to approach the 204 Kg goal.

[3] Flexible in design for both size and shape to fit the specific spacecraft.

tion valves and electrical controls (not shown in the figures) maintain the proper relationships of the working fluid temperatures and regulate the electric power output of the system. Table 8.11 provides some of the characteristics of the DIPS flight development system.

Summary

Radioisotope generators have compiled a truly remarkable record in space missions, consistently exceeding their design goals and projected lifetimes. Rigorous safety standards have been established and met in an effort to protect the biosphere against possible radiological hazards. In fact, aerospace nuclear safety considerations largely govern the design of these radioisotope generators. Their prime limitation is cost, due mainly to the high cost of the radioisotopic (plutonium-238) fuel. However, RTGs represent an enabling power supply technology, an electrical energy sine qua non, that can satisfy space mission power requirements where no other practical engineering alternatives exist. High reliability, long lifetimes, the ability to operate in severe environments and at great distances from the Sun—these are all factors that make the radioisotope power supply an ideal choice for selected space missions where power levels of up to a few kilowatts-electric are needed.

References

1. G. L. Bennett, J. J. Lombardo, and B. J. Rock, "Space Nuclear Electric Power Systems," Advances In Astronautical Sciences, Volume 44, American Astronautical Society, 1980.
2. G. L. Bennett, J. J. Lombardo, and B. J. Rock, "Development and Use of Nuclear Power Sources for Space Applications," *The Journal of the Astronautical Sciences*, Vol. XXIX, No. 4, Oct.-Dec. 1981, pp. 321–342.
3. "Viking '75, SNAP/Viking Lander System, Final Safety Analysis Report, Vol. I," Teledyne Isotopes Report ESD-3069-15-1, Timonium, Md.
4. "SNAP-19/Pioneer F Safety Analysis Report, Vol. I, Reference Design Document," Teledyne Isotopes, INSD-2873-42-1, Timonium, Md., December 1970.
5. R. T. Carpenter, "Status of U.S. Radioisotopic Space Power Systems," Isotope and Radiation Technology, Vol. 9, No. 3, pp. 330–344.
6. G. P. Dix, "Advances in the Safety of Space Nuclear Power Systems," 2nd International Symposium on Power From Radioisotopes, Madrid, Spain, June 1972.
7. G. L. Bennett, J. J. Lombardo, and B. J. Rock, "U.S. Radioisotope Thermoelectric Generator Space Operating Experience (June 1961–December 1982)," 18th Intersociety Energy Conversion Engineering Conference, Orlando, Fla., 21–26 August 1983.
8. G. R. Grove, J. A. Powers, N. Goldenberg, D. P. Kelly, and D. L. Prosser, "Plutonium-238 Isotopic Heat Sources: A Summary Report," Mound Laboratory Report MLM-1270, Miamisburg, Ohio, 7 June 1965.
9. P. Ronklove and V. Truscello, "Long-Term Tests of a SNAP-19 Thermoelectric Generator," Intersociety Energy Conversion Engineering Conference Proceedings, 1972, pp. 186–193.
10. C. J. Goebel and L. R. Putnam, "SNAP-19 Performance Update for Pioneer and Viking Missions," 14th Intersociety Energy Conversion Engineering Conference, 5–10 August 1979, pp. 1476–1479.
11. J. A. Van Allen, "Pioneer's Unfunded Reach for the Stars," Aviation Week & Space Technology, 12 April 1982, p. 11.
12. Wayne M. Brittain, "SNAP-19 Radioisotope Thermoelectric Generator Thermal/Electrical Integration With the Viking Mars Lander," Intersociety Energy Conversion Engineering Conference Proceedings, 1971, pp. 1189–1199.
13. G. Stapfer and V. Truscello, "The Long-Term Behavior of SNAP-19 Generators Which Use the TAGS Thermoelectric Material," Intersociety Energy Conversion Engineering Conference Proceedings, 1971, pp. 1184–1189.
14. B. M. French and S. P. Maran, *A Meeting With the Universe*, NASA, EP-177, 1981.
15. A. A. Pitrolo, B. J. Rock, W. Remini, and J. A. Leonard, "SNAP-27 Program Review," Intersociety Energy Conversion Engineering Conference Proceedings, 1969, pp. 153–170.
16. W. C. Remini and J. H. Grayson, "SNAP-27/ALSEP Power Subsystem Used in the Apollo Program," Intersociety Energy Conversion Engineering Conference, 1970 Proceedings, pp. 13-10 to 13–15.
17. "TRANSIT RTG Final Safety Analysis Report, Vol. I," TRW report TRW(A)-11464-0491, March 1971.
18. H. Lurie and S. Rocklin, "Transit RTG–A Status Report," Intersociety Energy Conversion Engineering Conference, 1970 Proceedings, pp. 15–111 to 15–116.
19. P. Ronklove, "Performance Testing of a Transit Generator at JPL," Intersociety Energy Conversion Engineering Conference Proceedings, 1973, pp. 52–57.
20. "Multi-Hundred Watt Radioisotope Thermoelectric Generator Program, Final Safety Analysis Report for the MJS-77 Mission," General Electric Document No. 77SDS4206, January 1977.
21. "Multi-Hundred Watt Radioisotope Thermoelectric Generator Program Final Safety Analysis Report," LES 8/9, General Electric Document No. GEMS-419, March 1975.
22. "Power for Space Experimentation," General Electric Pamphlet GEA-9326(8-74)1M.
23. L. Garvey and G. Stapfer, "Performance Testing of Thermoelectric Generators Including Voyager and LES 8/9 Flight Results," Intersociety Energy Conversion Engineering Conference, 1979 Proceedings, pp. 1470–1475.
24. C. E. Kelly, "The MHW Converter (RTG)," Intersociety Energy Conversion Engineering Conference, 1975 Proceedings, pp. 880–886.
25. V. Shields, "Long-Term Behavior of Silicon Germanium Thermoelectric Generators," Intersociety Energy Conversion Engineering Conference, 1981 Proceedings, pp. 315–320.
26. E. C. Snow, R. W. Zocher, I. M. Grinberg, and L. E. Hulbert, "General-Purpose Heat Source Development Phase II–Conceptual Designs," Los Alamos National Laboratory Report LA-7546-SR, November 1978.
27. A. Schock, "Design, Evolution and Verification of the General-Purpose Heat Source," Intersociety Energy Conversion Engineering Conference, 1980 Proceedings, pp. 1032–1042.
28. G. L. Bennett, J. L. Lombardo, and B. J. Rock, "Nuclear Electric Power for Space Systems: Technology Background and Flight Systems Program," Intersociety Energy Conversion Engineering Conference, 1981 Proceedings, pp. 362–368.
29. "Nuclear Powered Organic Rankine Systems for Space Applications," Proceedings of the AFOSR Special Conference on Prime-Power for High Energy Space Systems, Norfolk, Va. 22–25 February 1982.
30. "Dynamic Isotope Power System Applications Study Results," General Electric Co., Space Division, April 1979.
31. G. L. Sorensen, R. E. Niggemann, E. C. Krueger, R. C. Brouns, and F. A. Russo, "Status Report of the Dynamic Isotope Power System," Intersociety Energy Conversion Engineering Conference, 1979 Proceedings, pp. 1396–1400.

Chapter 9

Uranium-Zirconium-Hydride Reactor Power Plants

Introduction

From the very start of the Space Age it was recognized that ever-increasing amounts of electrical power would be required to support more ambitious space system developments. Consequently, in 1957 active development of nuclear reactors commenced with the development of a uranium-zirconium-hydride (U-ZrH) reactor system [1]. The first design, designated SNAP-2, involved a reactor with a three kilowatt-electric mercury Rankine conversion system. This program demonstrated the U-ZrH concept. It also led to the start in 1958 of a 500 watt-electric power plant using thermoelectric converters, called SNAP-10A. The SNAP-10A, the only U.S. nuclear reactor to fly in space, was successfully flight tested in 1965. U-ZrH reactor development continued in 1960 with the initiation of the SNAP-8 program, a 30-60 kWe power plant, incorporating a higher power level mercury Rankine cycle. The major features of these early U-ZrH reactor systems are shown in Table 9.1. With a change in emphasis in the space program following the Apollo lunar landings, development activities on all these systems were terminated in the early 1970s.

Six U-ZrH reactors were built to provide integral tests of technology at various program stages. Table 9.2 [2, 3] summarizes SNAP reactor test experience, while Fig. 9.1 [4] and Table 9.3 describe major power system component experience. In this chapter we will examine the major engineering features of the SNAP-2, SNAP-10A, and SNAP-8 systems. These now-terminated programs still provide a rich technical heritage for future nuclear power systems planning.

SNAP-2A

Starting in 1957, SNAP-2 system development extended over approximately a decade. This system represented the engineering synthesis of a zirconium-hydride reactor with a mercury Rankine cycle dynamic conversion system. Table 9.4 summarizes the major technical accomplishments of this program.

SNAP-2 was the precursor for both the SNAP-10A and SNAP-8 concepts. At the start of this program

*Units fueled **Table 9.1** U-ZrH reactor power plant summary.

	SNAP-2	SNAP-10A	SNAP-8
Power (kW_e)	3	0.5	35
Design lifetime (yr)	1	1	1
Reactor power (kW_t)	55	30	600
Efficiency (%)	9	1.6	8
Reactor outlet temperature (K)	920	810	975
Reactor type	U-ZrH$_x$ thermal	U-ZrH$_x$ thermal	U-ZrH$_x$ thermal
Primary coolant	NaK-78	NaK-78	NaK-78
Power conversion	Rankine (Hg)	Thermoelectrics (SiGe)	Rankine (Hg)
Boiling temperature (K)	770	—	850
Turbine inlet temperature (K)	895	—	950
Condenser temperature (K)	590	—	645
Hot junction temperature (K)	—	770	—
Cold junction temperature (K)	—	595	—
Radiator temperature (K)	590	595	575
Radiator area (m^2)	11.1	5.8	167.2
System unshielded weight (kg)	545	295	4,545
Specific weight (kg/kW)	182	590	130

Table 9.2 SNAP reactor test experience.

	SNAP-2		SNAP-10A		SNAP-8	
	EXPERIMENTAL REACTOR (SER)	DEVELOPMENTAL REACTOR (S2DR)	FLIGHT SYSTEM FS-3	FS-4	EXPERIMENTAL REACTOR (S8ER)	DEVELOPMENTAL REACTOR (S8DR)
Critical	September 1959	April 1961	January 1965	April 1965	May 1963	June 1968
Shutdown	November 1960	December 1962	March 1966	May 1965	April 1965	December 1969
Thermal Power (kW_t)	50	57	38	42	600	600–1000
Thermal Energy (MWh)	225	273	383	41	5,100	4,350
Electric Power (watts)	—	—	402	650 BOL 535 EOL	—	—
Electric Energy (MWh)	—	—	4.0	0.6	—	—
Time at Power and Temperature	1,800 hr at 920 K 3,500 hr above 750 K	2,800 hr at 920 K 7,700 hr above 750 K	417 days at approx 800 K	43 days above 785 K	100 days at 600 kW_t at 975 K; 1 yr above 400 kW_t at 975 K	429 hr at 1,000 kW_t; 1,670 hr at 965 K; 6,680 hr above 810 K

there were many challenging technical issues to be resolved. Table 9.5 describes some of these issues and the year in which they were resolved or satisfied. The mercury corrosion rate was regarded as one of the main problems to be considered. With mercury as the working fluid, special materials are needed throughout the power conversion system and the "crud" generated by such corrosion processes must be filtered or separated out to prevent possible deposition in critical flow passages. Other technical issues included (1) protecting the alternator stator windings from liquid mercury; (2) compensating for the poor lubricating properties of mercury so that the load-carrying capacity of the bearings would be adequate; (3) cavitation-erosion damage within the pump and bearings; and (4) the critical design tolerances and clearances associated with a dynamic conversion system for space applications. Engineering questions pertaining to the nuclear reactor included fuel fabrication, hydrogen retention, and coolant pumping [4].

Figure 9.2 is a schematic diagram for a 3-5 kWe SNAP-2 unit. The nuclear reactor heat source was cooled by the primary coolant loop which contained the liquid metal eutectic NaK-78, pumped by a thermoelectric DC conduction electromagnetic (EM) pump. Thermal energy was transferred to the secondary loop, the Rankine cycle loop using mercury, through the boiler. The EM pump obtained its electric current from a chromel-constantan themocouple that operated between the reactor outlet temperature and the boiler preheat temperature. On the secondary loop side of the boiler the mercury became a superheated vapor and then expanded through the turbine. When the mercury vapor left the turbine it was condensed in the radiator and was then returned to the boiler by the mercury pump.

In the SNAP-2 design the rotating components of the power conversion machinery were mounted on a single, common shaft. This design feature created a conversion system that had one moving part, supported on liquid mercury bearings. The assembly was enclosed in a hermetic housing that prevented the loss of the mercury working fluid (see Fig. 9.3). The very heart of the power conversion unit was this combined rotating unit (CRU). Mercury vapor entered at a temperature of 895 K and a pressure of 790 kPa, expanded through the two-stage axial flow impulse turbine that in turn drove the three-phase, six-pole permanent magnet alternator. This alternator was a permanent magnet device with a sealed stator. It delivered 5 kilowatts-electric at 1,800 Hz. The shaft of the CRU rotated at 36,000 rpm. The mercury working fluid exhausted at a pressure of 60 kPa and

Table 9.3 Major component experience for U-ZrH reactors.

	Number Units Tested	Total Test (Hrs)	Maximum Temperature (K)	Maximum Duration Single Test (Hr)
ZrH Reactor	6	43,000	975	10,000
NaK Pumps	59	212,000	920	20,000
NaK Volume Accumulators	50	100,000	670	10,000
Control System Actuators	68	241,000	880	24,400
Bearings	93	241,000	1,090	15,500
Radiation Shields	13	91,000	755	10,000

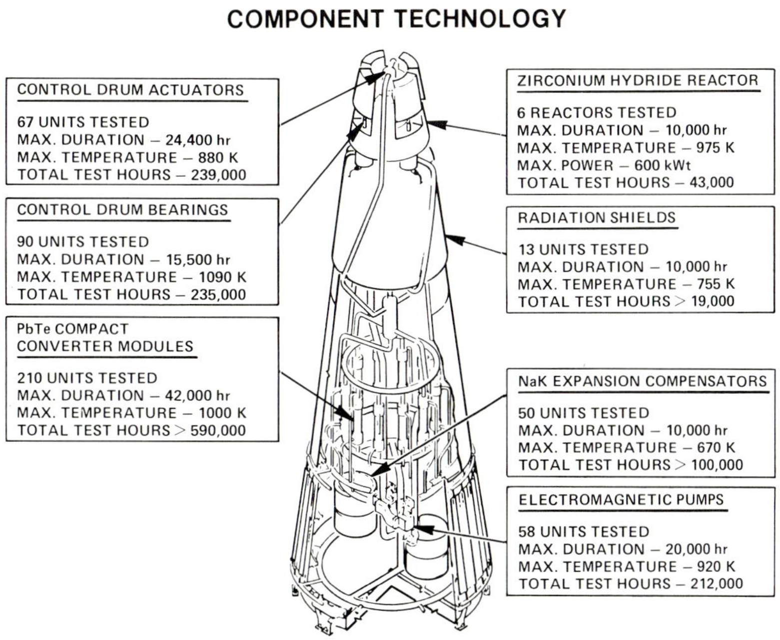

Fig. 9.1 U-ZrH component technology status. *From "ZrH-Reactor-Thermoelectric Space Power System Program Review with the DOD/AEC Ad Hoc Space Power Working Group," Rockwell Report AI-BD-AEC-73-27, 11 September 1973.*

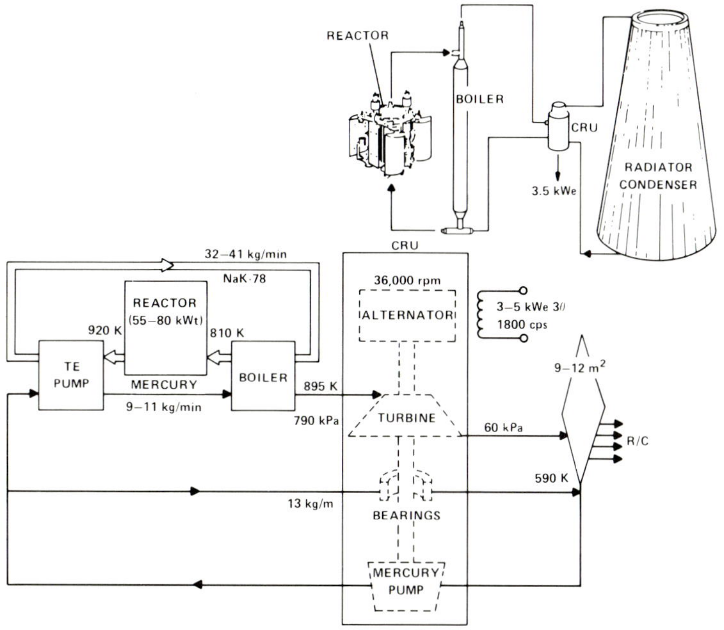

Fig. 9.2 SNAP-2 mercury Rankine system schematic [5].

Table 9.4 Major technical accomplishments of SNAP-2 system development.

- Uranium-zirconium-hydride reactor concept developed to operational nuclear heat source status
- Performance characteristics of a mercury Rankine dynamic energy conversion system demonstrated
- Feasibility of using a high-speed, single-shaft, working fluid-lubricated and hermetically sealed turboalternator assembly verified
- Startup, steady-state operation, and shutdown of a Rankine cycle space power plant demonstrated

cooled the alternator as it flowed through the unit and over the finned stator housing. The shaft itself was supported by two mercury-lubricated journal bearings, the lubricant for which was supplied by the on-shaft centrifugal mercury pump.

A boiler was used to transfer thermal energy liberated in the nuclear reactor to the mercury working fluid, which was superheated prior to entering the turbine. A single pass, counterflow design was used. Because the

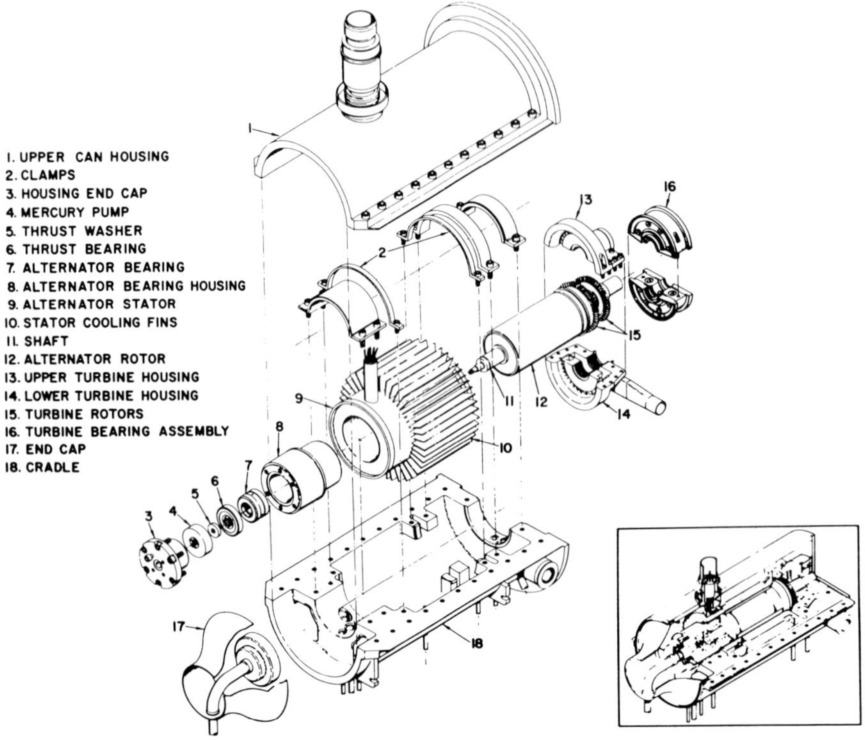

Fig. 9.3 Combined rotating unit components for the SNAP-2 system [5].

boiler needed to function in the microgravity environment of outer space, internal swirl wires were introduced to accommodate better heat transfer. Table 9.6 summarizes the final SNAP-2 system boiler parameters. The final boiler configuration involved four 9.5 mm diameter tubes for the mercury. These were 0.5 mm wall thickness and 5.8 m long. The shell length was 1.4 m with a 10 cm diameter straight cylinder configuration. The boiler had a dry mass of 9.5 kg and a 13.6 kg wet mass.

The NaK liquid metal pump was originally designed

Table 9.5 Technical issues associated with SNAP-2 system development.

Technical Problem Area	Primary Demonstration or Verification	Year Resolved
Criticality of U-ZrH Reactor	Zero Power Critical Experiment	1957
Fabrication of Enriched Uranium ZrH Fuel Elements	SER Fuel Elements	1958
Load Capability-Stability of Mercury-Lubricated Bearings	Prototype Bearing Component Tests	1958
High Speed Operability of Turbo-alternator Assembly	500-hr CRU[a]-1 Test	1959
920 K Power Operation of Reactor	SER Power Demonstration	1959
Zero-Gravity Boiling and Condensing	OFF-BASE Flight Test	1962
Protection of Alternator from Mercury Attack	CRU-V with Evacuated Stator Operation	1964
Long-Life Capability of Turbine, Pump, and Alternator	6,600-hr CRU-IVM-3 Test	1965
Mercury Corrosion Endurance Effects	2,500-hr CRU-V-1A Test (4,700-hr CRU-V-5C Test)	1963 (1966)
Stable Operation of Mercury Rankine	PSM[b]-1 Test	1964
Automatic Orbital Startup of Mercury Rankine System	PSM-3 Test	1965

[a]CRU = Combined Rotating Unit
[b]PSM = Power System Module

Table 9.6 SNAP-2 system boiler parameters.

Secondary (Mercury) Loop Side	
Mercury flow rate (kg/min)	9.1
Outlet temperature (K)	895
Inlet temperature (K)	645
Outlet pressure (kPa)	827
Mercury pressure drop (kPa)	414
Primary (NaK) Loop Side	
NaK flow rate (kg/min)	32
Inlet temperature (K)	920
NaK pressure drop (kPa)	1.7

that changed the relative position of a part of the beryllium reflector. The reflector also contained separate safety elements (see Fig. 9.5).

The core assembly consisted of 37 fuel elements that were arranged in a triangular array on 3.2 cm centers. As a result, the core configuration formed a right hexagonal cylinder about 20 cm across the flats, 23 cm across the corners, and 25 cm in length.

The fuel material was zirconium hydride plus approximately 10 weight percent (10 wt %) of enriched uranium alloy. This fuel mixture contained 90.08 wt % $ZrH_{1.87}$ and 9.92 wt % U. The hydrogen-to-metal ratio was 1.79. The core material volume was 7,000 cubic centimeters, with 3.12% by volume uranium metal and 96.88% by volume zirconium hydride.

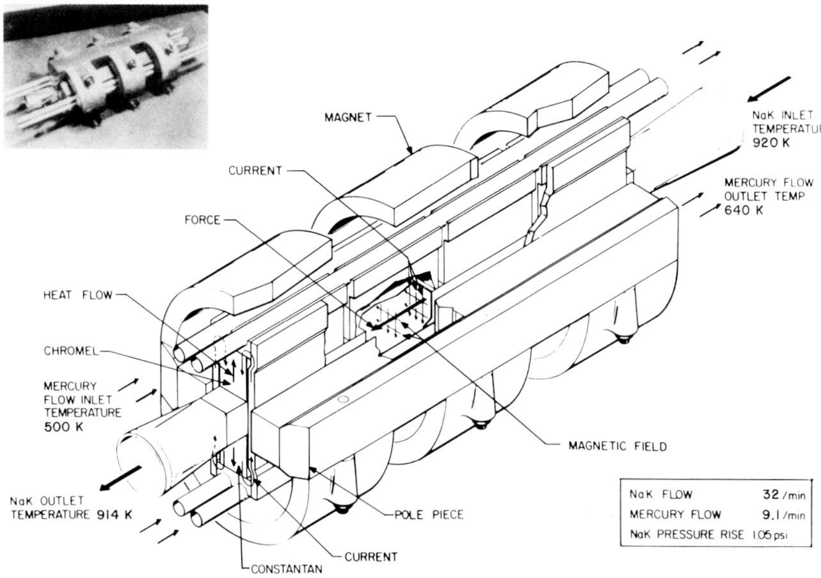

Fig. 9.4 SNAP-2 thermoelectric-electromagnetic NaK pump [5].

as a rotating member of the combined rotating unit. This so complicated the design of an already complex unit that a separate electromagnetic NaK pump unit was eventually chosen (see Fig. 9.4). This thermoelectric EM pump had the advantages of containing no moving machinery, supporting complete liquid metal sealing, providing a built-in electromagnetic flowmeter, and serving as a preheater to the mercury boiler. The pump combined a DC conduction pump and a thermoelectric (TE) generator in a single, all-metallic unit. The cromel and constantan TE elements extracted heat from the hot, circulating NaK and converted a portion of this thermal energy directly into electricity, which was supplied to the conduction pump. Since thermal energy was being rejected from the NaK loop to the mercury loop, the inherently low efficiency of the NaK EM pump proved to be acceptable.

Two reactors were built under the SNAP-2 program, the Experimental Reactor (SER) and the Developmental Reactor (S2DR) [5]. These systems were thermal reactors: fueled with fully enriched uranium-235, hydrogen-moderated, and containing a beryllium reflector. The power level was established by control drums

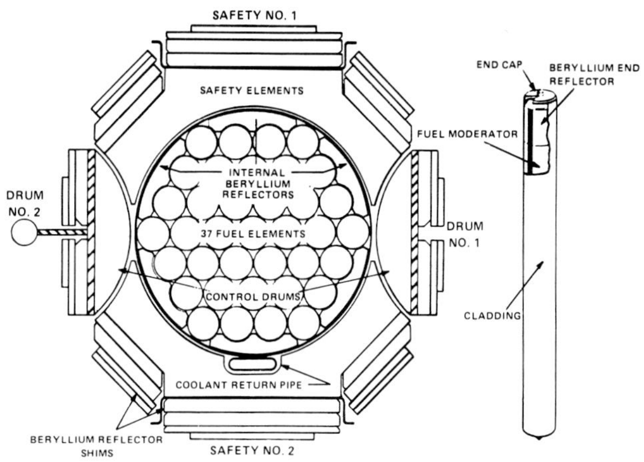

Fig. 9.5 Fuel rod assembly and core cross section for SNAP-2 reactor [5].

The fuel material was fabricated in a solid cylindrical rod, 25 cm long and 3.1 cm in diameter. The fuel element, in turn, was assembled by canning the fuel rod in a 3 cm outside diameter (0.3 mm wall thickness) Hastelloy N cladding tube. On each end of the fuel element there was a 3.8 cm long, 3.1 cm diameter beryllium oxide cylinder that served as an end reflector. The inside surfaces of the cladding tube and the end caps were coated with a ceramic barrier that inhibited the diffusion of hydrogen from the fuel element. This ceramic material contained samarium oxide to prepoison the reactor with equilibrium samarium. The fuel elements were sealed by welding the end caps.

The fuel elements, some 33.6 cm in overall length, were supported and held laterally by top and bottom grid plates. Coolant flow holes (9.5 mm diameter) in these grid plates were localized in line with the core coolant passages formed by each set of three adjacent fuel elements.

The reactor coolant entered the core vessel through the lower grid plate at the center of the bottom head. It then passed through the core into the plenum chamber between the top grid plate and the top vessel head. The coolant exit line originated at the center of the top head. It then made a U-bend and went along the side of the core vessel through a slot, machined in one of the beryllium safety elements. At a flow rate of 32 kg/min, the coolant experienced a total pressure drop of 1.2 kPa in passing through the core vessel. There was 86 kg of NaK in the primary coolant system.

The coolant flow passages were tricusp-shaped channels formed by three adjacent fuel elements—except for the outer row where the flow channels were formed by two adjacent fuel elements and the inner surfaces of the internal beryllium side pieces. The fuel element heat transfer area (over the 25 cm length) was 9,385 cm^2. Consequently, at 50 and 65 killowatts-thermal power level, the average heat flux was 53.3 and 69.3 kW/m^2, respectively. The power density had a peak-to-average ratio of 1.29 in the radial direction and 1.26 in the axial direction. The overall peak-to-average power ratio was 1.63. The peak heat flux was 85.1 kW/m^2 at 50 kW_{th} power level and 112.9 kW/m^2 at 65 kW_{th}.

An inside reflector, consisting of six radial reflector beryllium side pieces, was located within the core vessel. The outside reflector included two control drums with a total reactivity worth of 4.69$. Each control drum had individually driven, constant-speed, reversible motors capable of 0.56 degree per second or a maximum reactivity insertion rate of 2.1¢ per second. The safety elements were attached to the safety element drive yoke by electromagnets used for scramming the reactor. The SNAP-2 reactor could be scrammed in 200 milliseconds with 10.5$ negative reactivity insertion.

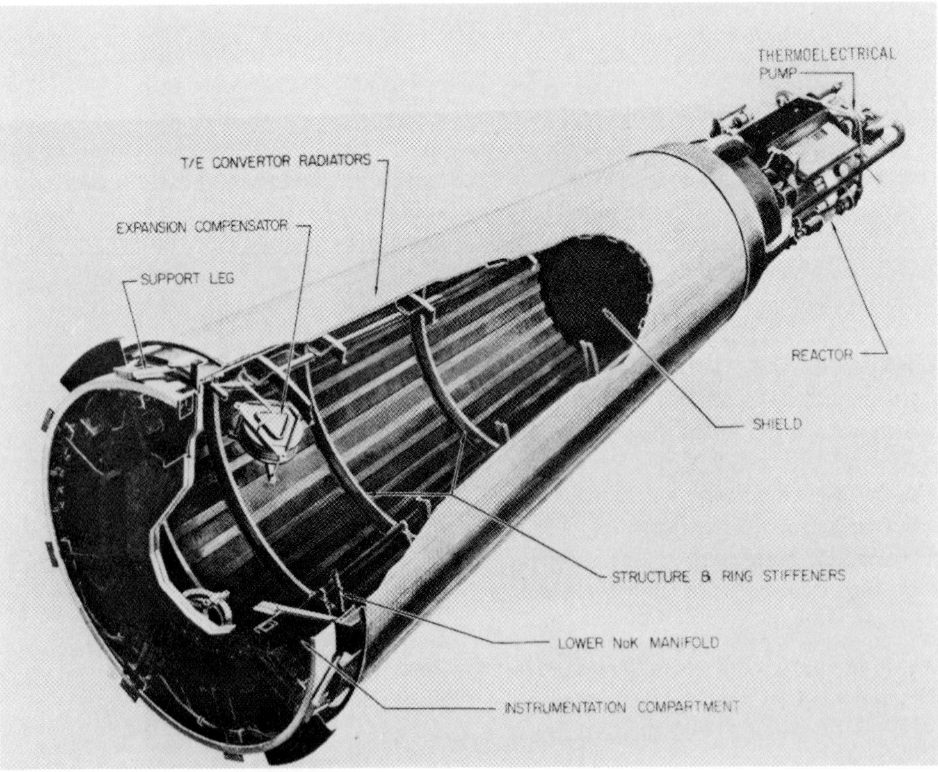

Fig. 9.6 SNAP-10A system [6].

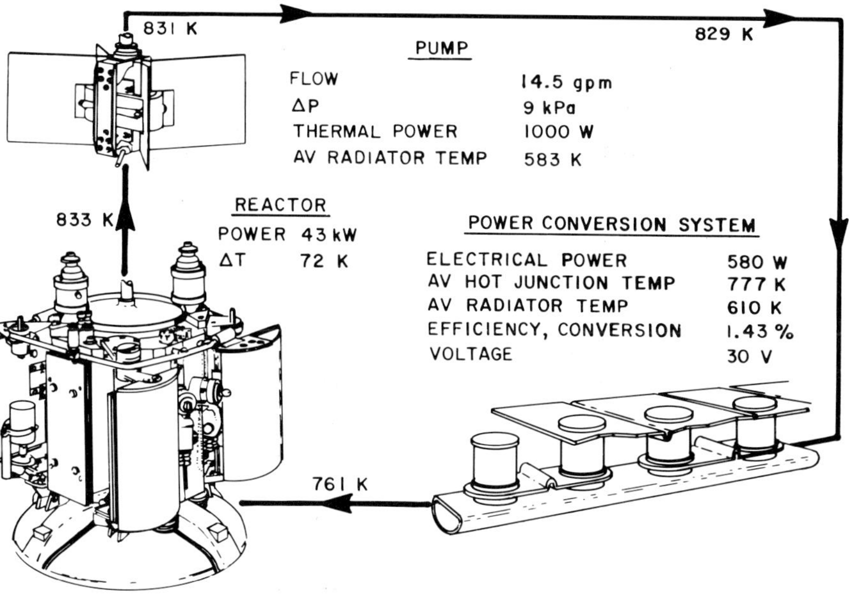

Fig. 9.7 SNAP-10A power conversion system [6].

Two SNAP-2 reactors, the SER and S2DR systems, were tested. The SER was operated at 50 kilowatts-thermal for 1,877 hours with a core outlet temperature of 920 K; and at 50 kilowatts-thermal for 2,290 hours with a core exit temperature less than 920 K. In addition, SER was also operated at less than 50 kW_{th} and less than 920 K for 1,868 hours. The equivalent time at 50 kW_{th} operation was 4,493 hours or some 187 days. There was both power-dependent and temperature-dependent long-term performance degradation. The power-dependent reactor losses were attributed to fuel burnup and fission product poisoning including samarium and xenon buildup. The samarium buildup was calculated to be the largest power-dependent factor. Temperature-dependent losses were attributed to the loss of hydrogen from the fuel rod, a phenomenon apparently caused by hydrogen dissociation from the matrix material and its subsequent diffusion through the cladding. The successful operation of the SER system permitted the S2DR reactor development to proceed with technical confidence.

The S2DR reactor was operated at 30.5 kW_{th} and 745 K for 1,150 hours to simulate SNAP-10 reactor conditions. It was also operated at 55 kW_{th} at a temperature of 920 K for 2,060 hours and at 30.5 kW_{th} at 810 K for 1,544 hours. During the 21 months of S2DR testing, there were no reactor failures. The success of this experiment supported the conclusion that SNAP-2 reactor operation for one year at 55 kW_{th} and a 920 K outlet temperature was feasible.

SNAP-10A

On December 30, 1960 a decision was made to proceed with the design of a convection-cooled reactor system based on SNAP-2 technology. Designated as SNAP-10, it was designed to provide 500 watts of electric power (at a nominal 28 volts) at the end of one year of life. Major SNAP-10A design features included liquid metal heat transfer, a DC electromagnetic pump, and thermoelectric power conversion with a cold junction serving as the heat rejection area to outer space.

The major components of the SNAP-10A system are shown in Fig. 9.6. The thermal reactor contained fully enriched uranium-235 fuel and a zirconium hydride moderator. It was cooled by circulating NaK. Silicon germanium (SiGe) thermoelectric modules were used to generate electricity and the waste heat was rejected to space. Approximately 5.8-m^2 of radiating surface was necessary. A thermoelectric-powered DC conduction pump was used to transport the liquid metal working fluid (NaK) to these TE elements (see Fig. 9.7).

The eutectic mixture, NaK-78, was chosen for thermal energy transfer because of its low melting point and low vapor pressure at reactor operating temperatures. Heated NaK exited the reactor core at 833 K and was circulated at a rate of 0.8 liter per second. The DC conduction pump consisted of a permanent magnet and an integral thermelectric (PbTe material) power supply. This integral TE converter operated between the NaK outlet temperature and a cold junction temperature established by a heat rejection radiator. The corresponding temperature differential (ΔT) was 167 K. As shown in Fig. 9.7, the hot NaK was then circulated through the SiGe TE power conversion system and back into the reactor core. Bellows-type expansion compensators were attached to the return leg. These devices

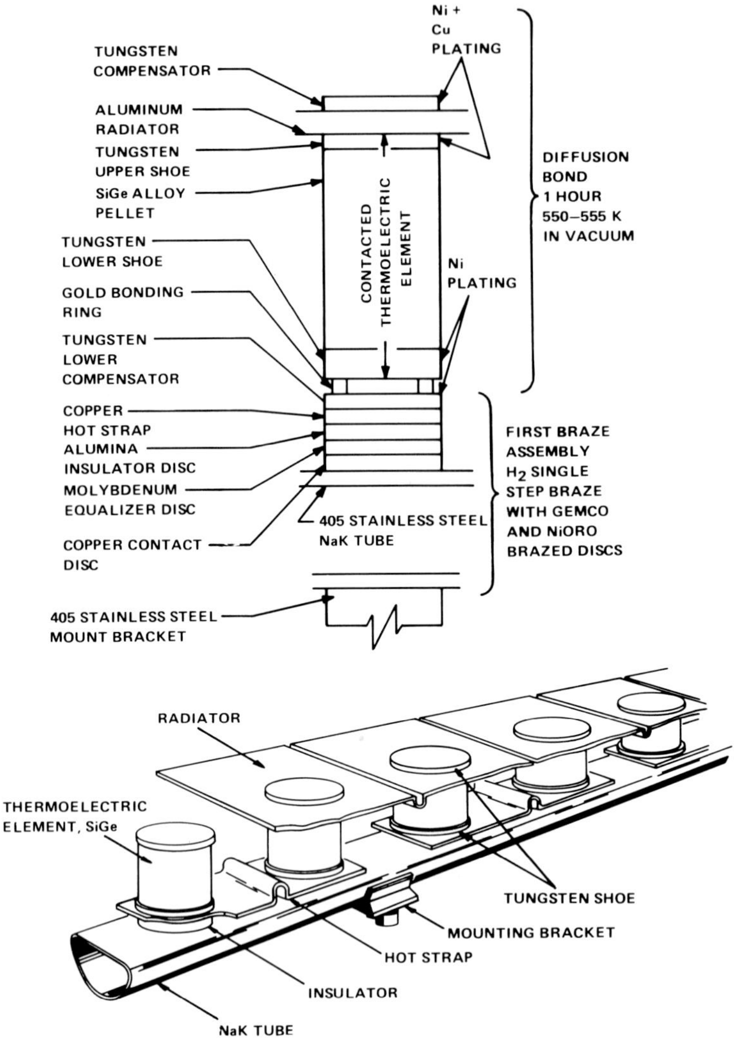

Fig. 9.8 SNAP-10A thermoelectric converter module (top) with construction details (bottom) [4].

pressurized the orbiting system to 34.5 kPa and accommodated thermal expansion of the NaK working fluid during reactor startup.

SiGe TE material was chosen over PbTe for the SNAP-10A system for the following reasons. SiGe is stable at tempertures above 1255 K versus PbTe TE material which sublimes above 700 K, requiring the use of an encapsulant when operated in elevated temperature regimes. Second, stable low-resistance electrical contacts to SiGe can be made by metallurgical bonding. Fabrication experience with PbTe material resulted in only unstable higher resistance mechanical contacting. Finally, the mechanical properties of SiGe alloy were more uniform and less design-restrictive.

The NaK flow was divided equally among 40 tubes. There were 72 SiGe thermocouple elements along each tube, for a total of 2,880 individual TE elements. Fig. 9.8 shows the arrangement and details of the converter. The temperature differential (ΔT) across the TE material was approximately 170 K from the hot to the cold SiGe junctions. The radiator had a total effective area of 5.8 m^2 and an average power dissipation capability of approximately 16.4 W/cm^2 at a mean temperature of 590 K. All materials were brazed or metallurgically bonded to ensure a sound structure and a thermally conductive stack.

The SNAP-10A reactor (see Fig. 9.9 and Table 9.7) was based on SNAP-2A design. It contained 37 uranium-zirconium fuel elements enriched with uranium-235. Each fuel element was clad in Hastelloy N tubing which had a ceramic hydrogen barrier applied to its inner surface. Upper and lower grid plates positioned these fuel elements in the 22.6 cm diameter, 41 cm high core. The grid plates were held by a fixed ring at the bottom and hold-down springs at the top. In SNAP-2 design, beryllium reflectors were used in the void space between the close-packed hexagonal fuel-

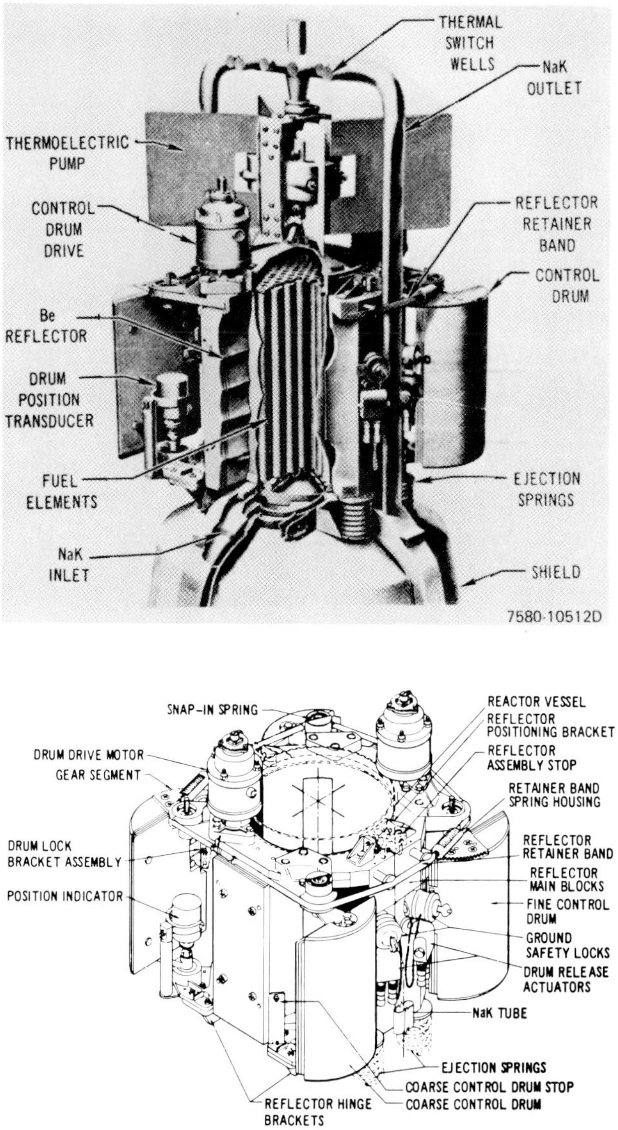

Fig. 9.9 SNAP-10A reactor (top) and control and reflector assembly (bottom) [*4*].

element array and the circular reactor vessel. The top reactor vessel head served as both a vessel closure and restraining structure for the core hold-down springs. Four drums, situated in the beryllium reflector surrounding the reactor, provided coarse and fine control. Reactor control was achieved by moving the cylindrical pieces outward, enchancing neutron leakage from the core.

Flight operations were initiated by a single command from the ground. This command released the locking pins holding the four control drums, applied power to the startup controller, and armed the heat shield eject temperature switch circuit. Two of the control drums were spring loaded and immediately drove to the full-in position; the two remaining drums were automatically inserted by the control system at a rate of 1.0°/300 sec. Approximately seven hours were required to reach criticality, during which time the system remained with the heat shields in place to protect the liquid metal (NaK) working fluid from freezing in the space environment.

After criticality was reached, a nominal power excursion occurred that was terminated by the reactor's inherent negative temperature coefficient of reactivity. With the continued insertion of reactivity, the reactor gradually approached full power. When the temperature of the liquid metal working fluid reached 408 K, the thermal heat shield switches released the heat shield halves and preloaded springs caused their ejection from the power plant. Upon an increase in reactor temperature, the thermoelectric pump, located on the reactor pressure vessel head, immediately bootstrapped flow. When the system reached full operating temperature, the ramp insertion of reactivity was terminated by the action of a temperature switch which sensed reactor outlet temperature. This temperature switch functioned like a thermostat by allowing an occasional reactivity insertion to maintain the reactor outlet temperature about the switch's setpoint. The SNAP-10A power plant was designed to have no moving parts except for the reactor's control system.

Driven by a system design requirement to minimize shield weight, a shadow shield configuration was chosen

Table 9.7 SNAP-10A system design parameters.

System Characteristics	Flight System (1965)
Electrical power, minimum (W)	533
Reactor thermal power, nominal (kW)	43.8
Reactor type	
Fuel	ZrH-U^{235}
Reflector	Be
Design life (years)	1
Radiator area (m^2)	5.8
Overall package length (m)	3.5
Mounting base diameter (m)	1.3
Coolant	NaK-78
Coolant flow rate (gpm)	13.3
Reactor outlet temperature (K)	827
Reactor inlet temperature (K)	790
Power conversion system	
Material	SiGe
Voltage	30.3
Mass (kg)	
Reactor	125
Shield	98
Converter	70
Pump	9
Expansion Compensator	13
Piping and NaK	20
Structure	38
Instrumentation and Compartment	48
Heat Shield	15
TOTAL	436

(see Fig. 9.6). The shielding specification limited radiation doses at the power unit-spacecraft interface to approximately 5×10^{12} fast neutrons and 4×10^{7} rads of gamma radiation in one year. All spacecraft components were located in the conical shadow of this shield to eliminate undesirable radiation scattering. The shield itself was located directly below the reactor and used a cold-pressed lithium hydride material that was reinforced with stainless steel honeycomb contained in a Type 316 stainless steel casing. The shield weighed 98 kg. It was designed to withstand a maximum temperature of 712 K at the attachment to the TE converter radiator and a minimum temperature of 579 K near the surface exposed to space.

This program achieved its objectives through the on-orbit testing of the SNAP-10A reactor in the SNAPSHOT mission. SNAPSHOT, flown in 1965, demonstrated that a nuclear reactor could be successfully launched into and operated in the space environment; and that nuclear reactors could be used safely in space without exposing the terrestrial population or biosphere to undue risk.

Prior to launch, the entire system was checked thermally, using electric heaters in a simulated space environment. The reactor was not, however, operated at full power to avoid fission product buildup that would have inhibited prelaunch access to the unit and severely impaired launch pad activities. Instead, the reactor was neutronically checked in a series of criticality tests that did not involve the creation of a significant quantity of fission product nuclides. In fact, launch pad activities at Vandenberg Air Force Base in California were very much the same for the SNAP-10A payload as for other payloads. The SNAP-10A system was mated to the Atlas-Agena vehicle in the gantry complex, and all subsystems were given a prelaunch check. The reactor itself was not made critical on the pad. The time required to mate components, perform functional checks, and complete booster preparation closely followed patterns experienced with other, non-nuclear payloads. No unique operational constraints were introduced by the presence of the SNAP-10A system and no unusually elaborate aerospace ground handling equipment or procedures were required to support the SNAPSHOT launch. Finally, since the "cold, clean" reactor did not pose a radiation hazard on the pad, normal working access at the launch site to either the payload or its Atlas-Agena launch vehicle was not restricted.

On 3 April 1965 at 1:24 P.M. Pacific Standard Time, SNAPSHOT was launched by an Atlas-Agena vehicle and achieved an orbit with an apogee of 1,328 km and a perigee of 1,295 km. It is interesting to note that the target orbit was 1,296 km circular. At 5:05 P.M. PST the start command was given to the SNAP-10A system. Criticality was reached at 11:15 P.M. PST and full power was obtained at 1:45 A.M. PST on April 4 [6].

After approximately six days at full power, the on board automatic control system was deactivated by ground command. From that time on, the SNAP-10A system was controlled by its inherent negative temperature coefficient of reactivity.

Then, on May 16, 43 days after startup, the SNAP-10A reactor system shut down. All indications are that the reactor performed normally and that this shutdown was caused by an apparent failure in the voltage regulator in the Agena vehicle.

Despite this premature shutdown, SNAPSHOT represents an extremely successful experiment. The entire reactor subsystem performed properly and very predictably. Criticality was achieved at the reactivity levels anticipated, indicating that the launch environment produced no adverse effects on the reactor. Furthermore, there were no signs of self-welding problems nor failures in the high temperature bearings of the reactor control elements. The thermoelectric power conversion system proved reliable, and heat rejection to space occurred a bit more efficiently than anticipated, increasing the power output by 3 percent. The liquid metal coolant loop was not adversely affected by the launch environment, especially launch vibrations, and operated as predicted. In addition, the system apparently did not experience any micrometeoroid punctures of its 0.5 mm thick tubing, and there was no detectable deterioration of applied emissivity surfaces. The launch shock and

Table 9.8 SNAP-8 reactor characteristics [4].

	S8ER	S8DR
Design power level (kW_t)	600	600
Design life (hr)	10,000	12,000
Primary coolant	NaK-78	NaK-78
Number of fuel elements	211	211
Fuel loading (kg of 93.15% enriched U^{235})	6.56	8.2
Core vessel outside diameter (cm)	23.7	23.5
Reflector (thickness of Be, cm)	7.6	13.0
Control drums	6	6
Reactor outlet temperature (K)	977	977
Reactor inlet temperature (K)	866	866
Average power density (kW/liter)		37
Core mass flow (kg/sec)	6.1	6.1
Mean prompt neutron lifetime (μ sec)	6.7	8.4
Effective delayed neutron fraction	0.0077	0.0080
Thermal flux ($n/cm^2/sec$)	2×10^{12}	3×10^{12}
Medium fission energy (eV)	0.21	0.15
Clean, wet excess reactivity ($)		14.4
Control drum worth ($)		21.1
Max. cladding temp at design conditions (K)	1,055	1,015
Max. core temp at design power (K)	1,112	1,080

vibration loads were actually a factor of approximately 1.5 under the qualification levels. Finally, the fast neutron radiation levels were about a factor of four higher on the vehicle centerline and a factor of four lower on the outer vehicle than predicted. The gamma radiation level toward the aft end of the Agena vehicle was about two orders of magnitude lower than anticipated, because of scattering and attenuation by the vehicle structure and components within it.

As a complement to the SNAPSHOT mission, a duplicate of the flight system, called the FS-3, was operated in a shielded vacuum chamber on the ground for 10,000 hours. Post-test examination of this system revealed that there were no apparent wearout modes or incipient failures present.

SNAP-8

A second generation of SNAP nuclear power plants was initiated in 1960 to support anticipated increasing demands for spacecraft power, stimulated by enhanced launch vehicle capabilities that supported larger payloads in orbit. The SNAP-8 system was planned to have a 30 kWe power level with capability to expand to 60 kWe, a 10,000 hour operational lifetime, and a reactor outlet temperature of the working fluid of 975 K. This reactor was based on the SNAP-2 concept of a liquid metal (NaK) cooled, uranium-zirconium-hydride reactor that was then coupled to a mercury working fluid Rankine cycle power conversion system. Two reactors were developed and tested as part of the SNAP-8 program: the SNAP-8 Experimental Reactor (S8ER) from May 1963 through April 1965 and the Developmental Reactor (S8DR) from January 1969 through December 1969. Table 9.8 lists the reactor characteristics of these two systems [4].

Fig. 9.10 is a schematic illustration of the four loop SNAP-8 power plant [7]. The reactor was coupled to the power conversion loop at the mercury boiler. The primary loop, containing a eutectic mixture of sodium and potassium (called NaK-78) transferred thermal energy from the reactor. The liquid metal working fluid was circulated by a specially designed NaK pump. The second or mercury working fluid loop consisted of the boiler, turbine-alternator assembly, condenser, and boiler feed pump. Excess heat was removed by the third loop, another pump-driven NaK loop which coupled the condenser and flight radiator assembly. The fourth loop used an organic fluid and low temperature radiator. It lubricated the bearings of the rotating components in the mercury loop and provided coolant for the alternator, pumps, and electrical controls. There was also an auxiliary loop (not shown in Fig. 9.10) for

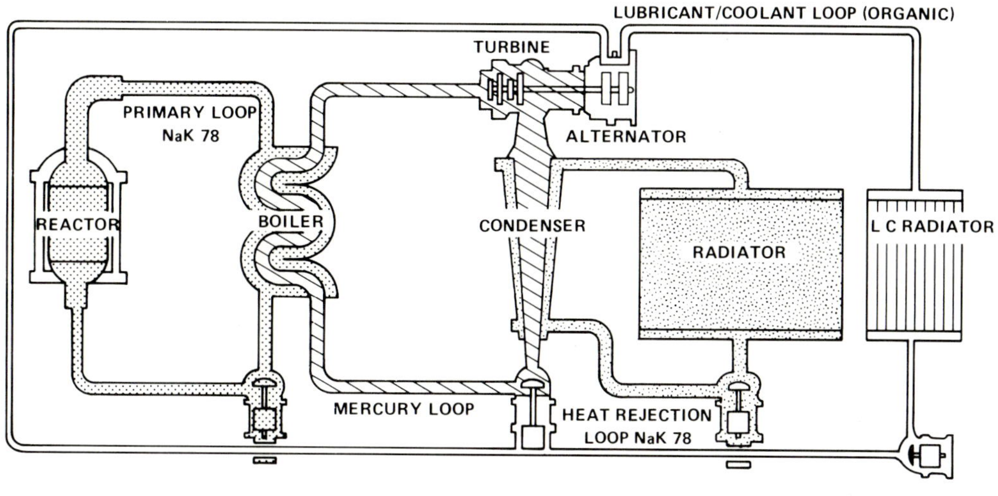

Fig. 9.10 SNAP-8 four loop power plant schematic [7].

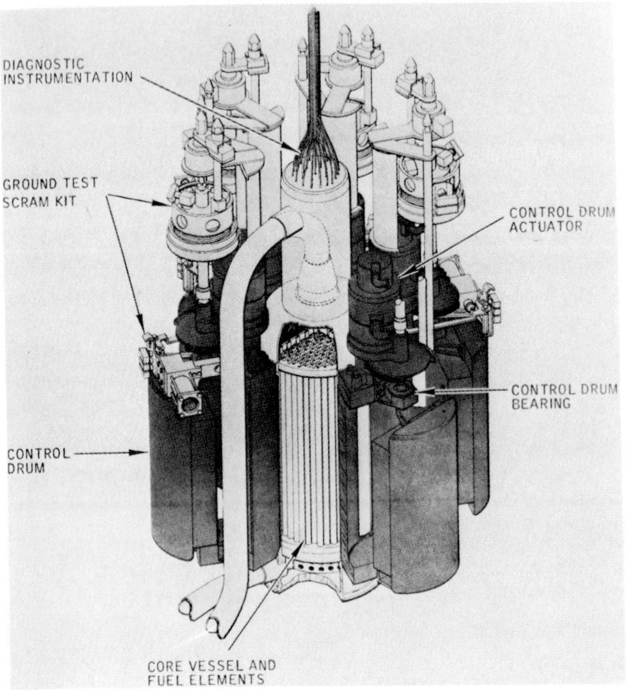

Fig. 9.11 S8DR ground test assembly [*4*].

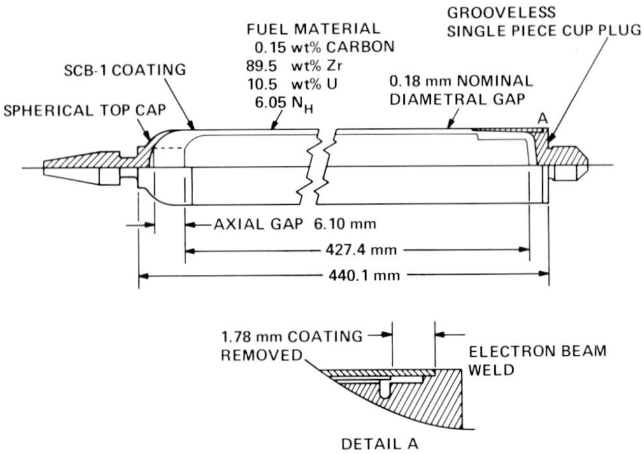

Fig. 9.12 Details of S8DR fuel element [*4*].

startup. It coupled the two NaK loops and provided an intermediate thermal load for the nuclear reactor prior to startup of the mercury loop.

The S8ER system was designed to produce 600 kilowatts of thermal power, with a coolant outlet temperature of 975 K. During testing, this reactor generated over 5.1 million kilowatt-hours of energy—a performance that included 100 days of operation at 600 kW-thermal and 975 K outlet temperature, and 365 days operation at more than 400 kW-thermal and 975 K outlet temperature. These tests demonstrated static and dynamic stability, as well as tolerance to rapid transients. The core was subjected to some 115 rapid changes in power (i.e., 100 kW-thermal in one minute). Post-test core disassembly and fuel element analyses indicated, however, that 80 percent of the 211 fuel elements experienced cladding cracks. Detailed examination of selected fuel elements revealed that these cracks resulted primarily from high temperature swelling of the hydrated fuel, combined with the low ductility of irradiated Hastelloy N cladding material. This phenomenon was attributed to excessive fuel swelling experienced in the core at temperatures greater than 1035 K.

The S8DR reactor was a modified S8ER design. The modifications of the core consisted of: (1) changing the fuel element hydrogen permeation barrier to a thermal cycle resistance coating; (2) increasing radial and axial fuel rod-cladding clearances; (3) redesigning fuel element end caps to alleviate stress cracking; and (4) modifying the coolant flow profile.

The S8DR reactor assembly (see Fig. 9.11) contained 211 fuel elements, each 1.34 cm in diameter and 42.7 cm long (see Fig. 9.12) [4]. The center fuel region contained a hydrated zirconium-uranium alloy with 10.5 wt % U. The hydrogen concentration was 6.05×10^{22} atoms/cc. The axial hydrogen gap was 6.1 mm, while the radial gap was 0.09 mm. A 0.05 mm ceramic coating was included on the inside of the 0.29 mm thick Hastelloy N cladding and it served as a hydrogen diffusion barrier. Included in this ceramic was 58 mg of a burnable poison (Sm_2O_3), which compensated for reactivity loss due to fission product buildup. As shown in Fig. 9.13, the S8DR fuel elements were arranged in a triangular lattice on 14.6 mm centers. The core was

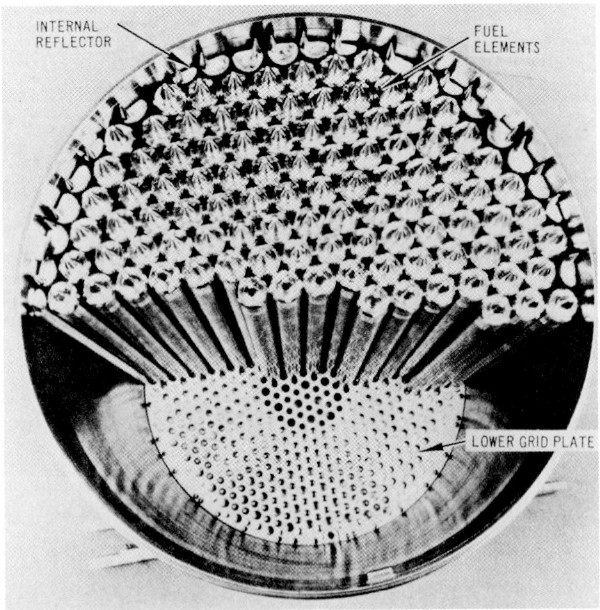

Fig. 9.13 Partial fuel array in SNAP-8 reactor core vessel. *Courtesy of Atomics International (AI)-AEC-13071.*

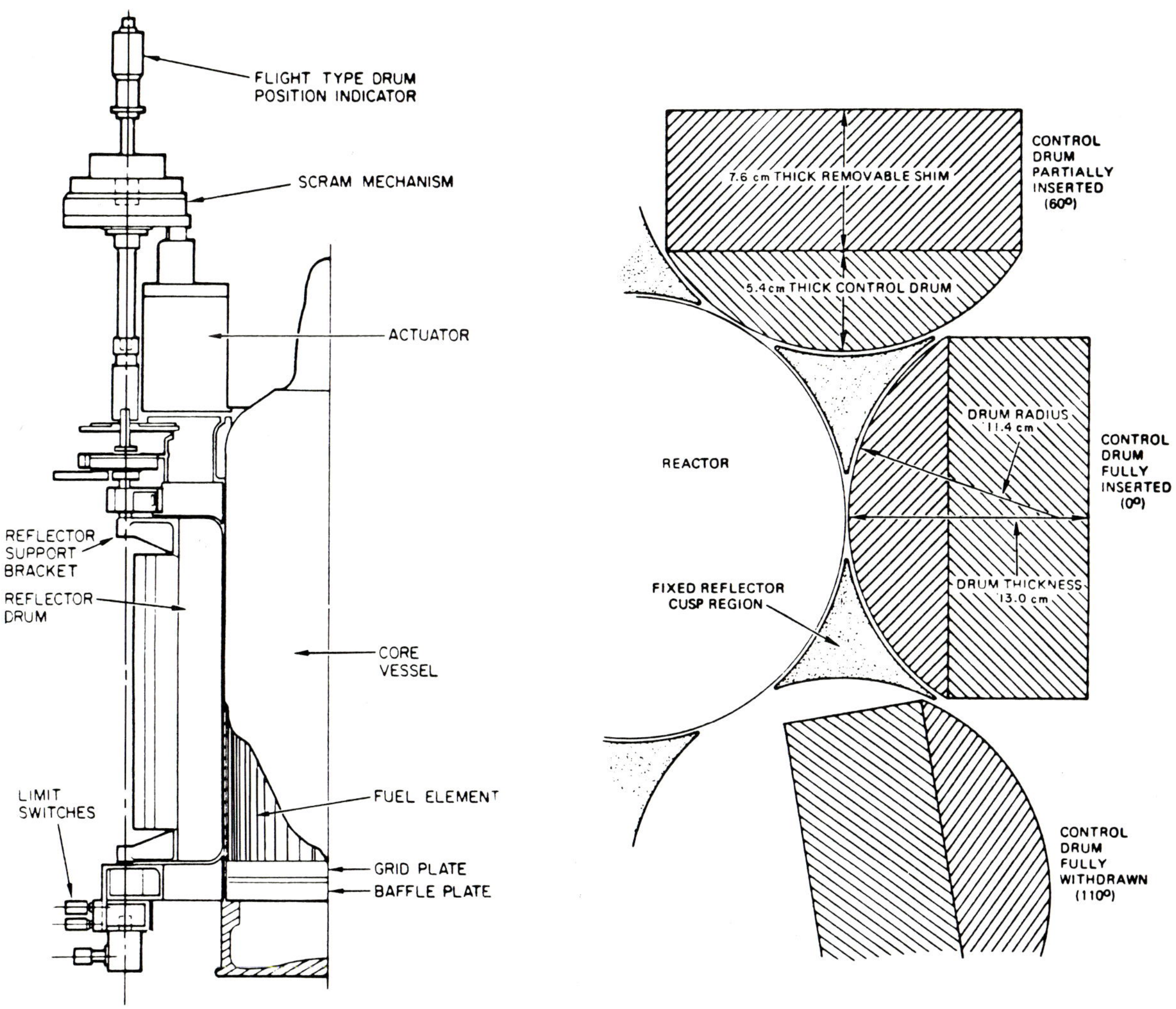

Fig. 9.14 Reflector and control drums for SNAP-8 development reactor [*4*].

cooled by the axial flow of NaK between the fuel elements.

The reactor core was located in a right circular cylinder pressure vessel that consisted of a lower plenum head, an intermediate cylinder, and an upper plenum head. These components were fabricated with Type 316 stainless steel. The lower plenum head had a flat external bottom and a convex 45.7 cm radius internal bottom surface. This plenum head had internal brackets for mounting the flow distributor, the baffle plate, and the lower grid plate. An external mounting ring was used to couple the core vessel to the support structure. The intermediate cylinder, with a nominal wall thickness of 2.3 mm, was welded to the upper and lower plenum heads. This component had internal lugs to support and position the upper grid plate and external brackets on which were mounted the external reflector and control system. The upper plenum head contained a coolant outlet nozzle, 3 holddown lugs for the upper grid plate, 3 support lugs for the thermocouple bridge, and 27 thermocouples to measure coolant temperature at the reactor outlet. The core vessel was designed for operation at a pressure of 241 kPa with an inlet temperature of 894 K and an outlet temperature of 1005 K.

Upper and lower grid plates were used within the vessel to position the fuel elements. A flow distributor and baffle plate were located beneath the lower grid plate to break up the flow coming into the lower plenum, to prevent vortexing, and to provide a constant pressure profile across the lower face of the flow baffle plate. There were 42 internal reflector pieces in the core to fill the void space between the hexagonal core array and the cylindrical vessel. Thirty were made of beryllium oxide, clad in Type 316 stainless steel, while the balance were solid Type 316 stainless steel.

The reactor vessel was surrounded by a beryllium reflector, as shown in Fig. 9.14. This reflector was divided into two stationary and six movable sections. Reactor control was provided through the movable sections. The reflector assemblies were attached to and

Table 9.9 S8DR nuclear characteristics.

Clean, wet, excess reactivity ($)	14.4
NaK reactivity worth ($)	0.25
Total isothermal temperature coefficient of reactivity	− 0.32¢/K at 295 K − 0.36¢/K at 920 K
Power coefficient of reactivity	− 0.035¢/kW
Rise to 600 kW power and 920 K core average temperature ($)	− 2.15
Reactivity inventory (12,000 hours) ($)	
Hot, end of life excess (600 Kw)	5.45
Samarium burnout	2.90
Fission product poisoning	
Xenon	− 1.05
Samarium	− 1.02
Other	− 1.17
Axial hydrogen redistribution	− 0.65
Hydrogen loss	− 5.15
^{235}U burnup	− 0.66
Control drum worth ($)	
Total—6 drums	21.06
Single drum	3.90
Maximum differential worth	5.75¢/deg

supported by brackets that were an integral part of the reactor core vessel. The drums were shaped as segments of right-circular cylinders with 11.4 cm radii. The basic drum had a length of 38.9 cm and a maximum thickness of 5.4 cm. A 7.6 cm beryllium shim was attached to the back of each drum to make up the total thickness of 13.0 cm. All beryllium surfaces were anodized to provide a high emissivity. The reflector assembly was designed to limit the beryllium temperature to less than 1005 K. Each control drum rotated on a pair of upper and lower self-aligning bearings. These bearings were of a journal type, designed for operation at 894 K. The control drums were designed to compensate for the reactivity changes listed in Table 9.9.

The control drum actuator was an eight-pole reluctance motor that produced torque by the interaction of magnetic forces between the stationary teeth on the fixed poles and the notched teeth cut into the cylindrical rotor. The field coils for the poles were sequentially energized to produce a revolving magnetic field that turned the rotor. The stator-pole tooth alignment was such that, as each pole was energized, the rotor teeth were in a position to provide maximum torque. When not energized, the rotor was held in position by a brake mechanism. The control drum drives operated in conjunction with a scram mechanism that would rapidly rotate the control drums outward, when an electromagnetic clutch was released.

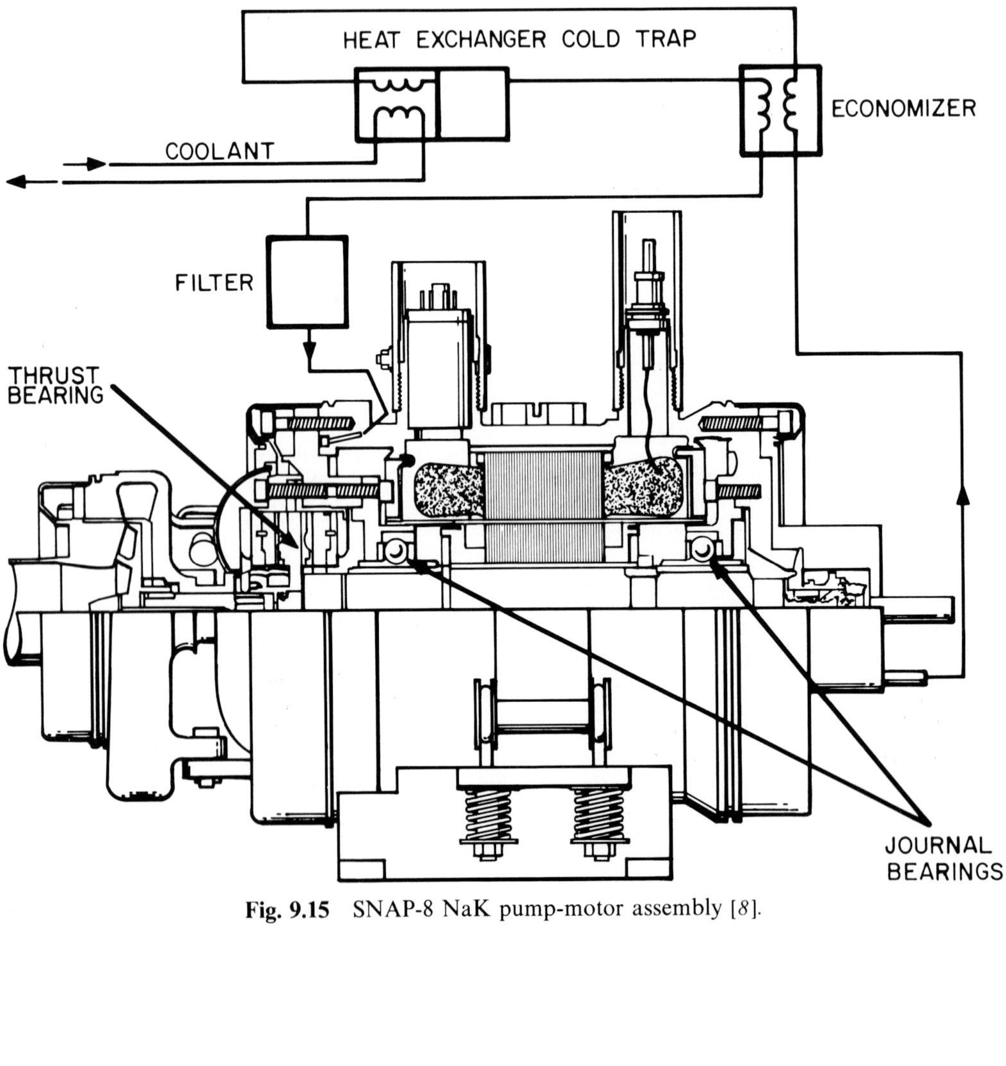

Fig. 9.15 SNAP-8 NaK pump-motor assembly [8].

Operation of the S8DR system generated 4.3 million kilowatt-hours, a performance with 1,670 hours at 600 kW-thermal and an outlet temperature of 975 K, and 5,575 hours at 600 kW-thermal and an outlet temperature greater than 920 K. The fuel elements exhibited failure modes similar to those experienced by the S8ER fuel elements, with 72 out of 211 elements developing cladding cracks typical of stress-rupture phenomenon. These failures were attributed to excessive cladding strain, caused by fuel swelling resulting from core temperatures in excess of design values. The probable source of this high temperature environment was entrainment of surge tank cover gas. This entrainment process was coupled with two standing, stable vortices in the inner plenum. The vortices tended to agglomerate small entrained gas bubbles and periodically release large bubbles that passed through the core in the region of maximum fuel element damage. Stable clustering also occurred, causing additional flow decreases and, consequently, fuel element temperature increases. This entire process resulted in fuel element temperatures that were estimated to be over 55 K hotter than the (maximum) design value [4]. Once again, such over-temperatures were attributed to thermohydraulic phenomena within the core. Other reactor components operated satisfactorily.

The SNAP-8 boiler was a single-pass counterflow heat exchanger that consisted of seven mercury containment tubes in a single NaK tube outershell, made of Type 316 stainless steel. Because of its superior mercury corrosion resistance and good mercury "wetting" properties, tantalum was selected as the mercury containment material. A non-flowing layer of NaK, contained in Type 321 stainless steel tubing, surrounded each tantalum tube. The primary coolant loop working fluid (NaK) flowed over the outside of these Type 321 stainless steel tubes. The nonflowing layer of NaK protected the tantalum from corrosion caused by interaction with the impurity materials in the primary loop NaK.

Two pumps were used to circulate the NaK working fluid through the power conversion system. One of these pumps was located in the primary loop, the other in the heat rejection (third) loop. A special, single pump was designed and developed to suit each purpose. The pump motor assembly (see Fig. 9.15) was a self-contained, hermetically sealed unit. This unit combined on a single shaft: a centrifugal pump, a hermetically sealed motor, an internally lubricated coolant recirculating pump, and NaK-lubricated bearings. A recirculating loop was integrated with the pump motor assembly. This loop contained a "coldtrap" system, a heat exchanger, and a NaK filter. No static or dynamic shaft seals were used in the NaK motor. Isolation of the NaK in the closed recirculating loop from the NaK in the primary loop was achieved through the use of close-clearance annulus around the shaft between the pump and motor. A feature of this design concept, a design in which a high thermal gradient existed between the pump and the motor, is that the recirculating fluid remained relatively uncontaminated by the primary loop working fluid. It should be noted that the primary loop NaK could contain a much higher oxide level than could be tolerated in the bearings and motor. Any oxides migrating through the annulus into the recirculating loop were trapped out.

A three-phase, 208 volt, 400 cycle squirrel-cage induction motor was used to drive the pump. The rotor and stators were canned with thin cylinders of Inconel. The rotor shaft was supported on two tilting pad journal bearings in assemblies of four pads per bearing. This design was selected because of its ability to accept misalignment, its inherent stability, and its low frictional characteristics.

The mercury turbine-alternator assembly in the secondary loop consisted of two subassemblies: the turbine and the alternator which were bolted together at a common flange. Turbine power was transmitted to the alternator by a splined drive on a flexible quill shaft (see Fig. 9.16). The turbine was a cantilevered, four-stage axial-flow type. The first and second stages were partial admissions, and the third and fourth stages were full admissions. Preloaded, angular-contact ball bearings were used with an over-hung turbine, and a space-seal system was interposed between the turbine exhaust duct and the outboard turbine bearings. The alternator was a hermetically sealed, three-phase, 220/208 V, 400 cycle, radial air-gap, homopolar inductor. It featured a brushless solid rotor straddle-mounted on angular-contact ball bearings. The turbine-alternator was lubricated and cooled by organic polyphenyl-ether.

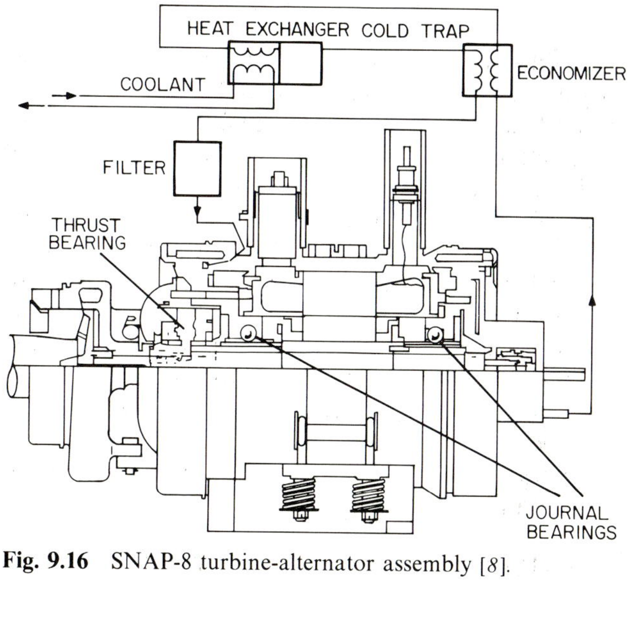

Fig. 9.16 SNAP-8 turbine-alternator assembly [8].

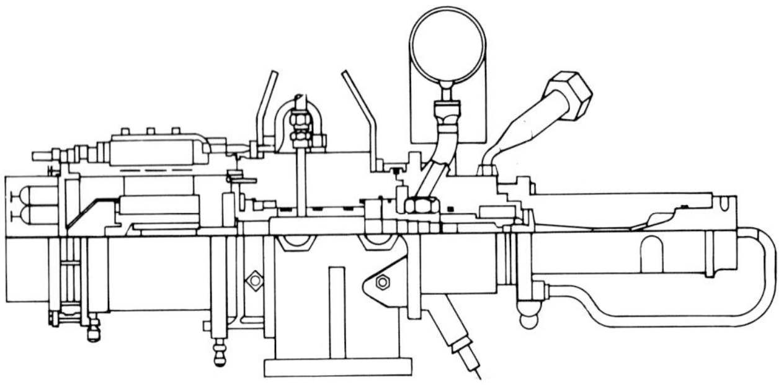

Fig. 9.17 SNAP-8 mercury pump-motor assembly [*8*].

The mercury pump motor assembly was a self-contained system with a "seal-to-space" (a mechanical shaft seal discharging to vacuum) (see Fig. 9.17) [8]. The pumping system consisted of a centrifugal pump, low-leakage dynamic shaft seals, and an induction motor. All of these components were located on one shaft, supported by angular-contact ball bearings lubricated with polyphenyl-ether. The motor and shaft seals were cooled with the lubricant fluid. Integral with this assembly were a jet-booster pump and lubricant-coolant valves along with their associated plumbing. The jet-booster pump was used to improve the suction performance of the centrifugal pump. The centrifugal pump was a conventional design. The volute was a partial-emission design and the back vanes of the pump impeller were used to balance the axial thrust.

The SNAP-8 condenser was a counterflow, tube-in-shell heat exchanger. The mercury working fluid was condensed in 73 tapered tubes. The tapered tubes were designed to maintain a high vapor velocity through the length of the condenser, sustaining a continual movement of condensate droplets toward the liquid-vapor interface in a microgravity environment. NaK was the shell-side coolant and it flowed through the condenser as part of the heat rejection system.

In the SNAP-8 program considerable operating experience was gathered during groundbased testing of the S8ER and S8DR reactors. The S8ER was operated for over one year at a power level between 400 and 600 kW-thermal and a reactor outlet temperature of 975 K. The S8DR reactor was operated for 1,670 hours at 600 kW-thermal and a reactor outlet temperature of 965 K and for 6,688 hours at 600 kW-thermal and reactor outlet temperatures greater than 810 K. In reviewing the S8DR test data, we should keep in mind that even though the S8DR reactor developed a significant number of cracked fuel elements, appreciable operating life still remained when it was shut down. In fact, it has been predicted that this reactor could have operated for more than 16,000 hours [9].

A major operating difficulty experienced by the S8DR system was that fuel element mechanical instability, coupled with larger than anticipated core coolant flow bypass, caused excessive fuel element temperatures in selected core locations. This condition, in turn, resulted in greater than expected fuel swelling, which subsequently led to fuel element cladding cracking.

Since the U-ZrH reactors used integral moderated fuel, the fuel element cladding must be lined with a ceramic coating that acts as a hydrogen permeation barrier. (An integral moderated fuel is a fuel that is a mixture of uranium with hydrided metallic zirconium, which is equilibrated by a low hydrogen overpressure, approximately 330 mm of H_G at temperature.) Excessive deformation or cracking of the cladding degrades or fractures the hydrogen barrier. This permits a higher than predicted loss of hydrogen that ultimately results in a far-shortened operating lifetime.

The SNAP-8 power conversion system components were well on their way toward being proven items [10, 11]. All the major components were operated at least 10,000 hours without maintenance. A "breadboard" configuration had also been run successfully for 8,700 hours [8]. The turbine-alternator was operated for a total of 10,823 hours at system-rated load conditions and 35 start-stop cycles. During the last approximately 5,000 hours of this test, a 0.5 percent loss in turbine-alternator aerodynamic efficiency was observed. This loss was attributed primarily to a reduction in the third-stage nozzle area due to deposits of mass transfer products and also due to rotor blade leading edge erosion and corrosion damage. These conclusions were based on post-test inspection of the turbine-alternator, flow-testing of the nozzle assemblies, and measurements of blade erosion.

The mercury boiler feed-pump was tested at rated conditions for 12,227 hours and 109 start-stop cycles.

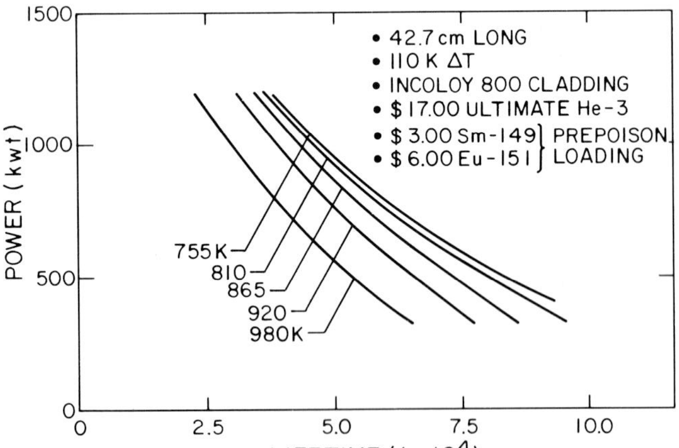

Fig. 9.18 Reference ZrH reactor performance [*4*].

Hydraulic performance tests were conducted at intervals to determine whether changes had occurred. The hydraulic-electric efficiency of the pump was the same at the end of the test as at the beginning.

The SNAP-8 NaK pump was tested as a complete assembly for 10,363 hours and 786 start-stop cycles. The hydraulic and electric performance of the pump was identical throughout this endurance test. Significant cavitation damage, however, had occurred in the impeller back vane hub area. This damage was so severe that some local penetrations of the hub wall had occurred.

We can make a general conclusion concerning the life expectancy of rotating components based on these performance tests—namely, that five years could be achieved with the exception of the mercury pump and impeller. The condition of seals must also be carefully evaluated, and some improvement in the turbine rotor might also be necessary. As shown in Fig. 9.18, the lifetime of ZrH reactor systems can be increased considerably by reducing the temperature [12].

In conclusion, we can say that long-life U-ZrH reactor systems, such as five year lifetime at 920 K, are obtainable. A number of ways have been examined to improve the fuel elements and the coolant flow distribution in the core. For example, the use of fins on the fuel elements would reduce temperature differences across single fuel elements and maximize channel-to-channel coolant mixing.

Improvements of the fuel elements are also possible, such as the substitution of Inconel 800 for the Hastelloy N. When compared to the Hastelloy N, this material would provide improved strain capability without a great loss of relative strength. In addition, a method was developed to produce a highly adherent hydrogen barrier coating on the Inconel 800, using a much simpler process than possible on the Hastelloy N material. Finally, the hydrogen content of the uranium-zirconium-hydride fuel material can also be increased to provide greater lifetime capability.

References

1. James H. Van Osdol, W. B. Thompson, and O. S. Merrill, "Uranium Zirconium Hydride Reactor Space Power Systems," International Astronautical Federation XXVIIth Congress, Anaheim, Calif., 10-16 October 1976, Paper No. 76-256, 250 Rue Saint-Jacques, 75005 Paris, France, p. 6.
2. "A Management Summary of the ZrH Reactor-Thermoelectric Space Power System Program," Atomic International Report AI-BD-AEC-73-23, Canoga Park, Calif.
3. "10 to 75 kW_e Reactor Power Systems for Space Applications," Rockwell International Report No. N652T1140013, 24 March 1976, Canoga Park, Calif.
4. D. W. Staub (Ed.), "SNAP Programs Summary Report," Atomics International Report No. AI-AEC-13067, 30 July 1973.
5. A. A. Jarrett, "SNAP-2 Summary Report," Atomics International Report No. AI-AEC-13068, 27 July 1973.
6. H. M. Dieckamp, "Nuclear Space Power Systems," Atomics International report issued September 1967, Canoga Park, Calif.
7. J. N. Hodgson, R. G. Germer, A. H. Kreeger, "SNAP-8—A Technical Assessment," Proceedings of Intersociety Energy Conversion Engineering Conference, 1967, pp. 135–143.
8. J. N. Hodgson and R. P. Macosko, "A SNAP-8 Breadboard System—Operating Experience," Intersociety Energy Conversion Engineering Conference (1968), pp. 338–351.
9. J. K. Asquith and D. G. Mason, "Zirconium Hydride Space Power Reactor Design," Proceedings of Intersociety Energy Conversion Engineering Conference, 1972, pp. 572–580.
10. J. R. Pope and A. Stromquist, "SNAP-8 Rotating Machinery Components—10,000 Hour Endurance Achievement," Proceedings of Intersociety Energy Conversion Engineering Conference, 1970, pp. 11–67 to 11–70.
11. H. Derow, "SNAP-8 Power Conversion System Breadboard Assembly—Materials Evaluation After 8,700 Hours Operation," Proceedings of Intersociety Energy Conversion Engineering Conference, 1970, pp. 11–24 to 11–27.
12. R. L. Detterman, L. M. Maki, R. F. Wilson, and D. E. Reardon, "Technology Status—ZrH Power Reactors," Intersociety Energy Conversion Engineering Conference (1970), p. 12–26 to 12–32.

Bibliography

J. G. Asquith, D. G. Mason, and S. Stamp, "Zirconium Hydride Space Power Reactor Design," Intersociety Energy Conversion Engineering Conference Proceedings (1972), p. 572–580.

G. E. Berg, "Final SNAPSHOT Performance Report," Rockwell International Report NAA-SR-11934 (15 August 1966).

G. F. Burdi, "SNAP Technology Handbook, Vol. I, Liquid Metals," Rockwell International Report No. NAA-SR-8617, Vol. I (11 August 1964).

W. Corliss, "SNAP Nuclear Space Reactors," U.S. Atomic Energy Commission, Library of Congress Catalog No. 66-62772 (1966).

H. Derow, "SNAP-8 Power Conversion System, Breadboard Assembly-Materials Evaluation After 8700 Hours of Operation," Intersociety Energy Conversion Engineering Conference Procedures (1970), p. 11–24 to 11–27.

L. E. Donelan, "Space Nuclear System Sheathed Electrical Cables Summary Report," Rockwell International Report No. AI-AEC-13089 (June 1973).

L. E. Donelan, "Zirconium Hydride Reactor Control Drive Actuator Development Summary Report," Rockwell International Report AI-AEC-13080 (June 1973).

J. P. Dooley and J. P. Beall, "The Disassembly and Post Operation Component Examination of the SNAP-10A FS-3," Rockwell International Report NAA-SR-12504 (15 September 1967).

R. E. Dyer, "Disassembly and Post Operation Examination of the SNAP-8 Experimental Reactor," Rockwell International Report NAA-SR-11520 (1965).

R. R. Eggleston (Ed.), "SNAP Programs Bibliography," Rockwell International Report AI-AEC-13066 (30 June 1973).

L. M. Fead, L. D. Felton, and V. L. Rooney, "SNAP-8 Experimental Reactor Operations and Test Results," Rockwell International Report NAA-SR-10903 (28 June 1965).

J. P. Hawley, "Water Immersion Safety for SNAP Reactors," Rockwell International Report NAA-SR-11361 (30 September 1967).

J. P. Hawley and R. A. Johnson, "SNAP-10A FS-3 Reactor Performance," Rockwell International Report NAA-SR-11397 (15 August 1966).

J. N. Hodgson and R. P. Macosko, "A SNAP-8 Breadboard System —Operating Experience," Intersociety Energy Conversion Engineering Conference Proceedings (1968), p. 338–351.

P. H. Horton and W. J. Kruzeka, "Zirconium Hydride Reactor Control Reflector Systems Summary Report," Rockwell International Report AI-AEC-13078 (June 1973).

H. G. Hurrell, F. Boecker, Jr., and K. S. Jefferies, "Startup Testing of the SNAP-8 Power Conversion System," Intersociety Energy Conversion Engineering Conference Proceedings (1970), p. 11–28 to 11–34.

C. S. Mak and J. R. Pope, "Operation Experience with NaK-Lubricated Journal and Thrust Bearings for a NaK Pump-Motor Assembly," Intersociety Energy Conversion Engineering Conference Proceedings (1970), p. 11–71 to 11–73.

J. M. Otter, K. E. Buttrey, and R. P. Johnson, "Aerospace Safety Program Summary Report," Rockwell International Report AI-AEC-13100 (June 1973).

Staff, "SNAP-8 Summary—S8ER and S8DS," Rockwell International Report NAA-SR-Memo-12099 (30 August 1966).

D. W. Staub, "SNAP-10A Summary Report," Rockwell International Report NAA-SR-12073 (25 March 1967).

O. P. Steele, III, and E. J. Donovan, "Design Summary Reflector Drive System" Rockwell International Report TI-652-220-006 (10 October 1972).

A. W. Thiele and S. Berger, "SNAP-8 Summary Report," Rockwell International Report AI-AEC-13070 (June 1973).

G. M. Thur, "Design Changes in SNAP-8 for Reduced Reactor Operating Temperatures," Intersociety Energy Conversion Engineering Conference Proceedings (1970), paper. no. 709137.

G. M. Thur, "SNAP-8 Power Conversion System Assessment," Intersociety Energy Conversion Engineering Conference Proceedings (1968), p. 329–337.

J. H. Van Osdol, "Design and Integration of Zirconium Hydride Reactor Power Systems," Rockwell International Report AI-AEC-13072 (June 1973).

J. H. Van Osdol, W. B. Thompson, and O. S. Merrill, "Uranium Zirconium Hydride Reactor Space Power Systems," International Astronautical Federation, XXVIIth Congress (October 1976), paper no. 76-256.

F. H. Welch, E. C. Phillips, J. G. Asquith, and N. F. Davies, "Lithium Hydride Technology, Vol. I, Properties of Lithium Hydride and Corrosion Studies," Rockwell International Report NAA-SR-9400, Vol. I (8 May 1964).

Chapter 10

Solid-Core Nuclear Rocket Program

Introduction

The basic secret of space travel and extending human presence throughout heliocentric space is energy, immense quantities of energy. The energy leverage of nuclear fission will ultimately allow us to develop spaceships that can sweep across the Solar System, carrying people and sophisticated machines to the farthest reaches of the Sun's domain.

The rocket engine operates according to the basic principles expressed in Newton's third law of motion: for every action there is an equal and opposite reaction. In a rocket, hot gases created by chemical combustion or from the application of nuclear energy sources flow through the throat of the rocket nozzle, expanding and developing thrust. A simple nuclear rocket system is illustrated in Fig. 10.1. It consists of a single propellant tank (usually cryogenically stored hydrogen), a nuclear reactor energy source, and turbopump machinery to move the propulsive working fluid into the reactor, where it is then heated and exhausted at very high temperatures through the nozzle assembly. This expansion process of the heated propellant delivers thrust to the rocket vehicle assembly and its payload.

The performance of a rocket is measured in terms of the specific impulse which is defined as

$$I_{sp} = \frac{\text{thrust}}{\text{mass flow rate of propellant}} = \frac{F}{\dot{m}} \qquad (10.1)$$

where F is the thrust in newtons (N) or pounds-force (lb_f)
$\dot{m}$ is the mass flow rate of propellant in kilograms per second (kg/s) or pounds-mass per second (lb_m/s).

We have used the International System of Units (SI System) throughout this book. However, to avoid possible confusion, it is also necessary here to briefly introduce the traditional engineering or English unit system, since it is also frequently encountered in propulsion. In the SI unit system eq. 10.1 has the following dimensions

$$I_{sp} = \frac{F}{\dot{m}} = \frac{\text{newtons}}{\text{kilograms/second}}. \qquad (10.2)$$

Since one newton equals one kilogram-meter per second squared ($1N = 1\ kg\text{-}m/s^2$), the basic units for specific impulse are

$$I_{sp} = \text{meter/second (m/s)}. \qquad (10.3)$$

In the traditional engineering or English system of units, however, the term specific impulse frequently

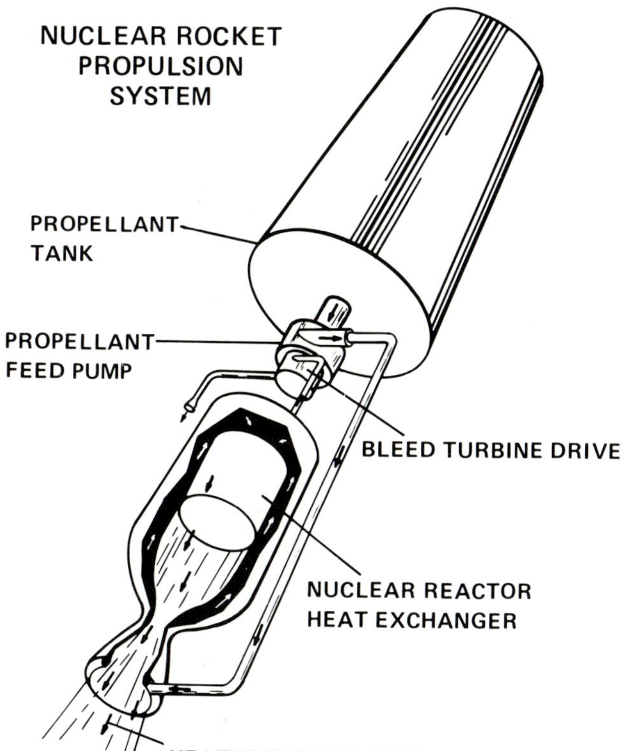

Fig. 10.1 Typical nuclear rocket propulsion module. *Courtesy of NASA and Department of Energy (After J. M. Taub, UC-33).*

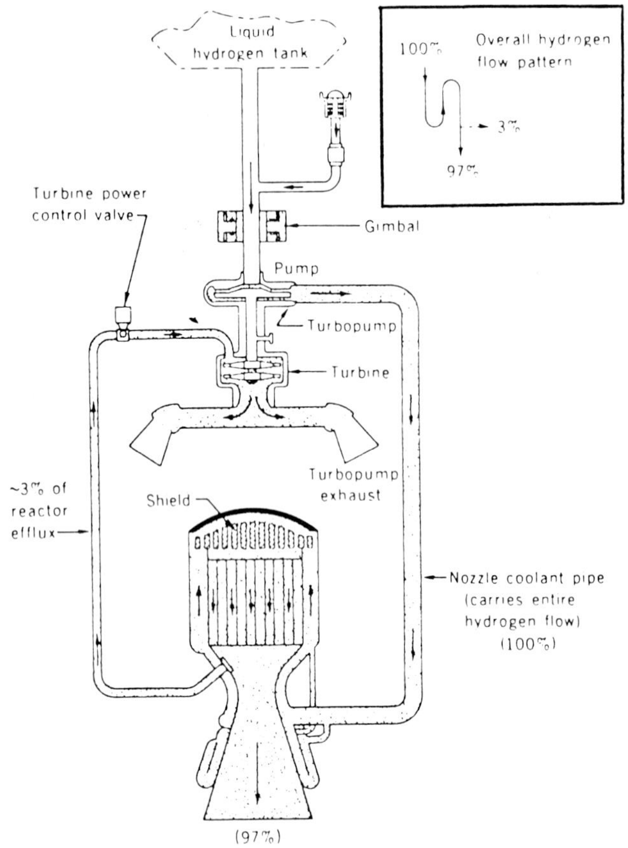

Fig. 10.2 A "hot bleed cycle" nuclear rocket reactor engine in which a small fraction of the reactor power is used to drive the propellant turbopumps. *Courtesy of Department of Energy.*

takes on a slightly different dimensional format. Again, from eq. 10.1

$$I_{sp} = \frac{\text{thrust}}{\text{mass flow rate}} = \frac{lb_f}{lb_m/s}. \tag{10.4}$$

Recalling Newton's second law of motion,

$$F = (1/g_c)\ m\ a \tag{10.5}$$

$$(1\ lb_f) = (1/g_c)\ (1\ lb_m)\ a$$

where a is the acceleration due to gravity (ft/s^2)
g_c is a universal dimensional constant (32.1739 $ft\text{-}lb_m/lb_f\text{-}s^2$).

At sea level on the surface of the Earth, a = 32.1739 ft/s^2 and eq. 10.5 reduces to 1 lb_f = 1 lb_m, that is, one pound-force is equal to one pound-mass. However, this relationship is only valid at sea level on Earth! Some confusion occurs in interpreting eq. 10.5, when this condition of equality is forgotten or ignored. Often this relationship between one pound-force and one pound-mass is used to simplify the expression found in eq. 10.4, namely

$$I_{sp} = \frac{lb_f}{lb_m/s} = \text{seconds.} \tag{10.6}$$

This simplification, strictly speaking, is valid only where 1 lb_f = 1 lb_m, that is, the sea level on the surface of the Earth. But, specific impulse is frequently described in terms of "seconds" in the traditional engineering system units.

Rocket engine performance can be further expressed as

$$I_{sp} = \frac{F}{\dot{m}} = A\ C_f\sqrt{(T_c/M)} \tag{10.7}$$

where I_{sp} is the specific impulse (m/s)
F is the thrust (N)
$\dot{m}$ is the mass flow rate of propellant (kg/s)
A is a performance factor related to the thermophysical properties of the propellant
C_f is the thrust coefficient which is a function of the nozzle parameters
T_c is the chamber temperature (K)
M is the molecular weight of the exhaust gases.

An inspection of eq. 10.7 quickly reveals that the specific impulse, and therefore the engine performance, increases as we make the chamber temperature (propellant gases) hotter and as we decrease the molecular weight of the exhaust gases. Consequently, hydrogen (H_2), as the lowest molecular weight gas (M = 2), is the working fluid of choice for nuclear rockets. In order to increase the quantity of hydrogen propellant that can be carried by a certain volume nuclear rocket, the hydrogen is usually stored as a liquid at cryogenic conditions. The size of individual propellant storage tanks is Earth-to-orbit launch vehicle limited, until on-orbit construction supported by a permanently manned space station allows the creation of space vehicles of unlimited size. In the interim, several of these tanks could also be launched separately from Earth and then assembled into clusters on orbit, again creating nuclear-staged propulsion systems of virtually any size.

During operation of a typical solid-core nuclear rocket (see Fig. 10.2), the hydrogen propellant is pressurized and fed through the nuclear reactor by a turbopump. This high pressure working fluid regeneratively cools the nozzle and reflector on its way to the reactor. Some hydrogen, bled from the propellant flow system at a number of points, such as the thrust chamber or reactor outlet region, can also be used to run the turbopumps. However, the main quantity of propellant flows through the reactor core, which, as configured here, contains solid fuel elements arranged in a hexagonal matrix. Holes in the fuel elements are used as coolant passages. The hydrogen propellant, heated to temperatures of 2,400 to 2,700 K, is then expanded and

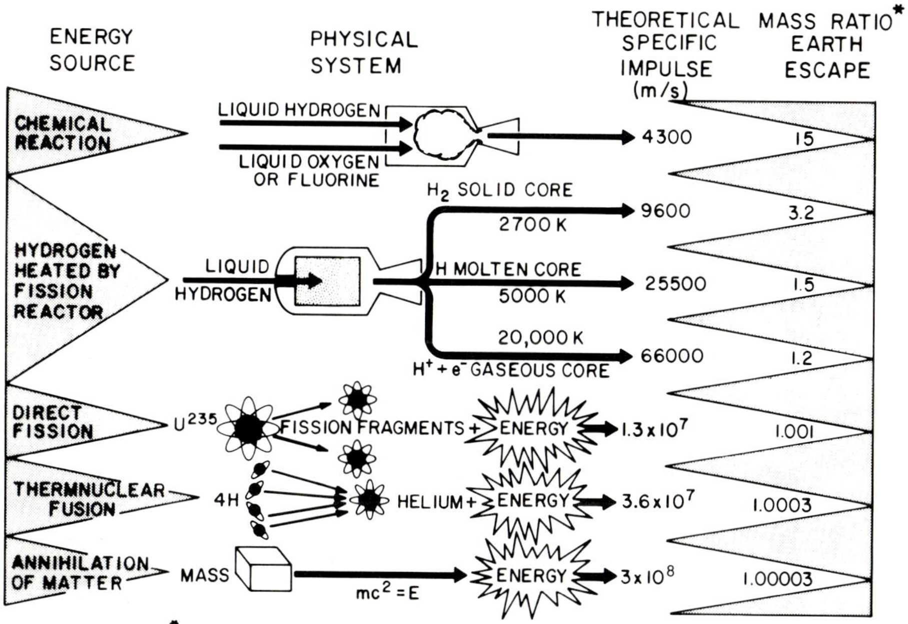

Fig. 10.3 Propulsion performance for various energy sources. *After J. M. Taub, UC-33.*

exhausted through the nozzle.

As shown in Fig. 10.3, solid-core nuclear rockets provide about twice the specific impulse of the best chemical rockets. In fact, the hydrogen-cooled, solid-core nuclear rocket was selected for extensive development in 1955 with one of the major program objectives being (in time) to provide propulsion for a manned expedition to Mars. During the 1960s, nuclear rocketry was viewed as a necessary part of such interplanetary expeditions, since the use of high specific impulse nuclear propulsion devices would greatly reduce the overall transit times to Mars. This, in turn, would alleviate strain on life support system reliability and also reduce the initial masses that had to be delivered to low Earth orbit at the start of the expedition. For example, the ratio of take-off mass from the Earth's surface to the final mass that achieves Earth escape velocity is a factor of about 15 for a chemical rocket and about 3.2 for a launch vehicle using a solid-core nuclear rocket upper stage. The nuclear rocket program undertaken by the United States continued until 1973. Although technically successful, as will be seen in this chapter, its termination was the result of changing priorities in the national space program.

Figure 10.4 illustrates the nuclear reactors that were part of the Rover Program, the name given to the overall United States nuclear rocket development program. This effort started with research reactors named Kiwi, whose objectives were to establish basic nuclear rocket reactor technology and to develop sound design concepts. For example, in the Kiwi series of reactors, high temperature fuels were demonstrated. These reactors were also the first to operate with liquid hydrogen (LH_2). They were automatically controlled, using drums in the reflector. Although thermal-hydraulic vibrational problems occurred early in the Kiwi program, a solution was achieved following several cold flow and experimental reactor tests. The Kiwi program culminated in the Kiwi-B4E reactor operating at over 1,890 K for 11.3 minutes and at 2,005 K and 937 megawatts (MW) for 95 seconds.

NUCLEAR ROCKET REACTORS

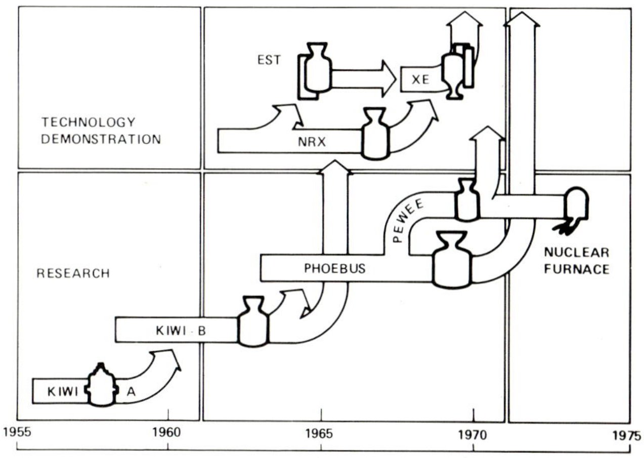

Fig. 10.4 Nuclear rocket reactor development series. *From G. H. Farbman and R. E. Thompson, "Applications of Nuclear Rocket Technology to Light Weight Nuclear Propulsion and Commercial Nuclear Process Heat Systems," AIAA Paper No. 75-1261.*

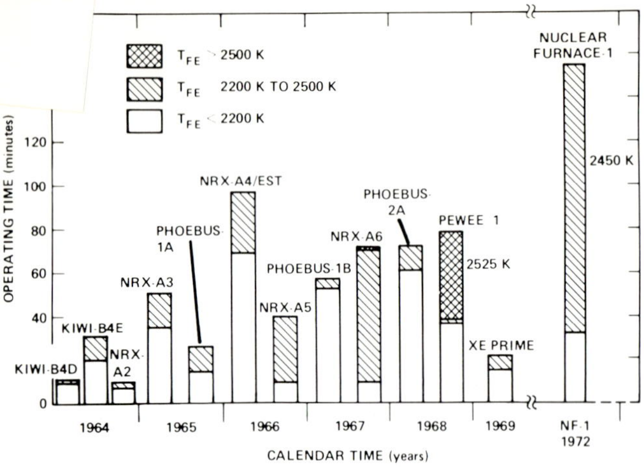

Fig. 10.5 Temperature performance demonstrated during rocket reactor tests. *After David S. Gabriel, "Nuclear Propulsion in the United States," 1972.*

The Kiwi reactors led to the NRX series of developmental reactors. Their goal was to demonstrate a specific impulse of 7,450 m/s (760 seconds) for 60 minutes in a 1,100 MW reactor. The thrust level was 245 kilonewtons (kN) [55,000 lb_{force}]. These objectives were exceeded in the NRX-A6 test, which ran for 62 minutes at 2,220 K and 1,100 MW with only 11¢ reactivity loss.

A further series of research reactors, called Phoebus, was also developed. Their objectives were to increase the specific impulse to 8,085 m/s (825 seconds), increase power density 50 percent, and increase the power level to 4,000–5,000 MW. These capabilities were demonstrated in the Phoebus-1B and Phoebus-2A tests. The Phoebus-2A test operated for over 12 minutes above 4,000 megawatts-thermal and reached a peak power of 4,100 MW_t.

The Pewee and Nuclear Furnace (NF) reactor series objectives were to demonstrate high temperature and longer life fuel elements. Pewee-1 ran at 2,555 K and 514 MW for 40 minutes. NF-1 ran at 2,450 K and 54 MW for 109 minutes.

In the engine development test program, an experimental engine was used to: (1) determine system characteristics during startup, full power, and shutdown conditions; (2) evaluate control concepts; and (3) qualify the engine test stand operations in a downward-firing configuration with simulated altitude conditions resembling operation in outer space. These objectives were met or exceeded in the NRX/EST and XE test programs. XE, a prototype flight engine system, was tested in the simulated space environment and performed some 28 starts and restarts. In this engine, non-nuclear flight components were demonstrated along with a flight-type reactor. Flight engine system designs were initiated, but not completed. The objectives were to come up with a suitable flight design that could operate for 10 hours and 60 cycles with a reliability of 0.997. The full-flow topping cycle had been selected.

The major emphasis of the reactor development efforts was to increase the reactor outlet temperature (since the specific impulse is proportional to the square root of the temperature), and also to increase the operating time of the reactor. The success of this part of the Rover Program is depicted in Fig. 10.5. Temperature levels at the core or fuel exit (T_{FE}) of over 2,500 K and operating times over two hours were demonstrated. The cumulative time-at-power is shown in Fig. 10.6. As the designs were perfected, there was an accelerated increase in the time-at-power.

Major events in the Rover Program are summarized in Table 10.1. The decision to proceed with the program was made in 1955, following several years of studies, and the first reactor was tested in 1959. Industrial contractors were brought into the program in 1961 and a reactor in-flight test program was started. The in-flight test program lasted two years, at which point it was decided to concentrate on engine development. By the end of 1964, the Kiwi program had demonstrated full-power operation, overcoming vibrational problems. The first fueled demonstration reactor, NRX-A2, was run in 1964 at 1,100 MW for about five minutes. Then, in 1965, as a safety test, a reactor was intentionally (but with great difficulty) put on a very fast transient to observe what would happen when it destroyed itself. Mechanical distintegration, not a nuclear explosion,

REACTOR AND ENGINE SYSTEM CUMULATIVE TEST TIME

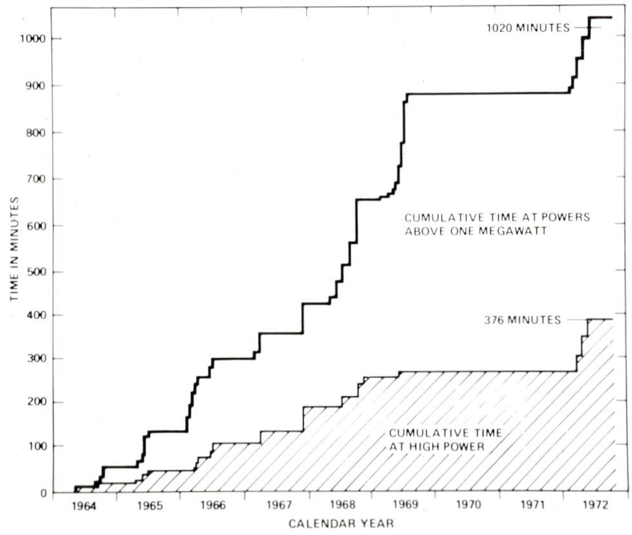

Fig. 10.6 Cumulative time-at-power during rocket reactor testing. *After David S. Gabriel, "Nuclear Propulsion in the United States," 1972.*

Table 10.1 History of the Rover Program

Date	Event
1955	Following several years of nuclear rocket studies, nuclear rocket program initiated as Project Rover at Los Alamos National Laboratory. Concept to be pursued was solid core, H_2 cooled, reactor expanding gas through a rocket nozzle.
July 1959	First reactor test, Kiwi-A, tested at 70 MW for 5 minutes.
Oct 1960	Proof-of-principle tests (Kiwi-A series of 3 reactors) completed.
July 1961	Industrial contractors (Aerojet-General for rocket engine and Westinghouse Electric Corporation for reactor) selected to perform rocket development phase. Reactor in-flight tests (RIFT) program initiated.
1963	RIFT program canceled.
1961–1964	Kiwi-B series of 1,000 MW reactor tests included five reactors plus several cold-flow unfueled reactors to resolve vibration problems and demonstrate design power.
May–Sept 1964	First full power test, Kiwi-B4D, at design power with no indications of core vibrations. Also demonstrated restart capability.
Sept 1964	NRX-A2, first tests of the NERVA (NERVA = *n*uclear *e*ngine for *r*ocket *v*ehicle *a*pplication) reactor, reached full power of 1,100 MW for about 5 minutes.
Jan 1965	Kiwi-B-type reactor deliberately placed on fast transient to destroy itself as part of safety program.
June 1965	The prototype of a new class of reactors, Phoebus-1A, was run at full power for 10.5 minutes.
Mar 1966	The NRX/EST, first rocket engine breadboard powerplant, operated at full power (1,100 MW) for 13.5 minutes.
Dec 1967	The fifth fueled NRX reactor in the NERVA engine series exceeded the design goal of 60 minutes at 1,100 MW.
June 1968	The Phoebus-2A—the most powerful nuclear rocket reactor ever built—ran for 12 minutes above 4,000 MW.
Dec 1968	Pewee set records in power density and temperature operating at 503 MW for 40 minutes at 2,550 K and core power density of 2,340 MW/m^3.
Mar 1969	The first down-firing prototype nuclear rocket engine, XE-prime, was successfully operated at 1,100 MW.
1969	Saturn V production suspended—prime launch vehicle for NERVA.
June 1972	In the 44 MW Nuclear Furnace (NF-1), fuel was demonstrated at peak power densities of approximately 4,500 MW/m^3 and temperatures up to 2,500 K for 109 minutes.
Jan 1973	Nuclear rocket program terminated. Judged a technical success but changing national priorities resulted in cancellation decision.

occurred.

A new class of reactors, aimed at a power level range of 4,000–5,000 MW, was also started in 1965. The first prototype breadboard engine system, called the NRX/EST, operated in 1966 and many successful startups and shutdowns were achieved. By 1967, the final reactor in the NRX series had operated, meeting the established goals. The highest power steady-state reactor ever built was operated in the Phoebus 2A test in 1968. It performed above 4,000 MW for 12 minutes. Also in 1968, the Pewee-1 reactor test established a core power density record. The year 1969 witnessed the downward-firing of the first prototype nuclear rocket engine, including simulated space environment. The configuration tested was similar to a flight engine. However, in 1969, production of the Saturn V launch vehicle was also suspended. Since the Saturn V vehicle was the prime launch vehicle for the nuclear rocket, this cancel-

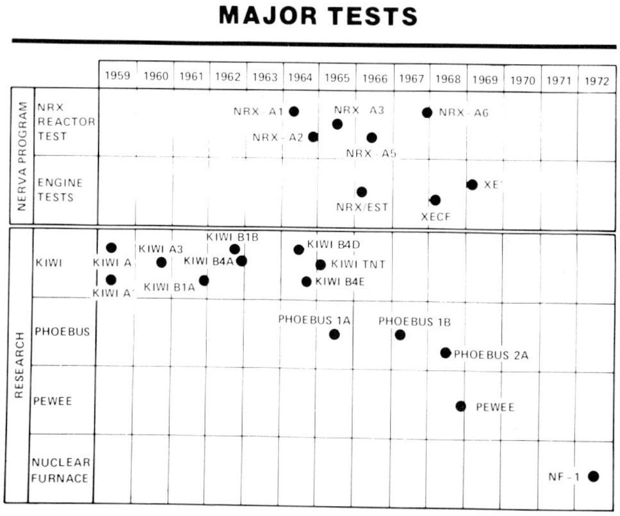

Fig. 10.7 Major tests accomplished during the Rover Program. *Courtesy of Los Alamos National Laboratory and Westinghouse Electric Co.*

Table 10.2 Performance of various Rover reactor systems.

	KIWI-4BE	NRX-A6	Phoebus-2A	Pewee-1
Reactor power (MW)	950	1,167	4,080	507
Flow rate (kg/s)	31.8	32.7	119.2	18.6
Fuel exit average temperature (K)	2,330	2,472	2,283	2,556
Chamber temperature (K)	1,980	2,342	2,256	1,837
Chamber pressure (MPa)	3.49	4.13	3.83	4.28
Core inlet temperature (K)	104	128	137	128
Core inlet pressure (MPa)	4.02	4.96	4.73	5.56
Reflector inlet temperature (K)	72	84	68	79
Reflector inlet pressure (MPa)	4.32	5.19	5.39	5.79
Periphery and structural flow (kg/s)	2.0	0.4	2.3	6.48

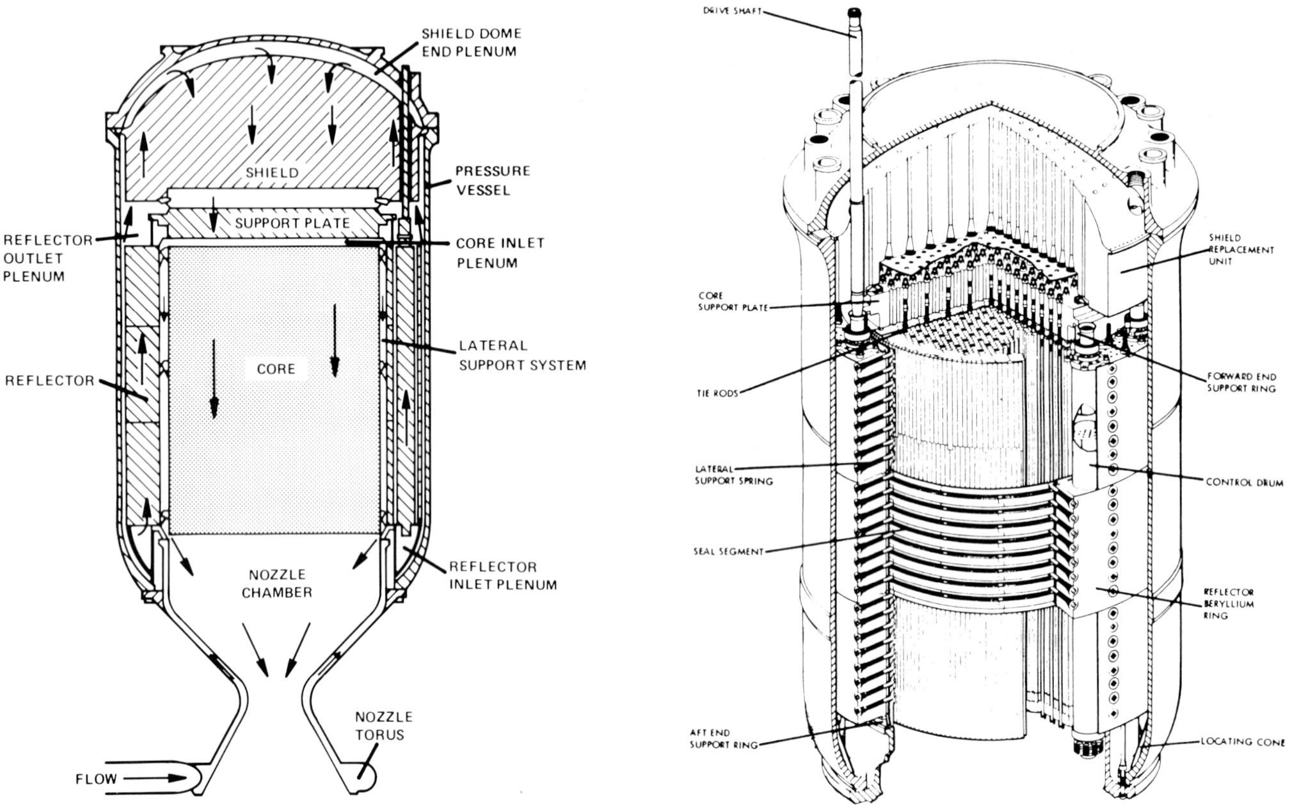

Fig. 10.8 Schematic (left) and cutaway (right) views of Rover nuclear rocket reactor. *From Westinghouse Astronuclear Laboratory, "NRX-A6 Test Predictions," WANL-TME-1613, November 1967.*

lation heralded a national decision to abandon the human exploration of Mars as a direct follow-on to the Apollo lunar landings. In 1972 a final reactor test was conducted in a nuclear furnace (NF) system. This test demonstrated up to 109 minutes operation at power densities of approximately 4,500 MW/m^3 and temperatures of over 2,500 K. Then, in 1973, the nuclear rocket program was terminated—a technically successful program, but also the victim of changing national space program priorities (see Fig. 10.7).

Reactor Development

The reactor design goals in the Rover Program were to maximize the core exit temperature of the hydrogen propellant, to increase longevity (initially one hour, then ten hours operation), to minimize hydrogen corrosion of the nuclear fuel, and to minimize fuel breakage from vibrational and thermal stress. Figure 10.8 presents schematic and cutaway views of the Rover nuclear rocket reactor. It was an epithermal, graphite-moderated, hydrogen-cooled reactor, using enriched (93.15%) uranium-235 as the nuclear fuel. The fuel elements were hexagonally shaped, with 19 cooling channels per element. The core was supported by a cold end support plate with a structural tube arrangement tied to the support plate. Power flattening was achieved by varying the fuel loading, while flow distribution was

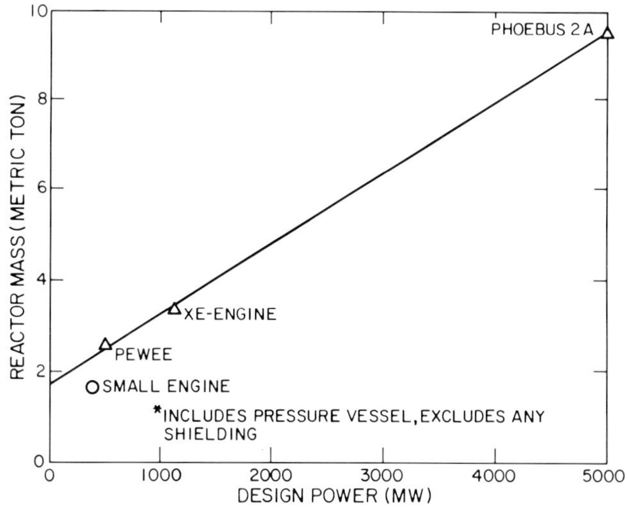

Fig. 10.9 Mass of various Rover Program reactors. *Courtesy of D. Koenig, Los Alamos National Laboratory.*

controlled by orifices in the core inlet plate. The reactor core was surrounded by a beryllium reflector, in which were located rotating drums that also contained a neutron absorber (boron carbide).

Table 10.2 describes some of the more significant Rover reactor tests. These tests covered a range from about 500 to 4,100 MW and involved flow rates of about 19 to 119 kg/s. The difference between the chamber temperature and the fuel exit average temperature depended on the amount of propellant flow used to cool the periphery. In many of the experimental reactor tests, a large amount of propellant flow was used, so that the fuel exit temperature rather than the chamber temperature is really more indicative of projected flight performance. For example, as shown in Table 10.2, Pewee-1 achieved an exit temperature of 2,556 K.

Figure 10.9 shows the mass of the reactor system as a function of its power level. A small rocket engine reactor at 365 MW had a mass of about 1,600 kg (including the pressure vessel); while the Phoebus-2 reactor, designed for 5,000 MW, had a mass of approximately 9,500 kg. By comparison, for approximately 6 times the mass, approximately 14 times the power was achieved.

A major technology effort was expended in the Rover Program on developing and improving nuclear fuels. The original Kiwi-A and Kiwi-B through the Kiwi-B4D reactors used highly enriched uranium dioxide (UO_2) fuel, extruded in a carbide matrix. The particle size was approximately 4 micrometers in diameter and its density was approximately 10.9 kg/m^3. The fuel melting temperature was 2,683 K. The demonstrated operating limit of this design series was 20 seconds at 2,127 K. The major engineering problem of concern was dimension control, because in the 1,875 to 2,275 K temperature regime, the UO_2 reacted with the carbon, converting to UC_2 with the evolution of CO and loss of carbon from the fuel element.

The Kiwi-B4E, Phoebus, Pewee, and NRX-A reactor tests used beaded UC_2 particles with pyrolitic graphite coatings as protection against oxidation and storage (see Fig. 10.10). The particles were 50 to 150 micrometers in diameter with a 25 micrometer thick coating. The fuel elements used a graphite matrix with niobium carbide (NbC) coatings to protect against hydrogen corrosion. In later reactor designs, the NbC was changed to zirconium carbide (ZrC). This fuel demonstrated one hour operation at temperatures between 2,400 K and 2,600 K. The major problems encountered with the beaded UC_2 particle coated fuel involved the large difference between the coefficients of thermal expansion of the graphite matrix and the NbC coatings. Excessive carbon loss after one hour occurred in the 2,375 to 2,575 K temperature range.

In the nuclear furnace (NF-1), composite (U, Zr)C particles coated with ZrC were used (see Fig. 10.10). This fuel performed for 109 minutes at 2,450 K at peak power densities of approximately 4,500 MW/m^3. Projections indicated that with matched thermal conductivity this fuel would be good for four to six hours at 2,500 to 2,800 K. The major problem observed in the NF-1 test was still some cracking from radiation damage. Several fuel elements of small, pure carbide particles (U, Zr)C were also included in the NF-1 experiment. These fuel elements exhibited extensive cracking at the power levels occurring in the NF-1 experiment.

A problem encountered with all the fuel configurations was called "mid-range corrosion," as shown in Fig. 10.11. The fuel operated at much higher temperatures toward the chamber end of the core than the front

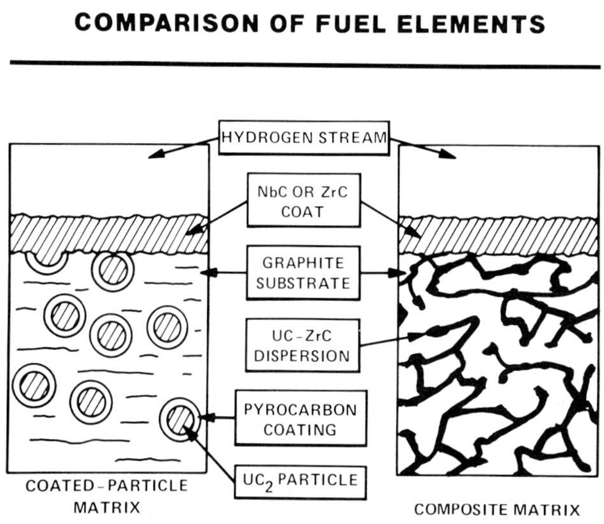

Fig. 10.10 Comparison of Rover Program fuel elements. *After David S. Gabriel, "Nuclear Propulsion in the United States," 1972.*

FUEL ELEMENT ENVIRONMENT AND MASS LOSS FROM PEWEE AND NF-1

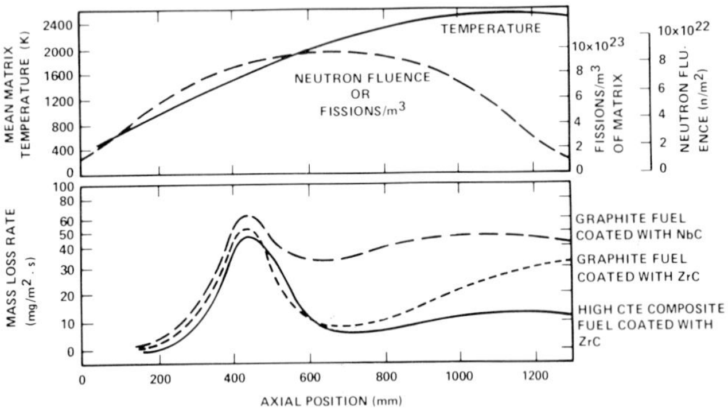

Fig. 10.11 Fuel element environment and mass loss from Pewee and NF-1 fuels. *After Luther L. Lyon, "Performance of (U, Zr)C-Graphite (Composite) and of (U, Zr)C(Carbide) Fuel Elements in the Nuclear Furnace 1 Test Reactor," Los Alamos National Laboratory report LA-5398-MS, September 1973.*

end. The fuel was processed to accept the high chamber end temperatures and the front end of the core had low mass loss rates because of the low temperatures found there. In the mid-range, however, there tended to be the highest mass loss rates, because temperatures were now appreciable and the neutron flux level was high, yet the temperatures were still far from the region at which the fuel was processed. On the hot end of the core, the temperatures matched the processing temperatures and the neutron flux levels tended to be low. As can be seen in Fig. 10.11, the result was lower mass losses, even for graphite fuel coated with NbC. The advantage of the zirconium carbide is depicted in Fig. 10.11, as well as the advantage of going to a composite fuel coated with ZrC.

Figure 10.12 illustrates the anticipated lifetimes at various operating temperatures for the three fuels of interest. If a ten hour life is desired, the reactor would have to operate around 2,200–2,300 K with a graphite matrix fuel; while with a composite fuel, this could go up to almost 2,400 K; and with a carbide fuel, up to approximately 3,000 K. For one hour operation, the graphite matrix fuel could be operated at over 2,500 K, the composite at 2,700 K, and the carbide possibly as high as 3,300 K.

Figure 10.13 is another way of depicting the progressive improvements achieved in Rover Program fuel element development. During nuclear rocket reactor operation, fuel can be lost from the core because of hydrogen corrosion. The control drums in the reflector are used to compensate for this. Starting with the NRX-A2 and NRX-A3 reactors normalized to unity, by the end of the NRX series the fuels had improved to the point that there was essentially three-tenths the relative corrosion rates as had occurred at the beginning of the NRX series. The development fuel for the Pewee-2 (a system which was never constructed) would have reduced this to less than one-tenth the relative corrosion rate.

Fuel structures were another major problem in the Rover reactor development. The early Kiwi series quite literally fell apart from vibrations. These vibrations were induced by thermal-hydraulic interactions. By the end of the Kiwi series, however, an adequate structural support system had been demonstrated. Improvements

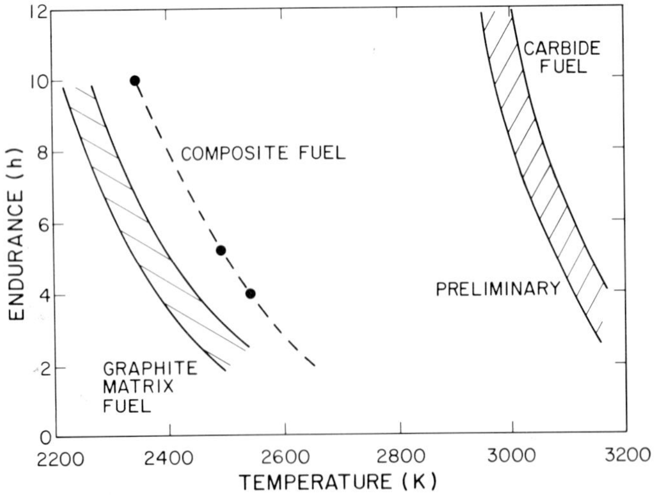

Fig. 10.12 Anticipated lifetimes of various Rover reactor fuels. *From David S. Gabriel, Statement to Committee on Aeronautics and Space Sciences, U.S. Senate, 1973.*

PROGRESS IN REDUCING HYDROGEN CORROSION OF FUEL ELEMENTS

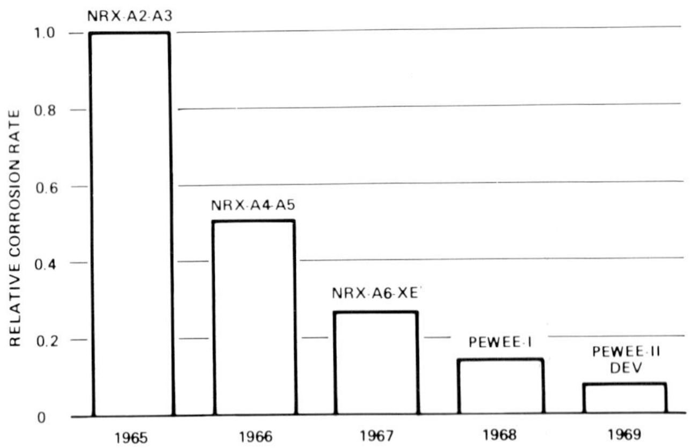

Fig. 10.13 Progress in reducing hydrogen corrosion of fuel elements. *Courtesy of Los Alamos National Laboratory and Westinghouse Astronuclear Laboratory.*

Table 10.3 Summary of NRX reactor structure anomalies.

Source	Significant Structural Anomalies	Phenomenon Causing/ Allowing Occurrence	Resulting Technique, Criteria or Geometry Change (Corrective Action)
NRX-A1	1. Control drum binding	1. Deflection of control drum and outer reflector due to thermal gradients	1. Complete deflection analysis of all interacting components for all reactor conditions indicated; increase bore diameter in outer reflector
	2. Pressure vessel-outer reflector support ring interference	2. Deflection due to thermal gradients	2. Deflection analysis for startup, steady state and shutdown required.
	3. Reflector cracks	3. Contraction due to thermal gradients on nozzle end ring	3. Finite element analysis technique was final technique prior to A6
NRX-A2	1. Aluminum barrel tab bent	1. Combination of barrel pre-test axial movement, shrinkage, and inner reflector axial contraction	1. None indicated, not considered serious
NRX-A3	1. Support block cracks, aft c' bore fillet to inner flow holes missed on A2 same cond.	1. Stresses due to thermal gradients and mechanical loads at termination of startup ramp	1. Initiated first finite element computer analysis study (This technique was developed extensively)
	2. Control drum rubbing	2. Deflection of sector and drum due to thermal gradients during cool-down after flow initiation	2. Transient analysis required, drum material removed
NRX-A4	1. Cracking of support blocks fillets also peripheral and axial cracks	1. Stresses due to thermal gradients	1. Reduction of start-ramp on A5, also planned test holds, also partial lobes removed, c'bore radius was increased
	2. Flaring and splitting of liner tube ends	2. Excessive growth due to overheating	2. Reduce liner tube length
	3. Light rubbing of control drum	3. Deflection due to thermal gradients	3. Analysis techniques improved, drum longitudinal slots added
NRX-A5	1. Support blocks peripheral and axial cracks	1. Stresses due to thermal gradients and mechanical loading	1. Serious problem at end of reporting period
	2. Liner tube flaring	2. Cluster components overheating caused liner tube expansion which coupled with tapered sleeve restraint caused flaring	2. Reduced liner tube length
	3. 1/3 of fuel elements broken	3. Reduced strength due to excessive corrosion plus inflow gradients and end of life	3. Increase emphasis on coating technology
	4. Hot buffer filler strip extensive breakage	4. Stresses due to thermal gradients	4. Eliminate hot buffer
NRX-A6	1. Beryllium reflector ring cracked	1. Stresses due to thermal gradients	1. Components tests verifying 3-D finite element analysis technique
	2. Support blocks cracks	2. Stresses due to thermal gradients	2. A predictable problem requiring extended efforts for similar components in the flight engine era of NERVA
	3. Cracking of peripheral composite cups and one tungsten cup	3. Stresses due to high non-symmetric transverse temperature gradient	3. Non-expected, protection cup performance not impaired

in the structural system continued to be made with the overall results listed in Table 10.3. It should be noted that what structural anomalies existed were identified after the tests and these anomalies did not cause a termination of power. At the end of the NRX-A6 test, some cracking was found in the beryllium reflector ring, in support blocks, in the peripheral deposit cups, and in one tungsten cup. These cracks were believed due to thermal gradients.

Rover Engine Development

Rover engine development was keyed to maximizing specific impulse, meeting various defined thrust levels, minimizing mass, and increasing longevity from an initial goal of one hour to ten hours. In addition to attempting to continuously improve the reactor fuel operating temperatures, a number of design changes were also made in the reactors for better performance. For example, the early reactors used tie-rods attached to the cold end core support plate that was cooled by hydrogen discharged into the nozzle chamber. This tended to lower the rocket specific impulse, because the tie-rod coolant was at a fairly low temperature relative to the hydrogen passing through the main fuel elements. In later reactors and in the Rover engine systems designed for space flight, tie-tubes were substituted for tie-rods. These were arranged as regeneratively cooled heat exchangers, with the coolant discharging into the core inlet. Also in the early Rover reactors, a high core peripheral flow rate was used to protect the outer section of the core. This flow was steadily decreased to an almost negligible amount in order to increase the temperature of the hydrogen discharged into the nozzle chamber. A final optimization involved the use of a regenerative-heat exchanger-cooled peripheral system.

The engine cycle was also changed to increase the specific impulse. In the XE experimental engine, the hot bleed cycle was used. This cycle extracted some of the exit chamber fluid, mixed it with fluid from the reflector outlet, and used it to drive the turbine. Because of the reduced pressure, the fluid could not be reintroduced into the main flow stream and, consequently, the turbine discharge was exhausted to space at a relatively low temperature when compared to the main nozzle chamber conditions. This significantly reduced the overall specific impulse of the engine. In the evolution of the engine to a full flow, or "topping cycle," the turbine used fluid from the tie-tube outlet and the turbine working fluid was then discharged into the core inlet. This made it possible to significantly raise the specific impulse of the system. The temperature at this point was much lower than in the hot-bleed cycle, so that a much higher turbine flow rate was needed and higher pump discharge pressures were experienced.

DESCRIPTION OF "SMALL ENGINE" NUCLEAR ROCKET DESIGN [71724 N (16125 lb)]

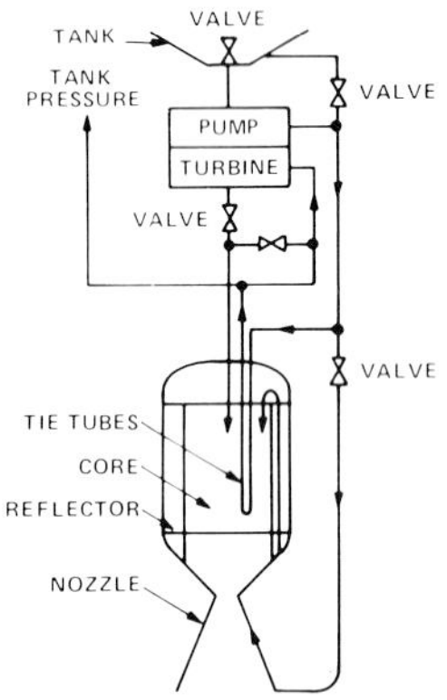

- HYDROGEN PROPELLANT
- FULL FLOW TOPPING CYCLE
- SINGLE-STAGE CENTRIFUGAL PUMP AND SINGLE-STAGE TURBINE
- REGENERATIVELY COOLED METAL-CORE SUPPORT ELEMENTS (TIE TUBES)
- RADIATION SHIELD OF BORATED ZIRCONIUM HYDRIDE
- 6 CONTROL-DRUM ACTUATORS
- 5 VALVES AND VALVE ACTUATORS
- REGENERATIVELY COOLED NOZZLE, AREA RATIO = 25:1
- UNCOOLED NOZZLE SKIRT, EXIT AREA RATIO = 100:1
- UNCOOLED NOZZLE SKIRT HINGED AND ROTATABLE
- OVERALL ENGINE LENGTH =
  3.1 m (123 in.) WITH SKIRT FOLDED
  4.4 m (174 in.) WITH SKIRT IN PLACE
- TOTAL MASS = 2550 kg (5620 lb)

Fig. 10.14 Schematic of small engine nuclear rocket design. *After F. D. Durham, LA-5044-MS, Los Alamos National Laboratory.*

Table 10.4 lists parameters for various Rover engine designs. The experimental XE′ was successfully tested. The NERVA engine was being designed for flight at the 337 kilonewton (kN) (75,000 lb_f) thrust level; while the small engine was being designed for a 72 kilonewton (16,000 lb_f) thrust level. The XE′ engine would have had a flight specific impulse of about 6,960 m/s (710 s); the NERVA flight engine could have improved this to 8,085 m/s (825 s); and the small engine could have achieved a specific impulse of 8,575 m/s (875 s) had these nuclear rocket engine programs been conducted through the flight stage. The power levels for the various engine systems were as follows: 1,140 MW for XE′, 1,556 MW for NERVA, and 367 MW for the small engine. The higher specific impulse levels are reflected in the chamber temperatures that went from 2,270 K (demonstrated in the XE′) to 2,695 K for the small engine.

In effect, the small engine (see Fig. 10.14) represented an accumulation of all of the knowledge gained in the

Table 10.4 Rover Program engine characteristics.

	XE′	NERVA	Small Engine
Thrust (kN)	245	337	72
Specific impuse (m/s)	6,970	8,085	8,575
Thermal power (MW)	1,140	1,566	367
Turbopump power (MW)	5.1		0.9
Turbopump speed (rpm)	22,270	23,920	46,950
Pump discharge pressure (MPa)	6.80	9.36	6.03
Engine flow rate (kg/s)	35.9	41.9	8.5
Chamber temperature (K)	2,270	2,360	2,695
Chamber pressure (MPa)	3.86	3.10	3.10

DESCRIPTION OF "SMALL ENGINE" REACTOR

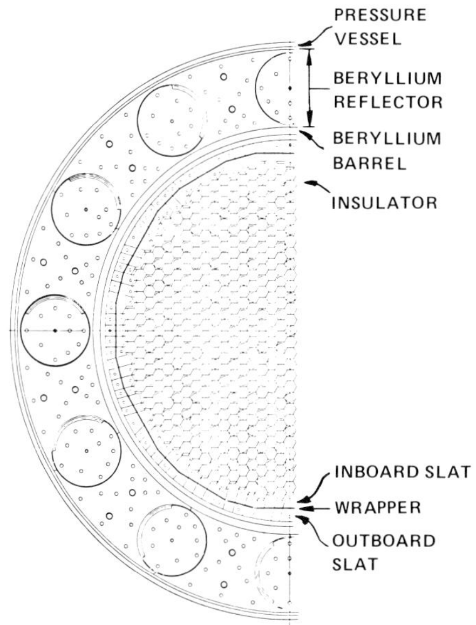

- PRODUCES 365 MW
- 564 HEXAGONALLY SHAPED (UC-ZrC) C COMPOSITE FUEL ELEMENTS
- 241 SUPPORT ELEMENTS CONTAINING ZrH NEUTRON MODERATOR
- 19 COOLANT CHANNELS PER ELEMENT
- CORE PERIPHERY CONTAINS AN OUTER INSULATION LAYER, A COOLED INBOARD SLAT SECTION, A METAL WRAPPER, A COOLED OUTBOARD SLAT SECTION, AND AN EXPANSION GAP
- REFLECTOR IS BERYLLIUM BARREL WITH 12 REACTIVITY CONTROL DRUMS
- CORE SUPPORT ON COLD END BY AN ALUMINUM-ALLOY PLATE. SUPPORT PLATE RESTS ON REFLECTOR SYSTEM
- REACTOR ENCLOSED IN ALUMINUM PRESSURE VESSEL
- CAPABLE OF 83 K/s TEMPERATURE TRANSIENTS

Fig. 10.15 Reactor core for small engine nuclear rocket. *After F. D. Durham, LA-5044-MS, Los Alamos National Laboratory.*

nuclear rocket program. It used hydrogen as the propellant; the full-flow engine topping cycle; and a single-stage centrifugal pump with a single-stage turbine. It had a regeneratively cooled nozzle and its support elements were tie-tubes. A radiation shield of borated zirconium hydride was incorporated above the reactor. This was used mainly to reduce neutron heating of the hydrogen propellant tank situated above the engine, although it would also have provided shielding for the payload and astronaut crew. Reactor control was achieved with six control drum actuators with the control drums being located in the beryllium reflector. The small engine had five valves and actuators. The small engine valving included a propellant tank shutoff valve (PSOV), a nozzle control valve (NCV), a turbine series control valve (TSCV), a turbine bypass control valve (TBCV), and a cooldown control valve (CCV). The PSOV was located at the bottom of the propellant tank and provided a tight seal against propellant leakage when the engine was not in use. The NCV adjusted the flow split between the nozzle coolant tubes and the tie-tubes. The TSCV was used to isolate the turbine during preconditioning and cooldown. The TBCV regulated the amount of the flow to the turbine and, consequently, the turbopump speed and flow rate. Finally, the CCV was used to regulate hydrogen flow for decay heat removal following engine operation. It was also used in conjunction with a small pump to provide prepressurization for the tank.

The small engine used a regeneratively cooled nozzle with an area ratio of 25 : 1. The nozzle was followed by an uncooled skirt section that extended the overall nozzle out to an area ratio of 100 : 1 (exit area to throat area). The uncooled nozzle skirt was hinged to facilitate packaging in the launch vehicle—a design arrangement that also provided room for a larger propellant tank. The overall engine length was 3.125 meters with the nozzle skirt folded or 4.415 meters with the nozzle skirt in place. The small engine had a total mass of 2,550 kg.

The small engine reactor core (see Fig. 10.15) was designed to produce 365 MW. There were 564 hexagonally shaped (UC-ZrC)C composite fuel elements. Each fuel element had 19 coolant channels. There were 241 support elements, containing zirconium hydride (ZrH) as a neutron moderator. The core periphery included an outer insulator layer, a cooled inboard slat section, a metal wrapper, a cooled outboard slat section, and an expansion gap. The core was supported by an aluminum alloy plate on the cold end with the support plate resting on the reflector system. The reactor was contained in an aluminum pressure vessel. A beryllium barrel containing 12 control drums surrounded the core. The small engine reactor was designed for 83 degrees kelvin per second (K/s) temperature transients. Figure 10.16 summarizes various flow rate, pressure, and temperature points throughout the engine system. The chamber temperature was about 2,700 K.

Figure 10.17 provides additional details about the small engine fuel modules. The fuel provided both the energy for heating the hydrogen and the heat transfer surface to accomplish propellant heating. It was enriched (93.15%) uranium-235 in a composite matrix of UC-ZrC solid solution and carbon. The channels were coated with zirconium carbide to protect against possible hydrogen reactions. The tie-tubes transmitted the

"SMALL ENGINE" STATE POINTS AT DESIGN CONDITION

CHAMBER PRESSURE	310 N/cm^2
CHAMBER FLOW RATE	8.51 kg/s
CHAMBER TEMPERATURE	2696 K
REACTOR POWER	367 MW
SPECIFIC IMPULSE	8580 m/s
THRUST	7297 N
NOZZLE FLOW FRACTION	44.9%
TURBINE BYPASS FLOW FRACTION	11.8%
TURBOPUMP SPEED	4917 rad/s
TURBOPUMP SHAFT WORK	0.93 MW
PUMP EFFICIENCY	65%
TURBINE EFFICIENCY	80%
NOZZLE VALVE AREA	2.63 cm^2
TURBINE CONTROL VALVE AREA	3.02 cm^2

STATE POINT DESCRIPTION	FLOW RATE (kg/s)	PRESSURE (MPa)	TEMPERATURE (K)
PUMP INLET	8.51	0.12	17.0
PUMP EXIT	8.51	6.03	19.8
TIE TUBE MANIFOLD INLET	4.05	5.72	20.3
TIE TUBE FIRST PASS EXIT	4.05	5.38	56.9
TIE TUBE EXIT	4.05	5.02	428.9
SLAT MANIFOLD INLET	0.64	5.72	20.3
SLAT FIRST PASS EXIT	0.64	5.24	167.1
SLAT EXIT	0.64	5.02	431.5
TURBINE INLET	4.13	4.86	428.6
TURBINE EXIT MIXED	4.69	4.13	415.6
TURBINE BYPASS INLET	0.55	4.86	428.6
NOZZLE INLET	3.83	4.63	21.4
NOZZLE EXIT	3.83	4.21	240.4
REFLECTOR EXIT	3.83	4.21	294.9
SHIELD INLET	8.51	4.06	361.0
CORE INLET	8.51	3.96	370.1
FUEL ELEMENT EXIT	8.33	3.10	2728.0
CORE BYPASS EXIT	0.18	3.10	370.1
CHAMBER	8.51	3.10	2695.8

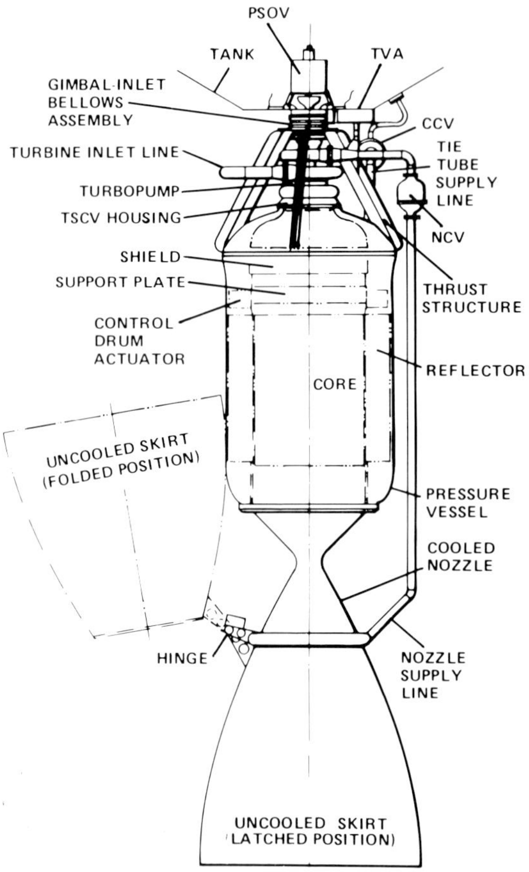

Fig. 10.16 Small engine design conditions. *After F. D. Durham, LA-5044-MS, Los Alamos National Laboratory.*

core axial pressure load from the hot end of the fuel elements to the core support plate. They also provided an energy source for the turbopump and contained and cooled the ZrC moderator sleeves. They consisted of a counterflow heat exchanger made of Inconel 718 and a zirconium moderator with zirconium carbide insulation sleeves. Their overall length was approximately 0.9 meter.

DESCRIPTION OF "SMALL ENGINE" FUEL MODULES

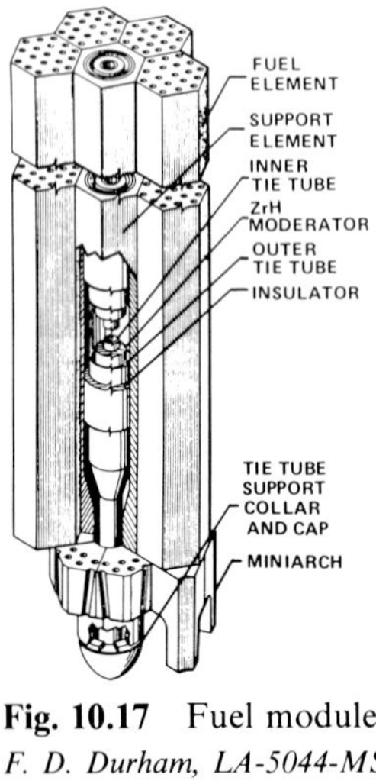

FUEL

- FUNCTION
  - PROVIDED ENERGY FOR HEATING HYDROGEN PROPELLANT
  - PROVIDED HEAT TRANSFER SURFACE
- DESCRIPTION
  - ^{235}U IN A COMPOSITE MATRIX OF UC-ZrC SOLID SOLUTION AND C
  - CHANNELS COATED WITH ZrC TO PROTECT AGAINST H_2 REACTIONS

TIE TUBES

- FUNCTION
  - TRANSMIT CORE AXIAL PRESSURE LOAD FROM THE HOT END OF THE FUEL ELEMENTS TO THE CORE SUPPORT PLATE
  - ENERGY SOURCE FOR TURBOPUMP
  - CONTAIN AND COOL ZrC MODERATOR SLEEVES
- DESCRIPTION
  - COUNTER FLOW HEAT EXCHANGER OF INCONEL 718
  - ZrH MODERATOR
  - ZrC INSULATION SLEEVES

Fig. 10.17 Fuel module for small engine nuclear rocket. *After F. D. Durham, LA-5044-MS, Los Alamos National Laboratory.*

Small engine component masses are listed in Table 10.5. The overall mass was 2,550 kg with the reactor

Table 10.5 Small engine mass estimates (kg).

Reactor core and hardware	868
Reflector and hardware	569
Shield	239
Pressure vessel	150
Turbopump	41
Nozzle and skirt assembly	224
Propellant lines	15
Thrust structure and gimbal	28
Valves and actuators	207
Instrumentation and electronics	159
Contingency	50
Total	2,550

The above total includes 100 kg for instrumentation cabling and electronics that are mounted on the stage rather than on the engine and 35 kg for the thrust vector actuators.

SHUTTLE-COMPATIBLE NUCLEAR STAGE

	STAGE MASS, kg (lb)	USABLE PROPELLANT MASS, kg (lb)
NUCLEAR STAGE (INCLUDING SMALL ENGINE)	17783 (39205)	12814 (28250)
PROPELLANT MODULE (18.3 m)	23181 (51105)	21265 (46880)

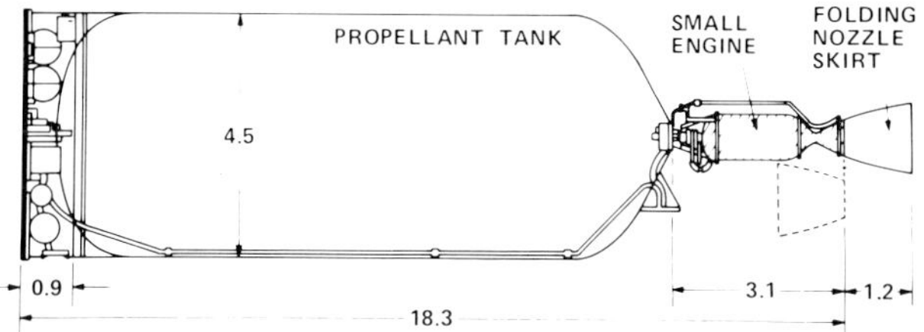

ALL DIMENSIONS IN METERS

McDONNELL DOUGLAS REUSABLE NUCLEAR STAGE CONCEPT

Fig. 10.18 Space Shuttle compatible nuclear stage propulsion module. *After F. D. Durham, LA-5044-MS, Los Alamos National Laboratory.*

(minus the shield) representing almost 1,600 kg of this amount and the turbopump only some 41 kg.

How might such a small engine nuclear stage with propellant tank fit within the Space Shuttle? is shown in Fig. 10.18. A modern small eng stage would have a mass of approximately which almost 13,000 kg would be propell tional propellant modules were placed in orbit separately, then this nuclear stage would have an added mass of about 23,000 kg with over 21,000 kg of hydrogen available for propulsion per propellant module. At a propellant flow rate of 8.5 kg/s, the baseline nuclear stage could operate for approximately 1,500 seconds. An additional propellant module would provide 2,500 seconds more operating time. Thus, a nuclear stage based on the small engine could support many interesting space missions in the Shuttle era.

We will now briefly discuss the state of development of some of the non-nuclear components in the Rover engine program. Figure 10.19 illustrates turbopump development. The turbopump pressurized the propellant for the engine feed system. The low propellant flow rates encountered with the small engine made it possible to run the shaft speed here at much higher rates than could be done in either the XE′ or NERVA engine turbopumps. The XE′ turbopump was a single-stage, radial exit flow, centrifugal pump with aluminum impel-

TURBOPUMP

- FUNCTION
  - PRESSURIZE THE PROPELLANT FOR THE ENGINE FEED SYSTEM
- DESIGN CONDITIONS

	XE′	NERVA	SMALL ENGINE
PUMP DISCHARGE PRESSURE (MPa)	6.69	9.36	6.03
PUMP FLOW RATE (kg/s)	35.8	20.9 – 41.7	8.5
TURBINE TEMP (K)	648	154	429
TURBINE FLOW RATE (kg/s)	3.32	19	4.1
SHAFT SPEED (rpm)	21989	23920	46952

- CONSTRUCTION (XE)
  - SINGLE-STAGE, RADIAL-EXIT-FLOW-CENTRIFUGAL PUMP WITH AN ALUMINUM IMPELLER, A POWER TRANSMISSION THAT COUPLES THE PUMP TO THE TURBINE, AND A TWO-STAGE TURBINE WITH STAINLESS STEEL ROTORS
- REACTOR TESTS EXPERIENCE
  - NRX/EST 8 STARTS INCLUDING 54.4 MINUTES AT HIGH POWER
  - XE-28 STARTS/RESTARTS INCLUDING RUNS TO RATED POWER
- POTENTIAL PROBLEMS
  - SHAFT SYSTEM BINDING AT BEARING COOLANT LABYRINTH IN XE TESTS (INCREASE CLEARANCE AND IMPROVE ALIGNMENT)
  - BEARING LIFE

Fig. 10.19 Turbopump. *Courtesy of Aerojet General Corporation.*

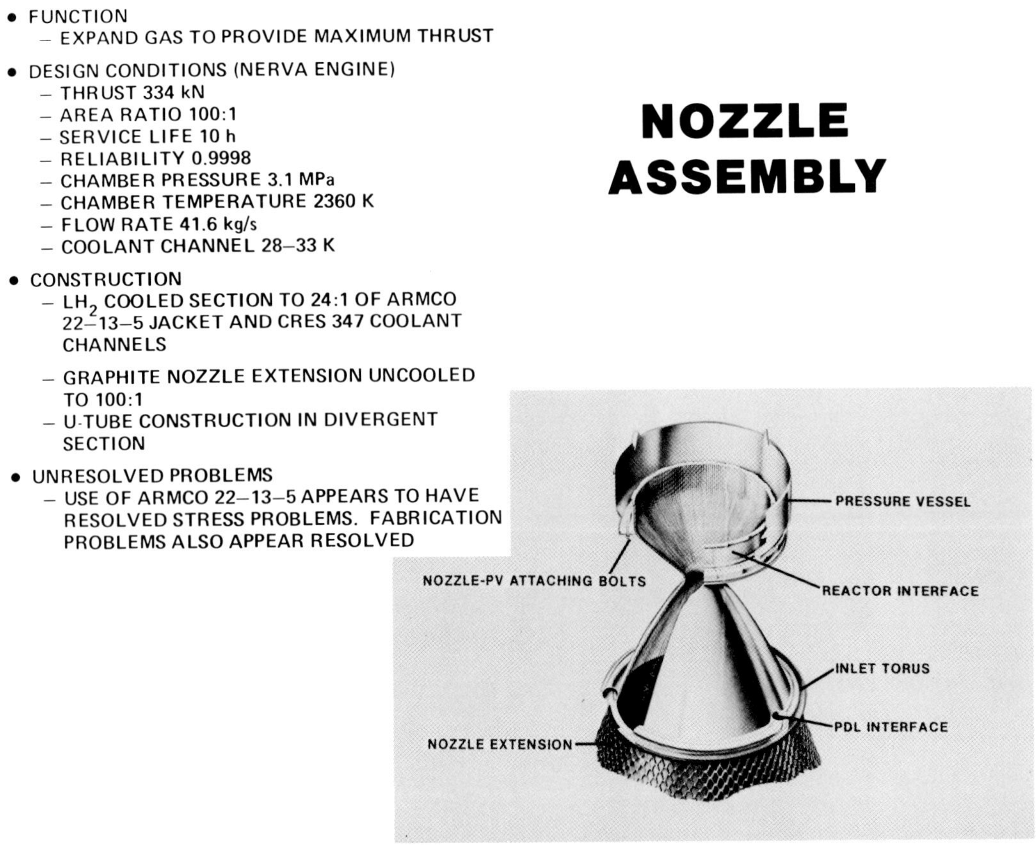

Fig. 10.20 Nozzle assembly. *Courtesy of Aerojet General Corporation.*

ler, a power transmission that coupled the pump to the turbine, and a two-stage turbine with stainless steel rotors. On the NRX/EST engine, this pump performed eight starts and operated 54.4 minutes at high power. In the XE′ engine, the turbopump performed 28 starts and restarts, including runs to rated power. One difficulty encountered in the XE′ tests was that the shaft system bound at the bearing. The solution was to increase clearance and to improve alignment. Bearings, in fact, probably represent one of the few life-limiting compo-

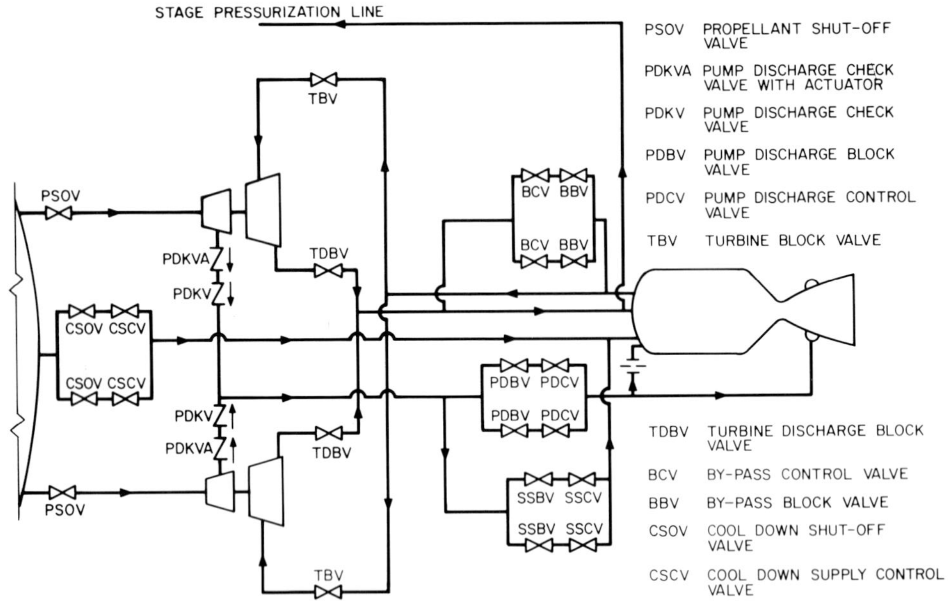

Fig. 10.21 NERVA engine schematic with parallel turbopumps and redundant engine valves. *After J. H. Altseimer et al., "Operating Characteristics and Requirements for the NERVA Flight Engine," J. Spacecraft, Vol. 8, No. 7, July 1971.*

Table 10.6 NERVA valve and actuator characteristics.

	Cooldown Control Valve (CCV)	Nozzle Control Valve (NCV)	Propellant Shutoff Valve (PSOV)	Turbine Bypass Control Valve (TBCV)	Turbine Shutoff Control Valve (TSCV)	Turbine Valve Actuator (TVA)	Control Drum Actuator (CDA)
Actuator size	small	small	large	large	large	large	small
Seat diameter, cm	2.54	2.54	15.24	3.175	11.43	—	—
Seat area, cm^2	5.06	5.06	182	7.93	103	—	—
Stroke	1.52 cm	1.52 cm	5.08 cm	1.52 cm	3.18 cm	± 0.044 rad*	3.14 rad
Load max. velocity	$0.15\ cm/s^{-1}$	$0.15\ cm/s^{-1}$	$0.5\ cm/s^{-1}$	$2.5\ cm/s^{-1}$	$0.2\ cm/s^{-1}$	$0.007\ rad/s^{-1}$	$0.2\ rad/s^{-1}$
Seating force, N	3,550	3,558	13,340	2,800	10,000	—	—
Plus max. ΔP at rated speed, $N\ cm^{-2}$	760	275	20.7	103	103	—	—
Lead screw diameter, cm	1.27	1.27	1.9	1.9	1.9	—	—
Lead screw advance, $cm\ rad^{-1}$	0.0809	0.0809	0.0809	0.0809	0.0809	—	—
Max stepping rate, steps s^{-1}	62	62	200	210	75	143	122
Motor gear ratio	50	50	50	10	50	380	917
Torque at speed $N - m$	—	—	—	—	—	3,250	34

*rad = radian
CCV and NCV are identical
PSOV and TCCV have same actuator

nents among the non-nuclear components of a nuclear rocket system. During the Rover Program, the solution to the bearing problem appeared dependent upon maintaining adequate cooling to reduce wear.

The nozzle assembly was used to expand the heated gas as it emerged from the reactor in order to provide maximum thrust. A typical nozzle assembly appears in Fig. 10.20. The design conditions for the NERVA flight engine included a thrust level of 337 kN, an area ratio of 100 : 1, a service life of 10 hours, a reliability level of less than 4 failures in 10^4 flights, a chamber pressure of 3.1 MPa, a chamber temperature of 2,360 K, a propellant flow rate of 41.6 kg/s, and operation of the coolant channels at temperatures between 28 and 33 K. The regeneratively cooled portion of the nozzle used an aluminum alloy jacket and stainless steel for the coolant channels. The graphite nozzle extension was uncooled out to an area ratio of 100 : 1. The major unresolved engineering problems in achieving a 10 hour operating life included stress problems in the aluminum alloy. Fabrication difficulties, however, appear to have been overcome in the program.

Various valves were required in the nuclear rocket system to control the hydrogen flow. These valves were binary valves, except for in-out control valves and check valves. Valve operating experience with a reactor was obtained in both the NRX/EST and XE′ engine tests. The major potential problems appeared to be from seal damage by contaminants, erroneous position indicators, and leakage from lip-seal tolerances.

The small engine design had just five valves. However, there was a desire in the design of the NERVA flight engine to increase system reliability by having two turbopumps (either one of which could provide full flow and pressure to the engine) and by having redundant valve configurations. In order to provide switching between the turbopumps and to provide high reliability by backing up each valve in case it should fail, some 26 valves were needed, as shown in Fig. 10.21. Table 10.6 lists NERVA valve and actuator characteristics.

Figure 10.22 depicts the pressure vessel and enclo-

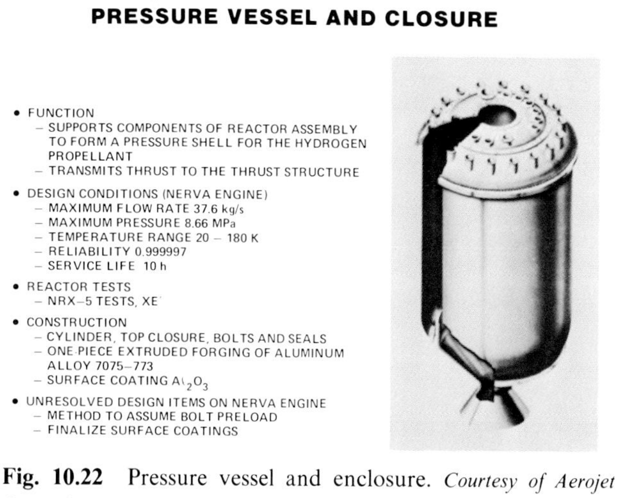

Fig. 10.22 Pressure vessel and enclosure. *Courtesy of Aerojet General Corporation.*

CHARACTERISTIC ROCKET ENGINE STARTUP

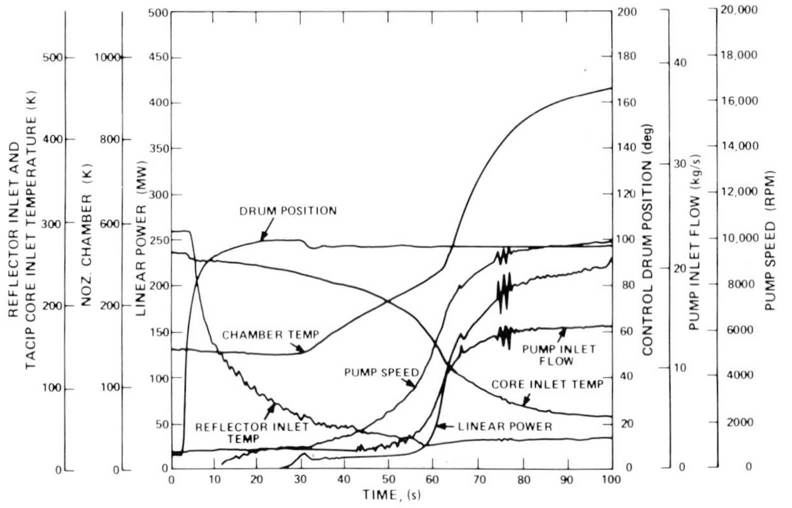

Fig. 10.23 Startup characteristics of a nuclear rocket.

sure. Its function was to support the components of the reactor assembly, to form a pressure shell for the hydrogen propellant, and to transmit thrust to the thrust structure. NERVA engine design conditions required a maximum flow rate of 37.6 kg/s, a maximum pressure of 8.66 MPa, temperatures ranging from 20 to 180 K, a reliability of less than three failures in 10^6 flights, and a service life of 10 hours. Similar designs were demonstrated in the five NRX tests and the XE′ engine. The pressure vessel consisted of a cylinder incorporating top closure with bolts and seals. A one-piece extruded forging of aluminum alloy 7075-773 was used with a surface coating of Al_2O_3. The major engineering features still under development at the time of program termination included optimum ways of assuring bulk preload and methods of finalizing the surface coatings.

Fig. 10.24 The ground experimental engine (XE) installed in Engine Test Stand No. 1 at the Nuclear Rocket Development Station in Nevada. *Courtesy of NASA.*

Controls and instrumentation were another major area of engineering development during the Rover Program. Pneumatic-type actuators were built for control drum actuators. These actuators were demonstrated on the XE′ engine without apparent degradation or anomalies. Thermocouples demonstrated performance at 2667 K for one hour without degradation. In addition, displacement, pressure, and vibration sensors were developed for several hours of operation. However, instrumentation with an accuracy of 1 percent or better required further development.

The startup characteristics of a nuclear rocket engine are shown in Fig. 10.23. For example, the engine components must be conditioned before high power can be reached. The turbopump, nozzle, reflector, and core inlet were all designed to operate at low temperatures. When the pump shutoff valve was first opened, the pump tended to vaporize the working fluid until sufficient fluid passed through the device, chilling it down to cryogenic conditions. The nozzle also tended to be a fluid choke point, as were the core and reflector inlets. Therefore, a certain amount of hydrogen must be passed through the system to remove thermal energy

Fig. 10.25 The Pewee 1 reactor being transported to a test location at the NRDS by the "Jackass & Western Railroad." *Courtesy of NASA.*

stored in the line, valves, and reflector. Once this chilling down process has been accomplished, the pump could be started and would operate normally. Approximately one minute of fluid flow was necessary to accomplish this function. During this period of time, the reactor could be brought up to a low power level. Reactor drums were programmed out rapidly, almost to the cold critical point, and then put on a slow transient. Once appreciable temperature was sensed in the chamber, the reactor could be switched over to closed-loop temperature control. This scheme did not require any neutronic instrumentation. When appreciable power had been achieved and the turbopump was running, engine heating could then be accelerated at the rate of 83 K/s. NRX/EST and XE′ engine experience indicate that the nuclear rocket engine can be controlled in a predictable and safe manner.

Nuclear Rocket Development Station (NRDS)

Major test facilities for the nuclear rocket program were located at the Nuclear Rocket Development Station (NRDS) on the Nevada Test Site (NTS). The NRDS was located on a desolate flat basin called Jackass Flats after some of the indigenous wildlife. Both reactors and complete engine assemblies were tested at the NRDS, which had three major test areas: Test Cells A and C and the Engine Test Stand Number One (EST-1) (see Fig. 10.24). These three test areas were connected by road and railroad to the R-MAD and E-MAD Buildings (Reactor and Engine Maintenance, Assembly, and Disassembly). Typically, a reactor or engine was assembled in one of the MAD buildings and then transported by the "Jackass & Western Railroad" to one of the test locations (see Fig. 10.25). This railroad was humorously referred to as the world's shortest and slowest. After a test, while the reactor or nuclear rocket engine was still highly radioactive, a heavily shielded railroad engine was used to tow it back to the proper MAD building for disassembly. The MAD buildings were primarily immense "hot cells" where engineers, protected by thick concrete and steel shields, could perform robot-assisted disassembly of the radioactive reactors and engines.

The reactor test facilities were designed to test a reactor in an upward-firing position, while the engine test facility (see Fig. 10.24) tested engines (but not complete rockets) in a downward firing mode. During a nuclear rocket test at the NRDS the desert basin became a literal inferno as the very hot hydrogen gas spontaneously ignited upon contact with the air and burned (with atmospheric oxygen) to form water. Unlike the familiar "pillar of fire" encountered with chemical rockets, during Earthbased testing a nuclear rocket engine's exhaust is essentially invisible—except perhaps for any incandescent impurities and the thermal aberrations it produces in the surrounding air (see Fig. 10.26).

Dual Mode System

A system to use a nuclear rocket reactor to generate electric power has been investigated. It is called a dual-mode rocket system. One mode involves the normal propulsion function, while the second mode involves the generation of electric power (at low to modest levels). Keeping the normal propulsion mode unchanged, the power production mode might be incorpo-

Fig. 10.26 The NRX-A6 engine being tested at the NRDS in December 1967. *Courtesy of NASA.*

DUAL-MODE NUCLEAR ROCKET

MODES

- HIGH-POWER ROCKET MODE OF 365 MW, 10 h LIFE, 2600-2660 K PROPELLANT TEMPERATURE, NEGLIGIBLE FUEL BURNUP
- LOW-POWER ELECTRICAL MODE OF 10-25 kW(e), 1 MW(t), ORGANIC RANKINE CYCLE DRIVEN BY THERMAL ENERGY FROM STRUCTURAL SUPPORT SYSTEM, USES PROPULSION MODULE SURFACE TO SUPPORT RADIATOR UP TO 10 kW(e)

ENGINE MODIFICATIONS

- STRUCTURAL SUPPORT SYSTEM ISOLATION VALVES FOR LOW-POWER
- CHANGE REACTOR DOME AND TIE TUBE CORE SUPPORT PLATE LINES FOR WIDER TEMPERATURE RANGE OPERATION TO STAINLESS STEEL FROM Al, AND ACTUATOR WINDINGS TO CERAMIC FROM POLYIMIDE

LIFETIME

- STUDY SHOWED TWO YEARS LIFE WILL NOT SIGNIFICANTLY AFFECT REACTIVITY CONTROL MARGIN

*STUDIED IN EARLY 1970's BY J. ALTSEIMER, L. A. BOOTH, "THE NUCLEAR ROCKET ENERGY CENTER CONCEPT" LA-DC-72-1262, 1972, BASED ON IDEAS OF JOHN BEVERIDGE

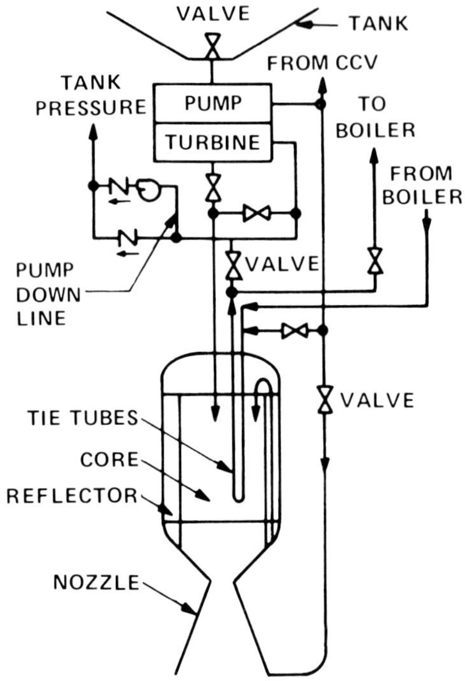

Fig. 10.27 Dual-mode nuclear rocket. *From L. A. Booth and J. H. Altseimer, "Summary of Nuclear Engine Dual-Mode Electrical Power System Preliminary Study," Los Alamos National Laboratory Report LA-DC-72-1111.*

rated into a nuclear rocket engine design by isolating hydrogen flow through the structural support system from the reactor chamber and nozzle sections and continuously recirculating it (see Fig. 10.27). With this type of scheme, some 10 to 25 kilowatts of electric power might be generated by a rocket reactor at the one MW-thermal operating level (see Fig. 10.28). The particular dual-mode configuration shown here involves an organic Rankine cycle. In the 10 kW_e power production version, the heat rejection radiator would be located along the surface of the power module. For higher levels of electric power generation, additional radiator surface area would be required. The major rocket engine modifications needed to implement the power production mode would be to add isolation valves for the low power gaseous hydrogen flow through the structural support system; to change the reactor dome and tie-tube core support plate and lines materials from aluminum to stainless steel in order to accommodate a wider range of operating temperatures; and to change the actuator materials from polyemid to ceramic materials. Engineering calculations indicate that over a two year lifetime there would be no significant degradation of the reactivity control margin.

Summary

The nuclear rocket program (see Fig. 10.29) demonstrated power levels as high as 4,100 MW and thrust levels of 930 kN (200,000 lb_f) with a hydrogen flow rate of 120 kg/s in the Phoebus-2A design. The highest equivalent specific impulse was achieved in the Pewee design, reaching 8,280 m/s (845 s) for an average coolant exit temperature of 2,550 K and a peak fuel temperature of 2,750 K. The minimum reactor specific mass, a value of 2.3 kg/MW, was reached with the Phoebus-2A system. The highest core average power density, some 2,340 W/cm^3, was obtained with the Pewee design, while the peak fuel power density was obtained in the NF-1 at 4,500 W/cm^3. The NF-1 also achieved the longest accumulated time at full power—some 109 minutes. The greatest number of engine starts and restarts was 28, accomplished in the XE′ program. In summary, basic nuclear rocket research and technology was completed during the Rover Program, a program terminated at the point of flight engine development. To have developed a flight system, it would have been necessary to verify the flight reactor and engine design, perform duration testing, and conduct reproducibility testing. In retrospect, however, there appear to be no technical barriers to the development of a successful nuclear rocket.

10 KW(e) REFERENCE DESIGN-STATE POINTS

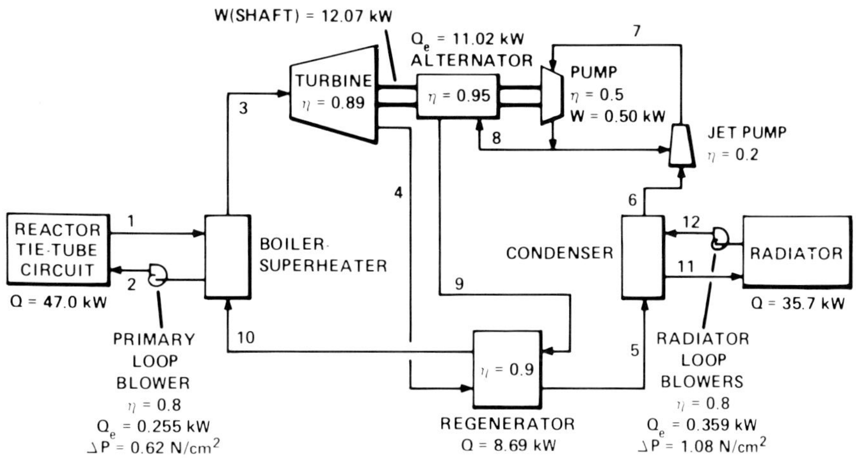

COMPONENT INLET	FLOW, (kg/s)	TEMP., (K)	PRESS., (N/cm^2)
PRIMARY LOOP – GH_2			
1 BOILER	0.0622	555.6	419.7
2 REACTOR	0.0622	504.0	420.0
POWER CONVERSION SYSTEM – THIOPHENE			
3 TURBINE	0.0891	550.0	140.0
4 REGEN VAPOR	0.0891	438.9	5.02
5 CONDENSER	0.0891	354.8	4.85
6 JET PUMP	0.0891	333.1	4.52
7 PUMP	0.178	334.5	14.90
8 ALTERNATOR	0.0891	342.1	155.6
9 REGEN LIQUID	0.0891	346.2	154.8
10 BOILER	0.0891	411.6	150.0
HEAT REJECTION LOOP GH_2			
11 RADIATOR	0.0798	355.2	419.5
12 CONDENSER	0.0798	304.6	420.0

POWER BREAKDOWN (kW)	
REACTOR THERMAL POWER	47.0
RADIATOR THERMAL POWER	35.7
ALTERNATOR OUTPUT	11.0
CONTROL SYSTEM	0.39
PRIMARY BLOWER	0.26
RADIATOR BLOWERS	0.36
NET ELECTRIC OUTPUT	10.0
TOTAL SYSTEM SPECIFIC MASS	48.6 kg/kW(e)

Fig. 10.28 Dual-mode nuclear rocket with a 10 kWe Rankine cycle. *From L. A. Booth and J. H. Altseimer, "Summary of Nuclear Engine Dual-Mode Electrical Power System Preliminary Study," Los Alamos National Laboratory Report LA-DC-72-1111.*

COMPARISON OF REACTORS TESTED IN ROVER PROGRAM

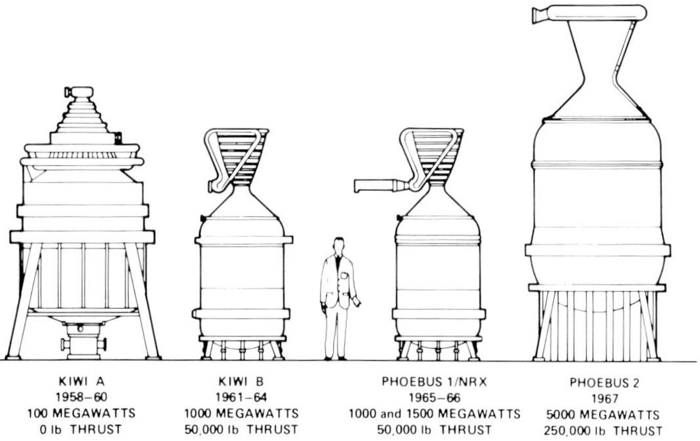

Fig. 10.29 Comparison of reactors tested in Rover Program. *From R. E. Schreiber, "The LASL Nuclear Rocket Propulsion Program," LASL Report LAMS-2036, April 1956.*

Bibliography

J. H. Altseimer, G. F. Mader, and J. J. Stewart, "Operating Characteristics and Requirements for the NERVA Flight Engine," J. Spacecraft, Vol. 8, No. 7, (July 1971).

J. H. Altseimer et al., "XE-Engine Systems Analysis Design Data Book," Aerojet General Report No. RN-S-0289A, (April 1967).

J. D. Balcomb, "Nuclear Rocket Reference Data Summary," LA-5057-MS, (October 1972).

J. H. Beveridge, "Feasibility of Using a Nuclear Rocket Engine for Electrical Power Generation," AIAA Paper 71-639, AIAA/SAE Joint 7th Propulsion Joint Specialist Conference, Salt Lake City, Utah, (14–18 June 1971).

L. A. Booth and J. H. Altseimer, "Summary of Nuclear Engine Dual-Mode Electrical Power System Preliminary Study," LASL Report LA-DC-72-1111.

D. Buden, "Operational Characteristics of Nuclear Rockets," AIAA Paper 69-515, AIAA 5th Propulsion Joint Specialist Conference, U.S. Air Force Academy, Colorado, (9–13 June 1969).

K. V. Davidson et al., "Development of Carbide-Carbon Composite Fuel Elements for Rover Reactors," LASL Report LA-5005, (October 1972).

J. DeStefano and R. J. Bahorich, "Rover Program Reactor Tests Performance Summary NRX-A1 Through NRX-A6," Westinghouse Astronuclear Laboratory Report WANL-TME-1788, (July 1968).

A. R. Driesner, "Summary of Disassembly and Post-Mortem Visual Observations of the Kiwi-B4E-301 Reactor," LASL Report LA-3299-MS, (July 1965).

F. P. Durham, "Nuclear Engine Definition Study Preliminary Report, Vol. I—Engine Description, Vol. II—Supporting Studies, Vol. III—Preliminary Program Plan," LASL Report LA-5044-MS, Vol, I, (September 1972).

W. J. Houghton and W. L. Kirk, "Phoebus II Reactor Analysis," LASL Report LAMS-2840, (April 1963).

M. T. Johnson, "NERVA Reactors," Proc. ANS 1970 Topical Meeting, Huntsville, Alabama, (28–30 June 1970).

L. D. P. King et al., "Description of the Kiwi-TNT Excursion and Related Experiments," LASL Report LA-3350-MS, (August 1966).

W. L. Kirk, "Advanced Nuclear Rocket Technology," Proceedings ANS 1970 Topical Meeting, Huntsville, Alabama, (28–30 April 1970).

W. L. Kirk, "Nuclear Furnace-1 Test Report," LASL Report LA-5189-MS, (March 1973).

L. L. Lyon, "Performance of (U, Zr)C-Graphite (Composite) and of (U, Zr)C (Carbide) Fuel Elements in the Nuclear Furnace-1 Test Reactor," LASL Report LA-5398-MS, (September 1973).

B. Misra, J. H. Altseimer, and G. D. Hart, "Hi-Flight Coolant Management Considerations for the NERVA Reactor Cooldown," Nuclear Technology, Vol. 12, (November 1971).

J. M. Napier, "NERVA Fuel Element Development Program Summary Report—July 1966 Through June 1972 Impregnation Studies," Oak Ridge Y-12 Plant Report Y-1852, Part 2, (September 1973).

C. M. Rice and W. H. Arnold, "Recent NERVA Technology Development," J. Spacecraft, Vol. 6, No. 5 (May 1969), p. 565.

J. C. Rowley, W. R. Prince, and R. G. Gido, "A Study of Power Density and Thermal Stress Limitations of Rover Reactor Fuel Elements," LASL Report LA-3323-MS, (July 1965).

R. E. Schreiber, "The LASL Nuclear Rocket Propulsion Program," LASL Report LAMS-2036, (April 1956).

R. W. Schroeder, "NERVA—Entering a New Phase," Astronautics and Aeronautics, (May 1968), p. 42.

J. M. Taub, "A Review of Fuel Element Development for Nuclear Rocket Engines," LASL Report LA-5931, (June 1975).

E. A. Warman, J. C. Courtney, and K. O. Koebberling, "Final Report of Shield System Trade Study," Vol. 1, Book 1, Aerojet General Report S054-023, SO54-CP090290-F1.

Chapter 11

Alternate Space Reactor Concepts

Introduction

Uranium-zirconium hydride (U-ZrH) reactors, such as the SNAP-2, -10A, and -8 systems, were NaK-cooled thermal reactors designed for operation around 900 K. The nuclear fuel was arranged in cylinders with pumped NaK coolant flowing between the fuel elements. In sharp contrast were the epithermal Rover nuclear rocket reactors that demonstrated operation at power levels in excess of 4,000 megawatts-thermal (MWth) and at temperatures near 2,450 K. The Rover reactors used uranium dicarbide (UC_2) fuel in a graphite matrix and thermal energy removal was achieved using hydrogen as the propellant. Power plant lifetimes were measured in hours as opposed to years.

A rich variety of alternate reactor designs was also created in the past to meet potential space power and propulsion requirements. Some of the most interesting of these concepts are discussed in this chapter. They include: the Rankine Cycle Liquid Metal Fast Reactors, the Medium Power Reactor Experiment (MPRE), the High Temperature Gas-Cooled Electric Power Reactor (710 Reactor Program), the Advanced Space Nuclear Power Program (SPR), fluidized bed reactors, thermionic reactors, and gaseous core reactors.

Rankine Cycle Liquid Metal Fast Reactors

From the early 1960s until 1973 active programs existed to develop liquid metal cooled, fast neutron spectrum reactors coupled with a potassium Rankine cycle power converter. This effort included the SNAP-50/SPUR component technology program (discontinued in 1965) and its follow-up program, called the Advanced Liquid Metal Cooled Reactor (which continued until 1973). The SNAP-50 program attempted to develop a 300–1,200 kWe power plant with a 10,000 hour lifetime. To achieve the lower end of this power generation spectrum, a 2.2 megawatt-thermal reactor was being designed along with a potassium working fluid Rankine cycle power converter, capable of producing 300 kWe. This same technology was to be used to achieve the upper end of the planned power generation regime, namely 1,200 kWe. In this case, the reactor would be enlarged to provide 8–10 MWth and four 300 kWe potassium Rankine cycle converter modules would be used to generate a total of 1.2 MWe [1–4].

Figure 11.1 illustrates the flow schematic for a 300 kWe potassium Rankine cycle advanced nuclear reactor, while Fig. 11.2 describes its nominal operating conditions [4]. A schematic of the SNAP-50 is shown in Fig. 11.3 [1]. This system has a fast neutron spectrum, approximately 0.35 MeV to 0.55 MeV. Highly enriched (93%) uranium-235, either as a carbide or nitride, was considered for the fuel. The nitride fuel form appeared to be the preferred choice. Columbium alloy was used in the pressure vessel, the core support structure, and fuel cladding, while titanium was used for parts of the support system at temperatures below 810 K. Lithium was used as the primary loop working fluid. It entered the reactor, flowing upward through an annular passage that was formed by the outer boundary of the core and the pressure vessel wall. Thus, the core boundary and the pressure vessel were cooled by the lithium before it entered the center of the core from the upper plenum.

A unique design problem encountered with this lithium-cooled core was the need to melt the lithium before operating the reactor. To accomplish this, the pressure vessel was enclosed by a columbium alloy jacket that provided a channel for high temperature helium to preheat and maintain the core and pressure vessel at 645 K prior to its powered operation. This jacket also provided the inert gas atmosphere needed for oxidation protection of the pressure vessel. Thermal insulation was provided outside the preheat jacket around the upper and lower regions of the pressure vessel to minimize the temperature of the support structure, the reflector, and bearings and components below the reactor.

Six pivoted reflector segments were arranged so that in the fully closed position they completely surrounded the reactor core. These segments consisted of a colum-

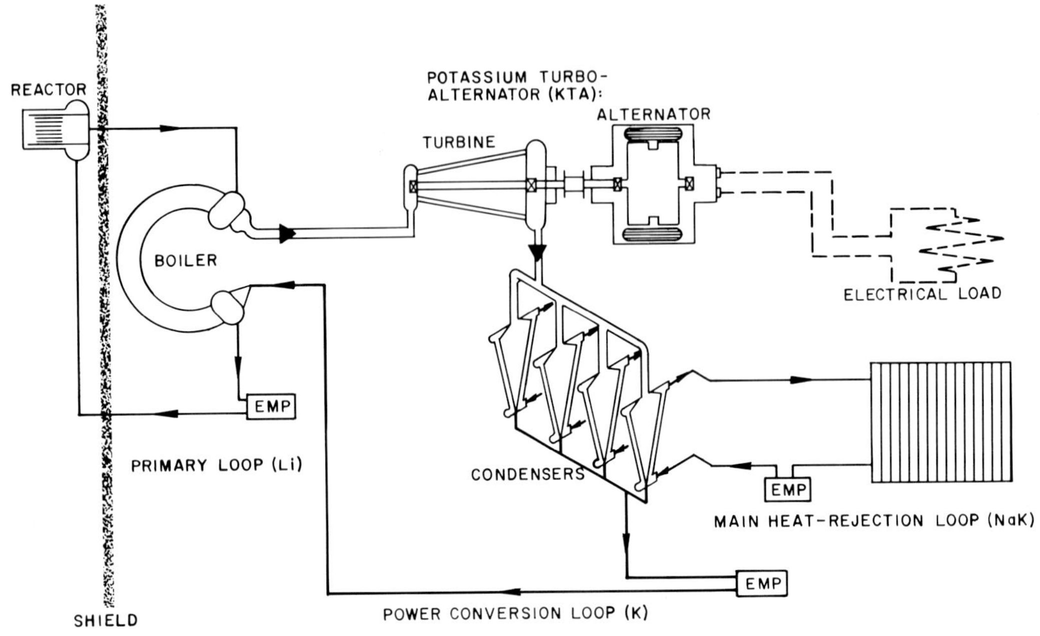

Fig. 11.1 Schematic diagram for advanced potassium Rankine cycle nuclear power plant [*4*].

bium alloy container filled with blocks of beryllium oxide. They were connected to the reactor and reflector support structure by means of flexure bearings at each end. Reactor control was provided by positioning the reflector segments to vary neutron leakage from the core. Additional details about the construction of the

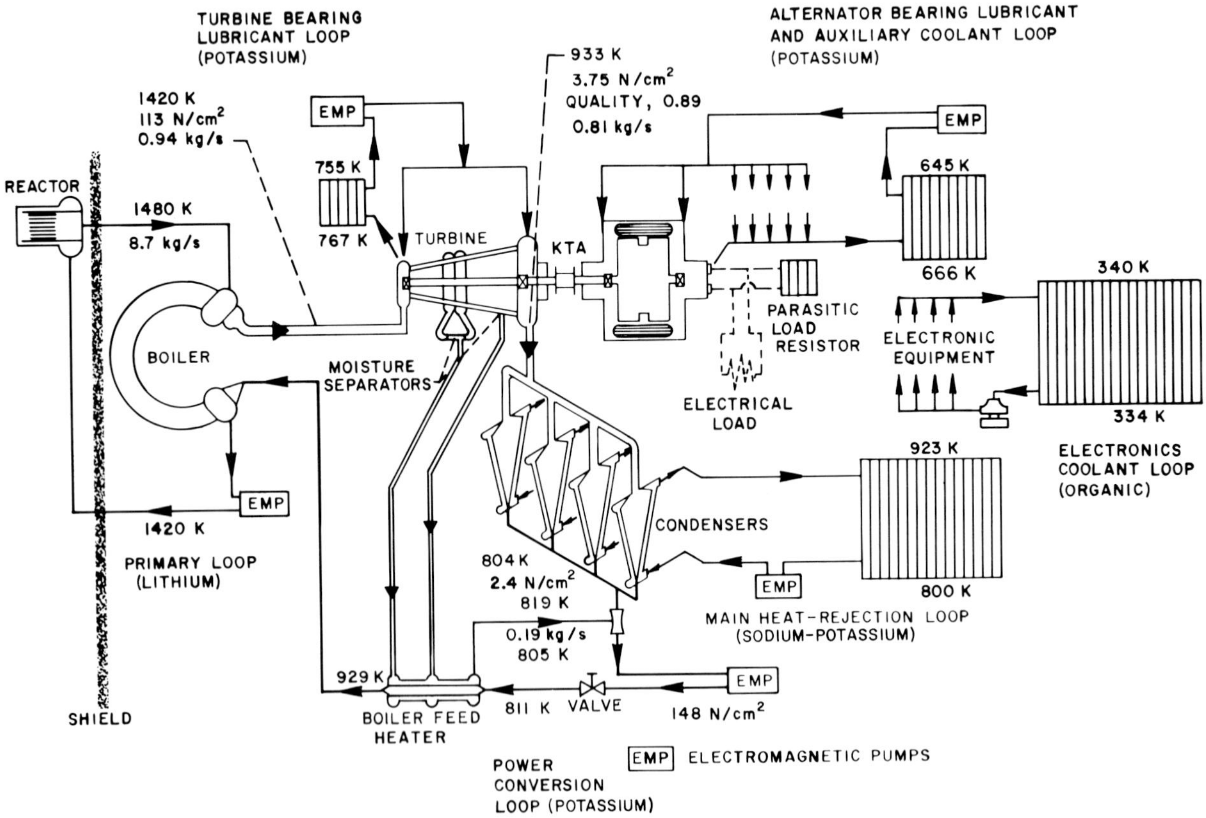

Fig. 11.2 Nominal operating conditions for an advanced 300 kWe potassium Rankine cycle nuclear reactor [*4*].

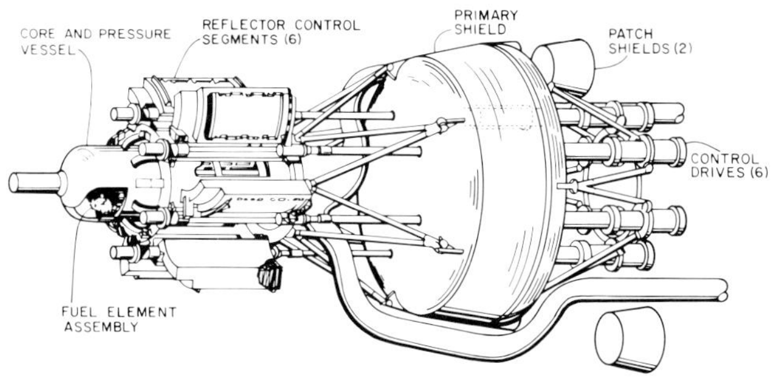

Fig. 11.3 SNAP-50 (PWAR-20) reactor and shield configuration [1].

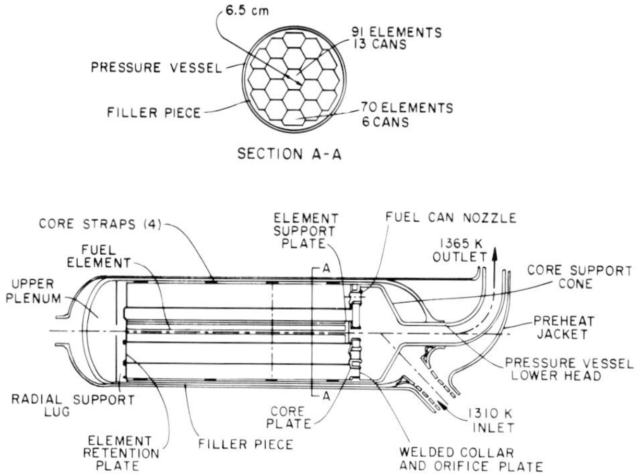

Fig. 11.4 SNAP-50 (PWAR-20) reactor core and pressure vessel [5].

SNAP-50 reactor core are presented in Fig. 11.4 [2, 5].

Extensive development was performed on both uranium nitride (UN) and uranium carbide (UC) as candidate fuels. For example, a UN fuel test exhibited no adverse effects after nearly 6,000 hours at an average cladding temperature of 1,350 K and a burnup of 1 percent of the total number of uranium atoms. In addition, UN fuel performed well at temperatures between 1,420 K and 1,480 K for 1,690 hours at 1.2 percent burnup. (Fuel swelling was less than 1 percent and gas release less than 12 percent.) For the UN fuel, the swelling rate remained quite low, as burnup rates progressed beyond 1 percent, but a sudden surge of gas release occurred between 1.5 and 1.6 percent burnup. The calculated gas pressure associated with this release, which was about 20 percent of the total gas produced, approached the strength limit of the cladding material. Burnups of the UN fuel beyond 1.6 percent would, consequently, require stronger cladding materials, such as tungsten-rhenium (W-Re) or lower gas release rates [3]. Uranium carbide fuel demonstrated good performance to 2,500 hours, prior to an equipment failure. However, UC exhibited higher swelling rates than did UN.

Non-nuclear components were extensively developed and tested for up to 10,000 hours. However, complete systems were never tested. Electromagnetic (EM) pumps, with their relative simplicity of design, were used in both the primary and boiler feed loops. A cutaway sketch of the boiler feed, helical induction pump appears in Fig. 4.4. This pump is analogous to an

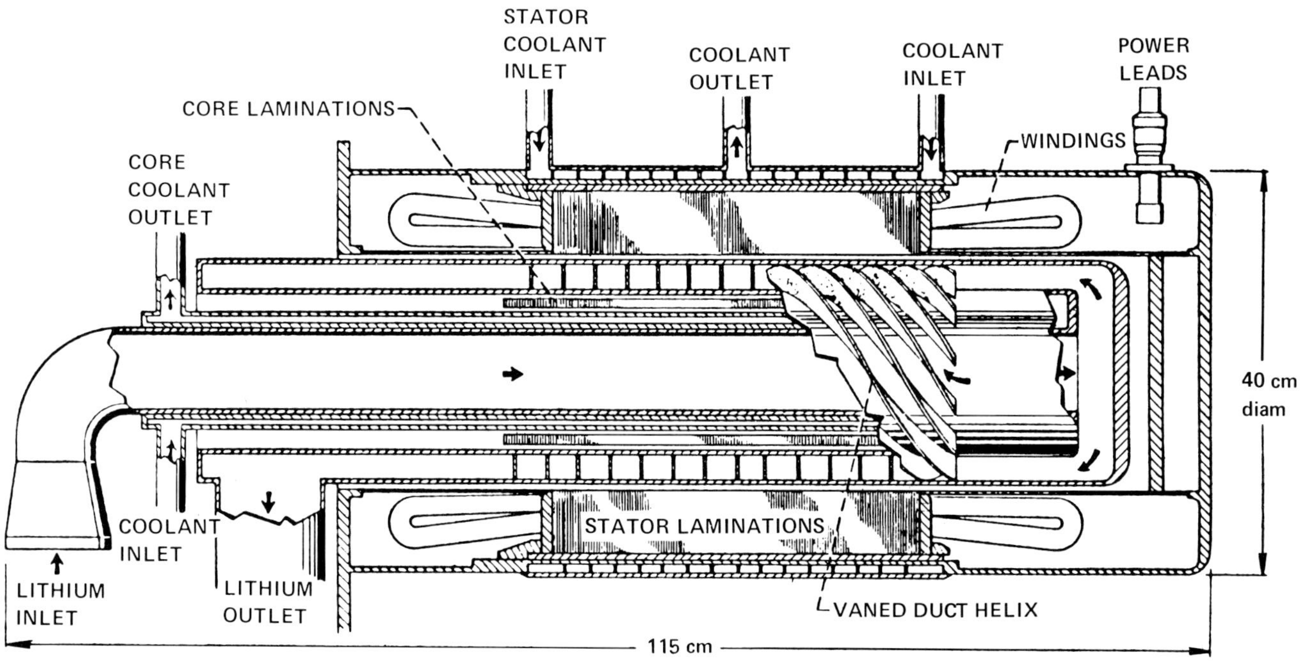

Fig. 11.5 Cutaway view of a lithium working fluid, helical induction pump for an advanced Rankine cycle reactor system [4].

Table 11.1 Electromagnetic pump design specifications.

	Primary-loop	Boiler-loop
Fluid	Lithium	Potassium
Fluid inlet temperature (K)	1,425	811
Flow rate (kg/sec)	13.6	1.47
Developed pressure rise (N/cm^2)	14	166
Net positive suction head (N/cm^2)		4.8
NaK coolant inlet temperature (K)	700	730
Coolant temperature rise (K)	310	280

induction motor except that here the rotor has been replaced by a liquid-metal-filled annular passage. Liquid metal (e.g., potassium) enters the pump at the right into an annular passage. Then, the rotating magnetic field imposes a force on the liquid, causing it to rotate circumferentially in the annulus. With flow constrained within the helical channel in the annulus, the fluid is then forced to move from one end of the pump to the other. The static pressure of the fluid increases along the helical channel with the high pressure fluid making a 180 degree turn at the end of the annulus and exiting the pump at the right through the center pipe. A lithium working fluid version of this EM pump, used for the primary loop, is illustrated in Fig. 11.5 [4]. The design is quite similar to the boiler feed EM pump but has an additional central coolant passage to meet higher operating temperature requirements. Table 11.1 summarizes the design specifications for these EM pumps [4]. Efficiencies of 19 percent (potassium) and 16 percent (lithium) were calculated [4].

The boiler transferred thermal energy from the primary reactor loop to the secondary power conversion loop. For this Rankine cycle, superheated potassium vapor at a temperature of 1,420 K was needed. The lithium temperature in the primary reactor loop was 1,480 K. The boiler had to be compact, provide stable performance during startup and at operating conditions, and also be structurally resilient to withstand system and environmental stresses over many thousands of operating hours.

As shown in Fig. 11.6, this boiler contained 31 tubes (approximately 230 cm long) which were housed in a 15.2 cm diameter shell. The potassium working fluid flowed in a single pass through the inside tubes, while the primary loop lithium flowed in the space between the tubes and shells in a countercurrent direction. This boiler was made of refractory metal, tantalum alloy T-111 (Ta-8W-2Hf). The boiler's arc shape facilitated packaging and minimized stresses. Each boiler tube contained a swirl-generating insert, such as a helical vane or wire coil. These inserts augmented boiling heat transfer processes and helped make the boiler's operation insensitive to gravitational effects. In the boiler, the potassium working fluid started as a liquid, experienced nucleate boiling, entered a transition boiling region, and finally emerged as a superheated vapor.

The superheated potassium then left the boiler and was expanded in the turboalternator. A representative turboalternator had a 10-stage axial-flow turbine flexibly coupled to a 10-pole homopolar alternator. The turbine and alternator rotors were each supported on a pair of potassium-lubricated, pivoted-pad journal bearings. A double-acting potassium pivoted-pad thrust bearing in each component supported forward and reverse loads. This unit operated at 19,200 rpm and produced a net electrical power output of 428 kWe. Dimensionally, the unit was 1.83 m long and 0.61 m diameter at the alternator and had a mass of 590 kg [4]. One major design consideration centered around keeping the "moisture" of the potassium in the turbine as low as possible (in Chapters 4 and 5 the moisture of a thermodynamic mixture was defined to be the fraction of mass present in the liquid phase). If, for example, the moisture of the potassium working fluid reached 20 percent, catastrophic turbine blade damage would result from liquid droplet impact erosion. To overcome this problem an interspool liquid separator was designed. It swirled the entire flow exiting the first stage by the upper set of vanes, thereby centrifuging the liquid to the wall. A slot in the wall scavenged the liquid and some vapor and returned the condensed mixture to the boiler feed heater. A lower set of vanes straightened the vortex flow prior to its entry into the second turbine stage. All the blading, the rotor discs, and the tie-bolts were fabricated from the molybdenum alloy TZM (Mo-0.6 Ti-0.1 Zr-0.035 C).

Estimated masses for a representative 300 kWe advanced Rankine cycle space power system are presented in Table 11.2. This system had an efficiency of 19 percent and a specific mass of 32 kg/kWe (minus the electrical equipment).

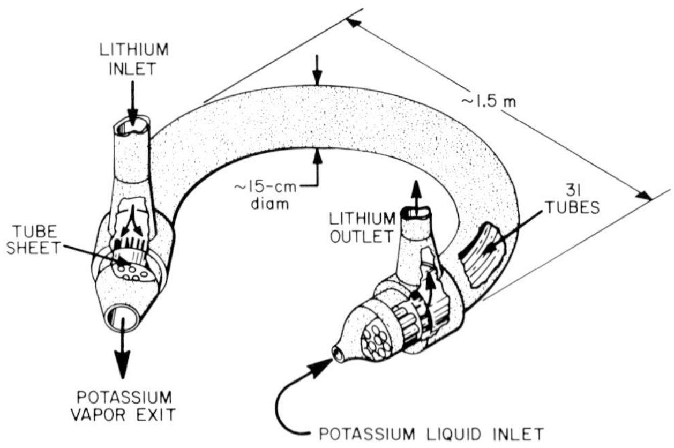

Fig. 11.6 Once-through 2 megawatt-thermal potassium boiler, advanced Rankine cycle reactor system [*4*].

Table 11.2 Mass estimates for nominal 300 kWe advanced Rankine cycle space power system.

Item	kg
Reactor (2 MW thermal, 3.8 at.% fuel burnup, 50,000-hr life) and reactor controls, drive, and structure	1,810
Primary loop (Li):	
Boiler	136
Pump	317
Pipe and insulation	145
Expansion tank	82
Lithium inventory (0.18 m^3) (72.5 kg)	(a)
Total weight	680
Power conversion loop (K):	
Turboalternator	590
Condensers (4)	110
Boiler feed pump	136
Condensate heat exchangers (2)	46
Separator	23
Potassium inventory control	46
Potassium injection tank	114
Potassium inventory	68
Vapor piping	114
Liquid piping	23
Total weight	1,270
Main heat-rejection loop (NaK):	
Radiator panels (4)	982
Pumps (4)	217
Expansion tanks (4)	181
Piping	195
NaK inventory	(a)
Total weight	1,575
Shadow shield (for unmanned spacecraft, 10° half-cone angle, ~46 m to payload)	2,700
Alternator coolant loop (K):	
Radiator panel	145
Pump	59
Expansion tank	23
Piping	36
Potassium inventory	(a)
Total weight	263
Turbine coolant loop (K):	
Radiator	34
Pump	23
Expansion tank	23
Piping	16
Potassium inventory	(a)
Total weight	96
Electronics cooling loop (DC-200):	
Radiator	124
Pump	11
Expansion tank	11
Piping	48
Inventory	46
Total weight	240
Electrical equipment:	
Speed control	46
Voltage regulator-exciter	92
Internal power conditioning	92
Parasitic load resistor	270
Interconnecting cable	23
Controls and instrumentation	23
Total weight	546
Structure (assumed to be 10 percent of engine weight)	890
Total estimated weight	10,070

[a] Included in above weights.

Extensive material and component testing had been performed on non-nuclear components. To be useful, such materials must resist the corrosive effects of lithium and potassium, exhibit long term strength above 1,365 K, and possess good weldability and ductility. It has been demonstrated in tests lasting beyond 10,000 hours that refractory alloys are compatible with alkali metals at temperatures above 1,475 K if careful procedures are followed. For example, corrosive attack by alkali metals is greatly accelerated in the presence of oxygen. Consequently, refractory alloys containing hafnium or zirconium as getters can be used to tie up the oxygen. Thus, the tantalum alloy T-111 (Ta-8W-2Hf) was used as the principal high temperature containment material for the primary (lithium) loop, the boiler, the secondary (potassium) loop and high temperature portions of the turboalternator. In March 1970, an endurance test of 10,000 hours was successfully completed with T-111 alloy in a corrosion study loop [6].

Heat transfer components were demonstrated in a prototype, single boiler tube and a three-tube NaK-cooled potassium condenser. Very stable operation was observed.

A successful 10,000 hour endurance test of a prototype boiler feed pump demonstrated the efficacy of EM induction pump technology. The pump operated at a potassium temperature of 810 K, a 1.5 kg/s flow rate, and a 1.6 MPa pressure. The stator coolant was NaK supplied at 700 K. The pump voltage was 135 volts at 60 cycles and the pump had a design efficiency of 16.3 percent.

A 5,000 hour erosion test and liquid extraction performance tests for a three-stage turbine using potassium vapor as the working fluid were considered to have verified design assumptions and to have indicated that a three to five year lifetime for the turboalternator was indeed achievable. The measured effectiveness values of

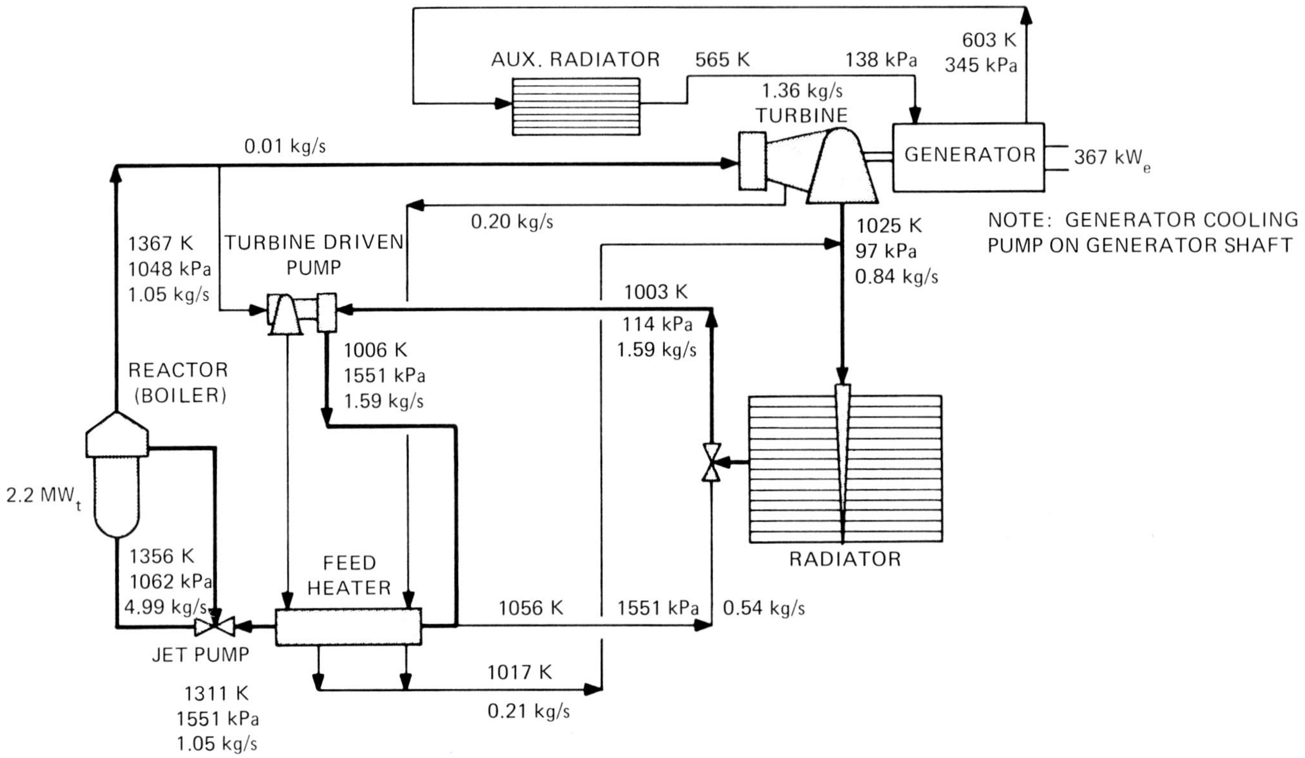

Fig. 11.7 Medium power reactor experiment (MPRE) showing single loop potassium Rankine cycle [7].

the rotor extraction device generally exceeded 30 percent and the effectiveness of the vortex separator approached 80 percent for separator vapor velocities of 9.3 m/s. The alternator bore seal was considered to be perhaps the component of least engineering certainty.

At the time of program discontinuation, high temperature Rankine cycle technology components had been demonstrated and a point was reached where a breadboard system could be tested with a high probability of success.

Medium-Power Reactor Experiment (MPRE) [7]

In 1958 work began on space nuclear power plants in the twenty to several hundred kilowatt-electric range with a minimum service life of 10,000 hours. This led to the design of a single-loop potassium Rankine cycle system. Activities were pursued on this concept until 1966. Initial technical questions centered around the compatibility of stainless steel with boiling potassium and the heat transfer capability of boiling potassium under forced convection for a reasonably small size core. By 1964, progress was sufficiently favorable to support the initiation of work on a reactor experiment of 1 megawatt-thermal, a power level corresponding to about 140 kilowatts-electric. When the program was terminated in 1966, both the one MWth reactor and its test facility had been designed. In addition, a series of component and system test rigs had also been designed, constructed, and operated to demonstrate and validate the design concept. Technical questions still to be resolved included the control of free-liquid surfaces under microgravity conditions and a reliable potassium vapor turbogenerator unit.

The MPRE (shown schematically in Fig. 11.7) was designed to be a major simplification over the commonly considered multiloop systems. Potassium was used as the working fluid, partially because it has sufficiently low neutron activation rates, so that it could be pumped beyond the radiation shield. Thermal energy was added to this working fluid in the reactor core, with boiling occurring in the upper segment of the core. This was considered acceptable because potassium has little effect on reactivity. Consequently, irregular potassium boiling in the core would not adversely affect nuclear power output. A vapor separator was located above the core. It was designed to give the existing potassium vapor a quality of approximately 99 percent. An expansion tank was located below the vapor separator.

In the MPRE design, potassium vapor exited the reactor and was expanded through the main turbine to generate electricity. A regenerative feed heater reduced the size of the radiator, increased cycle efficiency, and minimized thermal stresses in the boiler recirculation system. The potassium was condensed in the radiator.

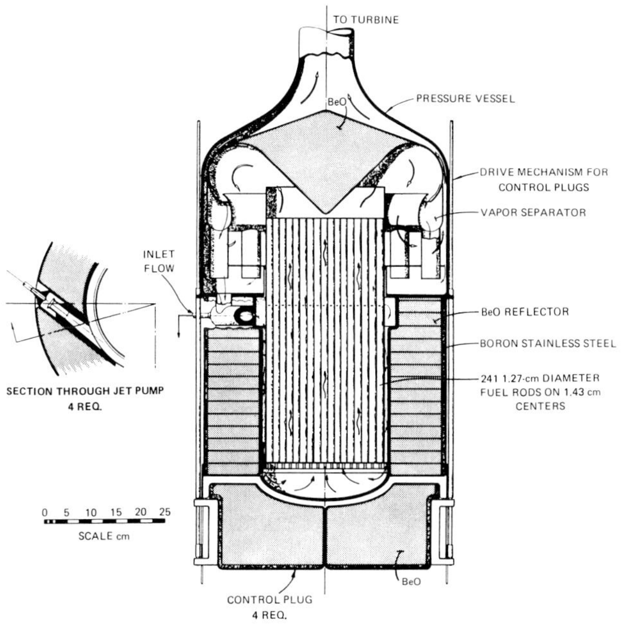

Fig. 11.8 Reactor schematic for MPRE [7].

Jet pumps, driven by a portion of the feed pump discharge flow, were installed near the condenser outlet manifold to facilitate removal of condensate from the condenser under microgravity conditions. A high heat transfer coefficient was achieved in the reactor through the use of a recirculating boiler effect in which the potassium liquid removed from the vapor separator was recirculated through the core by means of jet pumps driven by the boiler feed supply. To simplify the bearing and shaft system designs, an independent centrifugal feed pump (used to pressurize the fluid) was driven by a free turbine.

The MPRE reactor core (see Fig. 11.8) contained bundles of fuel rods with fully enriched UO_2 pellets in a stainless steel capsule (see Fig. 11.9) [7]. The 1.27 cm diameter fuel rods were arranged in an equilateral triangle pattern with a center-line spacing of 1.43 cm. This configuration created a coolant passage volume fraction of 0.283. A beryllium oxide (BeO) reflector surrounded the core. It was 8.26 cm thick on the side and 12.70 cm thick on the bottom. As shown in Fig. 11.8, the side reflector consisted of a stacked series of 3.05 cm thick BeO discs. As depicted in Fig. 11.9, a 7.6 cm long stack of BeO pellets were positioned above the UO_2 in each fuel rod and served as a portion of the top reflector. An additional 10.9 cm of void space above the BeO pellets provided a reservoir for fission gas accumulation. This void space also contained a piston and molybdenum spring to restrain axial movement of the BeO and UO_2 pellets. The lower portion of the BeO reflector was divided into four independent quadrants that could be moved axially to adjust reactivity and together were worth 5 percent in reactivity change (i.e., 5% in $\delta k/k$).

A thin, neutron absorbing layer of 0.1 percent boron steel surrounded the reflector. This thin layer of steel reduced the amount of neutron activation products and the need for gamma ray shielding. The layer of boron steel also tended to isolate the core from the environment in the event of a launch abort and reactor core immersion in water.

In the MPRE design, the potassium working fluid entered the reactor core through the bottom. It then rose between the fuel elements, emerged from the core at a vapor quality of about 20 percent, and entered the vapor separator where the liquid was skimmed off and discharged to the expansion tank directly beneath the vapor separator. The potassium vapor passed radially outward at the bottom of the vapor separator, then upward around the outer perimeter, and finally out the vapor exit pipe at the top. The liquid potassium entered a set of four jet pumps, operating in parallel beneath the expansion tank. These jet pumps extracted liquid potassium from the expansion tank and discharged it so that it returned to the core inlet.

Key MPRE parameters are listed in Tables 11.3 and 11.4. During the course of this program, many important technical issues were resolved. For example, the compatibility of stainless steel with boiling potassium was demonstrated in over 160,000 hours of component and systems tests, including 9,000 hours testing of reduced-scale mockups of the MPRE system. One such mockup operated for 2,500 hours. Calculations and critical experiments demonstrated that boiling potassium reactors have good nuclear stability and control characteristics. Boiling flow stability and flow distribution were demonstrated in electrically heated mockups. However, explosive boiling of the potassium was found to be a major problem, unless nucleation sites were introduced to maintain smooth nucleate boiling conditions. A lightweight, tapered-tube, direct condensing radiator was developed and shown to give a uniform flow distribution with good flow stability. A potassium

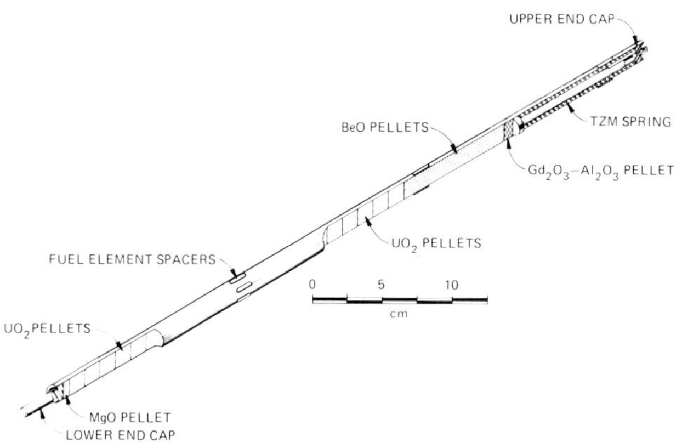

Fig. 11.9 Fuel element for MPRE [7].

Table 11.3 MPRE design parameters for stainless steel fuel element.

Reactor Operating Conditions	
Thermal Power (MW)	1
Coolant	Potassium (boiling)
Inlet Temperature (K)	1,105
Outlet Temperature (K)	1,111
Inlet Pressure (kPa)	207
Outlet Pressure (kPa)	200
Reactor Geometry	
Fuel Element Arrangement	Equilateral triangular pitch
Moderator	None in core
Diameter (equivalent) (cm)	23.3
Distance Across Corners (cm)	24.1
Length (cm)	29.5
Side Reflector Thickness (cm)	7.6
Fluid Flow Data	
Core Flow Passage Equivalent Diameter (cm)	0.5
Vapor Flow at Outlet (kg/s)	0.51
Liquid Flow at Core Outlet (kg/s)	2.02
Total Liquid Flow Entering Core (kg/s)	2.53
Vapor Quality Leaving Core (%)	20
Heat Transfer Data	
Power Density (core average) (W/cm^3)	79.5
Heat Transfer Surface Area (m^2)	2.8
Fuel Element Cladding Temperature (K)	
Average	1,119
Maximum (Hot Spot)	1,133
Fuel Central Temperature (K)	
Average	1,588
Maximum	1,755
Fuel Element	
Configuration	Rods
Fuel Form	UO_2 pellets in metal capsules
Number of Fuel Elements	241
Overall Length (cm)	50.5
Fueled Length (cm)	29.5
Fuel Enrichment (%)	93
Cladding Material	Type 316 stainless steel
Cladding Thickness (cm)	0.05
Fuel Element Outside Diameter (cm)	1.3
Burnup at End of Life (% of uranium atoms)	0.76
Core Composition, Volume Fraction	
Stainless Steel	0.11
UO_2	0.60
Void Inside Fuel Elements	0.01
Coolant Flow Passages	0.28
Control Elements	
Reactivity Control Method	4 movable BeO plugs (quadrants of the bottom reflector)
Control Plug Motion	Vertical displacement parallel to axis of reactor; total travel 20.3 cm
Vertical Height of Control Plugs (cm)	21.6
Sheath Material	Type 304 stainless steel
Combined Worth of Control Plugs ($\%\delta k/k$)	5.0
Pressure Vessel	
Material	Type 304 stainless steel
Shell Outside Diameter (cm)	26.7
Wall Thickness (cm)	0.64
Overall Length (with expansion tank and vapor separator) (cm)	86.4
Physics	
Mass of UO_2 (kg)	78.0
Mass of U-235 (kg)	63.8
Effective Delayed Neutron Fraction	0.0075
Median Energy For Fission (keV)	840
Prompt Neutron Lifetime (sec)	3.2×10^{-7}
Reactivity Worth of Control Plugs	0.05
Isothermal Temperature Coefficient of Reactivity $((1/k)\cdot(dk/dT))$ (K^{-1})	-1.6×10^{-5}
Fuel Reactivity Coefficient $(\Delta k/k)/(\Delta m/m)$	0.545
Loss in k, Fuel Burnup For 1 MW-year	0.0040
Reactivity Effect Of Potassium Filling Coolant Passages (relative to normal operating condition)	0.0014
Maximum Excess k From Control Plug Insertion	
Full Power	0.006
Cold, Clean	0.020
Neutron Flux at Outer Surfaces of Core ($neutrons/cm^2$-sec)	
Fast ($E = 0.9 - 10$ MeV)	0.41×10^{14}
Thermal ($E = 0 - 1$ eV)	3.4×10^{11}
Radiator	
Power To Be Dissipated (kW)	860
Emissivity of Treated Surface	0.92
Fin Efficiency	0.86
Area Required (m^2)	40.4
Radiator Diameter (m)	3.1
Radiator Height (m)	4.6
Vapor Quality (%)	91.7
Potassium Flow Rate (kg/s)	0.45
Number of Tubes	96

Table 11.4 Characteristics of a series of single-loop boiling potassium reactor systems (based on the MPRE concept).

Reactor Outlet Temperature (K)	1,110	1,110	1,365	1,365	1,365
Net Electrical Output (kW)	140	367	367	1,000	5,000
Reactor Thermal Output (MW)	1	2.2	2.2	6	30
Reactor Core Diameter (cm)	24.4	30.7	30.7	42.9	73.4
Reactor Mass (kg)	341	532	532	1,073	3,468
Radiator Height (m)	4.8	10.9	5.4	9.2	22.3
Radiator Diameter (m)	3.1	3.1	3.1	4.9	10.1
Radiator Mass Without Armor (kg)	423	843	529	1,441	7,180
Meteoroid Armor Mass (kg)[a]	205	314	241	918	6,455
Total System Mass (kg)[b]	1,305	2,602	1,910	5,080	25,385

[a]Armor designed to give 0.1% probability of a meteoroid penetration in 10,000 hours.
[b]Not including shield mass.

lubricated, tungsten carbide bearing was operated in a bearing testing rig for 4,000 hours with 350 starts and stops under load with no loss in performance or detectable surface damage.

The world's first potassium vapor turbine, employing potassium lubricated bearings, was successfully operated during the MPRE program. This operation included 4,000 hours of accumulated testing and 2,500 hours on one unit. Performance of typical jet pumps in the cavitating regime and pump design characteristics necessary for microgravity operation of a Rankine cycle system were also demonstrated. Four full-scale fuel elements were operated in-pile at design power and temperature for a total of over 30,000 hours with no sign of difficulty. In fact, one fuel element operated for over 14,000 hours—the equivalent of three years of normal service.

A parametric study of shielding showed that the effects of neutron and gamma scattering could be kept small in a cylindrical radiator concentric with the reactor. Bench tests, including startup and operation under microgravity conditions, were devised and the vapor separator, expansion tank, boiler, and condenser tested satisfactorily. In addition, a vapor separator-expansion

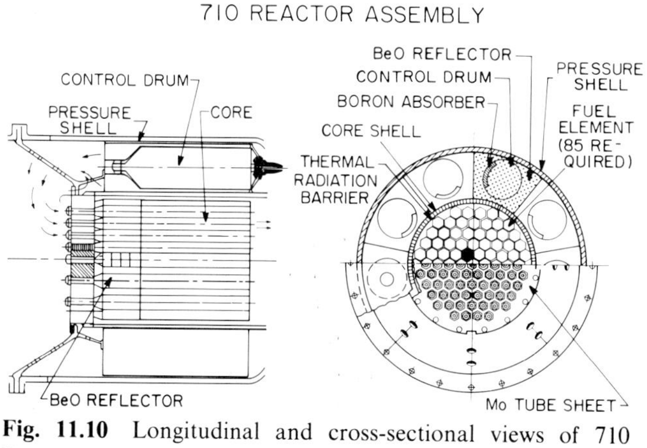

Fig. 11.10 Longitudinal and cross-sectional views of 710 Reactor [*8*].

tank was shown to operate fairly well in microgravity simulation tests conducted onboard a KC-135 aircraft. Finally, an electrically heated mockup (37 percent full-scale) of the MPRE was operated for over 3,600 hours with a potassium vapor, turbine-driven feed pump.

The next logical steps in the development of the MPRE system would have included: (1) testing fuel elements for up to 20,000 hours; (2) the intentional testing of a defective fuel element with a low fission product release rate; (3) the startup and microgravity operation of an electrically heated, potassium vapor power plant in space; and (4) the endurance testing of nuclear-fueled power plants on the ground.

High-Temperature, Gas-Cooled Electric Power Reactor (710 Reactor)

From 1962 until 1968, the 710 Advanced High Temperature Gas Reactor Program was conducted to develop and demonstrate a high performance, fast spectrum, refractory metal reactor as a heat source for mobile power and propulsion applications [8]. This program started out as a demonstration of both a closed-loop neon-cooled reactor for electric power generation and an open-loop hydrogen working fluid system for rocket propulsion. In 1963, the open-loop, rocket propulsion objective was deleted, while in 1965 the remaining program objective was reduced to a fuel element technology program.

The 710 Reactor was a fast spectrum, refractory metal reactor, using a hexagonal, cermet fuel element that had multiple tubular flow channels. Figure 11.10 is a schematic of the 710 Reactor [8]. Power flattening could be achieved by varying either the fuel loading or the hydraulic diameter between fuel elements or within fuel elements. The latter approach resulted in a smaller core size. The fuel elements were mechanically attached to a metallic front tube sheet, using a single point support. This tube sheet, in turn, was attached to the pressure vessel. The metallic front tube sheet also served as a reasonably efficient front reflector because of the

Table 11.5 Status of 710 reactor program fuel element testing (as of 14 October 1966).

Type of Test	Specimen No.	Specimen Geometry[a]	Cladding Material (at.%)	Temp. (K)	No. of Cycles	Time at Temp. (hr)
Static	ASH-7	P.L.-1 channel	W-30Re-30Mo	1,920	65	7,084
	ASH-11	P.L.-1 channel	W-30Re-30Mo	1,920	65	7,084
	E83W	P.L.-1 channel	W-30Re-30Mo	1,920	2	564
	P1W	P.L.-19 channel	W-30Re-30Mo	1,920	48	4,564
	P1T	P.L.-19 channel	T-111	1,920	5	2,000
	P2T	P.L.-19 channel	T-111	1,920	5	2,000
	P5T	P.L.-19 channel	T-111	1,920	5	2,000
	P3T	P.L.-19 channel	T-111	1,920	3	1,600
	P6T	P.L.-19 channel	T-111	1,920	3	1,600
	P7T	P.L.-19 channel	T-111	1,920	3	1,600
	P-20	P.L.-37 channel	Ta	1,920	64	4,500
Dynamic	FS1T	F.L.-1 channel	T-111	1,920	8	2,714
	F2T[b]	F.L.-19 channel	T-111	1,920	7	668
In-Pile	LT-710-1	P.L.-1 channel	W-30Re-30Mo	1,810	31	2,410
	LT-710-2	9 in.-1 channel	T-111	1,810	9	1,095

[a] P.L.—partial-length fuel element; F.L.-full-length fuel element
[b] This was a shakedown test for the "A" section of the first primary dynamic loop.

fast neutron spectrum. If a rear reflector was needed, a thicker header plate could have been used at the back of each fuel element. A radial beryllium reflector, consisting of cylindrical sections, surrounded the core. This reflector could be located inside or outside the pressure vessel. Reactivity control was obtained by attaching neutron-absorbing boron strips to these reflector cylinders and rotating them.

The 710 Reactor concept evolved from a desire for a high performance reactor that could be used for both power and propulsion. The high performance demands established the need for a high working fluid temperature and a high core power density. For Brayton cycle systems, a working fluid temperature of 1,475 K was selected for compatibility with uncooled refractory metal turbines. Rankine cycle systems had similar limits. Power densities of 80 to 500 W/cm^3 were used to meet one to five year system lifetimes. For nuclear rocket applications with hydrogen as the propellant, the appropriate range of gas temperatures was 2,475 to 3,000 K; the reactor power densities were on the order of 17,000 to 33,000 W/cm^3; while the operating life would be relatively short, ranging from one to ten hours. A cermet fuel was selected with a continuous refractory metal structure surrounding the individual fuel particles to provide partial fuel and fission product retention. It had high strength and good thermal conductivity. The cermet body was also enclosed with a bonded cladding. A tubular fuel element concept was considered advantageous, since tubes provide a flow channel independent of other structural elements, resist distortion, and can be fabricated to close dimensional tolerances. The choice of a refractory metal for a structural material to support and contain the fuel was based on the refractory metal's high melting point, low vapor pressure, sufficient ductility, ability to be fabricated to close tolerances, strength, and compatibility with nuclear fuels, such as uranium dioxide and uranium nitride. For example, tungsten melts at 3,655 K, a melting point temperature that is higher than UO_2 (3,090 K) and UN (2,900 K).

Uranium dioxide was initially selected as the 710 Reactor fuel, because it had the highest melting point of the nuclear fuels under consideration and the lowest vapor pressure. It was also nonreactive in the stoichiometric state with refractory metals, was readily available, and considerable technology existed for the material. Although uranium nitride has a lower melting temperature and higher vapor pressure at elevated temperatures than UO_2, UN also has greater high temperature creep strength, higher thermal conductivity (by an order of magnitude), higher nuclear density, and is more compatible with lithium (a candidate working fluid). These beneficial UN fuel characteristics result in a potentially smaller core size. Uranium nitride was, therefore, being considered for application in advanced reactor cores. Extensive development was performed on tungsten cermet fuel elements and various cladding candidates. The results of some of these efforts are described in Table 11.5. Consistent quality fuel was

Table 11.6 Summary of basic 710 reactor characteristics.

HIGH PERFORMANCE FUEL ELEMENT DESIGN

- Refractory Metal Cermet
  - High-temperature capability
  - Good strength
  - High thermal conductivity
  - Adequate ductility
  - Fabricability with high fuel loadings
  - Bonding of fuel to cladding
  - Low dimensional tolerances achievable in fabrication
- Wrought Refractory Metal Cladding
  - High-temperature capability
  - High strength
  - Good ductility
  - Positive fuel and fission product retention
  - Quality verification possible prior to fuel element assembly
  - Compatible with highest melting point fuels
  - Compatible with liquid metals and hydrogen
- Tubular Fuel Element
  - Dimensional stability
  - Perturbation alleviation (by use of multiple channels per element)
  - Flexibility of temperature flattening

SIMPLICITY OF REACTOR CONCEPT

- Single-Point Fuel Element Support
- Series Flow Without Bypass
- Integrated Front Support and Reflector
- Integrated Fuel Element and Rear Reflector (if desired)
- No In-Core Controls
- Only Fuel, Cladding, and Coolant in Core

RESULTANT FUNDAMENTAL FEATURES

- Ease of Temperature Flattening
- Small Temperature Perturbations
- Negative Doppler Coefficient
- Neutronics Insensitive to Structure
- Safe on Water Immersion
- Self-Shielding for Core Gamma Radiation
- Negligible Fission Product Poisoning

VERSATILITY OF APPLICATION

- Power Extensibility
- Coolant Variation
- Temperature, Power, and Life Flexibility

obtained with fuel particles encased in 40 volume percent tungsten (W). The thermal conductivity of UN is approximately ten times higher than that of UO_2. Small quantities of rhenium (Re) were found to significantly increase the ductility of the matrix. The claddings had similar thermal expansion coefficients. Analysis of the 710 Reactor fuel element indicated that the most likely cause of failure in power generation applications would be strains produced by fission product buildup. However, for nuclear rocket applications, where fuel burnups are relatively low and operating lifetimes are short, while pressure levels are high—the major stresses are produced by differences between the thermal expansion coefficients of the matrix and cladding under cyclic conditions. Table 11.6 lists the characteristics of the 710 Reactor and Table 11.7 provides reference design conditions for the Brayton cycle system [8].

Advanced Space Nuclear Power Program (SPR)

The Advanced Space Nuclear Power Program (SPR) was conducted between 1965 and 1968. Its main objective was to explore fundamental technologies for the optimum means of generating electric power in space at the hundreds of kilowatts to ten megawatt levels. Operating lifetimes for these advanced power systems were defined as tens of thousands of hours. In an effort to obtain the highest possible power-to-mass ratio, program emphasis was placed on materials capable of extremely high temperature operation.

For example, the SPR-4 system was a 2 MWth (375 kWe) Rankine cycle nuclear system using sodium heat pipes to transport core thermal energy to a boiler (see Fig. 11.11). This system consisted of the following elements: an integral reactor-boiler operating at 1,500 K (see Fig. 11.12); a nuclear shadow shield composed of two layers of lithium hydride separated by a layer of tungsten; a turbine-alternator unit; two heat pipe radiators (one operating at 1,035 K and the other at 770 K); and a single electromagnetic pump for circulating the potassium working fluid in the power conversion loop. The system had an overall length of approximately 10 m, permitting launch as an integral unit. The SPR-4 had an overall system efficiency of 18.8 percent and a 20,000 hour design lifetime.

The SPR-4 reactor utilized uranium nitride (UN) as the fuel, a tungsten-rhenium alloy (75% W–25% Re) as the structural material, and sodium as the heat pipe working fluid [9]. Uranium-235 enrichment was varied

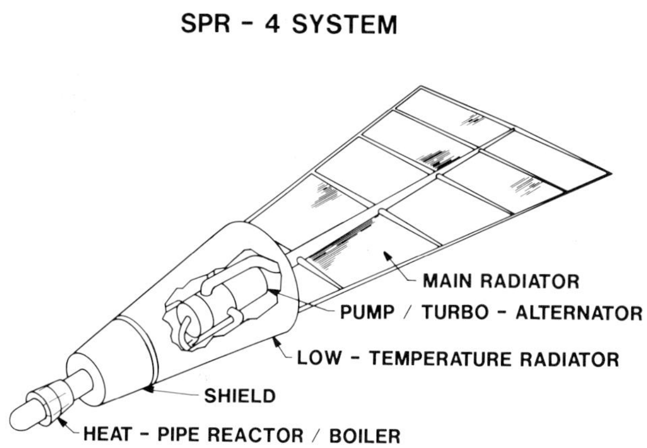

Fig. 11.11 SPR-4 space nuclear electric power plant.

Table 11.7 Operating conditions for various 710 reactor Brayton cycle systems.

	COOLANT		
General	He-Xe(40)	Neon	Argon
Total electrical power (kWe)	200	200	200
Total power thermal (MWth)	0.872	0.872	0.872
Reactor coolant flow rate (kg/s)	4.85	2.46	4.85
Reactor inlet gas temperature (K)	1,099	1,099	1,099
Reactor outlet gas temperature (K)	1,444	1,444	1,444
Reactor inlet pressure (MPa)	2.74	2.03	2.70
Reactor core pressure drop (%)	0.83	0.83	0.83
Core Thermodynamic and Nuclear Characteristics			
Channel inlet Reynolds number	14,900	7,170	18,260
Channel outlet Reynolds number	12,410	6,010	15,190
Average heat flux (watts/cm^2)	12.9	12.9	12.9
Max. average surface temperature (K)	1,505	1,581	1,631
Max. fuel burnup[a] (fissions/cm^3 of fuel)	8.8×10^{19}	8.8×10^{19}	8.8×10^{19}
Avg. core fuel burnup[a] (fissions/cm^3 of fuel)	4.6×10^{19}	4.6×10^{19}	4.6×10^{19}
Max. average element burnup[a] (fissions/cm^3 of fuel)	5.9×10^{19}	5.9×10^{19}	5.9×10^{19}
Max. fuel element[a] (nvt ($L_n > 1$ MeV))	7.90×10^{20}	7.90×10^{20}	7.90×10^{20}
Max. matrix power density (watts/cm^3) matrix	46.1	46.1	46.1

[a] For 10,000 hours operation

in two radial zones. Each of the 378 hexagonal fuel elements had a central hole through which a sodium heat pipe was inserted. In this design, two potassium boilers, one on each end of the cylindrical core, were integrated with the reactor (see Table 11.8 and Fig. 11.12) [9]. The potassium working fluid was boiled on the outside surfaces of the heat pipes (i.e., their condenser regions) which protruded into the boilers. Because of the heat pipe operating principle, thermal energy was transferred from the reactor core to the potassium boilers at essentially isothermal conditions. The potassium flow rate was varied to ensure that the working fluid departed the boilers as a saturated vapor. After the potassium vapor passed through the turbine, it was condensed in a large, low mass, flat radiator that contained 2,200 heat pipes (see Fig. 11.11).

Reactivity was regulated by means of a dual control system, consisting of a 2 cm thick, axially translating molybdenum reflector and centrally located, in-core liquid lithium-6 control circuits (see Fig. 11.12). Each system could independently control the SPR-4 reactor through its entire lifetime.

The SPR-5 concept was an out-of-core thermionic

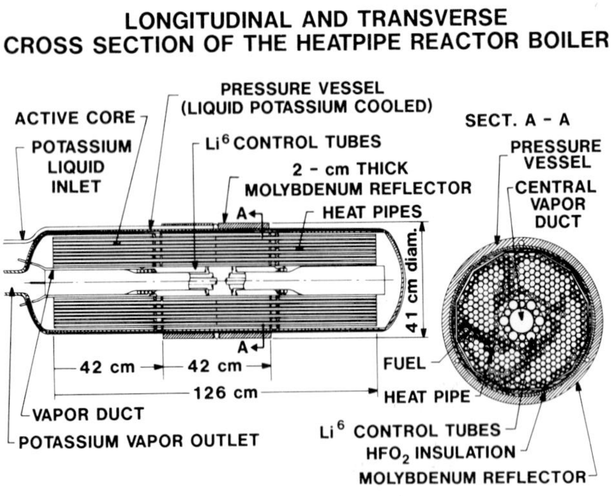

Fig. 11.12 Schematic diagram for the SPR-4 integral reactor-boiler.

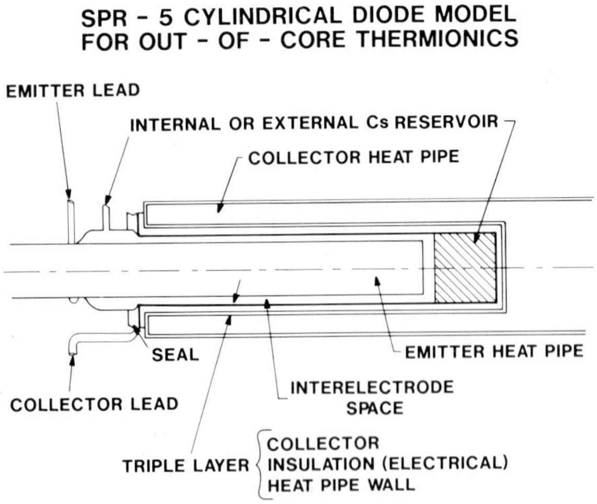

Fig. 11.13 SPR-5 cylindrical diode for out-of-core thermionic converter system.

Table 11.8 Operating characteristics of SPR-4 integral reactor-boiler.

Fuel Atom Burnup (%)	
—average	0.68
—maximum	1.08
Fuel Enrichment (fraction of ^{235}U)	
—average	0.93
—maximum	0.99
Average Fuel Power Density (W/cm^3)	100
Radial Peak-to-Average Power Density Ratio (at most severe transverse cross section)	1.18
Overall Peak-to-Average Power Density Ratio	1.59
Average Neutron Energy (Mev)	0.39
Prompt Neutron Lifetime (sec)	2.4×10^{-8}
Fueled Core	
—length (cm)	41.9
—mean outside diameter (cm)	33.9
Mean ^6Li Control Zone Outside Diameter (cm)	12.8
Potassium Vapor Return Duct Inside Dia. (cm)	6.3
Pressure Vessel Inside Diameter (cm)	36.2
Side Reflector	
—inside diameter (cm)	37.2
—outside diameter (cm)	41.2
Number of Fuel Elements	378
Number of Heat Pipes	756
Volume Fractions	
—fuel	0.616
—heat pipe metal	0.062
Intermediate Tube Spacer Volume Fraction	0.016
Heat Pipe Fluid Volume Fractions	
—heat pipe liquid	0.117
—heat pipe vapor	0.150
Swelling Allowance Volume Fraction	0.039
Total Fueled Core Volume (liters)	32.4
Total Heat Pipe Fluid Flow Rate (kg/s)	0.566
Heat Pipe Fluid Saturation Temperature (K)	1,500
Heat Pipe Fluid Saturation Pressure (bar)	11.0
Heat Pipe Inward Radial Heat Flux	
—average (W/cm^2)	46.4
—maximum (W/cm^2)	85.8
Maximum Heat Pipe Axial Heat Flux Based On Vapor Volume (kW/cm^2)	7.0
Nominal Peak Fuel Temperature (K)	1,533
Component Masses (kg)	
—shield	1,400
—reactor-boiler	900
—turbine-alternator	400
—main radiator	400
—pump	200
—miscellaneous	450
Total:	3750

system designed for 10 MWe output [10]. The reactor was cooled from one end by a set of silver-tungsten (Ag-W) heat pipes that extended the full length of the reactor and terminated shortly outside the cooled end. These core heat pipes transferred thermal energy to a second set of Ag-W heat pipes that were set in a matrix of high temperature shielding material, namely BeO with Eu_2O_3. The thermal energy removal system thus also doubled as the main shadow shield. At the other end of the shield, the heat was transferred by radiation across a void to a set of heat pipes that terminated as the emitters of an array of thermionic diodes (see Fig. 11.13). This void electrically isolated the emitters, while avoiding the use of high temperature insulation as much as possible. Waste thermal energy was removed from the diodes by lower temperature lithium-niobium (Li-Nb) heat pipes that extended to make up the main radiator.

Operating temperatures were selected to produce an overall system efficiency of 17 percent. Consequently, the SPR-5 reactor generated 60 MWth, which yielded a net electrical output of 10 MWe. The reactor coolant temperature was selected to be as high as material properties would permit. The selection of 2,273 K was based mainly on the results of a life test of the silver-tungsten heat pipe. This test showed negligible corrosion after 1,000 hours of operation at 2,223 K [10]. The overall SPR-5 system efficiency was selected to reflect the anticipated performance of a thermionic diode, operating at 2,000 K emitter temperature. The total specific mass of the system was estimated at 4-7 kg/kW.

The SPR-6 power plant concept is illustrated in Fig. 11.14 [11]. It consisted of a 57 MWth reactor, a nuclear radiation shield, and a potassium working fluid Rankine cycle power converter. Figure 11.15 depicts the various state points for the SPR-6 system [11]. In this design concept, high purity lithium was heated in a reactor and pumped through a heat exchanger that then boiled the potassium working fluid. Each of four

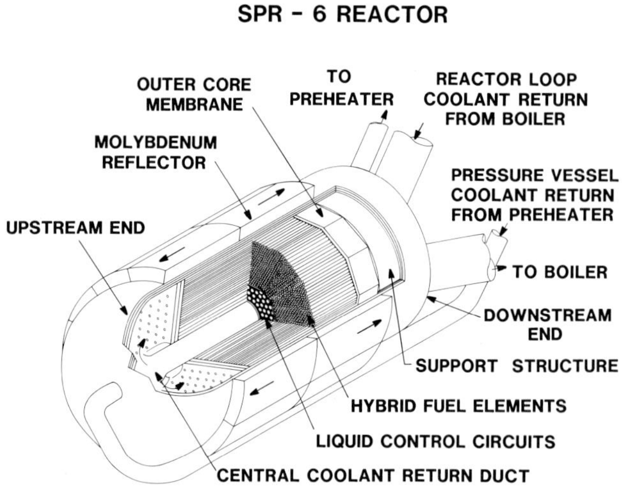

Fig. 11.14 Cutaway drawing of SPR-6 reactor.

turboalternator units extracted 2.5 MWe of useful power from the working fluid. Waste heat was rejected by means of a heat pipe radiator. The SPR-6 reactor was separated from the other system components by the lithium hydride (LiH) shadow shield. Reactor control could be obtained from either in-core liquid control circuits or by axial translation of the side reflector. Once a nearly neutral coolant mixture had been achieved in the reactor loop, each control method was independently capable of controlling the SPR-6 reactor.

The SPR-6 reactor used uranium mononitride (UN) as the fuel, lithium as the coolant, with 75% tungsten-25% rhenium alloy for the structural material. Compatibility of these materials at system operating temperatures was experimentally demonstrated. Characteristics of the SPR-6 reactor are given in Table 11.9 [11]. The SPR-6 core consisted of a closely packed bundle of 8,000 fuel elements and some 4,000 coolant channels supported on the downstream end by a 30-cm thick tungsten structure and on the upstream end by a spline attachment that provided for radial and axial growth. The downstream support structure also served as a radiation shield [11]. As shown in Fig. 11.14, the active portion of the SPR-6 core was an annular region that surrounded a central coolant return passage and a liquid control region. The core outer boundary formed a dodecagon, accommodating the fuel element hexagonal packing pattern. Thin, polygon-shaped tungsten-rhenium (75% W–25% Re) membranes enclosed the fueled core region at its inner and outer radii.

The boiler contained tantalum tubes that were arranged in a triangular configuration and separated from each other, permitting the liquid lithium coolant to circulate around the tubes. Pressure in the lithium loop was regulated to minimize the pressure differential across the boiler tube walls. To reduce neutron activa-

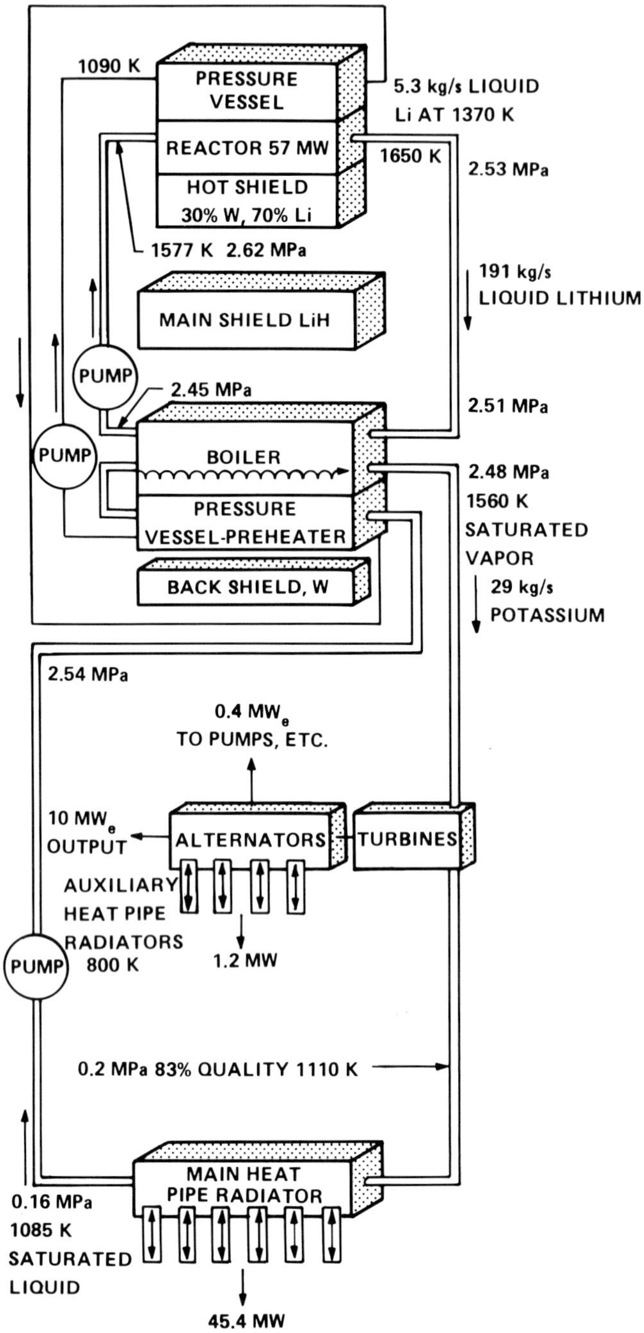

Fig. 11.15 SPR-6 system schematic.

tion of the natural potassium working fluid to acceptable levels, the boiler was located on the payload (i.e., protected) side of the radiation shield.

Waste thermal energy rejection was accomplished by means of a flat, rectangular heat pipe radiator. These heat pipes operated in parallel, providing radiator system redundancy. The potassium working fluid flow rate through the main heat pipe radiator was varied by adjusting the power to three mechanical pumps. A liquid metal storage expansion tank was needed in each loop to maintain the necessary system pressures.

Table 11.9 Characteristics of the SPR-6 reactor.

Fuel Atom Burnup (%)	
—average	2.1
—maximum	3.3
Fuel Enrichment (atom fraction of ^{235}U)	
—average	0.70
—maximum	1.0
Average Fuel Power Density (W/cm^3)	575
Radial Peak-to-Average Power Density	1.12
Overall Peak-to-Average Power Density	1.57
Average Neutron Energy (MeV)	0.25
Prompt Neutron Lifetime (ns)	81
Fueled Core	
—length (cm)	87.0
—mean outer diameter (cm)	62.1
Liquid Lithium Control Zone Outer Diameter (cm)	22.7
Reactor Coolant Return Duct Diameter (cm)	15.3
Pressure Vessel Inside Diameter (cm)	65.0
Side Reflector	
—inside diameter (cm)	67.4
—outside diameter (cm)	71.4
Fuel Radius (cm)	0.216
Fuel Element Cladding	
—inside radius (cm)	0.225
—outside radius (cm)	0.250
Coolant Tube Inside Radius (cm)	0.220
Number of Fuel Tubes	8,064
Number of Coolant Tubes	4,032
Fuel Volume Fraction	0.433
Cladding (plus coolant) Tube Volume Fraction	0.183
Primary Coolant Volume Fraction	0.234
Interstitial Fluid Volume Fraction	0.093
Swelling Allowance Volume Fraction	0.057
Total Fueled Core Volume (m^3)	0.228
Reactor Coolant Flow Rate (kg/s)	191
Reactor Coolant Outlet Temperature (K)	1,650
Reactor Coolant Temperature Rise (K)	73
Reactor Coolant Inlet Pressure (bars)	26.2
Reactor Pressure Drop (bars)	0.9
Core Reynolds Number	5.9×10^4
Reactor Coolant Velocity Inside Core (cm/s)	740
Average Heat Flux At Coolant Channel Surface (W/cm^2)	117
Average Centerline Fuel Temperature (degrees K above local coolant temperature)	46
Component Masses (kg)	
—shields	33,000
—main radiator	13,400
—alternators (including auxiliary cooling radiators)	7,200
—reactor	4,200
—boiler-preheater	1,800
—structure	1,500
—turbines	1,400
—reactor loop piping	1,000
—lithium coolant pump	1,000
—potassium working fluid pump	600
—lithium fluid	600
—power conversion loop piping	500
—potassium fluid	400
—controls	300
—pressure vessel cooling loop pump	100
—miscellaneous	3,000
Total:	70,000

A specific mass of 7 kg/kWe was made possible for the SPR-6 reactor concept by assuming a high fuel burnup capability, highly efficient heat pipe radiators, a compact radiation shield, and high strength alloys. An optimum reactor coolant temperature of approximately 1,650 K was selected for the system.

As part of the Advanced Space Nuclear Power Program, the vaporization behavior of UN powder at nitrogen pressures of 17,000, 13, and 0.4 pascals was experimentally investigated (see Fig. 11.16 [12]). It was found that back streaming of impurities from the sorption pumps was a problem. Consequently, the sorption pumps were baked at approximately 0.1 Pa.

Irradiation experiments involving UN fuel capsules were still in the early stages, when the program was terminated. Although none of these reactors were ever flown in space, UN fuel irradiation capsules had been exposed to 1.4×10^{22} fissions/cm^3 in 2,229 hours at 1,675–1,775 K with little or no apparent swelling (i.e., less than 1 percent), except for a fine grained UN

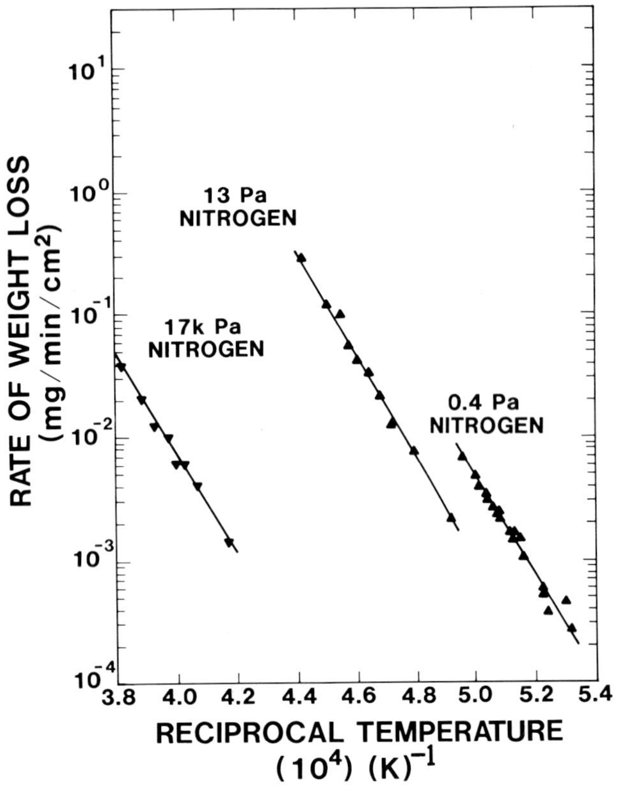

Fig. 11.16 Rate of mass loss for UN_{1-x} as a function of reciprocal temperature.

capsule that experienced approximately 3 percent swelling [13]. In addition, at program termination compatibility tests were also underway that exhibited negligible corrosion of the W-Re-Mo capsule with circulating lithium after 1,004 hours at temperatures near 1,675 K [13].

Fluidized Bed Reactors

Both rotating bed and fixed fluidized bed reactors have been considered for space nuclear applications [14–17]. The rotating bed reactor (RBR) has been investigated primarily for space nuclear propulsion, while the fixed bed reactor (FBR) has been proposed for power operation in space. RBRs offer potentially higher propellant temperatures, because of their simpler core structure and lower temperature gradients in the nuclear fuel versus solid core nuclear rocket designs (see Fig. 11.17). Specific impulse values on the order of 10,000 m/s appear achievable with carbide fuels.

The rotating fluidized bed reactor concept was initially proposed in 1960 and a modest development effort was carried out until 1973. As shown in Fig. 11.12, the UC-ZrC nuclear fuel, in the form of small 100–500 micrometer diameter spherical particles, is maintained in an annular arrangement by centrifugal force in a rotating cylindrical structure [14]. A rotor drive unit rotates a cylinder that consists of a porous material supported by a squirrel-cage type structure known as the frit. Hydrogen propellant first passes through the coolant in the rocket nozzle and then through the reflector. All or part of the propellant flow next passes through the turbine and then it enters the reactor core region. The now gaseous hydrogen flows radially inward through the cylindrical structure and annular core at a velocity sufficient to "fluidize" most of the nuclear fuel particle bed. The initiation of fluidization depends on such factors as the effective centrifugal force, the density of the bed particles, and the thickness of the bed. The heated hydrogen propellant then flows out through the nozzle. Predicted exit gas temperatures are on the order of 3,000 K with a core power density of 1,000 $MWth/m^3$ [14]. Control of the rotating bed reactor is achieved by means of drums in the reflector. These control drums would contain a moderator on one side and a neutron absorber on the opposite side.

The RBR itself is a thermal reactor with an external moderator/reflector made of beryllium or zirconium hydride. After moderation to low energies, neutrons are captured in the fuel bed, maintaining the chain reaction. The position of the neutron absorbers in the control drums is changed to achieve startup and to maintain desired reactor operating power levels. In the RBR core, only the nuclear fuel is at high temperature. Other core components, such as the frit, cylinder drive, and reflector, are actually at hydrogen coolant inlet temperatures. Consequently, high temperature strength and corrosion problems are potentially avoided in the

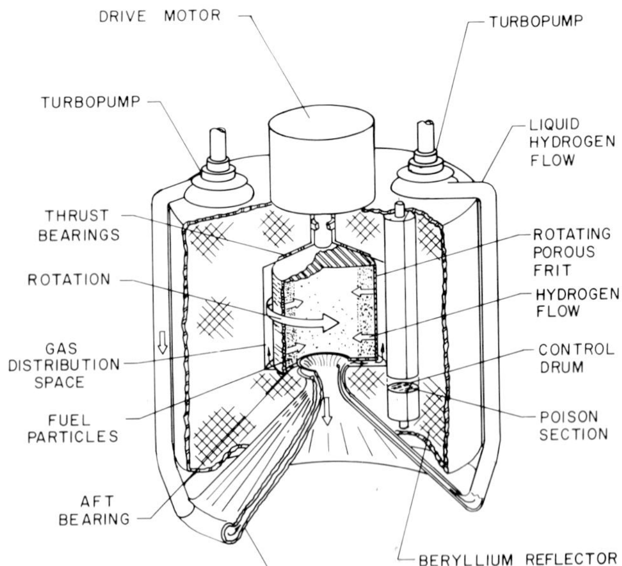

Fig. 11.17 Rotating fluidized bed rocket engine [14].

Table 11.10 Illustrative RBR design parameters.

Fuel	^{235}U
Bed internal diameter (cm)	63.5
Bed height (cm)	63.5
Fuel bed thickness (cm)	10.2
Reflector thickness (cm)	
Radial	30
Axial	30
Throat diameter (cm)	18
Overall height (cm)	123.5
Overall diameter (cm)	143.9
Critical mass (kg)	156
Bed voidage (%)	60
U concentration (atom %)	9.5
Chamber Pressure (MPa)	7.75
H_2 flow rate (kg/sec)	20
Power at $T = 3000$ K (MW)	1000
Reactor weight, including pumps and pressure vessel (Kg)	4,750

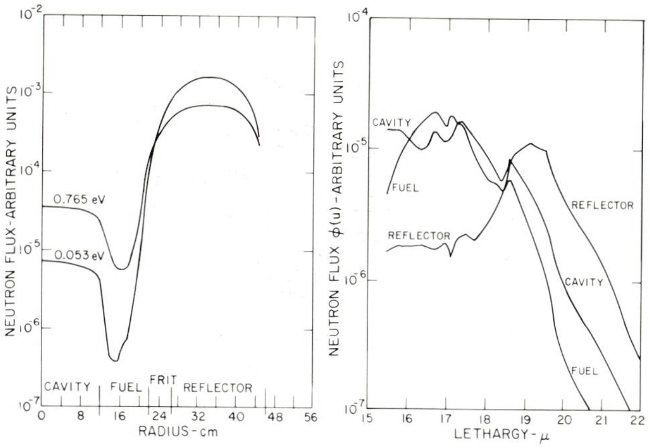

Fig. 11.19 Fluidized bed reactor analysis showing spatial variation of neutron flux and neutron spectra at various positions. *Courtesy of Brookhaven National Laboratory.*

RBR concept. Since the nuclear fuel particles are quite small, they can undergo rapid heatup and cooldown. Table 11.10 provides some typical rotating bed reactor design parameters. It is interesting to note that in this space reactor concept, core dimensions tend to be small, while reflector dimensions are relatively large [14–16].

A number of non-nuclear experiments have been performed in support of fluidized bed reactor concepts. These experiments investigated bed dynamics and obtained an empirical correlation between flow rate and bed fluidization. For example, a half-scale model (25 cm in height and diameter) was used in hydraulic tests. Spherical beads (either copper or glass) were employed to simulate nuclear fuel particles in these experiments, while ambient temperature nitrogen gas at approximately 1 megapascal pressure was used to simulate the hydrogen coolant. The nitrogen mass flow rate per unit area of the frit matched projected flow rates for a full-scale, full power rotating bed reactor. Selected test results are shown in Fig. 11.18 [14]. In this plot, the solid line represents the analytical prediction of pressure drop through a settled bed. As can be seen in this figure, such analytical predictions agree reasonably well with the experimentally observed results. As the gas flow rate is increased, the pressure drop through the settled bed increases until it fluidizes. After fluidization, the pressure drop remains level as the gas flow increases, although the void area in the bed also increases [14].

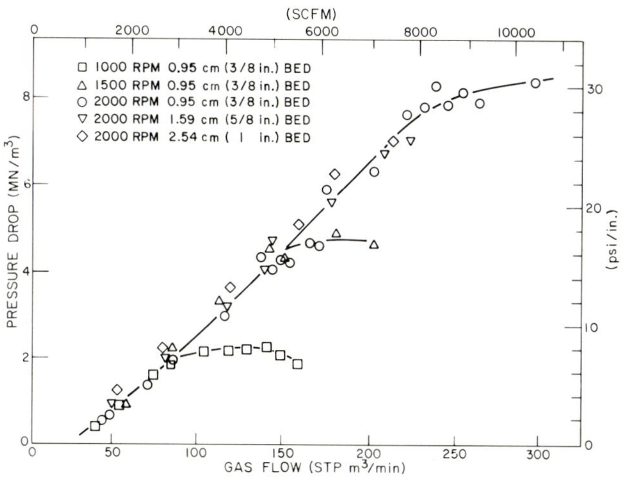

Fig. 11.18 Correlation of fluidization data for 500 micrometer glass beads (specific gravity = 2.5) [*14*].

Neutronic studies indicate that very small rotating bed reactors tend to be undermoderated, a condition requiring considerably higher critical masses than would be found in larger RBR systems. A reflector thickness of approximately 30 cm is desirable to minimize critical mass. The peak-to-average radial power ratio tends to be large with typical values of five, because of flux depression in the fuel bed. This high ratio, however, is readily accommodated. For example, the very large surface area of the small fuel particles supports thermal energy transfer to the working fluid. Figure 11.19 displays an analytical interpretation of the neutron flux and energy spectrum as a function of radial position in an RBR [15].

Fluidized bed reactor technology might also be considered for use as the heat source for electric power generation in space [16–17]. Its most likely application appears to reside in missions requiring great quantities of power for short (minutes to hours) periods of time. In a high power, short lifetime open-cycle power generation mode, the thrust nozzle (depicted in Fig. 11.17) would be replaced by a magnetohydrodynamic power converter. A closed cycle fluidized bed reactor system for power generation, on the other hand, would most likely involve the use of a fixed fluidized bed reactor

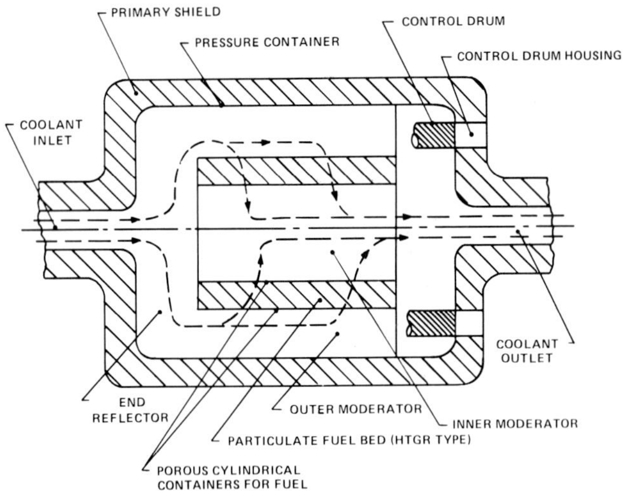

Fig. 11.20 Fixed bed reactor with outer and inner moderator regions. Control drums extend into the outer moderator/reflector region. Fuel bed dimensions are typically on the order of 45 cm in length and 30 cm in diameter [*17*].

(see Fig. 11.20) [17]. The operating temperatures of this type of fluidized reactor system are compatible with Brayton cycle component technology. For example, helium coolant temperatures would lie between 1,300 and 1,500 K. The nuclear fuel would be located in two porous frit plates with both an internal and external moderator/reflector. Enriched uranium fuel particles, derived from fuel technology efforts for terrestrial high temperature gas reactors (HTGRs) could also be utilized. These fuel particles would be typically 200 to 800 micrometers in diameter and would be coated with impervious ceramic materials. It is anticipated that such

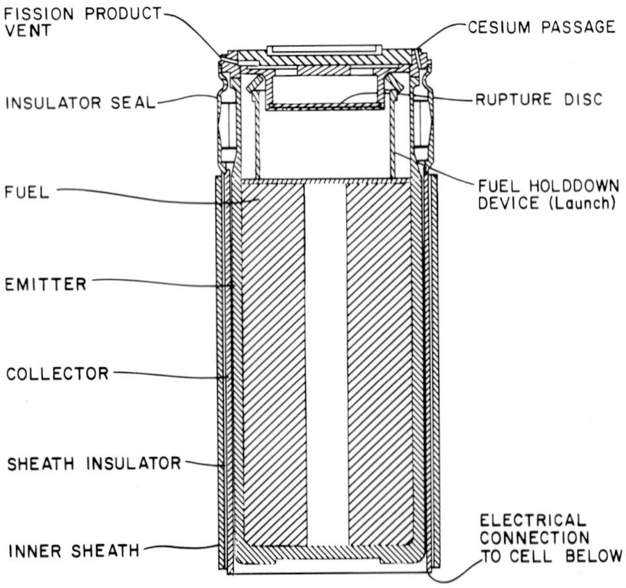

Fig. 11.21 Thermionic diode. *Courtesy of Gulf General Atomic.*

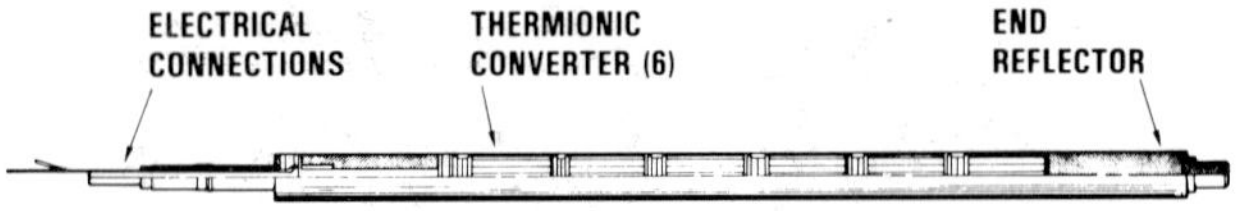

Fig. 11.22 Schematic of in-core thermionic fuel element [*18*].

spherical fuel particles could operate at 1,500 K for years with virtually complete fission product retention (less than 0.1 percent fission product loss).

At present, no nuclear-fueled rotating bed reactors have been tested. In any future development program, it would be possible to remove the old fuel and recharge the test reactor many times. This ease of refueling would, therefore, make it possible to maintain and reuse the fluidized bed engine system throughout a series of development and preflight ground tests without creating a residual fission product radiation hazard. As another aerospace nuclear safety feature, the nuclear fuel for such fluidized bed systems could be launched in separate, subcritical mass capsules, and then the system fueled only after a nuclear safe orbit (NSO) had been reached.

Thermionic Reactors

Thermionic reactor development was started in the mid-1950s with concentrated funding being placed from 1963 to 1973 on in-core thermionic conversion concepts [18–29]. Specific reactor system designs were prepared

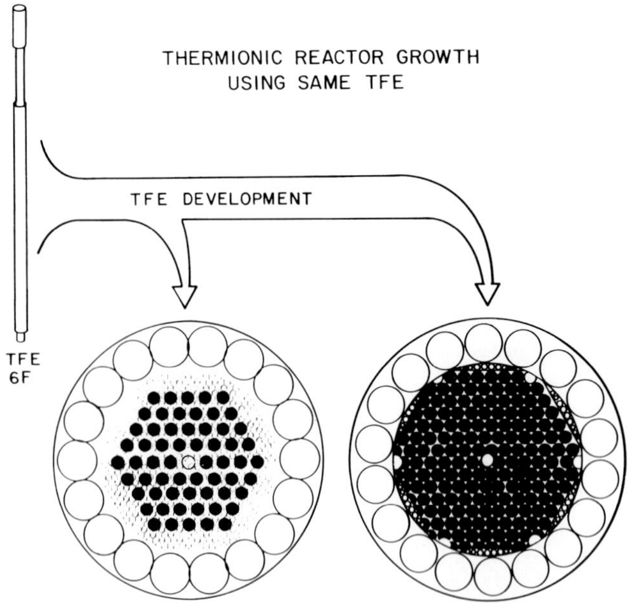

Fig. 11.23 Thermionic reactor core arrangements for 40–60 kWe and 100–250 kWe configurations [*18*].

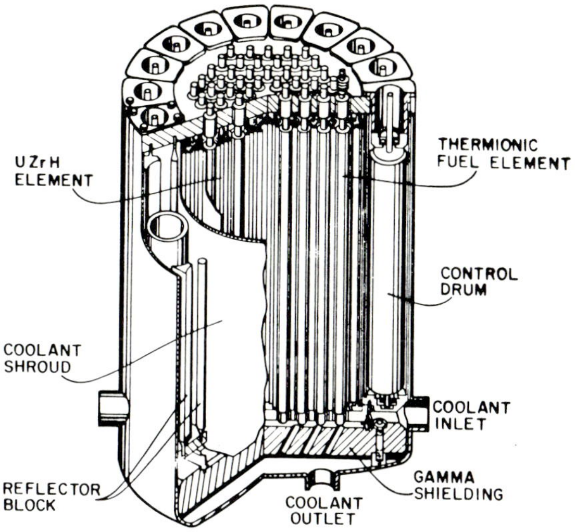

Fig. 11.24 Thermionic reactor schematic [18].

for many space applications including: a 5–10 kWe system for unmanned satellites; a 40 kWe system for manned space laboratories; and a 120 kWe system for nuclear electric propulsion. All of these power system concepts employed the same basic thermionic fuel element. There are, of course, several approaches to designing a thermionic reactor, some of which will be discussed later. We will focus the initial portion of this discussion, however, on the in-core "flashlight" thermionic reactor concept, since this was the most advanced approach.

The thermionic diode shown in Fig. 11.21 is the basic building block on which the major effort was concentrated as part of the in-core thermionic program. The fuel is arranged in the interior of the diode and is isolated from the core structure. The thermionic emitter, which surrounds the fuel and is designed to operate at up to 2,000 K, emits electrons that travel across the interelectrode gap to the collector. A cesium reservoir provides cesium atoms in the space between the emitter and collector. These cesium atoms are ionized partially by electron collisions and partially by contacting the hot emitter surface. The positive cesium ions help neutralize the space charge effect that would otherwise be built up by the flow of electrons. These positively charged cesium ions also prevent the electrons adjacent to the emitter surface from attaching themselves to both the emitter and collector surfaces, thereby lowering their work functions. The entire thermionic diode configuration is encapsulated in a sheath.

Much like a flashlight with its batteries in series, the thermionic diode was packaged six converters per fuel element (see Fig. 11.22) [18]. These fuel elements, which were about 2.5 cm in diameter by 40.6 cm long, were assembled into several different reactor core configurations.

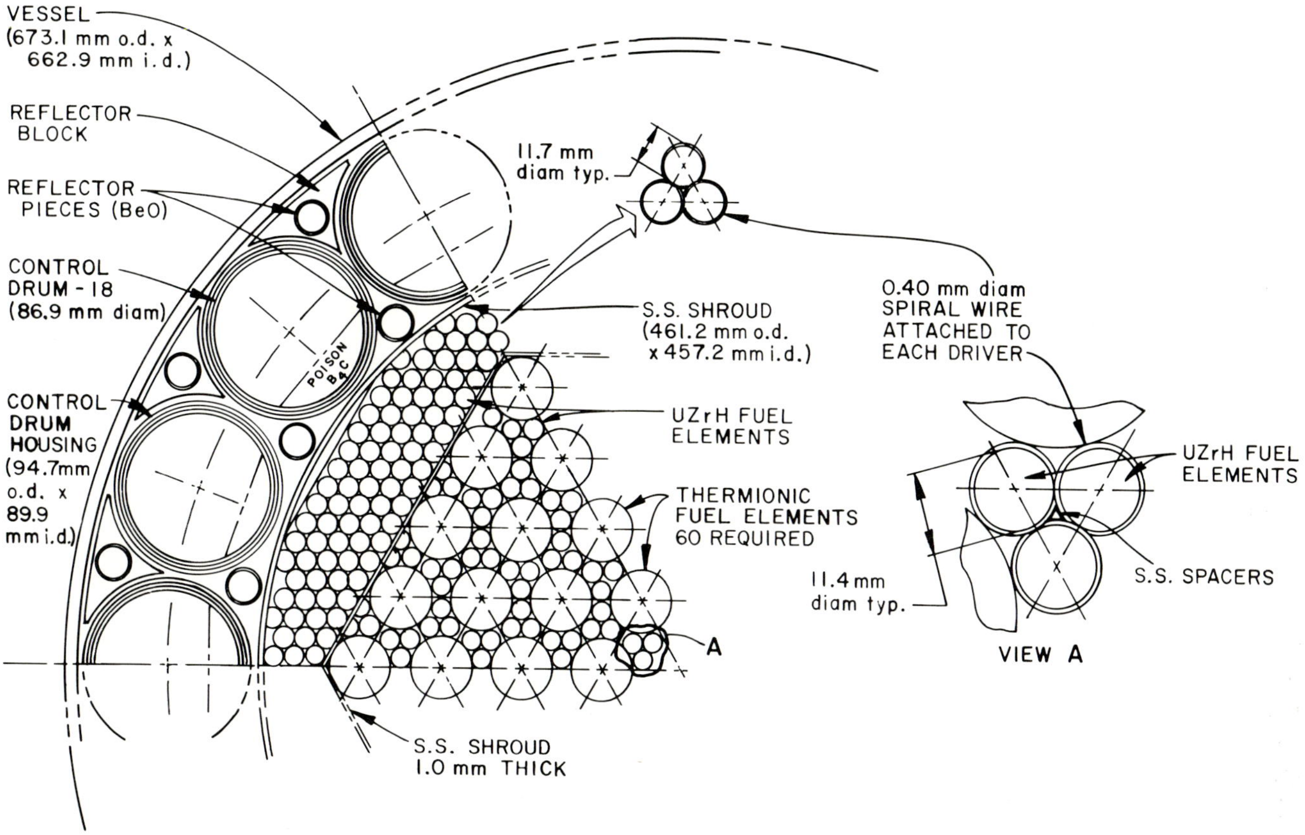

Fig. 11.25 Details of in-core thermionic reactor core configuration [18].

Table 11.11 Thermionic reactor performance parameters.

Reactor outside diameter (cm)	67.3	
Reactor length (cm)	96.5	
Reactor mass (kg)	987	
Number of thermionic fuel elements	60	
Number of U-ZrH fuel elements	813	
Number of control drums	18	
Control drum diameter (cm)	82.3	
	BOL	EOL
Gross electrical output (kWe)	53	58
TFE surface heat flux (W/cm^2)	24	29
Thermal power output (kW)	1,180	1,450
Coolant inlet temperature (K)	750	789
Coolant outlet temperature (K)	827	866

Representative core cross sections for 40–60 kWe and 100–250 kWe reactors are shown in Fig. 11.23 [18]. The solid circles represent the thermionic fuel elements. The 40–60 kWe moderated reactor contained 60 thermionic fuel elements. These were coupled with U-ZrH elements (the open circles in Fig. 11.23) similar to those developed for the SNAP-8 program. In this design, the thermionic fuel elements produced the desired electric output, while the U-ZrH fuel elements were added to achieve criticality and to moderate the neutron spectrum. The core was surrounded by a radial reflector region, containing control drums that were composed of beryllium oxide reflector material and boron carbide (neutron absorption) segments. The control drums, contained in dry wells within the reactor vessel, were rotated to achieve reactor control. If the boron carbide segments were turned toward the core, the reactor was shut down; if the boron carbide segments were turned away from the core, reactor power level increased.

The drive actuators were located above the radiation shield. A cross section of the 40–60 kWe thermionic reactor is shown in Fig. 11.24 and specific design details are provided in Fig. 11.25 [18]. The core consisted of an inner region with both types of fuel elements (i.e., thermionic and U-ZrH) and an outer region with only U-ZrH fuel elements. The U-ZrH fuel elements were operated under less demanding conditions than in the SNAP-8 system with their cladding temperatures about 110 K lower and the surface heat flux a factor of 2 to 4 lower.

The core and reflector regions were separated by a shroud that directed the incoming NaK coolant through the reflector region before it flowed back through the core. The coolant entered the side of the vessel near the bottom, was distributed in the small plenum area, and flowed up through the reflector region to remove neutron and gamma radiation heating. The coolant then entered the core region through openings in the shroud, was distributed in the region between the vessel head and the upper end of the fuel elements, and flowed down through the core. It passed through holes in the grid plate and through the shield block to the reactor outlet pipes. The coolant and vessel materials (NaK-78 and 316 SS) were the same as those in the reference SNAP-8 U-ZrH thermoelectric reactor system.

Significant performance parameters are provided in Table 11.11. Only 46 thermionic fuel elements were needed to meet end-of-life conditions, including an allowance for power degradation or failures.

The 100–250 kWe fast reactor design contained 162 thermionic fuel elements and was criticality limited using uranium-235 as the fissile fuel. This reactor produced 164 kWe at 23 volts. Losses in power distribution and conditioning equipment and in reactor coolant pumping requirements reduced the net power output to 120 kWe. The heat rejection system was designed to radiate 1,510 kWth at a temperature of 1,000 K. The radiator consisted of 1,620 heat pipes, containing sodium as the working fluid and employing niobium alloy for tubing. Waste thermal energy was carried to the radiator by NaK-78 coolant which was circulated in a loop by a linear induction pump.

For a 120 kWe configuration, the fuel burnup amounted to 1.6×10^{20} fissions/cm^3 in 20,000 hours. The component materials of the thermionic fuel element experienced a fast neutron fluence ($>$ 0.1 MeV) of 4.6×10^{21} nvt during the same period [21].

The major objectives of the thermionic fuel element development program were: (1) temperature and current capability of 2,000 K and 12 amperes/cm^2 of emitter surface; (2) lifetime of 2 years at 2,000 K and 12 amperes/cm^2 and 5 years at 1,800 K and 6 amperes/cm^2; (3) demonstration of reproducible thermionic converter performance levels in-pile; (4) demonstration that any electrical discharges which might occur were not destructive to the fuel element integrity and would not result in excessive power losses; and (5) qualification to anticipated launch vibration and shock environments.

Materials are the key to thermionic fuel element development. Tungsten was selected as the cladding material, mainly because of its compatibility with both the carbide and oxide fuel materials. In addition, its low vapor pressure, high melting point, and superior creep strength at high temperatures were distinctive advantages. Carbide fuels in the form of both 90 UC–10 ZrC and UC were being evaluated. High melting points (2,723 K for UC and 3,023 K for 90 UC–20 ZrC), high uranium concentrations (approximately 12 g/cm^3), and high thermal conductivities (approximately 0.2 W/m-K) made the carbides attractive as fuel materials. Difficulties existed, however, in their dimensional stability and thermionic converter performance.

Table 11.12 Summary of GGA thermionic in-pile life tests (as of 22 January 1973).

	Experimental Cells	Prototype Cells		Thermionic Fuel Elements			
	Mark VI	Mark VIIA	Mark VIIB	E Series (Two Cell)	F Series (One Cell)	F Series (Two Cell)	F Series (Six Cell)
Number of Tests	15	3	4	3	4	3	5
Total Test Hours	45449	4604	15915	27280	22152	16551	23972
Average Test Hours	3030	1535	3979	9093	5538	5517	4794
W/cm^2	6.3	5.2	3.7	4.5	3.8	4.0	2.6
Longest Test Hours	9754	2006	7881	12535	8560	8021	8062
W/cm^2	6.0	6.2	3.4	4.0	4.0	3.4	3.0
Failure Modes							
Carbide Fuel							
None	1	—	1	—	1	2	2
Short Circuit	1	—	–	—	2	—	1
Envelope Leak	10	2	2	1	—	—	1
Oxide Fuel							
None	1	—	—	1	—	—	1
Short Circuit	1	1	1	1	1	1	—
Envelope Leak	1	—	—	—	—	—	—

Oxide fuels were also evaluated. A stoichiometry of 2.002-2.007 was desired, based on the requirements that the uranium chemical activity should not be so high as to affect the structural integrity of the tungsten cladding during irradiation and the oxygen chemical activity should not be so high as to cause reaction with the tungsten cladding [21]. Niobium was selected for the collector material because its thermal expansion coefficient closely matched that of the Al_2O_3 layer used to electrically insulate the collector from the sheath. Niobium is also an excellent getter for oxygen and nitrogen and helps maintain a clean environment in the interelectrode gap. Copper was originally selected as the cesium reservoir material, but this choice was later changed to niobium in order to eliminate the need for a weld between the copper reservoir and the niobium collector. This weld had caused several previous converter failures.

Thermionic devices and fuel-clad capsules were tested in the Thermionic Test Reactor (TITR) at Gulf General Atomics [23]. Thirty-eight devices were tested in-core between 1962 and 1973. Table 11.12 summarizes the results of these tests. The in-pile thermionic devices were fueled with either carbide or oxide fuel. Considerable progress was made in improving converter lifetimes; for example, the longest tests had increased from the 1,000–2,000 hour range in 1965 to 10,000 hours by 1970. The number of test hours in a full-length fuel element (called 6F3) reached 8,062 at an average power output of 647 watts and 11,084 hours on a partial-length fuel element (called 2E2). In examining performance, oxide-fueled tests have consistently exhibited good initial thermionic performance and have maintained this performance throughout the tests. Carbide fuels, on the other hand, have tended to exhibit less stable performance during such tests. The longest UO_2 fueled tests, namely: IC-14 (9,754 hours), 2E2 (11,084 hours), and 2E1 (12,535 hours), had stable performance within ±10 percent. (In the case of the 2E1 test, a short occurred at 7,035 hours and its performance decreased approximately 15 percent over its lifetime.)

The behavior of other materials was also experimentally examined. For example, tests revealed that $W(WCl_6)$ emitters performed with a relative power some 8 percent better that those with $W(WF_6)$ emitters; while oxide-fueled devices performed approximately 4 percent better than carbide-fueled devices. In addition, cells with molybdenum collectors produced about 7 percent more power than those with niobium collectors. These changes in performance were attributed to the following factors: (1) leaks between the cesium or fission gas spaces and the containment gas (envelope failures); (2) a leak between fission gas and cesium spaces resulting in fission gas and fission products in the interelectrode space: (3) failure of the bonds in the collector-sheath tube structure, resulting in high collector temperatures; (4) contamination of the electrodes during fabrication; (5) electrode contamination due to fuel component diffusion through the emitter clad; and (6) interelectrode spacing changes.

Envelope failures were a major problem associated with braze joints. This problem was overcome by going to an all-welded design. Much of the fuel diffusion problem appears to have been solved by control of fuel stoichiometry and cladding structure. Further improvements were needed in dimensional stability to meet a 20,000 hour goal with a 1 millimeter cladding thickness.

The advantages of the in-core "flashlight" thermionic reactor concept included the following.

- The fuel element was similar to most other reactors.

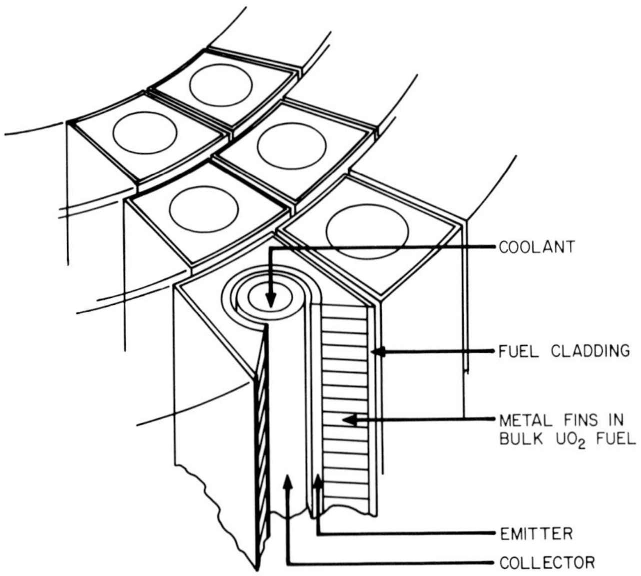

Fig. 11.26 Externally fueled, full-length thermionic diode modules [*22*].

- The high temperature components were thermally isolated from the core structure.
- Heat removal requirements were at conventional levels.
- Insulators were at a relatively low temperature.
- The concept was scaleable over a wide range of power levels.

Major difficulties encountered with this concept involved:

- Fuel swelling that shortened diode fuel life.
- Fast neutron damage at higher power and flux levels.
- Output voltage limitations imposed by insulator arcing considerations.
- The need for larger diameter emitters.
- Small values for the reactivity feedback coefficients.

This flashlight thermionic concept could be used in either a fast unmoderated reactor or a moderated thermal reactor. Selection of a moderated thermal reactor concept reduces both the minimum critical size of the core and the electric power output [24].

Another thermionic reactor concept involves an externally fueled diode, as illustrated in Fig. 11.26 [22]. Heat rejected from the collectors is removed by a liquid metal coolant loop or by heat pipes. Thermionic elements are connected in a series-parallel arrangement to build up the reactor output voltage and to provide greater system reliability. The high fuel volume fraction results in smaller core sizes than encountered with internally fueled thermionic reactor designs. Other advantages of this concept include the following.

- The insulator can be removed from the core.
- Fuel swelling may occur away from the cesium gap.
- Thermionic converter performance can be checked prior to reactor operation.
- The series-parallel electrical network provides flexibility optimizing system reliability.
- The system has a prompt expansion coefficient.
- Redundant cooling is possible.

There are disadvantages to the externally fueled diode concept.

- The concept has a growth limitation of a few hundred kilowatts-electric.
- It is difficult to add a moderator.
- The fuel element structure operates at emitter temperatures.
- Interelectrode spacing must be maintained over a long distance.
- The fuel elements must be electrically isolated from each other, while the temperature is too high to use an insulator.
- Arcing paths may exist between fuel elements.
- The large size associated with dioxide fuels could give rise to very high currents.

In-core thermionic concepts suffer, however, from the fact that the reactor and converters are physically combined or intertwined—a design relationship that results in heavier and larger reactors. Because of the proximity of the converters to the nuclear fuel, fuel swelling can result in an electrical short circuit in the converter should the emitter distort by as little as 0.5 mm [25]. In-core neutron damage to the alumina insulator also tends to limit thermionic diode lifetimes at high power, and therefore high neutron flux, levels. The develop-

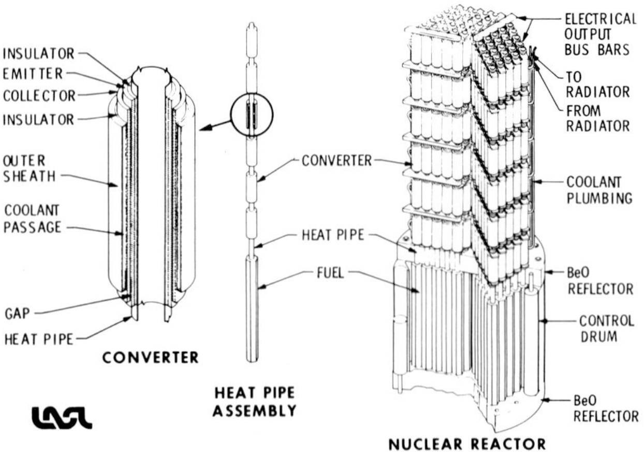

Fig. 11.27 Out-of-core thermionic reactor concept. *Courtesy of Los Alamos National Laboratory and Jet Propulsion Laboratory.*

Table 11.13 Out-of-core thermionic reactor operating characteristics.

Nominal electrical power output (kWe)	400
Thermal power level (MWth)	3
Thermionic converter efficiency (%)	15
Lifetime at full power (h)	75,000
Number of heat pipes	162
Nominal UO_2 content of fuel (vol %)	60
Average power density in Mo-UO_2 (MW/m^3 or W/cm^3)	78
Average power density in UO_2 (MW/m^3 or W/cm^3)	130
Power per heat pipe (kW)	18.8
Heat pipe axial heat flux (MW/m^2)	100
Heat pipe radial heat flux (MW/m^2)	0.91
Heat pipe temperature (K)	1675
Average fuel temperature (K)	1738
Maximum fuel ΔT (K*)	131
Maximum fuel temperature (K*)	1865
^{235}U burn-up (%)	5.8
Fission density in Mo-UO_2 (10^{20} fissions/cm^3)	6.3
Fuel volume swelling (%)	5
Reactivity change Δk due to fuel burn-up (%)	3.5
Reactivity change Δk due to thermal expansion (%)	1.5
Reactor Mass (kg)	
Mo-UO_2 fuel, total weight	372
^{235}U only	189
Mo heat pipes, 110 kg/m (to outer edge of reactor only)	60
BeO reflector	397
Control system	75
Reactor support structure	46
TOTAL	950

ment of insulators more resistant to neutron radiation damage would offer improved performance and lifetimes.

From 1973 until 1982, thermionic work has concentrated on out-of-core heat pipe concepts (see Fig. 11.27) [28]. These concepts separate the heat producing and power conversion functions by placing the energy conversion devices outside the reactor core and transporting core thermal energy to them. Most of this out-of-core thermionic effort has been focused on a configuration in which the converters are bonded directly to high temperature heat pipes. The reactor is similar in concept to the SP-100 heat pipe reactor discussed in Chapter 12. The major differences result from the need to operate the system at higher temperatures because of the thermionic converters. A design heat pipe temperature of 1,675 K results in a maximum fuel temperature close to 2,000 K. To reduce fuel swelling, hexagonally shaped fuel elements are used with the fuel in the form of UO_2 pellets in a molybdenum matrix that is brazed onto the heat pipe.

The nominal composition of the core fuel region is 60 percent UO_2 and 40 percent Mo. The operating characteristics for a 400 kWe out-of-core thermionic reactor are summarized in Table 11.13. Fuel swelling is estimated to be 5 percent.

A high temperature electrical insulator is used around the heat pipe, since it must not electrically short the system while transporting thermal energy. The high temperature insulating material must match the thermal expansion coefficients of the heat pipe. "Sealon" (a ceramic made from silicon, aluminum, oxygen, and nitrogen) was tested but failed to provide the necessary lifetimes. It is difficult to avoid degradation of such an insulator. Another technical problem was the inability to obtain the same performance at 1,650 K as at 1,850 K, despite a considerable research effort.

A modification of the out-of-core heat pipe thermionic reactor concept is shown in Fig. 11.28 [25, 29]. This design approach eliminates the high temperature insulator and uses radiative coupling. A large number (500–5,000) of small heat pipes helps to minimize temperatures in the reactor core, thereby reducing fuel swelling. Heat pipes are clustered around each thermionic converter. The cold inside electrodes are connected through a sheath insulator to heat rejection heat pipes that radiatively couple to the waste heat radiator.

Gaseous Core Reactors

Gaseous core reactors have historically received their primary impetus from space propulsion applications, because of their potential for large performance improvements over solid-core nuclear rockets. The fundamental concept for the gaseous core reactor is that of an externally moderated cavity assembly, containing the

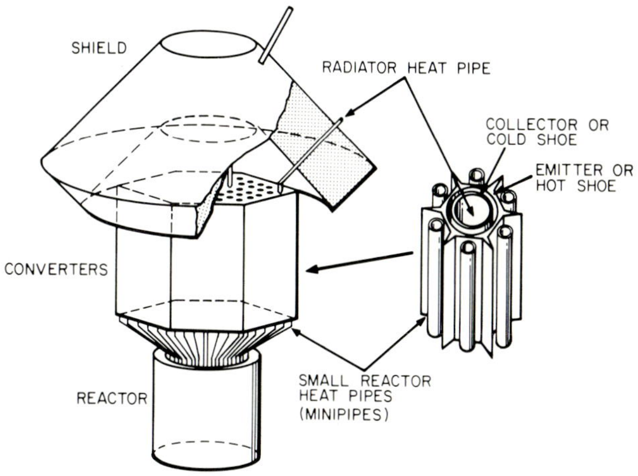

Fig. 11.28 Out-of-core radiation-coupled thermionic minipipe design [29].

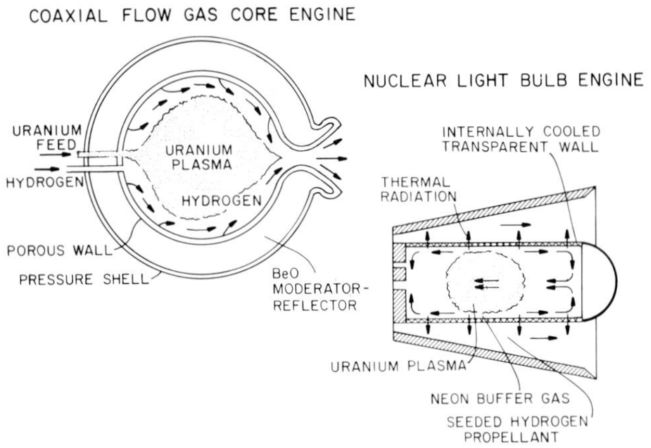

Fig. 11.29 Plasma core rocket concepts, including coaxial flow gas core engine (left) and nuclear light bulb engine (right) [*31*].

nuclear fuel in gaseous phase. For system temperatures less than a few thousand degrees kelvin (a temperature regime of interest in electric power generation), the nuclear fuel would be uranium hexafluoride (UF_6). However, for temperatures above about 5,000 K (a temperature regime ideal for nuclear rocket applications), vaporized uranium metal serves as the nuclear fuel in the form of a fissionable plasma. The gaseous nuclear fuel is separated from the cavity walls by hydrodynamic techniques, making it possible to operate systems at temperatures exceeding the structural limit temperatures of their containment materials. Power from the fissioning plasma can be extracted by convection and radiation heat transfer.

The original interest in gaseous core reactor was for advanced space propulsion systems, involving specific impulse values up to 50,000 m/s and nuclear fuel temperatures in excess of 10,000 K [30]. The uranium fuel in such systems would be in the plasma state. Many concepts were examined, with the coaxial flow system and the nuclear lightbulb system emerging as the favored approaches (see Fig. 11.29) [31].

The coaxial flow concept involves a low velocity inner stream of fissioning uranium metal plasma, surrounded by a very high velocity propellant gas (typically hydrogen) stream. The hot inner fuel plasma transfers thermal energy to the outer propellant stream by both convection and radiation heat transport. In this configuration, heat transfer occurs essentially unimpeded, but the propellant flow stream can also physically mix with the uranium plasma, leading to a loss of nuclear fuel.

In the nuclear lightbulb concept, the fissioning plasma is confined within a transparent (quartz) cell and is kept away from the cell's walls by the swirling flow of a tangentially injected buffer gas (e.g., argon). Here, energy transfer from the plasma to the propellant is by radiation heat transfer alone, since the two streams cannot physically mix. Even though the quartz cell effectively confines the nuclear fuel, radiation heat transfer is limited to selected regions of the electromagnetic spectrum, established by the transparency of the cell material and its resistance to nuclear radiation-induced "darkening" [31].

Both of these gaseous core reactor concepts share the common engineering problem of promoting effective radiation heat transfer to the flowing propellant, while avoiding thermally induced destruction of the cavity materials. Since neutrons and gamma rays from the fissioning plasma may not be totally absorbed by the propellant stream, nuclear radiation heating of the cavity walls must also be effectively handled. Table 11.14 provides representative preliminary performance estimates for a nuclear lightbulb engine with a cold beryllium reflector.

Another gaseous core reactor concept involves the use of a "mini-cavity," surrounded by a BeO moderator region and then a nuclear-fueled "driver" region (see Fig. 11.30 [32]). The uranium carbide fuel element driver region provides the majority of the neutrons

Table 11.14 Preliminary performance estimates for nuclear lightbulb rocket engine with cold beryllium reflectors.

Cavity Power (MWt)	189
Power to Propellant (MWt)	168
Reflective Liner Heat Load (MWt)	21
Be Reflector Heat Load (MWt)	7.5
Specific Impulse (m/s)	18,000
Number of Cavities	one
Thrust (N)	12,500
Engine Mass (kg)	4,620
Moderator (kg)	1,860
Pressure Vessel (kg)	1,830
Nozzles, Pumps, etc., (kg)	930
Cavity Length (m)	1.83
Cavity Diameter (m)	0.49
Cavity Volume (m^3)	0.34
System Pressure (MPa)	50.7
Fuel Material	^{233}U plasma
Buffer Vortex Fluid	neon
Reflector Material	B_e
Reflective Liner Material	B_eO
Liner Reflectivity	0.95
Fraction of Cavity Surface with Propellant Channels	0.25
Propellant Mass Flow (kg/sec)	0.68
Beryllium Coolant Outlet Temperature (K)	610
Effective Fuel Radiating Temperature (K)	8,300
Propellant Exit Temperature (K)	6,700
Cavity Power Density (MWt)	566
Space Radiator Mass (kg)	2,380
Total Mass (kg)	7,000

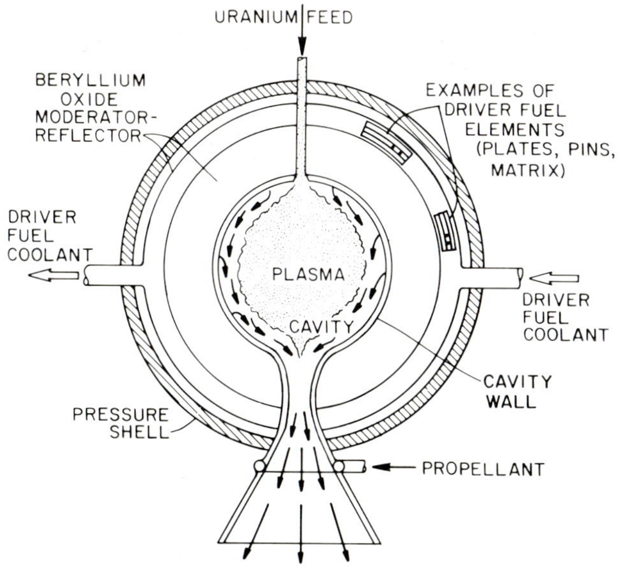

Fig. 11.30 Minicavity plasma core rocket engine [*32*].

needed for maintaining the chain reaction and substantially reduces the criticality dimensions of the gaseous cavity region.

Analytical and experimental studies have been undertaken for the coaxial flow concept [33]. These efforts examined hydrodynamic behavior and determined the rate at which nuclear fuel needed to be fed into a cavity to compensate for propellant stream losses. Vortex confinement of uranium hexafluoride gas at densities high enough to sustain nuclear criticality was experimentally demonstrated, as well as techniques for injecting UF_6 into confined vortex flows. Through both theory and experiments it was shown that the tangential velocity at the cavity wall must be significantly greater than the axial velocity of the buffer gas, in order to establish laminar flow at the fuel plasma. Vortex injection areas on the order of 90 cm^2 appeared necessary to produce a flow pattern with good confinement characteristics and laminar recirculation cells. The corresponding tangential injection velocity for the argon buffer gas was 360 cm/s, or about 18 times the maximum axial velocity of 20 cm/s [34].

Basic research on gaseous core reactors, demonstrating reactor physics and radiation heat transfer principles, has been completed. These efforts included a series of static UF_6 reactor experiments at temperatures between 300 and 500 K and power levels of 0.1 to 1.0 kW. These were followed by flowing UF_6 reactor physics experiments and flowing argon buffer gas physics experiments, which successfully demonstrated the confinement of a fissioning $U(93\%)F_6$ plasma inside an argon vortex. For example, on 10 May 1979 for a UF_6 flow rate of 1.8 g/s, a power level of 30 kW was sustained for approximately 120 seconds with excellent reactivity stability. Reactivity measurements indicated the achievement of a confined nuclear fuel mass of 35 g of UF_6 and a corresponding vortex residence time of 19 seconds. Similarly, for a nuclear fuel flow rate of 2.7 g/s, a power level of 10 kW was maintained for 140 seconds. In this case, a confined nuclear fuel mass of 50 g corresponded to a residence time of 18 s, indicating (as anticipated) that the confinement mechanism was independent of the flow rate [35]. The next logical technical step in the development of a gaseous core reactor system would be a reactor feasibility demonstration experiment. There are, however, no current plans for such a demonstration.

References

1. John E. Allen, "SNAP-50/SPUR Power Plant Requirements Report," Pratt and Whitney Aircraft, Middletown, Conn., Report No. CNLN-6291, 28 May 1965.
2. "Design No. 5 of SNAP-50/SPUR Reactor Test Systems," Pratt and Whitney Aircraft, Middletown, Conn., Report No. CNLM-5734, 9 July 1964.
3. L. M. Raring, "Fuel and Materials Highlights: SNAP-50 and Advanced Programs," Pratt and Whitney Aircraft, Middletown, Conn., Report No. CNLM-6155, February 1965.
4. J. A. Heller, T. A. Moss, and G. J. Barna, "Study of a 300-Kilowatt Rankine-Cycle Advanced Nuclear-Electric Space-Power System," NASA Technical Memorandum, NASA TM X-1919, November 1969.
5. H. J. Branch, "Preliminary Design of the 2 MWth Reactor and Shield (PWAR-20) for the SNAP-50/SPUR Power Plant," Pratt and Whitney Aircraft, Middletown, Conn., PWAC-445, 30 December 1964.
6. J. R. Peterson et al., "Status of Advanced Rankine Power Conversion Technology," 17th Annual Meeting of American Nuclear Society, Boston, Mass., 13–17 June 1971.
7. A. P. Fraas, "Summary of the MPRE Design and Development Program," Oak Ridge National Laboratory Report, ORNL-4048, June 1967.
8. General Electric Co., "710 High-Temperature Gas Reactor Program Summary Report," Cincinnati, Ohio, 1968.
9. John H. Pitts and C. E. Walters, "Conceptual Design of a 2 MWt (375 kWe) Nuclear-Electric Space Power System," *Journal of Spacecraft and Rockets*, Vol. 7, No. 11, November 1970, pp. 1282–1286.
10. Lawrence Radiation Laboratory, "Advanced Space Nuclear Power Programs Progress Report No. 5 (October through December 1966)," LRL Report C-92b, M-3679, 25 April 1967.
11. John H. Pitts and C. E. Walters, "Conceptual Design of a 10-MWe Nuclear Rankine System for Space Power," *J. Spacecraft*, Vol. 7, No. 3, March 1970, pp. 259–265.
12. Lawrence Radiation Laboratory, "Quarterly Report, Advanced Space Nuclear Power Program, October Through December 1967," LRL Report UCRL-50004-67-4, Part I.
13. Lawrence Radiation Laboratory, "Quarterly Report, Advanced Space Nuclear Power Program, January Through March 1968," LRL Report UCRL-50004-68-1.
14. J. R. Powell, T. E. Botts, and A. Hertzberg, "Applications of Space Power Beaming From Space-Based Nuclear Power Stations," 16th Intersociety Energy Conversion Engineering Conference, Proceedings, 1981, pp. 288–295.
15. H. Ludewig, A. J. Manning, and C. J. Raseman, "Feasibility of Rotating Fluidized Bed Reactor for Rocket Propulsion," *J. Spacecraft*, Vol. 11, No. 2, February 1974, pp. 65–71.
16. J. R. Powell, "Nuclear Power Systems," 1st Exoatmospheric Neutral Particle Beam Technical Interchange Meeting, Huntsville, Alabama, 17–18 November 1981.
17. T. E. Botts, J. R. Powell, J. L. Usher, and F. L. Horn, "Nuclear Reactor Using Fine Particulate Fuel for Primary Power in Space," 17th Intersociety Energy Conversion Engineering Conference, 1982 Proceedings, pp. 1403–1408.
18. J. W. Holland et al., "Thermionic Reactor Development," Intersociety Energy Conversion Engineering Conference, 1971 Proceedings, pp. 939–947.
19. A. J. Gietzen and W. G. Homeyer, "Thermionic Reactor Power Systems," Intersociety Energy Conversion Engineering Conference., 1972 Proceedings, pp. 556–558.
20. D. S. Beard, "Thermionic Reactor Technology—An Overview," Intersociety Energy Conversion Engineering Conference., 1971 Proceedings, pp. 933–938.
21. L. Yang and J. Chin, "Developmental Status of Thermionic Materials," Intersociety Energy Conversion Engineering Conference., 1972 Proceedings, pp. 1041–1049.
22. D. S. Beard, "Thermionic Fuel Development," Intersociety Energy Conversion Engineering Conference., 1969 Proceedings, pp. 131–136.
23. "Development of a Thermionic Reactor Space Power System," Gulf General Atomic Report No. Gulf-GA-A1208, San Diego, Calif., 30 June 1973.
24. L. K. Price, "Thermionic Reactor Concepts," Intersociety Energy Conversion Engineering Conference., 1969 Proceedings, pp. 122–130.
25. "Direct Conversion Nuclear Reactor Power Systems," Razor Associates, Sunnyvale, Calif., 1982.
26. D. R. Koenig and W. A. Ranken, "Reactor Design for Nuclear Electric Propulsion," 14th Intersociety Energy Conversion Engineering Conference, Boston, Mass., 1979 Proceedings, pp. 1418–1424.
27. W. M. Phillips, "Nuclear Electric Power System for Solar System Exploration," *J. Spacecraft*, Vol. 17, No. 4, (Article No. 79-1337R), pp. 348–353.
28. W. M. Phillips and E. V. Pawlik, "Design Considerations for a Nuclear Electric Propulsion System," 13th Intersociety Energy Conversion Engineering Conference, San Diego, Calif., 1978 Proceedings, pp. 218–223.
29. E. J. Britt and G. O. Fitzpatrick, "Direct Conversion Nuclear Reactor Power Systems," 17th Intersociety Energy Conversion Engineering Conference, 1982 Proceedings, pp. 1894–1901.
30. K. Thom, "High Grade Power from Fissioning Gases," NASA Report, 1976.
31. K. Thom and F. C. Schwenk, "Gaseous Fuel Reactor Systems for Aerospace Applications," AIAA Conference on the Future of Aerospace Power Systems, AIAA Paper No. 77–513, 1–3 March 1977.
32. Stanley Chow, "Mini-Cavity Plasma Core Reactors for Dual-Mode Space Power/Propulsion Systems," The Princeton University Conference: Partially Ionized and Uranium Plasmas, September 1976, pp. 217–223.
33. H. Weinstein, "Review of Coaxial Flow Gas Core Nuclear Rocket Fluid Mechanics," The Princeton University Conference: Partially Ionized and Uranium Plasmas, September 1976, pp. 123–129.
34. L. L. Lowry, "Gas Core Reactor Power Plants Designed for Low Proliferation Potential," Los Alamos National Laboratory Report No. LA-6900-MS, September 1977.
35. D. M. Barton et al., "Characteristic of Vortex Contained UF_6 Gas in a Beryllium Reflected Reactor," 1980 Annual American Nuclear Society Meeting, Las Vegas, Nev., 8–13 June 1980.

Chapter 12

Contemporary Space Nuclear Electric Systems

Heat Pipe Power Plant (SPAR/SP-100): Introduction

In 1979, to accommodate anticipated order-of-magnitude increases in space system power needs, the development of a new generation of space reactors was initiated. This power plant, an evolution from the Space Power Advanced Reactor (SPAR) program, was subsequently renamed SP-100. Because the space reactor power program initiated in 1983 also took the name SP-100, we will use the designation SPAR/SP-100 to distinguish the heat pipe power plant design. The major design objective of this generation space reactor is to provide 10-100 kW_e power unattended for a period of 7 to 10 years. The mass and size of the nuclear power plant are affected by the following: payload volume limits imposed by the launch vehicle; other components of the spacecraft on which the power plant will be used; and any constraints imposed by the orbital transfer vehicle used to place the nuclear-powered spacecraft in its final operational orbit. For example, the cargo bay of the Space Shuttle Orbiter is essentially a right circular cylinder, 18.3 m long and 4.6 m in diameter. If we consider the presence of an orbital transfer vehicle like the two-stage Inertial Upper Stage (IUS), the room available in the Shuttle's cargo bay is reduced to a cylinder 12.8 m long and 4.3 m in diameter. This upper stage configuration would be capable of delivering a payload of 2,270 kg to geosynchronous orbit. A power plant goal of 20-30 kg/kW has been established in anticipation of the development of advanced orbital transfer vehicles. Aerospace nuclear safety and system reliability have also been key factors incorporated into the design concept.

Description of SPAR/SP-100 Power Plant [1]

Figure 12.1 is a diagram of the power plant. The system consists of the nuclear subsystem (including the nuclear reactor heat source and radiation attenuation shield) and the power conversion system (including the thermoelectric converters and heat rejection devices).

The heat source is a solid core nuclear reactor that includes heat pipes for thermal energy transport. The heat pipe evaporator section is located in the reactor core, while the condenser section (supporting thermal energy extraction) is located within the electric power conversion equipment. A major advantage of this heat pipe-in-core design concept is that the need for mechanical pumps and compressors is eliminated. Furthermore, many parallel heat transport systems are incorporated into the reactor design, providing redundant thermal energy transport pathways and eliminating single point failure modes.

Advanced space mission requirements for high power levels, compact size, and long lifetimes favor the use of a fast neutron spectrum, highly enriched reactor that has a large inventory of fuel in a small volume. An abundant fuel inventory prevents large reactivity decreases due to fuel burnup. For example, in seven years, a 1,600 kilowatt-thermal (kW_t) reactor will burn up about 3.6 percent of the uranium-235 atoms in the fuel. Since this typical amount of fuel burnup is a very small

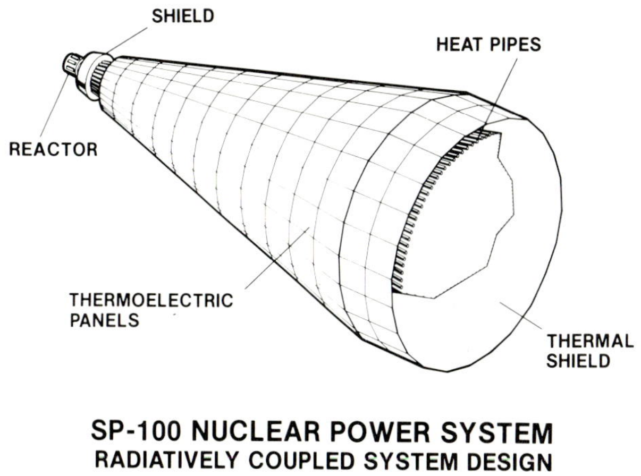

Fig. 12.1 The SP-100 nuclear power system, featuring a radiatively coupled design between the nuclear reactor subsystem and the thermoelectric power conversion subsystem. *Courtesy of Los Alamos National Laboratory.*

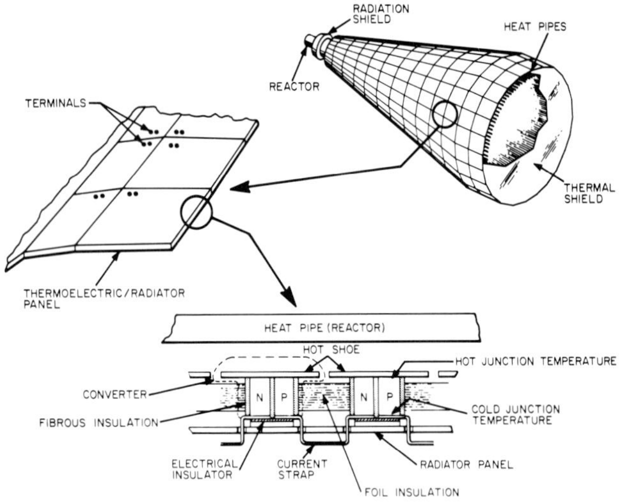

Fig. 12.2 Schematic of the SP-100 power conversion matrix. *Courtesy of Jet Propulsion Laboratory.*

percentage of the total fuel inventory, reactor criticality can be maintained throughout a mission. The operating temperature of the nuclear plant is constrained by both electrical conversion system material limits and fuel swelling considerations. This results in a reactor heat pipe temperature of 1,500 K. To support this operating temperature, the refractory fuel, uranium dioxide (UO_2), was selected—a choice based on materials compatibility with the heat pipes.

Power conversion for the SPAR/SP-100 concept is achieved using thermoelectric (TE) converters. Thermal energy is extracted from the reactor core by heat pipes that then radiate it to panels containing TE materials in an arrangement shown in Fig. 12.2. Hot shoe thermal collectors are used to concentrate this radiant energy. The thermal energy flows through the TE material by conduction. Thermal insulation is used to minimize thermal energy losses around the thermoelectric material. Thermal energy that cannot be effectively used due to second law of thermodynamics limitations is then rejected to space by radiation heat transfer. The heat rejection surface forms the cold shoe of the TE elements. By distributing a sufficient number of these TE elements over a wide area, the cold shoe, in effect, serves as the heat rejection radiator.

A special feature of the design is the elimination of the need for an independent afterheat removal system, whether accompanying planned or emergency reactor shutdown. This feature derives from the fact that heat pipes, as self-contained passive heat transport devices, will continue to function even at lower thermal energy input levels and in the absence of prime electric power. The heat pipes also possess sufficient out-of-core surface area to support radiation heat transfer rejection of the core afterheat.

Table 12.1 contains projected power plant performance parameters.

Table 12.1 SP-100 performance parameters.

Output Power (kW_e)	
Range	10–100
Nominal	100
Reactor Thermal Power (kW_t)	
Range	200–1,600
Reference Design	1,470
Design Life (year)	
Design Power	7
Total	10
Overall Dimensions	
Length (m)	8.5
Diameter (max)(m)	4.3
Radiator Area (m^2)	70
System Mass (at reference design)(kg)	
Reactor and Controls	450
Shield	790
Heat Pipes	460
TE Conversion and Circuitry	250
Thermal Insulation (incl. end panels)	290
Radiator	100
Power Control Subsystem	130
Interface Equipment and Structure	300
Total System Mass	2,770
Specific Power (W/kg)	36

Nuclear Subsystem

The nuclear subsystem is depicted in Figures 12.3 and 12.4. It consists of:

1. the nuclear reactor itself which serves as a thermal energy source;
2. the core heat pipes which transport thermal energy from the reactor to the TE converters;
3. the reflector which surrounds the core and scatters neutrons back into the fueled region;

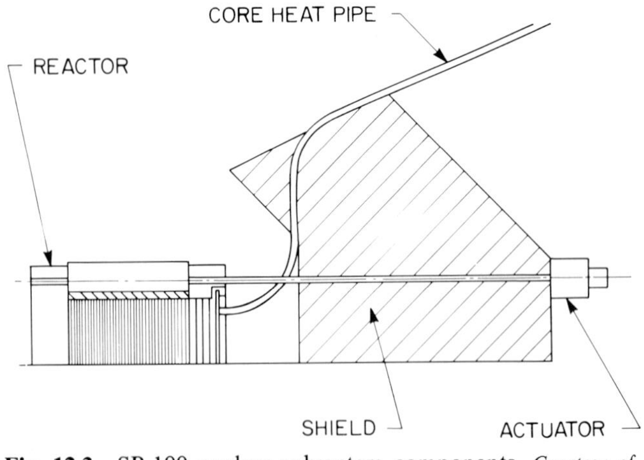

Fig. 12.3 SP-100 nuclear subsystem components. *Courtesy of Los Alamos National Laboratory.*

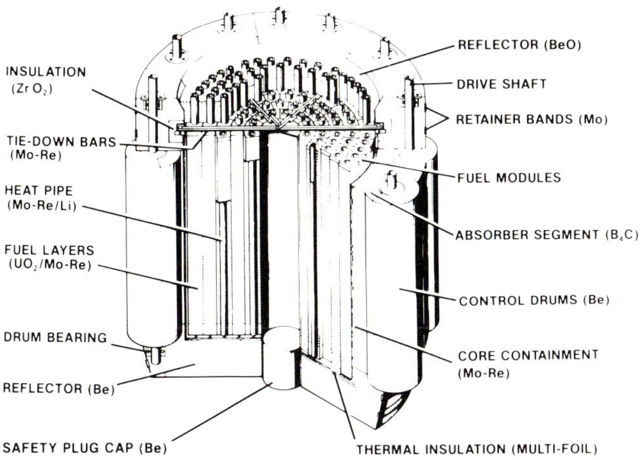

Fig. 12.4 Heat pipe space reactor. *Courtesy of Los Alamos National Laboratory.*

4. the shield (full or partial) which attenuates radiation levels at the payload or crew compartment; and
5. the control system which regulates the reactor power level.

The reactor has a core region composed of fuel modules in which the nuclear fuel is arranged in layers between circumferential fins that are attached to the heat pipes. Enhanced heat transfer is accomplished through the use of these fins. This design feature helps lower the operating temperature in the UO_2 fuel and also results in reduced fuel swelling. A barrel-shaped containment structure surrounds the core. This structure supports the fuel modules, but is not a pressure vessel. The containment barrel also provides non-compressive support for the multifoil insulation, which consists of multiple reflective insulation layers that reduce the core heat loss to an acceptable level. A beryllium reflector surrounds the core and reduces neutron leakage. Control drums, rotated by electromechanical actuators, are located within the reflector. Each of these drums contains a wedge of boron carbide (B_4C), a neutron absorbing material. Movement of this "neutron poison" wedge is used to establish reactor power level.

As shown in Figures 12.2 and 12.3, the core heat pipes extend from the reactor, bend around the radiation shield, and continue through the TE converter region. The radiation shield is located between the reactor and the payload. It serves as a barrier that substantially reduces the amount of reactor-generated nuclear radiation (e.g., neutrons and gamma rays) reaching the payload sensitive portions of the spacecraft or the astronaut crew. This shield contains primarily two types of materials: low atomic weight lithium hydride (LiH) to attenuate neutrons and high atomic weight tungsten (W) to shield against gamma radiation.

Reactor Core

Figure 12.4 is a cutaway view of the reactor core. It consists of heat pipe/fuel modules arranged in concentric rings. The portions of the heat pipes within the core have circumferential fins made of the same alloy as the heat pipes, namely molybdenum containing 13 percent rhenium (Mo-13%Re). Fuel tiles are positioned or sandwiched between these fins. The fuel itself is highly enriched uranium-235 in the form of uranium dioxide (UO_2). A void space appears in the center of the core, accommodating a neutron poison plug made of B_4C. This plug is inserted before the reactor is installed with its payload onboard the launch vehicle, e.g., the Space Shuttle, and remains in place as a safety feature during launch operations and orbital ascent. The poison plug keeps the reactor subcritical, even if it were immersed intact in water with all its heat pipes broken. Once the reactor and its spacecraft are on orbit and deployed out of the Shuttle's cargo bay, this B_4C plug is removed.

A typical fuel module is depicted in Fig. 12.5. The fins are an integral part of the heat pipe, while the fuel is fabricated as tiles and is sandwiched between the fins.

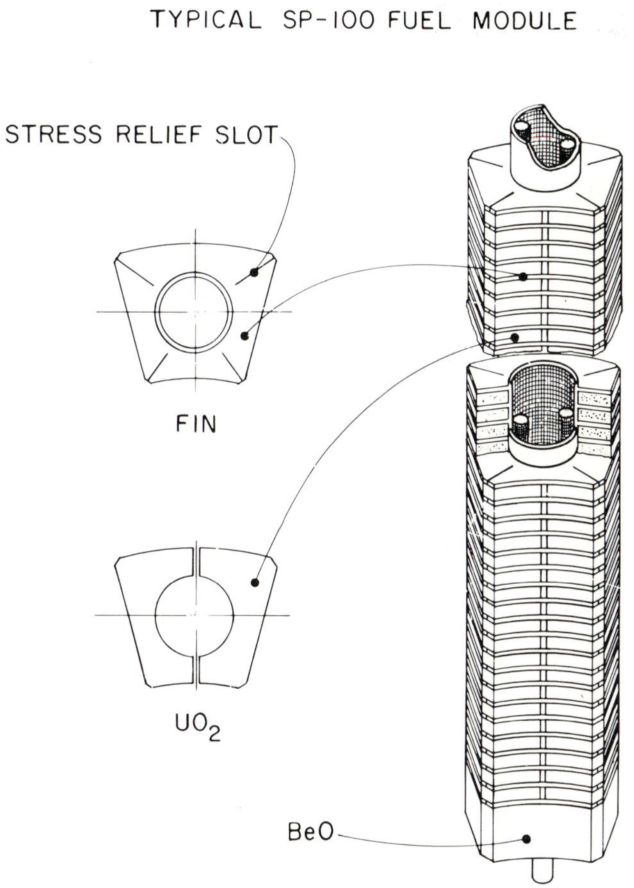

Fig. 12.5 Typical SP-100 fuel module. *Courtesy of Los Alamos National Laboratory.*

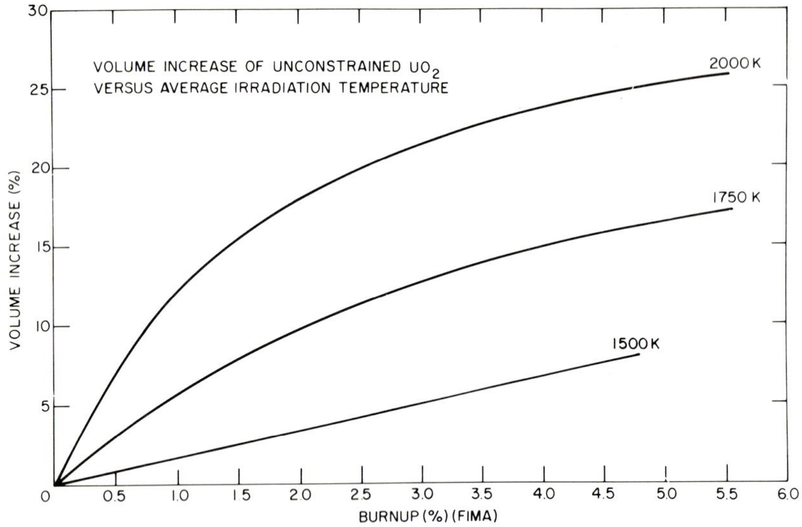

Fig. 12.6 Fuel swelling behavior of UO_2. *Courtesy of W. A. Ranken, Los Alamos National Laboratory.*

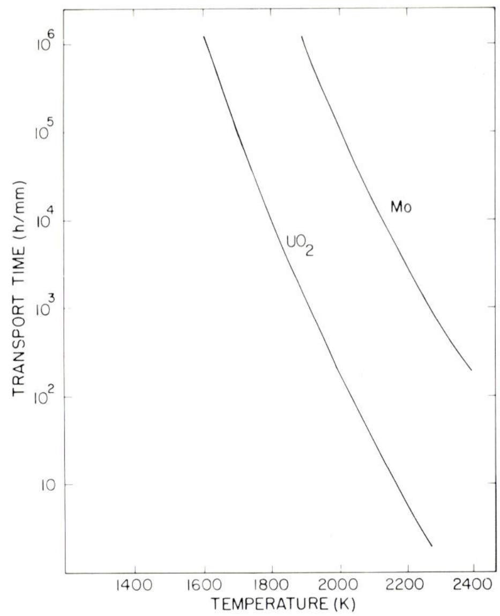

Fig. 12.7 Self-transport times for UO_2 and molybdenum. *Courtesy of Los Alamos National Laboratory.*

Distinct axial and radial power profiles exist in the core. Instead of varying the isotopic content of the fuel, the amount of fuel in each row of modules is varied, resulting in a uniform amount of heat generation per module. Consequently, each core heat pipe transports about the same amount of thermal energy. Axial temperature flattening is achieved by the natural heat transport action of the heat pipe itself.

Fuel swelling is both a function of temperature and operating time. Data on UO_2 fuel swelling are summarized in Fig. 12.6 [2,3,4]. However, the SP-100 design provides voids to accommodate fuel swelling. The thermal stress generated by fuel swelling must also be considered. By intentionally omitting an occasional fuel tile, the accumulation of axial stresses and core distortion are prevented.

Because of the high operating temperatures, core materials to some extent will evaporate and redeposit themselves. For example, the rate of redeposition of UO_2 and molybdenum is shown in Fig. 12.7 as a function of temperature.

A frequently asked question is what happens if a heat pipe fails? In the case of an individual heat pipe failure, transient heat transfer analysis indicates a peak temperature buildup will occur in a few minutes, because of poor thermal contact between adjacent fuel modules. Within 24 hours, the UO_2 will redeposit, closing the gap and reducing the fuel temperature to acceptable levels. Since core heat pipes are being designed to accommodate a 25 percent overcapacity, the heat pipes adjacent to an individual failed pipe will transport this additional heat load out of the core. In fact, an additional 25 percent design margin has also been included so that the heat pipe will operate at nearly isothermal conditions and to protect against performance degradation. Nevertheless, multiple failures of adjacent heat pipes are unacceptable, and the current design does not guarantee good thermal contact between fuel modules under these unusual circumstances.

The circumferential reflector is made of beryllium, while the reflector section that is penetrated by heat pipes is made of beryllium oxide (BeO). The reflector operates at a temperature of about 800 K and is cooled by radiation heat transfer to space from its outer surface. Of particular concern in reflector design is neutron irradiation-induced swelling and the subsequent loss of

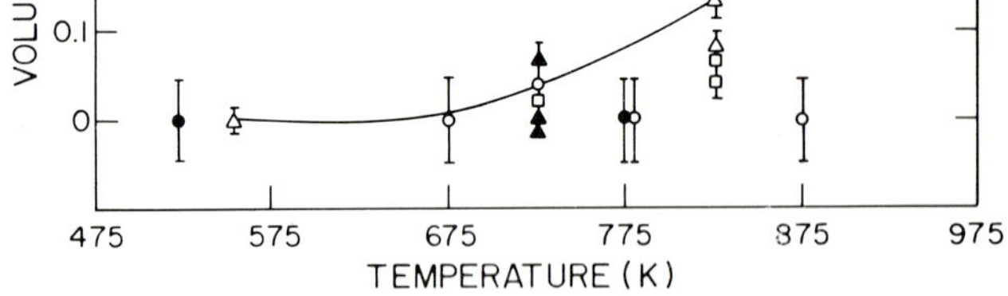

Fig. 12.8 Volume changes observed in irradiated beryllium as a function of temperature. (The reported values have been normalized to a dose of 10^{24} neutrons/m^2.) *Courtesy of Los Alamos National Laboratory.*

Table 12.2 Nuclear subsystem characteristics.

Thermal power (MW)	1.5
Heat pipe evaporator temperature (K)	1,500
Number of heat pipes	120
Heat pipe materials	
Container	Mo-13% Re
Working Fluid	Lithium
Getter	Hafnium, Zirconium
Maximum UO_2 temperature (K)	1,730
Fissions per 7 y (fis/cm^3)	8×10^{20}
Beginning-of-life reactivity (K_{eff})	1.05
Burnup 7 y (%)	3.6
Core diameter (cm)	33
Core height (cm)	33
Be reflector average thickness (cm)	9
BeO reflector average thickness (cm)	10
Reactor diameter (cm)	54
Reactor height (cm)	55
Nuclear subsystem mass (kg)	440
B_4C plug mass (kg)	3
Shield fluence at 7 y (nvt)	10^{12}
Gamma dose at 7 y (rad/Si)	10^6
Shield mass (kg)	790
Shield core half angle (Deg)	15
Shield axial thickness (cm)	80

thermal conductivity. The primary cause of this swelling is the formation of helium by (n, α) reactions. At temperatures above 575 K, this helium gas accumulates into bubbles and swelling proceeds as the bubbles grow. Swelling data are summarized in Fig. 12.8, normalized to a dose of 10^{20} nvt (10^{24} n/m^2). The term nvt stands for neutron-velocity-time and is a unit of neutron fluence, that is n/cm^2 or n/m^2. As the temperature is increased, no significant swelling occurs up to approximately 825 K. It is anticipated that helium bubble formation at high neutron fluences (greater than 10^{21} n/m^2) and elevated temperatures (above 825 K) will reduce the beryllium's thermal conductivity.

There are twelve cylindrical control drums within the reflector. Reactivity is increased or decreased by rotating the neutron absorbing material away from or toward the core, respectively. Each drum will be rotated by a stepping actuator. The actuators are located away from the core behind the shield for protection against both nuclear and thermal radiations. The reactor has built-in (approximately 5 percent) excess reactivity to compensate for control margin, fuel burnup, temperature defect, and operational contingencies. The B_4C is enriched to 90 percent ^{10}B to provide a reactivity worth of 12.5 percent. After seven years of operation, the reflector will increase in volume by about 2 percent.

The reflector consists of segments held together by molybdenum retainer bands (see Fig. 12.9) and is designed to disassemble on atmospheric reentry in support of aerospace nuclear safety requirements. These molybdenum bands provide mechanical integrity to the reactor during launch and orbital operations; yet they also rapidly oxidize in the event of atmospheric reentry.

Table 12.2 describes typical SP-100 nuclear subsystem performance parameters, including dimensions, fuel swelling, and burnup.

Core Heat Pipes

The heat pipe container material is made of a low rhenium content alloy of molybdenum, which was selected because of its excellent high temperature creep strength and its compatibility with the internal working fluid and with the uranium dioxide fuel. In addition, this particular alloy was chosen because it is ductile at low temperatures and can withstand the launch load environments. Properties of Mo-13% Re are shown in Table 12.3. The working fluid is lithium at 1,500 K. Lithium was selected because of its vapor pressure, high heat capacity, low operating pressure, and compatibility with the wall and wick materials. An artery wick design provides a low pressure drop, redundant wick structure. A circumferential distributive wick is also used to transport the condensing fluid to and from the arteries, while the arteries transport the lithium axially. Figure 12.10 shows the predicted behavior of a typical core heat pipe. The limitations on heat pipe performance in this particular design are the sonic limit and the capillary action that can be achieved with a given wick pore size.

Getter materials are included in the heat pipes so that residual fabrication impurities and impurities that diffuse through the walls during operation are absorbed and will not react with the container or wick materials.

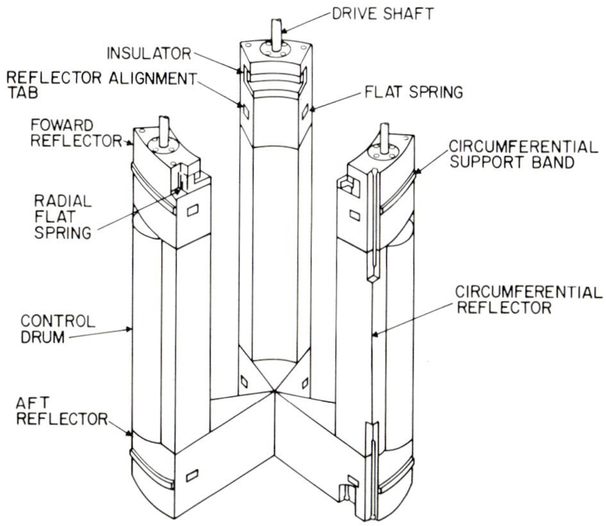

Fig. 12.9 Segmented reflector showing drums and banding arrangement that supports upper atmospheric dispersal upon reentry. *Courtesy of Los Alamos National Laboratory.*

Table 12.3 Selected properties of molybdenum-13% rhenium alloy.

Physical Properties	Molybdenum-13% Rhenium[a]
Density at 293 K (mg/m^3)	11.89
Thermal conductivity at 1,400 K (W/m K)	~70
Melting point (K)	2,810
Complete recrystallization in 1 h (K)	1,645
Chemical Compatibility	
Lithium at 1,500 K in heat pipe	No reaction experienced
UO_2	No reaction to 2,475 K
Mechanical Properties	
Ductile-brittle transition temperature (K)	140
Projected creep deformation (%)[b]	3×10^{-4}

[a] Electron beam melted alloys
[b] of 15.9 mm o.d. by 14.1 mm i.d. heat pipe after 7 yr at 1,400 K ($p \simeq 7$ Mpa)

Hafnium is used for this purpose because of its affinity for oxygen. The advantage of using hafnium is shown in Fig. 12.11, which compares the free energies of formation of various materials in the core. The compounds appearing at the lower part of the figure will be formed in preference to those appearing higher up.

Radiation Shield

Nuclear radiation shield designs were extensively developed in the SNAP program and the SP-100 design is based on that work. It uses lithium hydride (LiH) for neutron shielding and a layer of tungsten (W) to attenuate gamma radiation.

The operating temperature of the radiation shield must be at least 600 K to permit reabsorption of radiolytically decomposed hydrogen, thus preventing swelling. The shield, however, must also remain below 675 K to avoid hydrogen loss due to excessive thermal dissociation if the casing is punctured by meteoroids. To avoid single failure points, the LiH is encapsulated in a number of compartments, so that a pressure containment failure will deplete the hydrogen in only a small portion of the shield. The tungsten part of the shield will operate satisfactorily at the LiH temperature level.

Figure 12.12 shows shield mass as a function of radiation level, shield half angle, and distance from the center of the core.

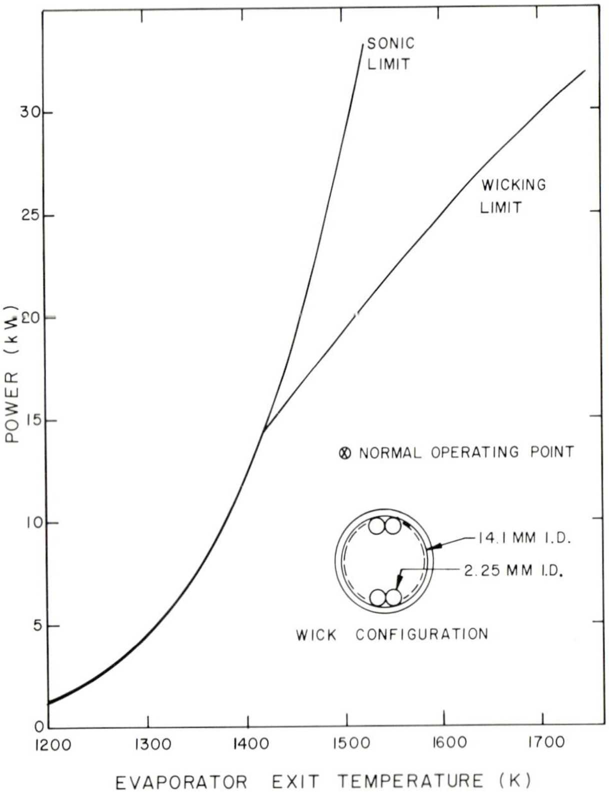

Fig. 12.10 Typical high temperature heat pipe performance curve. *Courtesy of Los Alamos National Laboratory.*

Reactor Control System

The current concept is to control the reactor by adjusting the 12 rotating drums that surround the core. These control drums will be operated under closed-loop control by functionally redundant reactor controllers. Safety criteria dictate that, to prevent overheating of high temperature system elements, the reactor controller will provide a power cutback capability. In addition, if an emergency shutdown situation is detected, a shutdown control function will be provided to scram the control drums.

During startup, reactor control is based on the temperatures of the core heat pipes. To avoid excessive power and temperature overshoots due to long time constants in transfer functions, the current startup logic sequence takes several hours. Faster startups are feasible, if desired.

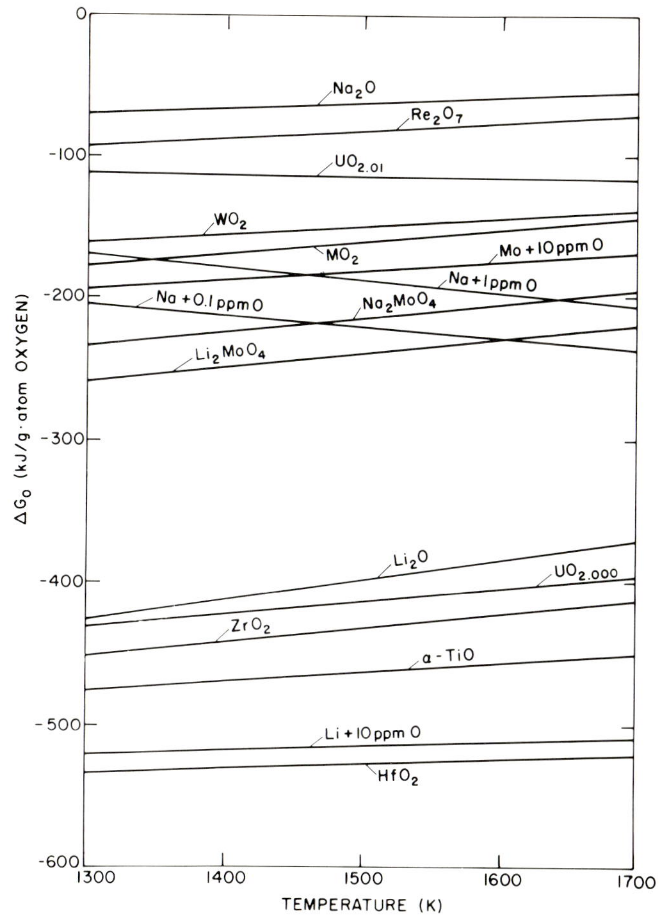

Fig. 12.11 Free energy of formation of various compounds. *Courtesy of Los Alamos National Laboratory.*

A representative startup scheme is as follows:

1. The control drums are ramped relatively quickly from shutdown to a position well below cold critical condition.
2. The drums are then ramped through this cold critical condition very slowly.
3. When the heat pipe temperatures reach the minimum controllable value, the system is switched to temperature control.
4. Then the power level is raised to increase the control temperature enough to permit normal operation.

To shut down the reactor, the drums are driven to their in-position. Since the core heat pipes will continue to function as a passive heat removal system even after shutdown, no additional afterheat removal system is needed.

Aerospace nuclear safety requirements are an important factor in the control system design and redundancy is also valuable in creating systems with good safety characteristics. The following safety features are incorporated in this design:

1. The control drums are pinned in their shutdown position with captive-key locks for all noncritical ground handling nuclear operations. However, it will be possible to unlock and operate one drum at a time for testing.
2. A central plug of B_4C is installed in the core for launch and orbital ascent. This neutron poison plug is then removed after the nuclear-powered payload has been deployed from the Space Shuttle Orbiter's cargo bay. A teleoperated device or an astronaut doing extravehicular activity (EVA) can perform this removal task.
3. Control drum actuators have brakes that hold the drums in position until the brakes are energized. Their motors do not have enough torque to drive the drums with the brakes engaged.
4. With twelve control drums, reactor shutdown can still be attained, even if a number of these drums were to become inoperative.
5. The system has redundant, parallel-operating, self-test reactor controllers that will ramp the control drums to their shutdown position. Each redundant controller has its own energy source and

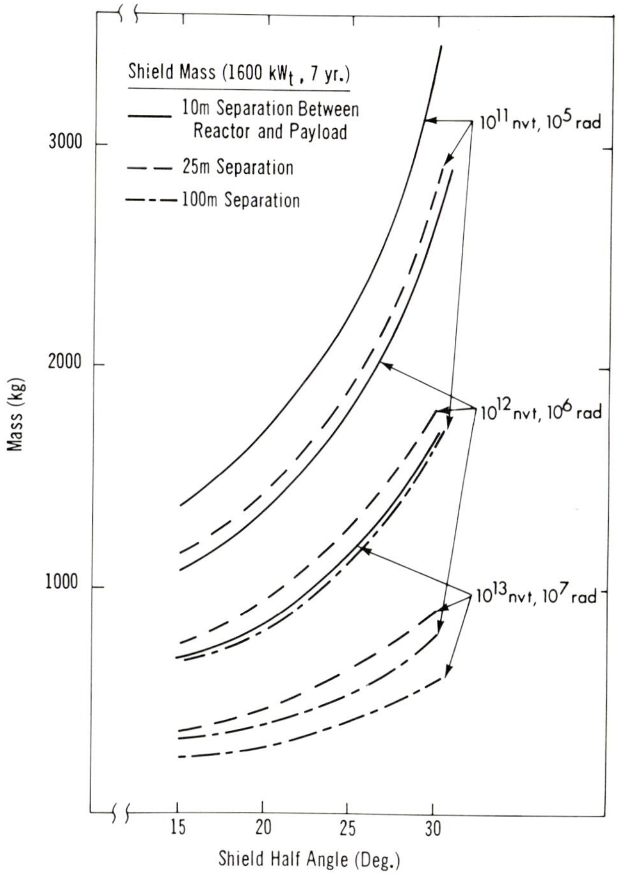

Fig. 12.12 Shield mass as a function of shield half angle for various radiation attenuation levels and separation distances. *Courtesy of Los Alamos National Laboratory.*

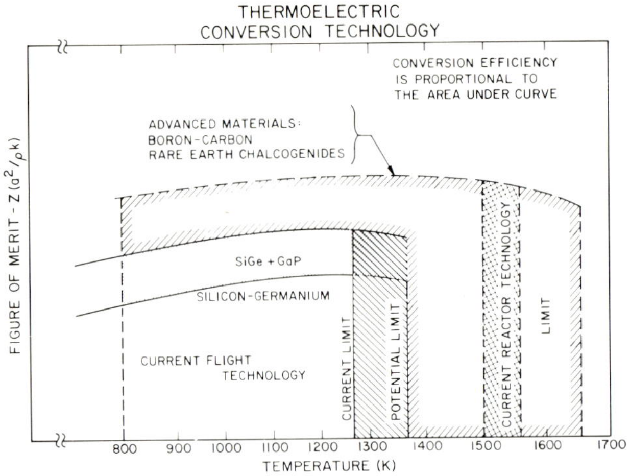

Fig. 12.13 Thermoelectric conversion technology figure-of-merit values. *Courtesy of Jet Propulsion Laboratory.*

does not share electronic components with the others. The system is capable of being reset to normal operation and restart after an emergency shutdown.

6. Finally, the control drums are spring-loaded—a fail-safe design feature that returns them to the shutdown condition in the event of the loss of electrical power.

Electrical Conversion Subsystem

The conversion of the thermal energy to electrical energy is accomplished through the use of thermoelectric (TE) converters. The conversion systems shown in Fig. 12.2 consist of:

1. thermoelectric converters that transform thermal energy from the reactor into electricity;
2. radiator panels that remove waste thermal energy from the TE converters and reject it to space;
3. thermal insulation that isolates the radiator panels from direct thermal contact with the nuclear energy heat source; and
4. interconnecting circuitry that links the electrical output of the TE converters with the power processing equipment.

Thermoelectric conversion performance depends on the choice of thermoelectric material. As shown in Fig. 12.13, the choice of TE material will significantly affect the power plant performance. The lower line in this figure represents current SiGe alloys. Improved versions of this TE material are now being developed in which GaP is being alloyed with the SiGe. Characteristics of these improved materials are given in the middle line. An overall gain of about 40 percent arises mainly from a reduction in thermal conductivity. The top curve represents potential improved TE converters using such materials as boron-carbon and rare earth compounds. These materials promise stability at even higher temperatures than alloys based on SiGe. Some of the factors that make these improved materials desirable are: (1) they have long-term stability at high operating temperatures; (2) they can be metallurgically bonded to the other materials making up the TE module; and (3) they possess acceptable strength characteristics. Sublimation is the major limitation to lifetime. This disadvantage, however, can be reduced through the use of suitable coatings.

The TE converters are arranged in panels. These panels form the heat rejection elements, eliminating the need for a separate radiator. The panels are arranged to form a frustum of a cone about 7 m long that begins at the end of the shield and grows in diameter until reaching a diameter of 4.3 m. At this point, the panels then extend in a cylindrical shape to the full radiator length. The minor diameter of this frustum is determined by the nuclear radiation shield, while the major diameter is controlled by the Space Shuttle Orbiter's payload bay limits. The total panel area is dimensioned to reject heat to space at the design cold side temperature. The amount of heat to be rejected is governed by the power level of the plant; the conversion efficiency of the TE material; and the operating temperatures of the hot and cold side of the TE converters.

The thermoelectric material is arranged in two legs—one doped as an *n*-type material and the other as a *p*-type material (see Chapter 5). These legs are also covered with a coating to reduce high temperature sublimation to acceptable levels. For example, in the multihundred-watt (MHW) radioisotope generator program, the SiGe converters were coated with silicon nitride (Si_3N_4). To achieve even higher operating temperatures and longer converter lifetime, an improved sublimation coating is needed. The goal is to reduce the hot side sublimation loss of SiGe to less than 1×10^{-7} g/cm^2-h. Silica glass appears to be a good candidate material.

Thermal energy from the heat pipes is transferred to the TE modules by radiation transfer directly from the heat pipes to the hot shoes. The hot shoe material should possess the following characteristics: good thermal conductivity, an ability to operate at 1,355 K, compatibility with the thermocouple materials, and light weight. The prime candidates are molybdenum and carbon/carbon composites. Because of lower weight, the use of carbon/carbon composites is favored. To support radiative coupling, the TE converters are distributed in panels, with the cold shoes serving as the heat rejection surface. No armoring is needed for micrometeoroid protection. The use of a carbon/carbon composite, with its high emissivity, as the shoe

Table 12.4 SPAR/SP-100 conversion subsystem parameters.

Electrical power delivered (kW)	100
Voltage delivered (Vdc)	174
Thermal losses (%)	10
Electrical losses (%)	10
Total subsystem efficiency (%)	6.8
Radiator area (m^2)	70
Thermoelectric hot junction temperature (K)	1,355
Cold junction temperature (K)	840
Average figure-of-merit (K^{-1})	1.0×10^{-3}
Power/couple (W)	0.5
Hot shoe material	graphite
Cold shoe material	carbon/carbon composite
Surface emissivity	0.85
Radiator temperature (K)	800

material eliminates the need for a special emissivity coating (such as zinc oxide or carbon black) which is needed on many conventional radiator surfaces. Thermal insulation is used to minimize heat flow leakage from the hot shoe to the cold shoe. Multifoil insulation, similar to that used in the reactor core to thermally isolate the core from the reflector, is used here. An additional quartz wrap insulation layer surrounds the TE material, protecting against electrical shorts. A 100 kW_e plant design uses 220,000 couples.

Important conversion subsystem parameters are given in Table 12.4.

SPAR/SP-100 Reactor Technology Status

The key technology development needed in the heat pipe reactor concept is demonstrating extended time behavior of the heat pipes and fuel modules. Sufficient short term testing experience exists to provide a high technical confidence that component performance requirements can be met. The heat pipe and fin material, Mo-13% Re, has been shown to be ductile at temperatures well below 200 K and at this low temperature exhibits excellent weld ductility. In fabrication of an in-pile irradiation experiment, fins were successfully machined onto a heat pipe. Compatibility experiments, involving the UO_2 and molybdenum alloy, indicate that O_2 diffusion is sufficiently low to avoid serious lifetime limiting effects. Additional work, however, is needed to fully characterize the alloy. This includes investigations of high-temperature strength (including long-term creep), low-temperature ductility, the effects of impurities, long-term grain growth phenomena, and the effects of neutron irradiation and fission fragments. Experimental heat pipes have exceeded the normal and contingency design performance goals. However, such experiments involved short term tests, so that endurance testing is now the next logical step in the heat pipe development. Lifetime testing of lithium heat pipes is shown in Table 12.5 [2]. The longest lithium heat pipe lifetime demonstrated to date is with a short unit fabricated from low-carbon, arc-cast (LCAC) molybdenum that operated for 25,400 h at 1,700 K with a heat transport of approximately 1 kW. The wall of this heat pipe experienced extensive grain growth, to the point that the grain diameter became several times the 1 mm wall thickness. The unit failed when the lithium leaked after the unit had been chucked up in a lathe for remachining of a support stem. The tests identified in Table 12.5 were all conducted at representative SPAR/SP-100 reactor core heat pipe temperatures and higher. Thus, the results indicate that longevity goals at 1,500 K are indeed achievable.

Fuel swelling data were presented in Fig. 12.6 as a function of temperature and burnup. These fuel data are based on sources that most nearly simulate SPAR/SP-100 conditions—that is, fuel in the temperature range of 1,500 to 2,150 K and with a low thermal gradient. The primary experimental data are shown in Fig. 12.14 [3]. With expected burnup for seven years

Table 12.5 Lithium heat pipe life test data.

Material	Temp (K)	Operation Time (h)	Reference	Remarks
W-26 Re	1,875	10,000	Busse, 1971	
TZM	1,775	10,526	Rouklove, 1968	Weld cracking
TZM	1,775	10,400	Eastman, 1969	Weld cracking
TZM	1,775	9,800	Eastman, 1969	Weld cracking
Nb-1Zr	1,775	9,000	Busse, 1966	
Mo (LCAC)	1,700	25,400	Los Alamos Nat. Lab., 1982	Large grain formation
Mo/Mo-Re Screen	1,500	1,700	Los Alamos Nat. Lab., 1984	Continuing
Mo-Re	1,500	5,000	Los Alamos Nat. Lab., 1984	Continuing
Mo	1,500	> 20,000	Los Alamos Nat. Lab., 1984	Continuing (Na fluid)

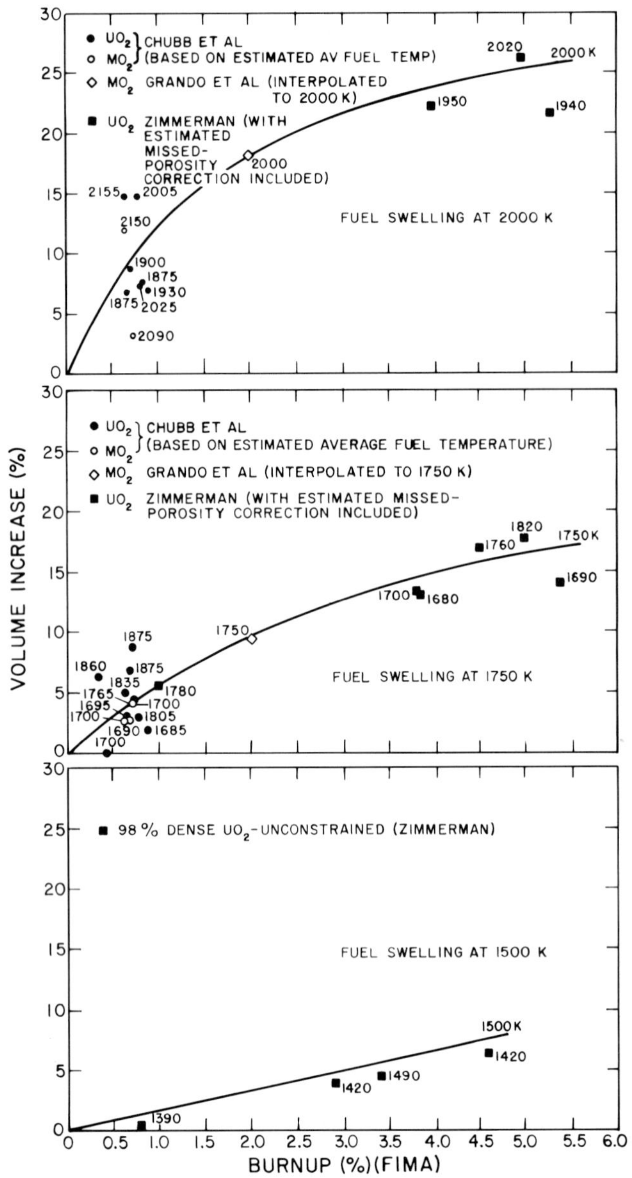

Fig. 12.14 Recommended fuel swelling data for SP-100 systems at 1,500, 1,700, and 2,000 K. *Courtesy of W. A. Ranken, Los Alamos National Laboratory.*

operation of 3.6%, these data cover the regions of design interest.

Other reactor components such as the reflector, reactivity control actuators, and shielding incorporate technologies from previous space reactor programs. The actuators, for example, in the S8DR development had a single unit run over 20,000 h.

Current SP-100 Program: Organization and Scope [4]

In 1983, the National Aeronautics and Space Administration (NASA), the Department of Energy (DOE), and the Department of Defense, Defense Advanced Research Project Agency (DARPA) entered into an agreement to fund a program also called SP-100 to jointly develop space reactor system technology. The goal is to permit a space power plant concept decision in 1985 to proceed with the development of ground engineering system. In order to facilitate this decision, mission definition studies, both civilian and military, are being carried out concurrently with system definition studies and technology work. The latter is to provide timely answers in critical technology areas. Special emphasis is also being placed on aerospace nuclear safety. Areas of technology work encompass reactor fuels and materials, power conversion, and an array of other aerospace technologies. The program also includes detailed planning for the ground engineering phase.

The management structure and rationale for the current SP-100 Program is given in Fig. 12.15. A senior-level Steering Committee from the three agencies provides overall progammatic guidance and direction. The Program Office is responsible for the execution of the program. Project management is charged with providing the detailed planning and execution of the technology development efforts. A Deputy Project Manager for Nuclear Technology provides nuclear technology, design, and engineering. In scope, the SP-100 Program has the DOD, DOE, and NASA charter for all space nuclear reactor power systems technology development.

Specific SP-100 Program objectives are:

a. Ensure that aerospace nuclear safety is fully considered.
b. Establish performance limits and satisfy military and civilian applications for space nuclear power in 100 kilowatt and multimegawatt regimes.

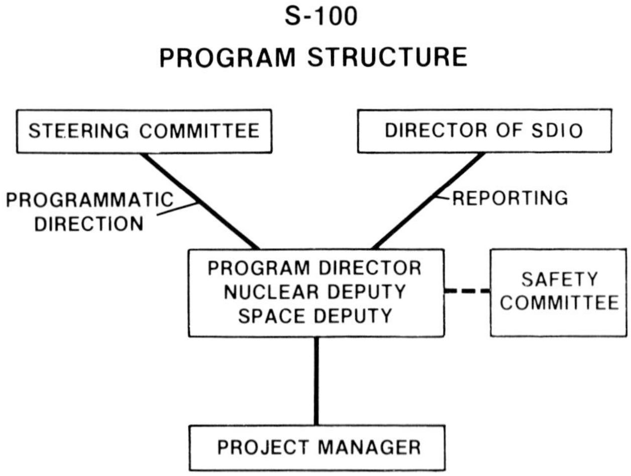

Fig. 12.15 SP-100 Program organization chart [4].

Table 12.6 SP-100 goals.

Performance	
Power output, net to user (kW_e)	100
Output variable up to 100 kW_e	
Full power operation (y)	7
System life (y)	10
Reliability (%)	
First system, 2 y	0.95
Growth system, 7 y	0.95
Multiple restarts	
Physical Constraints	
Mass (kg)	3,000
Size, length within STS envelope (m)	6.1
Interfaces	
Reactor induced radiation after 7 y operation, 25 m from forward end of reactor	
Neutron fluence (nvt)	10^{13}
Gamma ray dose (rads)	5×10^5
Mechanical	STS launch conditions
Safety	
Nuclear Safety Criteria and Specifications For Space Nuclear Reactors	

c. Assess and advance technology for a 100 kilowatt class space nuclear power system in support of military and civilian missions.

d. Evaluate the need for multimegawatt space nuclear power systems and perform long range technology development efforts, if reasonable needs exist in the next few decades.

e. Ensure an orderly and effective transition from technology assessment and advancement to engineering development and production for space reactor users as warranted.

The strategy of the SP-100 Program is to first identify potential users and space mission applications of large quantities of electric power, and, then, to determine power system functional requirements from such mission requirements. System concepts are then developed to meet these functional requirements, tempered by safety and other design constraints. Critical technical issues for these concepts will be identified and confirmed through analysis and experiment.

Current SP-100 Technology Approach

Following investigation of over a hundred potential space nuclear power system concepts, the field has now been narrowed to three candidate systems which appear to meet the goals in Table 12.6 with a reasonable balance of technical risk and development time.

One concept uses a fast spectrum, lithium-cooled reactor coupled to thermoelectric (TE) converters. The coolant is pumped with electromagnetic pumps. Thermal energy is radiatively transferred to thermoelectric elements to avoid excessive thermomechanical stresses on the TE converters. Waste heat is removed through a heat pipe distribution system and space radiator configuration. The principal technical uncertainties include the compatibility of fuel, clad, and lithium coolant; fuel properties in the temperature and burnup range of interest; and the performance and long-term behavior of the TE converters at high temperatures. Potential advantages of this design approach are the relative lack of moving parts and the use of similar (but not identical) space-mission-proven TE converters.

The second approach is an in-core thermionic system with a pumped sodium-potassium (NaK) eutectic coolant. Advantages of such a system include overall design simplicity, previous in-core data on thermionic elements for limited lifetimes, and relatively little mass and size growth to accommodate higher power levels. An unanswered question is whether the increased non-nuclear material and large thermionic gaps required in the core to sustain long-life operation also force the overall concept to be excessively massive in the lower power level regime. Specific technical issues include fuel swelling rates and mechanisms and insulator material lifetimes in the required temperature and flux regime.

The third approach uses a Stirling engine to convert reactor thermal energy into electricity. Since this dynamic cycle has potentially much higher conversion efficiency than the two static conversion approaches, reactor temperatures could be significantly reduced while generating the same electric power. Conversely, the use of the Stirling cycle could provide substantial growth capacity for a high temperature reactor. In this case the development issues involve the Stirling engine itself and include high temperature operation, long-life performance, the ability to provide high efficiency with a relatively low temperature drop, mass reduction, and dynamic balancing so that vibrations do not adversely impact space system operations.

These three concepts will now be discussed in more detail in the following sections.

High-Temperature Pin-Fuel Element Reactor With Thermoelectric Conversion [5]

Figure 12.16 depicts the high-temperature, pin type fuel element reactor with thermoelectric power conversion in its orbital configuration. The system contains a 12-sided cone structure with a 17 degree shield cone half angle. The reactor, which is a right circular cylinder approximately 1 meter in diameter and 1 meter high, is at the apex of the conical structure. The reactor has a

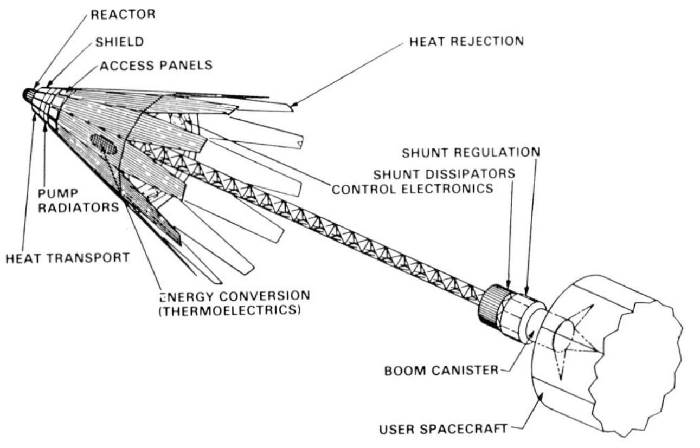

HIGH-TEMPERATURE REACTOR WITH THERMOELECTRIC POWER CONVERSION CONCEPT

Fig. 12.16 High temperature reactor with thermoelectric power conversion concept [5].

fast spectrum and uses lithium in the liquid state for heat transport. The nominal output power is 2 MW_t and the operating outlet temperature is 1,540 K. Control is achieved by using twelve rotatable drums. Each drum has a section of absorbing material and a section of reflective material to control the criticality level.

The reactor subsystem contains cylindrical shaped pins containing UO_2 pellets (with Nb-1% Zr-0.1% C clad), lithium coolant, a safety plug, a Mo-Re vessel, a BeO reflector, rotating B_4C absorber drums, thermoelectric EM coolant pumps, heat exchangers, and associated piping, accumulators, and core restraint. The fuel pins are sintered cylindrical UO_2 pellets with a BeO reflector column at each end. The whole column will be contained in a thin-walled cladding tube sealed at one end with a metal plug welded to the cladding and at the other end with a closed welded plug. A vent device is also included at the closed welded plug end to prevent excessive pressure buildup within the pin. The pin is filled with low pressure, high purity helium. At operating temperature the helium pressure will increase to a value approximately equal to the coolant system-pressure. The helium enhances heat transport from the fuel to the cladding, thereby minimizing fuel temperatures. Finally, a helical spring arrangement at the upper end plug prevents axial movement during launch.

A safety plug is used to guarantee subcriticality in the event of core water immersion following an aborted launch. The plug consists of a B_4C poison, BeO reflector, and tungsten and LiH shield sections. During launch, the neutron poison section is positioned in the core region; during space operations, the plug is moved so that the reflector section is in the core.

The core is contained in a cylindrical vessel with a central sleeve for the safety plug. The coolant flow path is baffled and 2 millimeters of foil insulation reduce thermal energy loss and protect the control drums and reflector against excessive temperatures.

The shield is mounted directly behind the reactor and consists of both gamma ray and neutron shielding materials. The gamma ray shield consists of multiple layers of tungsten designed so as to prevent radiation-induced thermal warping. The neutron shield consists of a series of axial sections with a thermal conductor between the axial sections. The thermal conductor carries the gamma ray and neutron-induced heat to the shield surface where it is then radiated to space. Temperature levels are maintained below 675 K.

Thermal transport is accomplished by means of a lithium working fluid that is circulated by a thermoelectrically powered EM pump. The thermoelectric converters for the pump are powered by the temperature drop between the working fluid and the pump radiators. This approach assures pumping of the working fluid as long as the reactor is at temperature and provides power for core afterheat removal.

Figure 12.17 shows the interface between the reactor coolant loop and a set of high temperature heat pipes. The heat pipes are used to radiatively couple energy from the reactor to the thermoelectric elements. The heat exchangers contain a thimble/finger interface that self-brazes in orbit as the power system is brought up to operational temperatures. Self-braze materials are used in the volume between the thimble and finger interface. These materials have the characteristic of a depressed melting point until the material becomes molten. In the molten condition, the additives in the braze material migrate away from the braze material and the material melting point elevates above the operating temperatures. Solidification results and a good thermal and mechanical joint is formed.

The thermoelectric elements are bonded to the internal surfaces of the heat rejection radiator panels. TE materials are expected to be carbide or sulfide com-

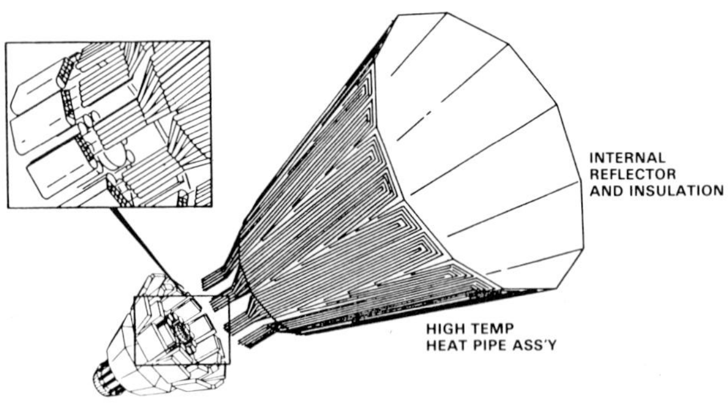

HEAT EXCHANGER INTERFACE FOR HIGH-TEMPERATURE REACTOR WITH THERMOELECTRIC CONVERSION CONCEPT

Fig. 12.17 Heat exchanger interface for high temperature reactor with thermoelectric conversion concept [5].

Table 12.7 Key parameters for high temperature reactor and thermoelectric conversion power plant.

Mass summary	Kg	Area summary	m^2
Reactor	525	Main body	
Shield	665	Fixed area	31
Heat transport	439	Deployed area	35
Energy conversion	307	E.M. pump	1.7
Radiation panels	363		
Structure & insulation	150	Power processing	
Aux. power & electronics	350	Shunt radiation	3.4
	2,799	Low temp radiation	8.6

Performance summaries

Temperature	K	Efficiency	.051
Reactor outlet	1,544	Thermal input power	2.21 MW
Conversion in	1,333	Rejected power	2.10 MW
Conversion out	864	Electric power	110 kW
Radiator	863		
Shunt radiator	1,000		

pounds. On the order of 60,000 bi-couple thermoelectric elements are needed. The heat rejection surfaces are beryllium sheets with titanium potassium heat pipes brazed to the surface to distribute and carry the thermal energy to the deployable panels. These panels provide an additional heat rejection surface. The deployable panels are thermally coupled by means of a heat pipe to heat pipe thermal joint that is very similar to the source heat pipe to heat exchanger thermal joint. For deployment of the panels, a bellow-like heat pipe section is mounted at the tail end of the heat pipes on the fixed panel.

Table 12.7 summarizes key parameters for this design concept and Fig. 12.18 provides projected scaleability values for this configuration. A figure-of-merit of 0.85×10^{-3} K^{-1} is needed to meet 100 kW_e system requirements.

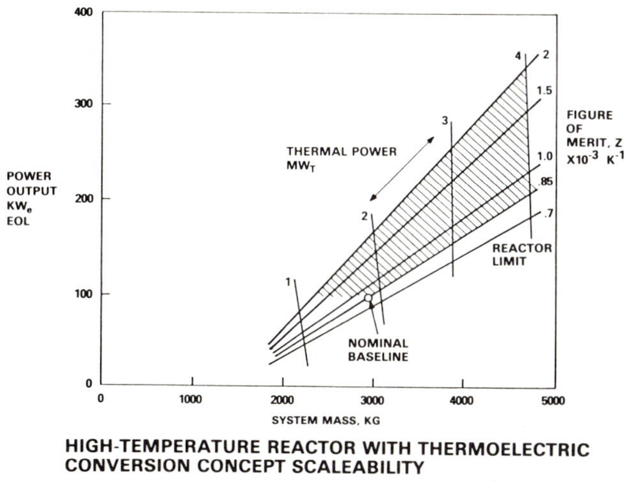

Fig. 12.18 High temperature reactor with thermoelectric conversion concept scaleability [5].

In-Core Thermionic Power System [6]

The general arrangement of the in-core thermionic space power system design is shown in Fig. 12.19. The design forms a conical frustum that is 5.8 m long with minor and major diameters of 0.7 m and 3.6 m respectively. The reactor-converter subsystem consists of the reactor, its reflector/control drums, and the neutron shield. The reactor contains the thermionic fuel elements (TFE) converters within a cylindrical vessel that is completely surrounded by control drums. Routing of the NaK primary coolant from the reactor vessel is accomplished with two sets of C-shaped, toroidal manifolds at the aft end of the reactor, each connecting to two sets of pipes that penetrate the vessel. The vessel penetrations are arranged so that the hot NaK leaves the reactor at the aft end and the cold NaK returns to the forward end. This helps minimize differential thermal expansion in the piping. The reactor is also surrounded by an array of long, thin cylindrical reservoirs that collect and retain the fission gases generated in the

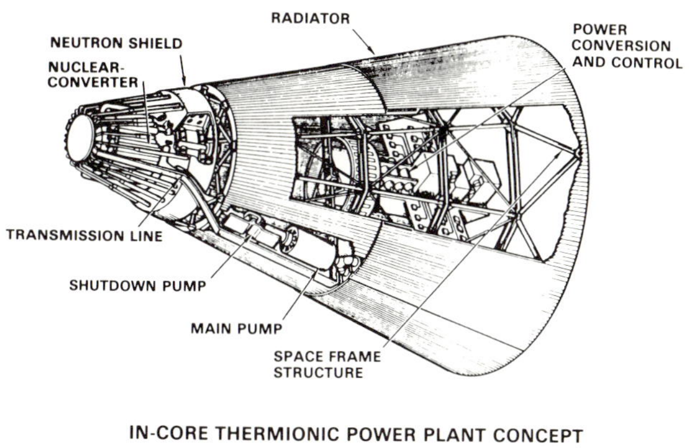

Fig. 12.19 In-core thermionic power plant concept [5].

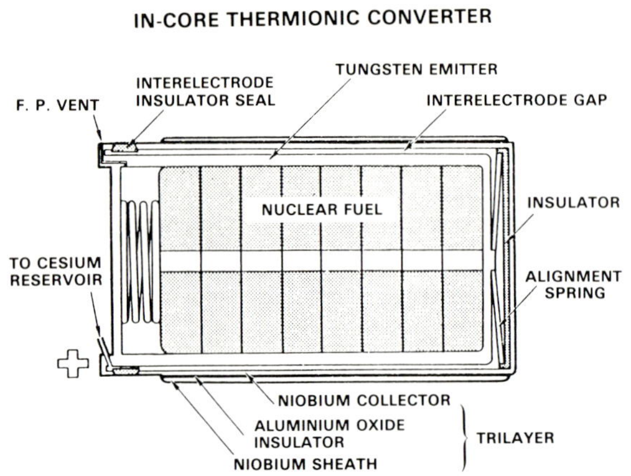

Fig. 12.20 In-core thermionic converter.

reactor core during the operating life of the system. The raw dc electric power generated by the thermionic process within the reactor is delivered to bus bars in the reactor head (at the forward end of the system). These bus bars are connected to the power conditioning module by two main transmission lines routed aft along the conical contour of the system assembly. Waste heat is removed from the primary reactor loop through the heat exchanger. The thermal energy is transferred through the heat sink heat exchanger to heat pipes that form the radiating surfaces for ultimate rejection to space.

The reactor vessel contains the reactor core, consisting of 176 TFEs, a grid plate to support TFEs at one end, a tungsten gamma shield, and eutectic NaK coolant. Each TFE is welded into the flat top head of the vessel. The TFE is allowed to move axially in the grid plate that restrains radial movement. Differential axial expansion is expected to be small, since the TFE sheath tubes and reactor vessel are both made of Nb-1% Zr alloy and their operating temperatures are nearly equal.

The TFE consists of six cells connected in series with end reflectors of BeO. Boron carbide neutron absorber is placed at both ends of the fuel element to reduce the thermal neutron flux in the coolant plenums, the gamma shield, and neutron shield. This reduces activation of the coolant, secondary gamma ray production, and nuclear heating of the lithium hydride shield.

The individual cells (see Fig. 12.20) are connected in series to build up voltage from the 0.4 individual cell voltage output. Electrical power is generated in the space between the duplex tungsten emitter and niobium collector. The electrical current output is conducted from one cell to the next through the tungsten stem of the emitter and the tantalum transition piece. The UO_2 fuel is held in place and supported during launch by a retention device designed to retract when the fuel expands on heating. The alignment spring at the base of the emitter centers the emitter in the collector, maintaining a uniform interelectrode spacing. This spring also restrains the emitter against launch vibration, preventing large displacements and limiting stresses in the thin stem at the other end of the emitter.

Fission gases are vented from the UO_2 fuel to prevent the buildup of pressures that would cause creep deformation of the tungsten emitter and close the interelectrode space. Most of the fission products released from the fuel are condensed or chemically combined on the baffles to the right of the fuel retention device, but noble gases are allowed to pass outward into the space between the outer tube and the outside of the cell. These noble fission gases are kept separate from volatile cesium by a ceramic-to-metal seal and the arrangement of passages through the emitter cap and transition piece.

The major parameters of the TFE are given in Table 12.8.

The neutron shield is a stainless steel container filled with lithium hydride. It is placed between the reactor and the remainder of the space system. Additional shielding from gamma radiation is provided within the reactor assembly by an integral tungsten plate.

Reactor control is provided by the rotation of the 20 cylindrical control drums surrounding the reactor. The drums are electrically driven in pairs by stepping motors installed on the aft side of the neutron shield. These motors are connected to drive shaft pinions which function inside sleeves that penetrate the neutron shield. Each drive pinion engages its respective control drum pair through appropriate sector gears. The stepping motors are cooled passively by radiation heat transfer to space through a gap between the neutron shield and the radiator.

The heat transport subsystem is a single loop that includes all of the NaK plumbing aft of the reactor, the heat sink heat exchanger, and the radiator. The 100 mm diameter NaK lines to and from the reactor are routed inside helical grooves in the outer surface of the neutron shield and then passed along the inside surface of the radiator to connect to the heat sink heat exchanger. The

Table 12.8 TFE/cell parameters.

Outside diameter, mm	33.38
Fueled/total length, mm	411.5/658.9
Number of cells	6
Cell length, mm	68.6
Emitter diameter/length, mm	27.94/55.8
Emitter clad thickness, mm	1.78
Fuel diameter/length, mm	24.1/50.8
Interelectrode spacing, mm	0.51, 0.25
Collector thickness, mm	0.94, 1.19

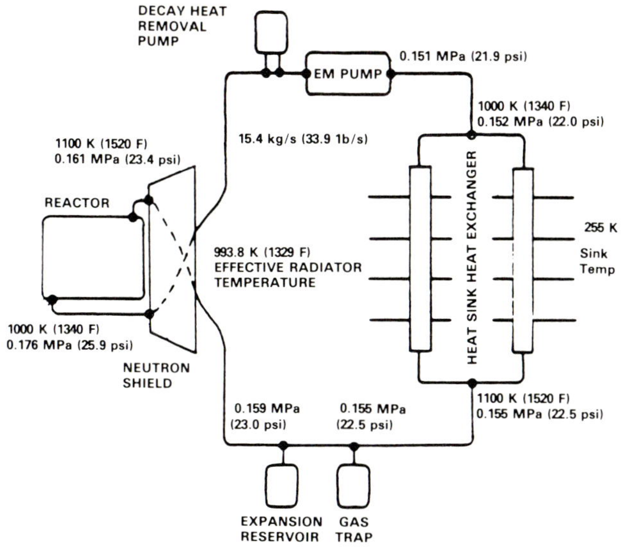

Fig. 12.21 Temperature and pressure at various points in an in-core thermionics system.

Table 12.9 In-core thermionics parameters.

Core/reactor diameter, mm	468/712
Core/reactor length, mm	411.5/940
Number of TFEs	176
Number in series connection	44
Power output at reactor terminals, kW_e	108.4
Output voltage, V	104.7
Conversion efficiency BOL/EOL	0.091/0.083
Thermal power BOL/EOL, kW_t	1,186/1,306
Coolant	NaK
Coolant inlet/outlet temperature at EOL, K	1,000/1,100
Reactor vessel material	Nb-1% Zr
Reflector material	BeO

Mass (kg)		
Nuclear converter subsystem		1,974
Reactor assembly	1,030	
Reflector/converter assembly	407	
Neutron shield assembly	537	
Heat transport subsystem		743
Coolant loop assembly	327	
HX assembly	60	
Radiator assembly	356	
Power conditioning and control subsystem		151
Space frame subsystem		128
System total		2,996

configuration of the NaK lines along the shield is helical, rather than linear, to avoid degradation of the shield performance due to neutron streaming in the pipe channels. The helical channels in the shield are also occupied by electrical transmission lines that are flattened in cross-section and are routed over the NaK lines to serve as meteoroid protection.

Electromagnetic pumping is used to circulate the NaK during normal operation and during shutdown periods when continuing NaK flow in the circuit must be maintained. Two electromagnetic pumps are provided in the cold leg of the NaK circuit: an annular linear induction pump, serving as the main pump, and a parallel thermoelectromagnetic pump (with a check valve) providing shutdown pumping capability.

The hot side of the NaK circuit contains an expansion reservoir and gas trap. The expansion reservoir accommodates the volumetric expansion of the NaK during startup and maintains sufficient static pressure in the NaK loop to suppress boiling. The gas trap, which is located in a parallel leg, utilizes a vortex NaK flow chamber separated from a gas reservoir by a semipermeable capillary barrier to remove and collect gases contained in the NaK stream.

The radiator comprises two finned-heat pipe panel assemblies that form a conical frustum when the panels are assembled on the radiator structure. The heat pipes follow the slant height of the cone and are deployed fore and aft of the heat sink heat exchanger, to which they are thermally coupled. The radiator also provides environmental protection for the equipment that it houses. The inside surface of the radiator is covered with multiple-layer foil-type insulation to maintain a thermal zone acceptable to the electrical equipment and support structure. This insulation combines with the panels of redundant heat pipes to serve as protection against meteoroids and space debris.

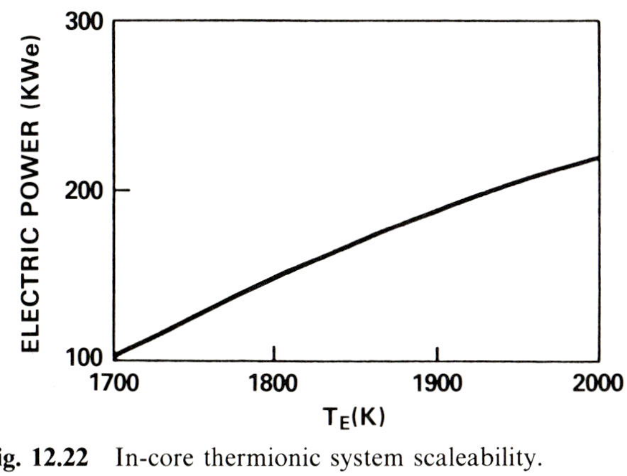

Fig. 12.22 In-core thermionic system scaleability.

Figure 12.21 summarizes key temperatures and pressures in the system. Table 12.9 provides reactor parameters and system mass. Figure 12.22 shows electric power growth versions of the in-core thermionic systems using higher emitter temperatures. Probably 2,000 K is the peak temperature for a tungsten emitter. Growth to the megawatt level is possible by adding more TFEs (see Fig. 12.23).

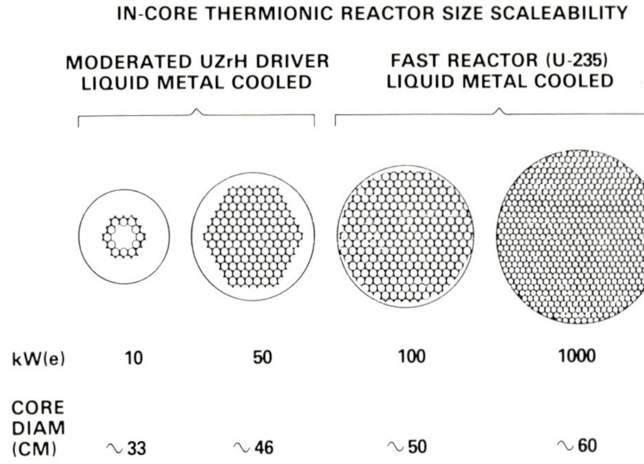

Fig. 12.23 In-core thermionic reactor size scaleability [5].

Low-Temperature Pin-Fuel Element Reactor With Stirling Cycle Conversion [5]

The general arrangement of the low-temperature pin-fuel element reactor with Stirling cycle conversion is shown in Fig. 12.24. This concept emphasizes the use of state-of-the-art stainless steel and UO_2 fuel pins with sodium as the working fluid. Temperatures can be as low as 900 K and still meet SP-100 goals. Such fuel pins have been developed for the breeder reactor program with 1,059 days of operation and 8.5 percent burnup demonstrated.

The reactor can be similar in design to the high-temperature reactor but utilizes lower temperature materials. In Fig. 12.24, the reactor is constructed as a separate module from the conversion subsystem. Four Stirling engines, each rated to deliver at least 33 kW_e, are included in this design concept. The design provides some redundancy in the event that a unit fails. Normally the Stirling engines operate at 75 percent of their rated power in producing a 100 kW_e output. Each engine contains a pair of opposed motion pistons that operate 180 degrees out of phase. This arrangement eliminates unbalanced linear momentum. Each engine receives heat from a pumped loop connected to the reactor vessel. An alternate arrangement would deliver thermal energy through an interface heat exchanger with heat pipes between the heat exchanger and engine. The heat is supplied to heater heads integral with the engine. Waste heat is removed from the cooler heads and delivered to a pumped liquid-to-heat pipe heat exchanger. The heat pipes, in turn, deliver the waste heat to the radiator, where it is then rejected to space.

The power output may be increased by adjusting the engine controls. Stroke control, employing a linear motor built into the engine displacer rod, provides one method of achieving the required control function. Fig. 12.25 illustrates how the engine power and efficiency vary as a function of stroke control. By limiting the stroke to about 80 percent of the full stroke, 75 percent of the full power can be obtained while the indicated efficiency is at the optimum value.

Fig. 12.26 provides performance curves for the Stirling system. A low temperature will meet the 100 kW_e goals. However, high power, growth systems favor combining the Stirling engines with higher temperature reactors both to minimize mass and to reduce the heat rejection surface areas.

Technology Status of SP-100 Subsystem [7]

Aerospace technology issues relate primarily to heat transport systems and the three candidate energy conversion technologies: thermoelectrics, thermionics, and Stirling engine technology.

Thermoelectric Technology

State-of-the-art thermoelectric materials, silicon-germanium (SiGe), are limited in both operating temperature and performance. System designs using SiGe have been evolved for SP-100, but only generate about 80 kW_e for current weight and volume design constraints. To meed the 100 kW_e power requirement will require addressing three key technology issues:

1. Development of high-temperature thermoelectric materials with a figure of merit in excess of $1 \times 10^{-3}\ K^{-1}$
2. Development of a lightweight converter
3. Development of a high temperature heat pipe with a high emissivity coating

The converter is radiatively coupled to the heat source, therefore, development of a stable, long-life, high-emissivity coating to enhance radiative heat transfer is required.

Samples of boron carbon (p-type) and lanthanum sulfur (n-type) thermoelectric elements have been fabricated, tested, and evaluated. Recent experiments with

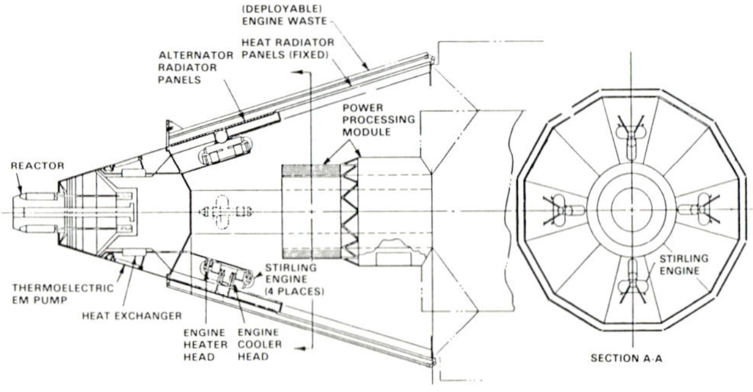

Fig. 12.24 Stirling engine conversion concept [5].

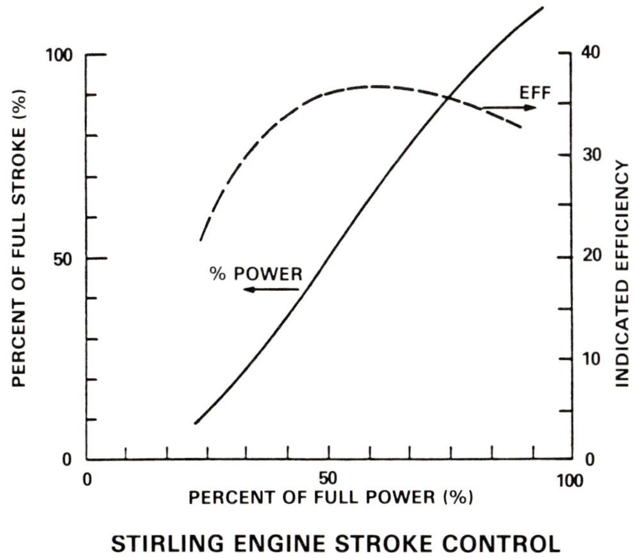

Fig. 12.25 Stirling engine stroke control.

metallurgically bonded couples indicate that with present material processing, a very-high-temperature, thermally stable (low vapor pressure) material is now producible. Material processing modifications have been identified that should enable fabrication of TE material with a figure-of-merit of greater than 1×10^{-3} K^{-1}. Couple fabrication shows compatibility between the thermoelectric materials and the hot and cold shoes. However, mismatches in thermal expansion coefficients must now be resolved.

Thermionic Technology

A considerable amount of work was performed in the 1960s and early 1970s on in-core thermionic systems. Two key technical feasibility issues for thermionic converters remain: (1) emitter instability at temperatures of 1,700 to 1,800 K (causing electrical shorting of the emitter and collector); and (2) neutron irradiation degradation of the electrical insulators (causing a short to ground). Experiments must be conducted to determine the lifetime of in-core thermionic fuel elements.

Improved thermionic performance can allow the interelectrode spacing to be increased from 0.25 to 0.50 mm while still obtaining the same power density and efficiency, thus increasing the time at which emitter distortion causes electrode shorting. One way to obtain this improved performance is to add a small quantity of oxygen to the surface of the collector. During operation, this oxygen is released to the emitter surface where it beneficially affects the emitter work function, thereby improving its performance. The UO_2-fuel in the tungsten emitter may represent an adequate supply of the oxygen necessary to obtain this performance improvement.

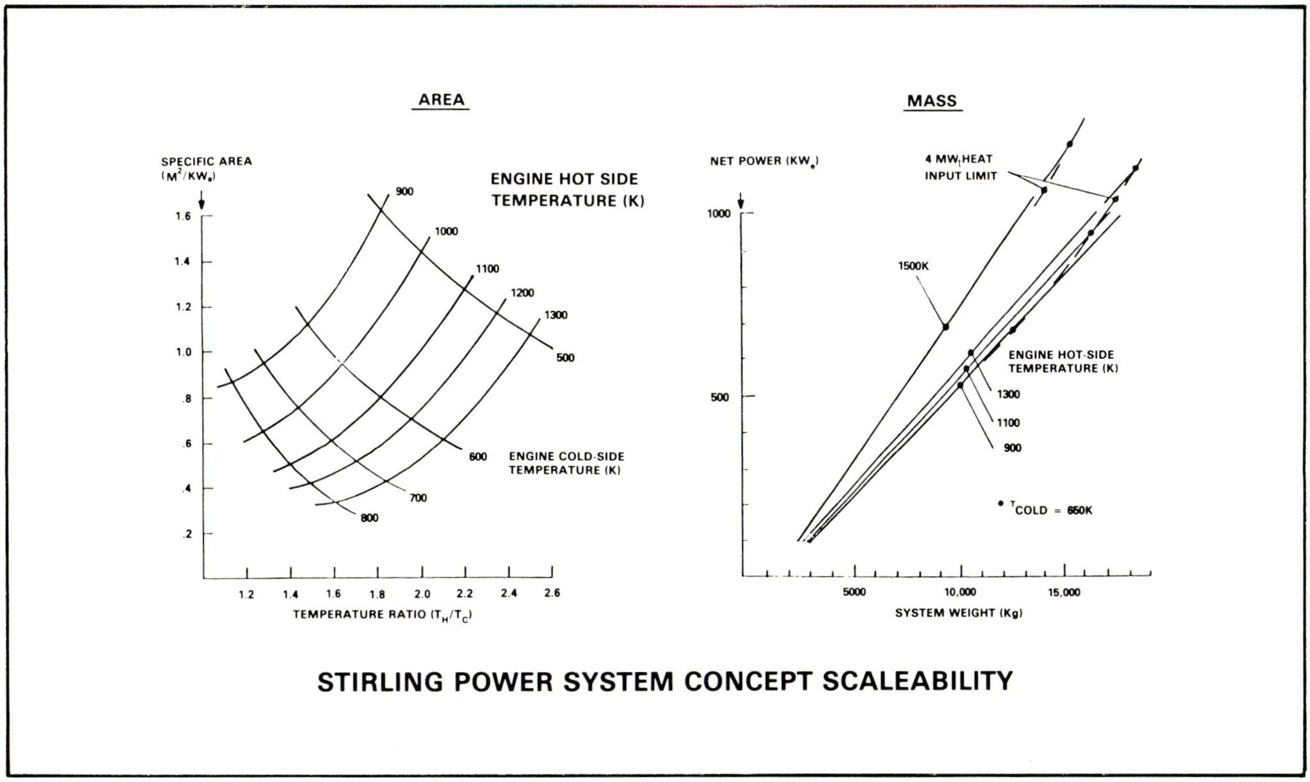

Fig. 12.26 Stirling engine power system concept scaleability [5].

Stirling Engine Technology

Free-piston linear alternator Stirling cycle engines for power conversion have the potential of operating at state-of-the-art reactor outlet temperatures. This feature can reduce the risk of long-life reactor operation and even eliminate the need for high temperature reactor development. Key technology issues for Stirling cycle engines include:

1. Demonstration of engine power at 25 kW_e range per module—a factor of ten above the largest engine built to date
2. Demonstration of high engine efficiency for heat source to heat sink temperature ratios of two
3. Fabrication of a dynamically balanced engine using internal gas bearings with an engine specific mass less than 8.0 kg/kW_e

Heat Transport Technology

Some of the concepts under consideration require heat transport systems too large for intact placement in the Space Shuttle cargo bay. Potential solutions to this problem could involve the use of separable mechanical joints with heat transfer across the interface and the use of a jointed or bendable heat pipe with continuous vapor and liquid transport across the joint. Flexible heat pipes capable of bending through 180° while at temperature have been operationally demonstrated.

Nuclear Technology

The major nuclear technology issues, mainly associated with the reactor and radiation shield subsystems, involve reactor materials compatibility, structural material development, and heat transport in high temperature systems. Another important feasibility issue, centered around the use of refractory metals in a space nuclear reactor, is the coolant/cladding/fuel chemical interaction. The rates at which oxygen or nitrogen will diffuse from the fuel (UO_2 or UN) through the cladding and into the coolant (typically Li or Na) must be quantified. Also, the consequence of a cladding failure allowing lithium to come into contact with UO_2 or UN must be determined. Thermodynamically, Mo-13% Re is stable with UO_2 as long as the oxygen-to-uranium (O/U) atom ratio is less than 2.05. Studies must be conducted to determine if this ratio remains below 2.05 throughout the reactor lifetime. Capsule and loop tests must also be performed to determine reaction rates and used to confirm analytical models for the behavior of the fuel/cladding/coolant system.

An understanding of the physical and mechanical properties of refractory alloys, both before and after irradiation, is also important for proper reactor system material selection. Materials, including Mo-Re, ASTAR 811C (a Tantalum alloy), W-Re, and Nb-Zr must be characterized as to their static and dynamic tensile behavior, fatigue behavior, fracture toughness, and high temperature creep. The ductile-to-brittle transition temperatures are important for launch environment behavior.

Liquid metals in heat pipes or pumped loops are the major means of energy transport among the various subsystems, including the reactor, the power conversion unit, and the radiator. Technical issues here are concerned with:

1. Startup from a frozen state
2. The thermal performance and lifetime of very long heat pipes
3. Transient behavior of heat pipes including possibility of dry-out
4. Lifetime and temperature limits of electromagnetic pumps

A basic feasibility issue in the use of high temperature pumped liquid metal heat transfer systems for primary heat transport from space reactors centers around the ability to restart the system from a frozen state. In space, equilibrium temperatures for most core cooling loops will be below the working fluid's solidification point at some location in the loop when the system is shut down. Restart will require melting of this "frozen" fluid. If thermoelectric electromagnetic pumps are used, startup will be complicated by the need to provide a heat source for the pump in order to initiate coolant circulation.

Because radiation coupling of large quantities of thermal energy from heat pipes requires the use of long condenser sections, the flow regime in the condenser is quite different than normally encountered, and operational parameters, such as the dynamic pressure recovery factor, are more difficult to predict. Subjecting the heat pipe to any thermal cycle (e.g., system shutdown) may induce wick dry-out and prevent the heat pipe from being restarted. This is potentially more significant with long pipes.

Materials Technology

Materials technology issues include materials for fuel cladding, reactor core structure, heat transport piping, heat exchangers and pumps, and power conversion high-temperature components such as insulators for thermoelectric and thermionic concepts. The structural materials selected must have design margin at operating temperatures and be capable of fabrication.

Refractory alloys will be required to meet the needs of the high temperature power conversion systems. Tantalum- and molybdenum-based alloys are prime candidates to meet temperature and creep strength constraints. They may also be suitable for the fuel cladding, although tantalum alloys require a diffusion barrier between the clad and the fuel. In contrast, the low temperature Stirling system concept can use existing stainless steel clad UO_2 fuel technology and super alloy or columbium alloy materials for engine construction.

Both the tantalum and molybdenum alloys could be used for high temperature static conversion systems with heat pipes. Weldability concerns, on the other hand, imply that molybdenum alloys may be less suitable for the reactor structure. Further research is consequently needed to determine the weldability of the molybdenum alloys and demonstrate that they can be used for piping throughout a particular energy conversion system or for the reactor structure itself.

References

1. "SP-100 Conceptual Design Description, No. 1," Jet Propulsion Laboratory, Pasadena, Calif., June 1982.
2. W. A. Ranken, "Status of High-Temperature Heat Pipe Technology," National Research Council Symposium on Advanced Reactor Systems, Washington, D.C., 20545, 15–17 November 1982.
3. Dale Dutt et al., "SP-100 Fuel Panel Report," Westinghouse, Hanford, WN, October 1982.
4. J. Ambrus, W. Wright, D. Bunch, "Review of the Tri-Agency Space Nuclear Reactor Power System Technology Program," 19th Intersociety Energy Conversion Engineering Conference, 19–24 August 1984.
5. "SP-100 Conceptual Design Study," General Electric Co., Valley Forge, Pa., Doc. No. 83SDS4262, 7 December 1983.
6. "Conceptual Design Studies in Support of SP-100 Program," GA Technologies and Martin Marietta, La Jolla, Calif., Report No. GA-C17354, 9 December 1983.
7. H. S. Davis, "Progress in Space Nuclear Reactor Power Systems Technology Development: The SP-100 Program," AIAA/NASA Space Systems Technology Conference, Calif., June 1984.

Bibliography

J. Angelo, Jr. and D. Buden, "Space Nuclear Power—A Strategy for Tomorrow," AIAA-81-0450, Proc. of 2nd AIAA Conference on Large Space Platforms, 2–4 February 1981, San Diego, Calif.

J. Angelo, Jr. and D. Buden, "Space Nuclear Reactors—Energy Gateway Into the Next Millennium," XXXII International Astronautical Congress, Rome, Italy, 6–12 September 1981.

D. Buden et al., "Selection of Power Plant Elements for Future Reactor Space Electric Power Systems," Los Alamos National Laboratory Report No. LA-7858, September 1979.

D. Buden, "The Acceptability of Reactors in Space," Los Alamos National Laboratory Report No. LA-8724-MS, April 1981.

D. Buden, "100-kW_e Nuclear Space Electric Power Source," *Electric Propulsion and Its Applications to Space Missions*, edited by Robert C. Finke, vol. 79 of Progress in Astronautics and Aeronautics, 1981.

D. Buden, W. A. Ranken, and D. R. Koenig, "Space Nuclear Reactor Power Plants," Los Alamos National Laboratory Report No. LA-8223-MS, January 1980.

D. Buden and J. Angelo, Jr., "Reactors for Nuclear Electric Propulsion," AIAA-81-0697, Proc. of AIAA/JSASS/DGLR 15th International Electric Propulsion Conference, 21–23 April 1981, Las Vegas, Nevada.

K. C. Cooper and R. G. Palmer, "System Tradeoffs in Space Reactor Design," 15th Intersociety Energy Conversion Engineering Conference, Seattle, Wash., 18–22 August 1980.

Gerald Fly, "A Compact, Efficient Thermoelectric Module for a Space Reactor," 16th Intersociety Energy Conversion Engineering Conference, Atlanta, Ga., 9–14 August 1981.

S. P. Girrens, K. L. Meier, and J. M. Dickinson, "Titanium Heat Pipes for Space Power Systems," 15th Intersociety Energy Conversion Engineering Conference, Seattle, Wash. 18–22 August 1980.

C. Grando, M. Montgomery, and A. Strasser, "Unrestrained Swelling and Fission Gas Release of Fast Reactor Fuels," Fast Reactor Fuel Element Technology, Edited by Ruth Tarmaker, American Nuclear Society, 13–15 April 1971, pp. 771–784.

Allen Harper, "Technical Report Study of Reactor Brayton Power Systems for Nuclear Electric Spacecraft," Air Research Manufacturing Co. of Arizona, Report No. 31-3321, Phoenix, Ariz. September 1979.

M. H. Kaplan, R. M. Trn, and D. Buden, "A Nuclear Electric Orbital Transfer Vehicle for the Shuttle Era," Paper No. 79-2109, Princeton/AIAA/DGLR 14th International Electric Propulsion Conference, 30 October–1 November 1979, Princeton, N.J.

Donald L. Keller, "Progress on Development of Fuels and Technology for Advanced Reactors During July 1970 Through June 1971," Battelle Columbus Laboratories, Report No. BMI-1918, Columbus, Ohio, July 1971.

J. E. Kemme, "Heat Pipe Design Considerations," Eleventh Heat Transfer Conference, 3–6 August 1969, Minneapolis, Minn.

D. R. Koenig and W. A. Ranken, "Reactor Design for Nuclear Electric Propulsion," 14th Intersociety Energy Conversion Engineering Conference, Boston, Mass., 5–10 August 1979.

L. B. Lundberg, "A Critical Evaluation of Molybdenum and Its Alloys for Use in Space Reactor Core Heat Pipes," Los Alamos National Laboratory Report No. LA-8685-MS, January 1981.

L. B. Lundberg, "An Evaluation of Molybdenum and Its Alloys," AIAA-81-1105, Proc. of AIAA 16th Thermophysics Conference, 23–25 June 1981, Palo Alto, Calif.

K. L. Meier, H. E. Martinez, and J. E. Runyan, "Development of Space Reactor Core Heat Pipes," 16th Intersociety Energy Conversion Engineering Conference, 9–14 August 1981, Atlanta, Ga.

F. C. Prenger and J. E. Kemme, "Performance Limits for Liquid-Metal Heat Pipes Containing Long Adiabatic Sections," 14th Intersociety Energy Conversion Engineering Conference, 5–10 August 1979, Boston, Mass.

F. C. Prenger, "Performance Predictions and Measurements for Space Power System Heat Pipes," Fourth International Heat Pipe Conference, London, England, 7–10 September 1981.

W. A. Ranken and D. R. Koenig, "Baseline Design of the Thermoelectric Reactor Space Power System," 14th Intersociety Energy Conversion Engineering Conference, 5–10 August 1979, Boston, Mass.

W. A. Ranken, "Experimental Results for Space Nuclear Power Plant Design," 15th Intersociety Energy Conversion Engineering Conference, Seattle, Wash., 18–22 August 1980.

W. A. Ranken, "Heat Pipe Development for the SPAR Space Power System," IVth International Heat Pipe Conference, London, England, 7–9 September 1981.

H. Zimmerman, "Investigations on Swelling and Fission Gas Behavior in Uranium Dioxide," Journal of Nuclear Materials 75, 1978, pp. 154–161.

Chapter 13

Space Nuclear Power Plant Design Considerations

Introduction

In response to the question, "What space nuclear power technology is best?" there is no simple answer. As shown in preceding chapters, a wide range of possible combinations of nuclear energy sources, power conversion equipment, and heat rejection techniques exists which can satisfy a given set of mission requirements. In addition, such factors as power level, mass, lifetime, reliability, development time, and cost must be optimally structured within the framework of aerospace nuclear safety requirements. Radioisotope power supplies appear most suitable for applications needing a few kilowatts electric or less. As described in Chapter 8, these systems have demonstrated technologies that support safe, long-life, high reliability operations in a variety of hostile extraterrestrial environments. Space nuclear reactors, on the other hand, become most appropriate for missions requiring tens of kilowatts of electric power and beyond. As described in Chapters 9 through 12, there are a number of very interesting reactor technology pathways that have been, are now being, or could be followed. The selection of an optimal space nuclear reactor design strategy is, however, not as simple and potentially clearcut as are the choices associated with RTG technology. For example, at different power levels, the mass-dominant component of a reactor system will change. At lower power levels (tens to a few 100 kW_e) the reactor and its radiation shield are the most mass-significant components in the power plant. But, as mission power levels reach and exceed megawatt proportions, the mass of the heat rejection system becomes dominant. Furthermore, regardless of the power level, a systematic design approach is needed in selecting reactor concepts, components, and subsystems, to minimize development risks, time and costs, while still satisfying mission requirements and aerospace nuclear safety criteria.

In this chapter we discuss the fundamental aspects of contemporary aerospace nuclear safety and emphasize its all-pervading influence on the design of any space nuclear power system, radioisotope or reactor. We also summarize the status of space reactor technologies, primarily as a function of mission power level. The status of RTG technology was treated extensively in Chapter 8 and will not be repeated here. Material presented in this chapter should help you develop a logical, technically sound framework for assessing the role and implementing the use of nuclear energy in space missions of tomorrow.

Aerospace Nuclear Safety

General Safety Philosophy

The principal safety guideline associated with the use of nuclear energy in space is to minimize the likelihood or consequences of interactions of radioactive materials with the terrestrial biosphere, so that any potential exposure levels are within the limits of internationally established standards. Stringent design factors and well-demonstrated operational procedures are used to protect both people and the Earth's environment under normal flight conditions and under any credible accident scenarios. Contemporary aerospace nuclear safety design philosophy for radioisotope heat sources involves the selection of the nuclear fuel form that has minimal biological effect and its containment under all credible accident scenarios. This safety philosophy has evolved over a period of time from the early approach of reentry burnup of the radioisotope fuel, to reentry dispersal, to intact reentry, to intact reentry and impact. (Reentry burnup involves the disintegration and vaporization of a spacecraft, satellite, or other manmade aerospace object as a result of aerodynamic heating by a planetary atmosphere.) At the same time, radioisotope fuels have also been made more resistant to the accident environment, evolving successively from ^{238}Pu metal, to $^{238}PuO_2$ microspheres, to ^{238}Pu solid solution molybdenum cermet discs and pressed PuO_2 (PPO) [1,2].

For space nuclear reactors, current aerospace nuclear safety philosophy requires that the reactor remain subcritical in all credible accident environments [3–5]. This prevents the generation of fission products and the release of radioactivity, before the reactor and its payload have been placed in a permitted orbit. Consequently, space nuclear reactors are launched in a subcritical, "cold clean" condition. The term "cold clean" means that there is no significant fission product inventory in the reactor core.

Aerospace Nuclear Incidents

To date, the United States has launched 19 space missions with radioisotope power sources. After four successful RTG launches, the Transit-5BN-3 failed to achieve orbit on 21 April 1964 due to a launch vehicle abort that was traced to conflicting guidance controller signals occurring during ascent. This guidance malfunction caused the launch vehicle to pitch improperly and orbital insertion of the payload was not achieved. Despite the ascent abort, however, the SNAP-9A RTG on board the spacecraft performed as designed for a launch/mission abort and it burned up on reentry into the Earth's atmosphere. The RTG's plutonium-238 metal fuel was injected into the atmosphere in the Southern Hemisphere at an altitude between 45 and 60 kilometers. Airborne and surface sampling was initiated following this abort and four months later plutonium dioxide (PuO_2) was first positively identified at an altitude of 32.9 km at 35 degrees south latitude. It was subsequently concluded that, as designed, the radioisotope fuel had completely burned up during reentry over the West Indian Ocean north of Madagascar [2, 6].

RECOVERY OF SNAP 19

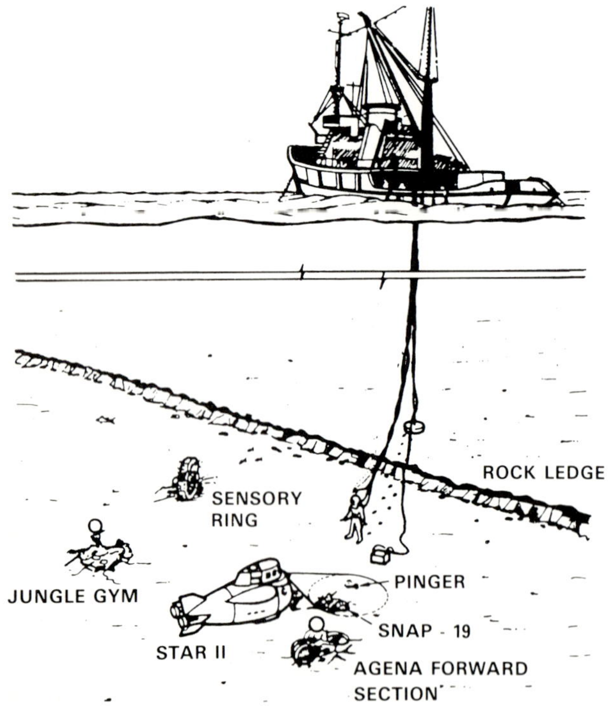

Fig. 13.1 Recovery of SNAP-19 RTG capsule. There were no adverse impacts on biosphere. *Courtesy of NASA.*

The second U.S. RTG aerospace nuclear incident occurred on 18 May 1968 and involved a SNAP-19 generator on board the Nimbus B-1 meteorological spacecraft. In this case, erratic behavior of the launch vehicle forced its intentional destruction by the Range Safety Officer when the vehicle and its payload were at an altitude of 30 km and travelling downrange from the Vandenberg Air Force Base launch site. Tracking data placed the impact point of the launch vehicle and spacecraft debris in the Santa Barbara Channel about 5 km north of San Miguel Island off the California coast. Here, the water depth is about 90 meters. The SNAP-19 generator was designed for intact reentry and had been tested in a marine environment. Since data indicated that the radioisotope fuel capsules were still intact and that they posed no immediate environmental or health problem, there was no immediate urgency to recover them from the ocean floor. In fact, the SNAP-19 generator was recovered from the Pacific Ocean five months later (see Fig. 13.1). The entire incident verified that the radioisotope fuel capsules of this design could remain in a marine environment for long periods of time following a launch/mission abort without concern for fuel release. Post-incident examination of the plutonium fuel capsules revealed that no detrimental effects were suffered from the destruction of the launch vehicle, impact in the ocean, or nearly five months residency on the ocean bottom. The graphite ablators surrounding the capsule were also intact [2, 6].

A third RTG aerospace nuclear incident involved the aborted Apollo 13 mission to the Moon in April 1970. In this event, the SNAP-27 fuel capsule, containing 44,500 curies of plutonium-238 oxide microspheres, reentered the Earth's atmosphere along with the *Aquarius* Lunar Module (LM) which had served as a translunar trajectory lifeboat for the in-flight stranded Apollo 13 astronauts. Enroute to the Moon, an oxygen tank had exploded in the Service Module. Following the near fatal explosion, astronauts Lowell, Swigert, and Haise powered up the *Aquarius*, battened down the crippled command ship *Odyssey*, and continued on a course around the Moon and back to Earth. For more than 90 hours these three men rode a lunar landing craft designed to accommodate just two astronauts for two days. Then, approaching Earth, they again fired the *Aquarius* (LM) engine to thread themselves carefully through a narrow reentry corridor, shifted to the lifeless *Odyssey* command module (CM), and cut loose both the damaged Service Module and their LM lifeboat. The three astronauts were recovered within 45 minutes of splashdown in the Pacific Ocean [7].

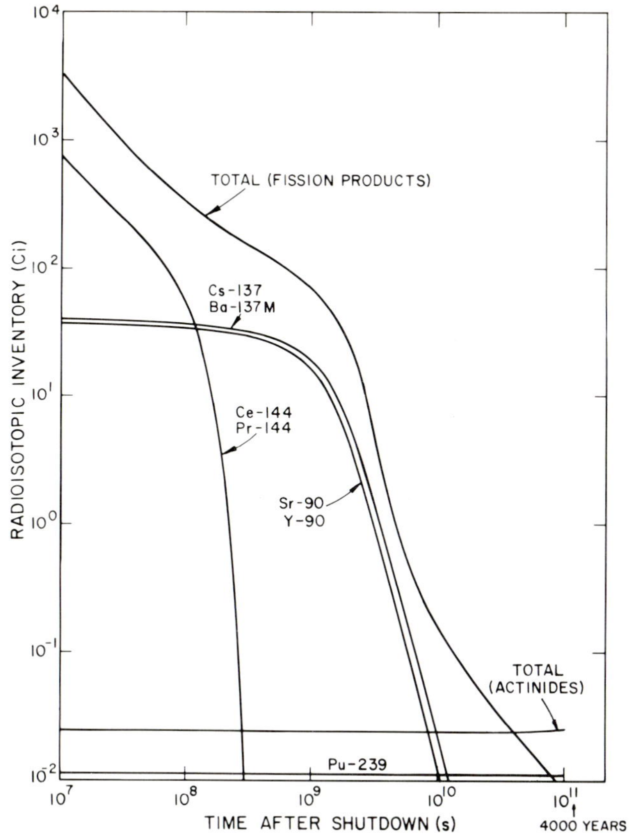

Fig. 13.2 Calculated radioisotope inventory of SNAP-10A reactor as a function of time after shutdown. The 34 kW_t reactor was launched in 1965, operated for 43 days, and currently resides in a shutdown mode in a 4000-year orbit [5].

The Apollo 13 SNAP-27 fuel capsule, on the other hand, was contained in a graphite fuel cask attached to the LM. Both reentered at approximately 122 km above the South Pacific Ocean. Atmospheric monitoring at several high and low altitudes in the area indicated that no nuclear fuel was released. Consequently, it was assumed that the SNAP-27 capsule impacted intact, as designed, in the deep ocean south of the Fiji Islands and now resides near the Tonga Trench in some 6 to 9 kilometers of water. There was no observable adverse effect on the biosphere as a consequence of this incident—indicating again the efficacy of the U.S. aerospace nuclear safety program.

The only U.S. reactor flown in space, SNAP-10A, was launched on 3 April 1965. This system was placed in a planned 4,000 year lifetime orbit. After successful startup and operation at power in its permitted orbit for 43 days, a series of spurious electronic signals accidentally shut down the SNAP-10A reactor. Because of aerospace safety design features, it was not possible to restart the reactor system using groundbased commands. Consequently, this reactor remains quiescent in its 4,000 year orbit. Figure 13.2 provides a calculated radioactivity inventory for the SNAP-10A core as a function of time after shutdown. Fifteen years after shutdown the fission product inventory is less than 100 curies, while after 100 years the core radioactivity level will be less than 0.1 curie. When the SNAP-10A finally reenters the Earth's atmosphere several millennia from now, the radioactivity level of its core will be negligible. Then, as designed, it will harmlessly disperse itself during a final fiery plunge through the Earth's upper atmosphere [2, 3, 5, 6].

The Soviet Union (U.S.S.R.) routinely flies spacecraft in low Earth orbits that include a nuclear reactor as a power source [8, 9]. When their low orbit mission is completed, the Soviets use a boost system to increase the orbital altitude of the reactor. By increasing the orbital altitude and, therefore, the orbital lifetime of a shutdown reactor, sufficient time is permitted before eventual reentry for the accumulated radioactivity in the core to decrease to acceptable levels. In the event that the nuclear power system falls from its operational (working) orbit after a failure of the boost system, the Soviet aerospace nuclear safety philosophy includes a core dispersal mode. Soviet space reactors are reported to be equipped with a backup safety system that will disperse the reactor core in such a way so that the radiation dose to people living in any potentially contaminated area will not exceed 0.5 rem (5 mSv) during the first year after reentry and surface contamination [8].

Soviet aerospace nuclear safety design was put to the test in a major accident in 1978—an international incident that focused world attention on the use of space nuclear power sources. On 24 January 1978, Cosmos 954 entered the Earth's atmosphere over Canada's Northwest Territories. A massive airborne and ground search and recovery program was undertaken under the direction of the Canadian Atomic Energy Control Board. This effort was named Operation Morning Light. A 124,000 square kilometer area of the Northwest Territories as well as selected portions of northern Alberta and Saskatchewan were identified for search. Debris was found in a 600 km tract stretching southward from Great Slave Lake. The Soviets stated that the reactor was designed to burn up on atmospheric reentry [8–10]. Table 13.1 summarizes the Cosmos 954 debris that was recovered. Although no large fuel particles were found, several large fragments with high radioactivity levels were discovered. For example, the four steel plate fragments appeared to be part of the periphery of the reactor core and may have been cylindrical in shape. Radioactivity levels in the beryllium pieces, believed to be part of the reflector, were mainly attributed to the activation products associated with impurities in the material. Some recovered fission products were believed to be the remnants of a cladding material.

Table 13.1 Summary of COSMOS 954 debris recovered in Canada.

Material	Size, Weight	Radiation Level When Found
Large fragments		
4 steel plate fragments	Largest 225 mm × 75 mm × 6.5 mm weighing 272 g	To 200 R/h on contact
41 beryllium rods (some with niobium sheaths)	100 mm × 20 mm diam. About 51 g each	600 mR/h to 100 R/h on contact; 30 to 150 mR/h at 1 meter
6 beryllium cylinders	250 mm × 100 mm diam. 3.6 kg each	5 to 15 R/h on contact 40 to 800 mR/h at 1 meter
Tubes, rods, plate (the "antlers")	6 rods about 1 meter long attached to plate. 20 kg.	To 15 R/h on contact Variable over the debris
Steel tube (the "stovepipe")	About 500 mm long, 360 mm diam. 2.5 mm wall. 18.2 kg	Non-radioactive
Other chunks, flakes, slivers	Variable	Most 10 to 30 R/h on contact one to 500 R/h
Small particles		
About 4000 recovered	a) spherical, 1 mm to less than 0.1 mm diam. Mass 5 mg to less than 100 μg	Most in the range 10 to 100 mR/h at 1 meter
	b) irregular, or flaky. Some several mm on a side and very thin and friable	

The largest fragment of Cosmos 954, actually two pieces, was a complex structure nicknamed the "antlers" (see Fig. 13.3). This structure consisted of six legs or tubes, three still attached to a circular base plate and three others that apparently broke loose on impact. The complex structure of the entire assembly appeared consistent with that of a reactor control function. The boron-rich link on the movable rods suggested that these were reactor control or shutoff rods and a large amount of lithium hydride would provide a neutron shield for the payload.

The small particles found appeared to represent the reactor fuel. These were scattered over 100,000 square kilometers. Their mean particle diameter as a function of distance is shown in Fig. 13.4. Near the reentry trajectory, these particles were about 1 mm in diameter, while those found 200 km to the south were typically 0.2 mm. There appeared to be two types of small particles. The majority were fuel particles that had a crystalline outer layer, which was analyzed as being uranium dioxide. A metallic core beneath this surface was apparently made up of a variable uranium/molybdenum alloy containing 5 to 15 percent or more molybdenum. The uranium was found to be nearly 90 percent enriched uranium-235. A second category of particles were low density, often flakey, thin, curved, or twisted. These particles contained some uranium alloyed with stainless steel or in some cases, niobium and tantalum.

Except for those small particles that drifted southward, it appears that most debris fell on Great Slave Lake or on unpopulated and little frequented areas to the northeast. There is reasonable assurance that all accessible, significantly sized radioactive fragments from Cosmos 954 were located and removed in the cleanup operations. The radioactivity inventory indicated that about 20 percent (or some 4 kg) of fuel had come back to Earth as of 1 April 1978. Various estimates made on the actual particles recovered in specific sections of the search area indicate that distribution of these particles may be as high as several hundred per square kilometer. The average was one particle every 4,000 square meters. The effects of the debris identified or observed on any part of the natural environment may be considered insignificant [10].

International Aspects of Space Nuclear Power

The reentry of COSMOS 954 focused world attention on the issues surrounding the use of nuclear power sources in space. In 1978, the Scientific and Technical Subcommittee of the United Nations Committee On the Peaceful Uses Of Outer Space (COPUOS) established a forum for these discussions in its fifteenth session [11]. Its Working Group on the Use of Nuclear Power Sources in Outer Space met from 1978 to 1981 and then reconvened in 1984. Since 1981, the Legal Subcommittee of the COPUOS has also addressed the issue of space nuclear power sources. Both the United States and the Soviet Union were major participants in these meetings and discussions. As major launching states, position papers from these two nations reflect contemporary aerospace nuclear safety practices.

Summary of U.S. Position Papers

In the position papers from the United States, it was pointed out that from the beginning of the U.S. space nuclear power program great emphasis has been placed

(A)

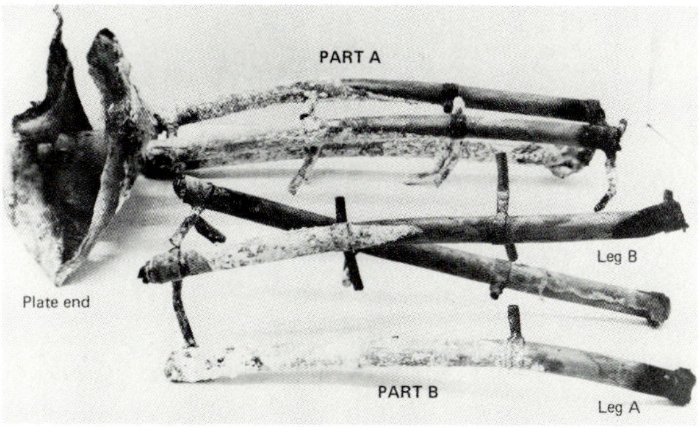

(B)

(C)

Fig. 13.3 COSMOS 954 debris. (A) Shows the "antlers" as discovered in the ice and snow on the Thelon River, Canada (29 Jan 1978); (B) shows the "antlers" as photographed at Canada's Whiteshell Nuclear Research Establishment (the white coating is hydrated lithium hydroxide); (C) is a cross section of one of the legs of the "antlers" [*10*].

on the safety of people and the protection of the environment [12]. The operational philosophy adopted for orbital missions requires that the normal lifetime in space be long enough to permit radioactive decay of either the radioisotope fuel or reactor core fission product inventory to a safe level prior to reentry into the Earth's biosphere. It is recognized, however, that space missions frequently use parking orbits from which to launch a spacecraft on a deep space or high Earth orbit trajectory.

Stringent design and operational measures are used to minimize the potential interactions of the radioactive materials with the global populace and to keep any such exposure levels within limits established by international standards [13]. For radioisotope power sources, this leads to a design philosophy of containment, immobilization, and recovery. For reactors prior to startup, the emphasis is on maintaining a subcritical configuration under all credible accident environments and to avoid the generation and/or release of any fission products. Therefore, one aerospace nuclear safety approach for a reactor system is to launch an appropriately shielded reactor in a subcritical configuration; to design it so as to prevent criticality at or after terrestrial impact, should the subcritical reactor reenter before startup; and to delay startup until the system has achieved an orbit of sufficient lifetime to provide for fission product decay. This design philosophy ensures minimal exposure of individuals to the nuclear material and results in general population exposure levels that conform to recommended international radiation protection standards (see Chapter 7). If the reactor is intended for use in a short duration orbit, the safety assessment would include the duration of reactor operation, the duration of the orbit, and a probabilistic risk analysis concerning the type of reentry and ultimate radionuclide disposition in the biosphere. Even the most reliable systems have associated with them a finite failure probability. Consequently, accident probability analyses, population dose assessments, and health effect evaluations must be undertaken in the development of flight approval criteria and the assessment of mission benefits versus potential risks.

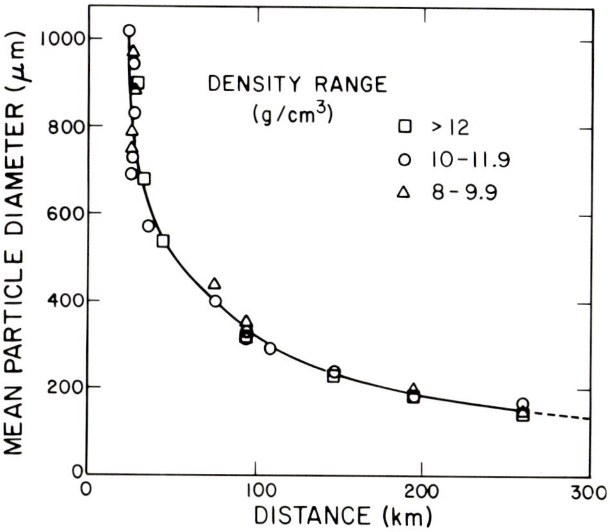

Fig. 13.4 Mean particle diameter of debris from COSMOS 954 versus distance from reentry trajectory [*10*].

The United States safety guidelines for plutonium-238 radioisotope heat sources were provided to the United Nations Working Group [13]. These safety objectives are to contain the solid radioactive materials and to limit the interactions of the radioactive materials with human beings and their environment. For normal operations, containment and interaction requirements are absolute, while for potential accident situations, the degree of interaction will be reduced to a practical level and the evaluation shall be based on probabilistic analyses. Under normal operating conditions, the heat source and its associated power system components are designed to be compatible with the radiation exposure limits set forth in Table 13.2. In controlled areas, these limits can be met by controlling exposure time, while in uncontrolled areas, an external shield in the form of a shipping container can be used.

Under accident conditions, the radioisotope heat source and/or its fuel containers are designed to limit individual external doses to 5 rem (50 mSv) per person per year at a reference distance of 10 meters from the centerline of the fuel or heat source fuel matrix. If a fuel release occurs, the internal dose to lung, bone, liver, or any critical body organ of any individual in the exposed population should not exceed 15 rem (150 mSv) per year at a distance of 10 meters in any direction from the point of release.

The radioisotope fuel form and its solid radioactive products are designed to be contained under all normal operating and accident conditions. In postulated accident scenarios, however, the heat source is designed and evaluated to the following limiting release objectives: in accident environments where the heat source is within the biosphere or lithosphere, the airborne release of fuel and/or its daughter products shall be limited to 0.1 picocurie per curie contained in the heat source and 0.01 microcurie per launch regardless of the fuel inventory involved, considering the fuel release probability to be unity.

For accidents where the radioisotope heat source is in the hydrosphere, the fuel is designed to be contained for at least three years, after which the release shall be controlled by the dissolution of the fuel and it shall be limited to a plutonium concentration of less than 3×10^{-5} curie per milliliter at a distance of 10 meters in any direction from the release point.

In addition, the radioisotope heat source is designed so that the subcriticality of the fuel is assured under all normal and accident conditions.

While detailed United States safety criteria for reactors were not expressly provided in the United Nations Working Group papers, this information can be readily found in reference 4. These aerospace nuclear safety design requirements include:

1. The reactor shall be designed to remain subcritical if immersed in water or other fluids, such as liquid rocket propellants, to which it may be exposed.
2. The reactor shall have a significantly effective negative power coefficient of reactivity.
3. The reactor shall be designed so that no credible launch pad accident, ascent, abort, or reentry from space resulting in Earth impact could result in a critical or supercritical geometry.
4. The reactor shall not be operated (except for zero power testing that yields negligible radioactivity at the time of launch) until a stable orbit or flight path is achieved and must have a reboost capability from low Earth orbit if it is operated in that orbit.
5. Two independent systems shall be provided to reduce reactivity to a subcritical state and they shall not be subject to a common failure mode.
6. The reactor shall be designed to ensure that sufficiently independent shutdown heat removal paths are available to provide decay heat removal.
7. The unirradiated fuel shall pose no significant environmental hazard.
8. The reactor shall remain subcritical under the environmental conditions of postulated launch vehicle explosions or range safety destruct actions.

Table 13.2 Normal space mission operations radiation exposure limits.

Individuals in Controlled Areas:

Type of Exposure	Condition	Dose (rem)
Whole body, head and trunk, active blood-forming organs, gonads, or lens of eye	Accumulated dose	5(N-18)*
	Calendar Quarter	3
Skin, thyroid, and bone	Year	30
	Calendar Quarter	10
Hands and forearms, feet and ankles	Year	75
	Calendar Quarter	25
Other organs	Year	15
	Calendar Quarter	5

*Where N equals age in years at next birthday.

Individuals in Uncontrolled Areas:

Type of Exposure	Annual dose (in rem) to critical individuals at points of maximum probable exposure	Average annual dose (in rem) to a suitable sample of the exposed population
Whole body, gonads, or bone marrow	0.5	0.17
Other organs	1.5	0.5

Note: 1 rem = 10 mSv (see Chapter 7)

A reentry core dispersal capability shall be provided for shortlived orbit missions. A shortlived orbit can be defined as one with an altitude such that reentry of the reactor would occur in less than 300 years. This capability shall be provided such that the reentry environment of the reactor (separate from the launch vehicle) will initiate and cause the required core dispersal with no active systems required to function.

In addition, an orbital altitude boost system shall be provided by the mission agency for shortlived missions. This boost system shall be capable of moving the reactor to a high orbit, following completion of its mission or upon mission failure.

Summary of the Soviet Position Paper

The Soviet Union position paper stated that progress in the investigation of outer space is "unthinkable without the use of nuclear power sources for thermal and electrical energy." Space research was considered to provide enormous benefits for all mankind and progress was impossible without the utilization of nuclear energy on board space vehicles. Consequently, equipping Soviet space vehicles with nuclear power sources is considered "socially justified" [8].

Nuclear power sources have recognizable dangers to the public in the event of an uncontrolled fall to the surface of the Earth. The main principle followed in the design of Soviet nuclear power sources is to preclude any uncontrolled return of the source to the surface of the Earth in the event of a failure after the spacecraft has been placed in orbit. For reactors, the system is made critical only after a working orbit has been reached. A system is also provided to boost the spacecraft to a higher, long-duration orbit. The boost vehicle would raise the reactor/spacecraft after the programmed flight is completed or if a malfunction occurs. If the orbital boost system fails to work and reentry into the Earth's atmosphere occurs, however, a radiation danger may exist. Then, a backup radiation safety system will ensure dispersion of the reactor core into particles so small that after they fall to Earth, the level of radioactive contamination of the land surface involves only a minimal risk to the population. International Commission on Radiological Protection (ICRP) standards are identified as acceptable dose-equivalent limits for population exposures, i.e., 0.5 rem (5 mSv) per year per individual (see Chapter 7) [8, 14].

U.S. Aerospace Nuclear Safety Review Process

Every United States nuclear-fueled power supply that is considered for use in space must undergo a safety review process. This process establishes that the potential risks associated with the nuclear energy source use are commensurate with the anticipated mission benefits. A formalized review process has been developed for evaluating the safety aspects of nuclear system launches. At the center of this process is the Interagency Nuclear Safety Review Panel (INSRP) comprised of representatives from the Department of Energy (DOE), the National Aeronautics and Space Administration (NASA), and the Department of Defense (DOD). These agencies are responsible for evaluating mission safety for each launch. DOD and NASA per-

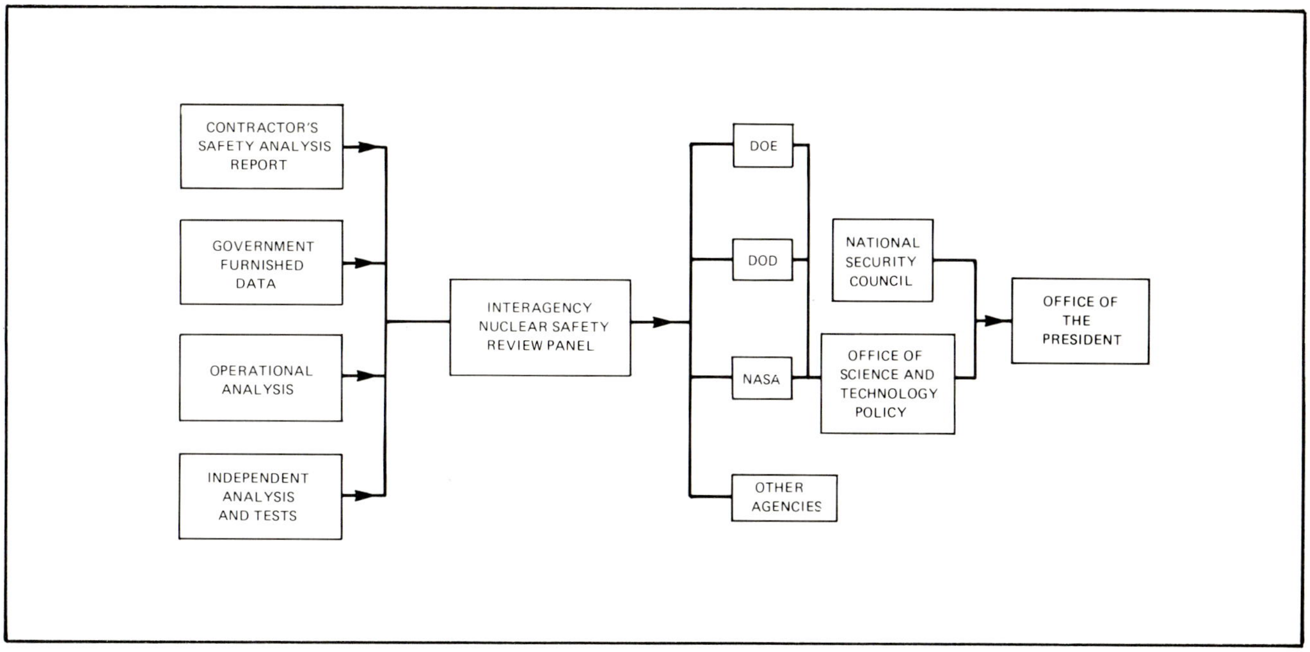

Fig. 13.5 United States safety review and launch approval process [2].

sonnel are involved because these two government agencies have safety responsibilities and expertise, both as launching organizations and as user organizations of space nuclear power. DOE has statutory responsibility for the safety of space nuclear power systems.

The evaluation process consists of the following elements:

1. The lead or sponsoring agency directs the manufacturer of the nuclear power system to write a Preliminary Safety Analysis Report (PSAR) or Updated Safety Analysis Report (USAR) describing all aspects of mission safety.
2. The particular version of the Safety Analysis Report is distributed to the members of the INSRP and each member agency conducts its own review and critique of the PSAR or USAR.
3. A meeting of the INSRP is held with member agencies and their mission hardware contractors (launch vehicle, nuclear fuel, power system, space vehicle, etc.) in attendance. The results of the independent reviews are presented and discussed at this meeting. Action items are generated to resolve any open questions or issues.
4. The power system contractor, with input from other agencies/contractors responsible for action items, writes a Final Safety Analysis Report (FSAR) taking into account the FSAR critiques and any appropriate new information.
5. Elements 2 and 3 are repeated with the Final Safety Analysis Report (FSAR).
6. The INSRP generates a Safety Evaluation Report (SER) that accompanies the request for Presidential approval of the launch.

Figure 13.5 shows the generalized sequence of events in this flight safety review process [2, 4, 15]. Because safety features are designed into U.S. nuclear power sources from the very beginning, this safety review process is actually an integral part of the overall flight system development and in no way constrains the overall mission schedule.

Reactor Technology Considerations

Because of the potentially unique and demanding requirements that will be imposed by future space missions, imaginative design and development efforts must also be undertaken to successfully create tomorrow's advanced space nuclear reactors. Some pacing design requirements include: minimum size and mass, extended operating lifetimes without refueling or maintenance, high reliability, and strict adherence to aero-

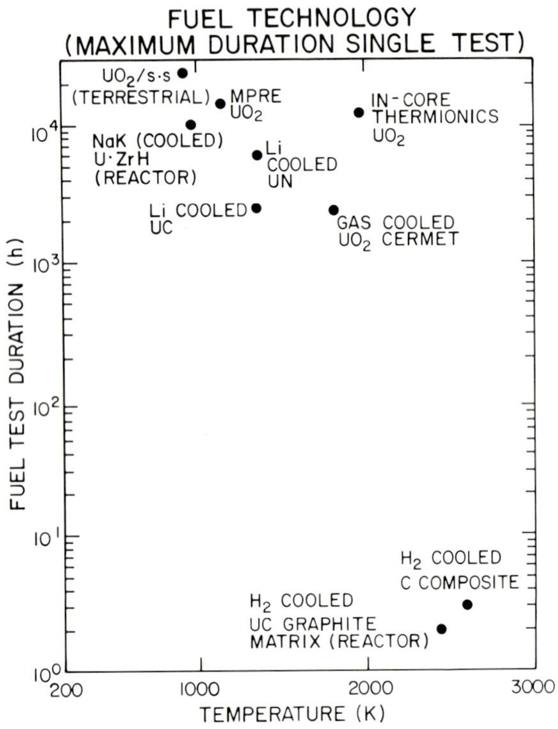

Fig. 13.6 Nuclear fuel technology status (based on maximum duration of longest test in series).

Table 13.3 Representative values for previous space power plants.

	SNAP-10A	SNAP-8	SNAP-50/Advanced Liquid-Metal-Cooled Reactors	In-Core Thermionics	710	SP-100
Power (kW_e)	0.5	35	300	120	200	100
Design Life (yr)	1	1	1	1	1	7–10
Reactor Power (kW_t)	30	600	2,000	1,670	870	1,600
Efficiency (%)	1.6	8	19	7	23	6.5
Reactor Outlet Temperature (K)	810	975	1,480	2,000	1,475	1,500
Reactor Type	U-ZrH, thermal	U-ZrH, thermal	UO_2, fast	UO_2, UC thermal, fast	UO_2, cermet, fast	UO_2, fast
Power Conversion	thermoelectrics	Rankine (mercury)	Rankine (potassium)	thermionics	Brayton (neon)	thermoelectrics
Specific Weight (kg/kW)	510	130	27	22	—	20–30

Table 13.4 Fuel element coating data.

Fuel	Reactor	Coolant	Coating	Maximum Fuel Temperature (K)
U-ZrH	SNAP	NaK	Hastelloy N	1,035
UC_2	Rover	H_2	NbC, ZrC	~2,700
UO_2	SPAR/SP-100	Li, Na	Mo-13% Re*	> 2,400
UO_2	710	Neon	T-111 W-30, Re-30, Mo	1,800
UO_2	MPRE	K	316 SS	1,135
UN	SNAP-50	Li	Cb-1, Zr-0.090C	1,350 to 1,500
UO_2, UC	In-core thermionics	—	W	> 2,200

*Heat pipe wall separates fuel from heat transport fluid.

space nuclear safety criteria. Previous space nuclear power plant programs (see Table 13.3) provide a technically sound and logical starting point. In fact, depending on system mass limitations and other design constraints, technologies to support the construction of power plants capable of providing tens of kilowatts-electric for a year or more can be considered reasonably well in hand. As operational lifetimes extend to a decade or more and power demands approach the megawatt-electric level, however, space nuclear power plant engineers quickly enter new design regimes. In this section we will briefly summarize established space nuclear reactor technologies and identify technology pathways that appear most reasonable for future development and exploitation.

Demonstrated Reactor Core Technologies

Once system power levels and operating temperatures are established, reactor design choices tend to be dominated by factors associated with fuel element technology, energy conversion technology, and heat transport and rejection technology.

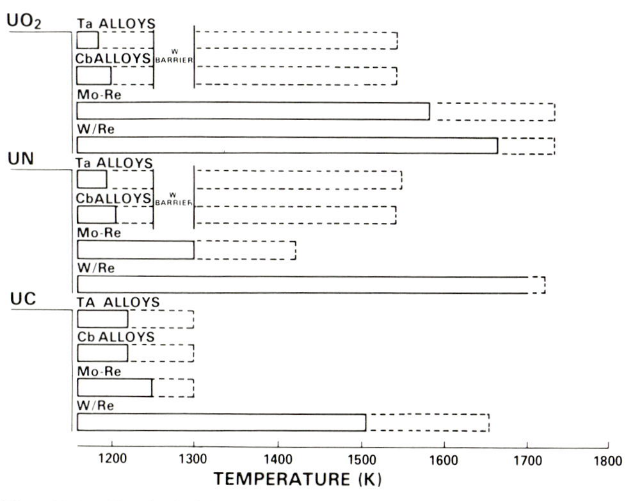

Fig. 13.7 Fuel-cladding compatibility summary [*16*].

Extensive fuel element development efforts have already occurred, primarily during the period 1965 to 1973. The two fuel element technology areas receiving the greatest focus were uranium-zirconium hydride (UZrH) fuel for the SNAP program and coated uranium dicarbide fuel for the Rover program. The SNAP program demonstrated fuel element technology capable of 10,000 hours operation at 975 K with NaK as the core coolant, while the Rover program demonstrated fuel technologies capable of approximately 2 hours operation at 2,450 K with hydrogen as the working fluid. The U.S. fuel element technology heritage for space reactors is summarized in Fig. 13.6.

Fuel element coatings serve as both a chemical barrier and a physical erosion barrier between the nuclear fuel and the core working fluid. In addition to being chemically compatible with the fuel and coolant, a good coating material must also exhibit stability at high temperatures and match the thermal expansion coefficient of the nuclear fuel. Table 13.4 is a summary of representative fuel element coating data.

For future space nuclear reactors, materials technology will be a pacing item, since fuel elements will typically be required to perform for seven to ten years at temperatures of 1,200 to 1,800 K. Figure 13.7 describes fuel-cladding compatibility for such refractory fuels as UO_2, UC, and UN [16]. At temperatures below 1,100 or 1,200 K, thermodiffusion rates are small and the choice of a fuel-cladding combination can be based on other factors rather than thermodynamic stability. As the operating temperature increases, however, thermodynamic stability becomes a significant consideration in the selection of a fuel-cladding combination.

Fuel swelling is another major consideration in selecting the fuel for a particular advanced reactor. There are three dominating factors affecting fuel swelling behavior: crystal lattice structure, burnup level, and temperture. An increase in temperature and burnup level will result in increased fuel swelling [17].

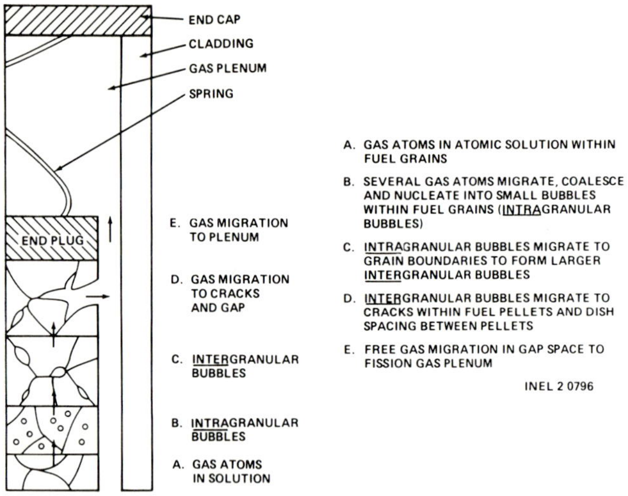

Fig. 13.8 Fission gas behavior in solid fuel [*17*].

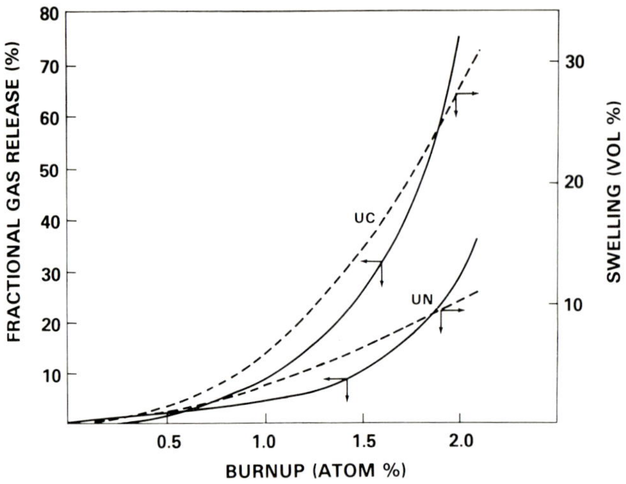

Fig. 13.10 UN and UC swelling and gas release data [*17*].

Fission gas induced fuel swelling is caused primarily by the growth of intergranular bubbles, entrapped at fuel grain boundaries. Figure 13.8 depicts the general nature of fission gas behavior in polycrystalline fuel [17]. Fission gas atoms, particularly xenon (Xe) and krypton (Kr), migrate to the grain boundaries of the fuel microstructure. These atoms migrate either individually or as small intragranular bubbles. Then, once at the grain boundaries, they form larger intergranular bubbles. Since the gas density is lower than that of a solid fission product, atoms residing in such gas bubbles occupy more volume than the solid fissile atoms they've replaced. Consequently, this bubble precipitation leads to fuel swelling. The accumulation of fission gas bubbles at grain boundaries stresses the grain structure, resulting in a grain boundary separation that eventually interlinks to form an interconnected network of pores.

Fuel swelling occurs prior to the development of a network of interlinked grain boundary pores. Once such an interlinked porosity has developed, fission gas release by means of "tunneling" to the fuel pellet exterior takes place. This results in a saturation of swelling [17]. Figures 13.9 and 13.10 summarize contemporary fuel swelling data as a function of fuel burnup [17]. Table 13.5 describes some of the major factors controlling fission gas behavior in polycrystalline fuel. Empirically established behavioral trends are described with the symbol ↑, representing a direct relationship between the variable under consideration and its effect on fuel swelling/fission gas release; the symbol ↓ denotes an inverse relationship. For example, increasing burnup

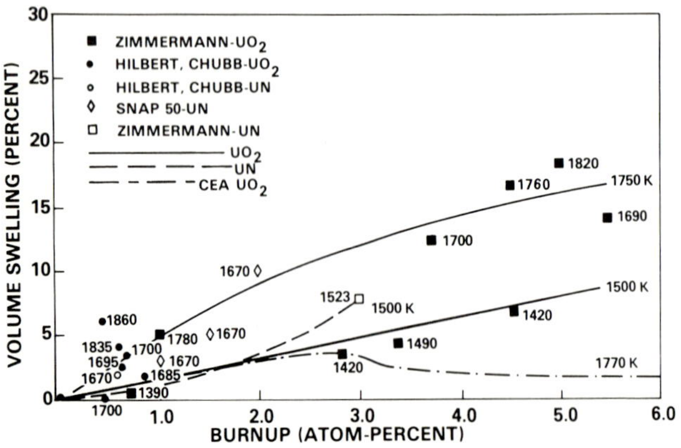

Fig. 13.9 Fuel swelling data [*17*].

Table 13.5 Comparison of UO_2, UC, and UN properties and parameters affecting fission gas induced fuel swelling and gas release behavior.

Properties:

Property	UO_2	UC	UN
Theoretical density, g/cm^3	10.97	13.63	14.32
Uranium density, g/cm^3	9.67	12.97	13.52
Melting point, K	3,120	2,660	3,030
Thermal conductivity at 1,300 K, W/cm-K	0.03	0.25	0.23
Crystal structure	Cubic-CaF_2 a = 5.47 Å	Cubic-NaCl a = 4.96 Å	Cubic-NaCl a = 4.89 Å

Swelling/Gas Release Trends:

Variable	Fuel Swelling	Fission Gas Release
Increased Microcracking	(↓)	(↑)
Increased Burnup	(↑)	(↑)
Increased Temperature	(↑)	(↑)
Increased Dislocation Density	(↓)	(↓)
Increased Grain Size	(↓)	(↓)
Increased Porosity	(↓)	(↑↓)[a]

(↑) Denotes direct relationship
(↓) Denotes inverse relationship
a. Medium initial porosity traps gas bubbles in fuel matrix, high initial porosity vents gas to pellet surface.

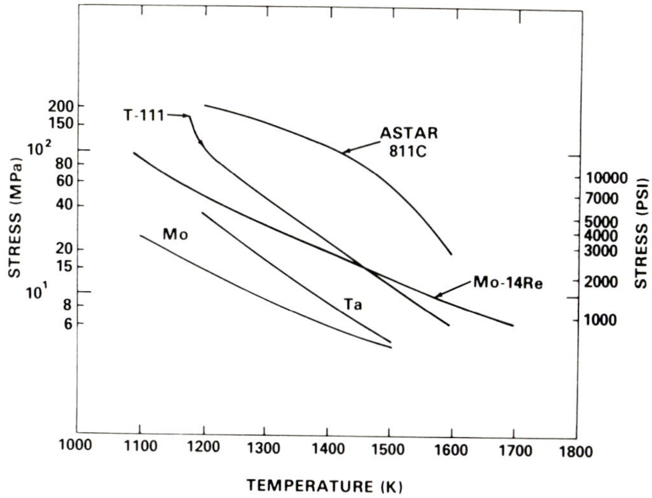

Fig. 13.11 Creep data estimates. *Courtesy of NASA.*

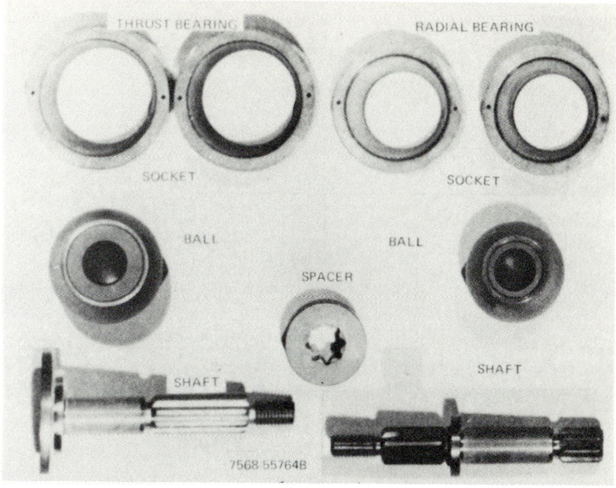

Fig. 13.12 SNAP-8 composite bearing [*18*].

also increases fuel swelling and fission gas release, while a larger grain size decreases both fuel swelling and gas release [17].

As depicted in Figures 13.9 and 13.10, for equivalent burnup, temperature, and fabricated porosity conditions, UC fuel experiences significantly more swelling than either UO_2 or UN fuel. On a relative scale, enhanced "breakaway" swelling has been observed for UC fuel temperatures above approximately 1,300 K, for UO_2 above approximately 1,500 K, and for UN above approximately 1,600 K. Fuel stoichiometry and impurity content can also exert significant influence on fuel swelling. For example, other factors being equal, hyperstoichiometric (nitrogen-rich) UN exhibits less swelling than hypostoichiometric (nitrogen-deficient) UN. The same can be said for UC fuel. In general, however, hyperstoichiometric UO_{2+x} undergoes greater swelling than hypostoichiometric (oxygen-deficient) UO_2, because of enhanced noble gas diffusivities in oxygen-rich UO_2 fuel [17].

The fraction of gas released gives rise to a cladding overpressure during its operational lifetime. Space reactor designers must ensure maximum fuel rod stability by optimizing the gas retention/fuel swelling characteristics of a particular fuel form versus the use of gas release (venting) and rod pressurization design techniques. This is especially challenging in the creation of high burnup, high temperature fuels that must operate reliably for many years in space.

Component structural materials will also limit the operating temperatures of future space reactors. Figure 13.11 provides creep estimate data for candidate structural materials. The material ASTAR-811C is a tantalum alloy with 8% W, 1% Re, 0.7% Hf and 0.025% C. T-111 is a tantalum alloy with 8% W and 2% Hf.

A variety of arrangements involving moveable elements have been used in the reflector for reactivity control. These control assemblies have included sliding blocks, rotatable half-cylinders, and rotating drums with neutron poison segments. All of these configurations employed some type of actuator to move the reactivity control element and all required bearings in the movable member.

The longest operation times for beryllium reflectors was demonstrated in U-ZrH fueled reactors [18]. These reflectors were cooled by NaK. For example, the SNAP-8 Demonstration Reactor (S8DR) reflector reached a neutron dose of 1×10^{20} nvt and a gamma dose of 1×10^{11} rads (10^9 grays). During steady-state operation, this reflector operated around 600 K. The beryllium was anodized to provide oxidation protection and emittance enhancement. In a one year long vacuum test, the reflector drive operated satisfactorily and no signs of self-welding or sticking were displayed by any component. Bearings made of a solid carbon-graphite ball (see Fig. 13.12) were operated successfully for 7,000 hours in the S8DR reactor test. Four individual bearing sets completed 12,000 hours of vacuum testing at 895 K and 0.001 Pa or lower pressure. In all, over 100,000 test hours were accumulated on the bearings.

Space qualified, longlife control actuators, as shown in Fig. 13.13, were most extensively developed in the SNAP program. Successful actuator operation was demonstrated as follows [19]: (1) SNAP-10A ground test: 2 actuators, 10,000 test hours; (2) SNAP-10A flight: 2 actuators, 1,000 test hours in space; and (3) S8DR: 6 actuators, 6,400 test hours. The SNAP-10A actuators were designed to provide 8.5 N-cm torque at 615 K and the S8DR actuators, 26.1 N-cm of torque at 810 K. Furthermore, during development testing, one of the S8DR actuator units was tested without failure for over 20,000 hours.

S8DR Actuator After Acceptance Testing

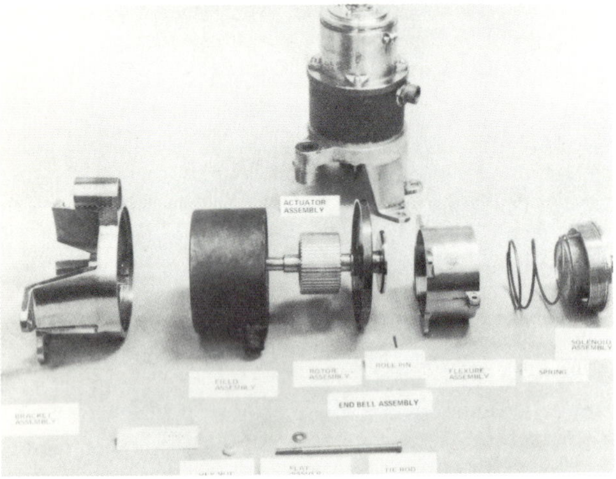

Indirect Drive 1.8-Degree Step Solid Rotor Actuator

Fig. 13.13 SNAP-8 actuator: (A) an external view; (B) components [*3*].

The longest life demonstration of radiation shielding technology for space nuclear reactors is associated with the SNAP program. Within material constraints, similar shielding technology can be applied to future space reactors. For the SNAP-10A program, five cold-pressed lithium hydride (LiH) shields were fabricated. One of these shields was used in the SNAP-10A flight system. Another was operated successfully for 10,000 hours as part of the reactor test assembly in the FS-3 ground test. Following completion of this test on 15 March 1966, visual examination of the LiH block revealed only slight surface darkening and one out of eight broken Hastalloy springs. The spring failure was attributed to fatigue. Dimensional examination of the LiH block revealed that it had remained within the accuracy of the measurement (i.e., $\pm 1.2\%$). Furthermore, lattice parameters and density measurements of the LiH samples removed from the ground test block were found to agree very well with values for unirradiated LiH [20]. Activation analysis of the stainless steel vessel indicated that the top and bottom of the shield experienced a fast neutron fluence of 2.6×10^{18} nvt. It was, therefore, concluded that the cold pressed LiH block had withstood the rigors of the reactor experiment without noticeable damage, but that improved support was needed to avoid damage during launch.

Evaluation methods for fabricating LiH shield shapes has led to a melting and casting process instead of the cold pressing and machining technique. This melting and casting technique is a faster, cheaper, and more versatile shield fabrication process that results in structurally more reliable shields because the LiH can be solidified in the shield vessel intimately surrounding all internal structural members, penetrations, etc.

In the SNAP-8 system shield design, lithium enriched in the isotope lithium-7 was used to reduce neutron-induced heating. To reduce cost, however, this was done in only one segment of the shield. The SNAP-8 shield assembly is shown in Fig. 13.14 [20]. In the 7,000 hour SNAP-8 system test, the LiH shield was exposed to a maximum neutron fluence of approximately 10^{19} nvt and to temperatures ranging from 365 to 500 K.

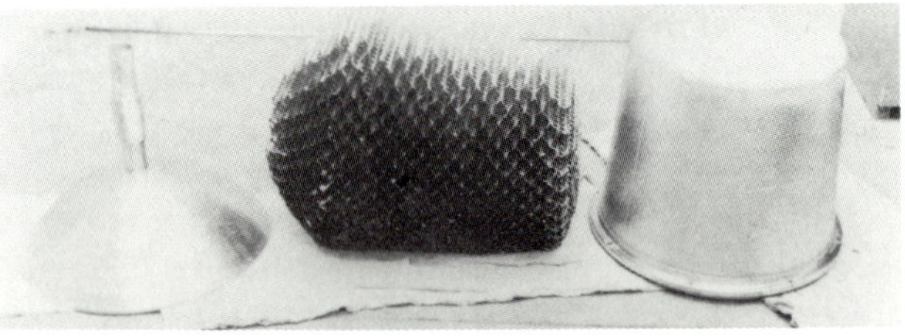

Fig. 13.14 SNAP-8 shield: (A) shield vessel, shaped honeycomb, and vessel bottom with filler tube and bellows attached; (B) instrumented SNAP-8 shield [*20*].

Post-test examination of the shield vessel showed that it was clean and that the chromic oxide emissivity coating was intact except for small spalling where the top head and the side wall met. The top of the vessel was estimated to have bulged about 1.6 cm, but no bulging was observed on the side walls. Examination of the LiH under the wafer revealed that the shield material was hard and crystalline, a condition typical of cast LiH.

Depending on the working fluid and heat transport methods chosen, a variety of components are needed to cool the reactor core and transfer thermal energy to power conversion elements. For example, gas-cooled reactors require a gas pressurization component. This device is usually combined with the power conversion components in the form of an integral design turbo-alternator-compressor. Such units have been operated for up to 38,000 hours without failure. The key to device longevity has been the development of gas bearings that eliminate physical contact or rubbing between solid surfaces during operation.

Liquid metal cooled reactors, on the other hand, require pump and boiler elements. The pump, whether electromagnetic or turbopump, pressurizes the core coolant loop and transports the liquid metal working fluid to a heat exchanger/boiler element that then transfers thermal energy to the power conversion loop. Electromagnetic pump and boiler configurations have been tested for 10,000 hours, using lithium at 1,480 K as the working fluid. NaK working fluid has also been tested for 10,000 hours.

Finally, except for the heat pipes themselves, heat pipe-cooled reactors require no additional elements to achieve core cooling. The ability to meet a single heat pipe thermal energy transport goal of 17 kilowatt-thermal at 1,500 K has been demonstrated in limited duration tests, while heat pipe endurance tests lasting as long as 25,400 hours at 1,700 K have been performed at the 1 kW_t level. Longevity tests at both high power and high temperature are now being initiated.

Power Conversion Technologies

Thermoelectric (TE) converters are the most extensively demonstrated thermal-to-electric power conversion technology for use with space nuclear power systems. As discussed in Chapter 8, certain units, such as the RTGs on board the Pioneer 10 and 11 spacecraft, have operated successfully in space for over a decade. In fact, all space nuclear power systems flown to date by the United States have used TE converters. Figure 13.15 summarizes the figures-of-merit for the major TE materials used in the United States space nuclear flight programs [21]. The tellurides have higher figure-of-merit values, but are also limited in their useful temperature range. Silicon germanium (SiGe) converters, on the

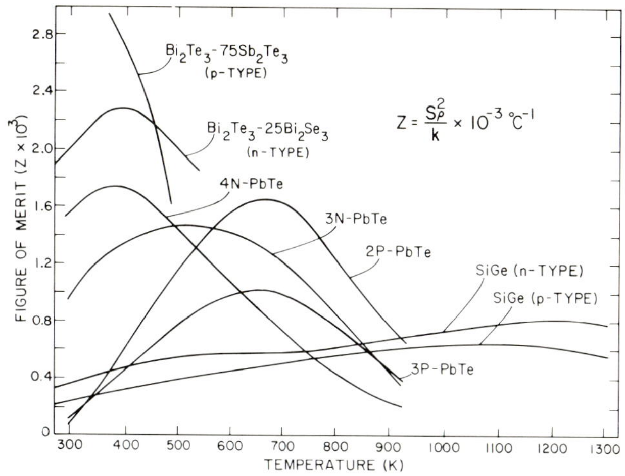

Fig. 13.15 Figures-of-merit for selected thermoelectric materials. *From C. W. Wood, "High Temperature Thermoelectric Energy Conversion, Part II-Materials Survey," Jet Propulsion Laboratory, 1983.*

other hand, can be used at higher temperatures and are now the principal TE material for operational systems. Development work is being performed on SiGe alloys and on rare earth sulfides and boron carbides to improve converter efficiencies over the plain SiGe materials.

From 1965 to 1973, thermionic converter development was concentrated on the in-pile thermionic reactor configuration [22, 23]. Tungsten, niobium, and aluminum oxide were evaluated as the emitter, collector, and insulator materials, respectively. Tests were conducted both in reactors and out-of-pile. The accumulated results of approximately 50 in-pile and out-of-pile tests indicated that about one to one and one half years performance can be anticipated for in-pile devices, while a five year life occurred for out-of-pile devices. The most significant of these test data are presented in Table 13.6. Out-of-pile thermionic devices (e.g., LC-9) have a demonstrated performance of 12 percent efficiency for 46,647 hours operation at an emitter temperature of 1,970 K; while the longest operating in-pile device (2E-1) functioned for 12,534 hours at an emitter temperature of 1,820 K. A reduction of voltage loss in the plasma and a reduction of the arc drop must be achieved in order to improve thermionic device performance. At present, the arc drop (or potential loss in the plasma) is about 0.5 to 0.6 volt. It is possible, in theory, to reduce this loss to about 0.1 volt, but up until now it has been difficult to put theory into practice.

A prototype Brayton cycle for space power applications has been successfully operated for 38,000 hours with the hardware still in good working condition at the end of the test. The working fluid was an inert gas mixture of helium and xenon. The turbine was made of 713 C steel and was operated at an inlet temperature of

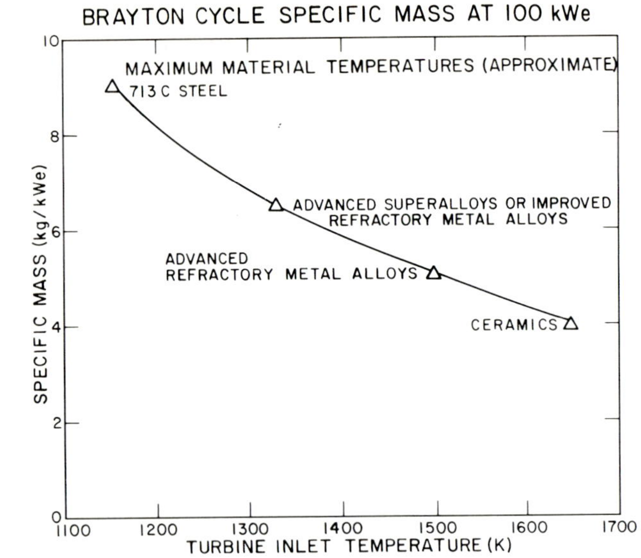

Fig. 13.16 Projected Brayton cycle performance for various turbine materials.

1,145 K. An efficiency of approximately 25 percent was achieved.

Superalloy materials and gas-bearings were being developed in a program that coupled the Brayton cycle with radioisotope heat sources for space power generation at the 1.3 kilowatt-electric level. The operating temperature of this system was also 1,145 K. Further Brayton cycle improvements are possible with the use of refractory metals that permit operation at 1,500 K and with ceramics at 1,650 K. Figure 13.16 summarizes projected Brayton cycle performance levels. As operating temperatures increase, major developments will be needed in turbomachinery and heat exchangers.

Extensive work has been done at the component and subsystem loop levels in developing the Rankine cycle for space power applications. Table 13.7 summarizes these Rankine cycle efforts.

An organic working fluid Rankine cycle was under development in the Dynamic Isotope Power Systems (DIPS) program. At completion of the program in December 1980, this system had demonstrated a DC efficiency of 16.6 percent and an AC efficiency of 18.5 percent. The organic Rankine cycle operated at a turbine inlet temperature of 615 K and a power level of 1.3 kilowatts-electric. Technology demonstrations involved an accumulation of more than 11,000 hours of testing with no component failures. This included a 2,000 hour endurance test on the final ground demonstration unit.

Table 13.6 Summary of thermionic converter tests.

Type of test	Device Designation	Fuel	Emitter Temp. (K)	Average Electrode Power (W/cm^2)	Testing Time (hours)	Burnup (10^{19} Fission /cc)	Fast Neutron Fluence (10^{20} nvt) (> 0.1 MeV)	Failure Mode of Converter
1. Out-of-pile unfueled	LC-9	—	1,970	8.0	46,647	—	—	Still operable
2. Out-of-pile fueled	LC-11	90 UC-10 ZrC-4W	1,870	6.5	18,569	—	—	Still operable
	LC-3	W-UO_2 cermet	2,000	8.9	10,406	—	—	Electrode short by Ti getter ring
3. Capsule irradiation	V-2C	90 UC-10 ZrC-4W	1,873	—	11,000	30	—	—
	FC-3	UO_2	1,910	—	10,244	12	9	—
4. In-pile MK6 experimental converter	I-4	UO_2	1,840	6.0	8,754	31	4	Envelope leak
5. In-pile MK7B prototype converter	C11	90 UC-10 ZrC	1,840	3.4	7,881	6.8	3	Emitter crack
6. In-pile MK7A partial fuel element	2E-1	UO_2	1,820	4.0	12,534	24	9	Emitter swelling
	2E-2	UO_2	1,820	5.6	11,084	22	8	Still operable
7. In-pile MK7B partial length fuel element	1F-1	UO_2	1,780	4.0	8,560	6.6	4	Emitter swelling
8. In-pile MK7B full length fuel element	6F-2	90 UC-10 ZrC-4W	1,780 1,950	2.3	7,685	6.6	4–8	Emitter cracking
	6F-3	UO_2	1,740 1,820	3.0	8,062	7.3	2.7–5.9	Still operable

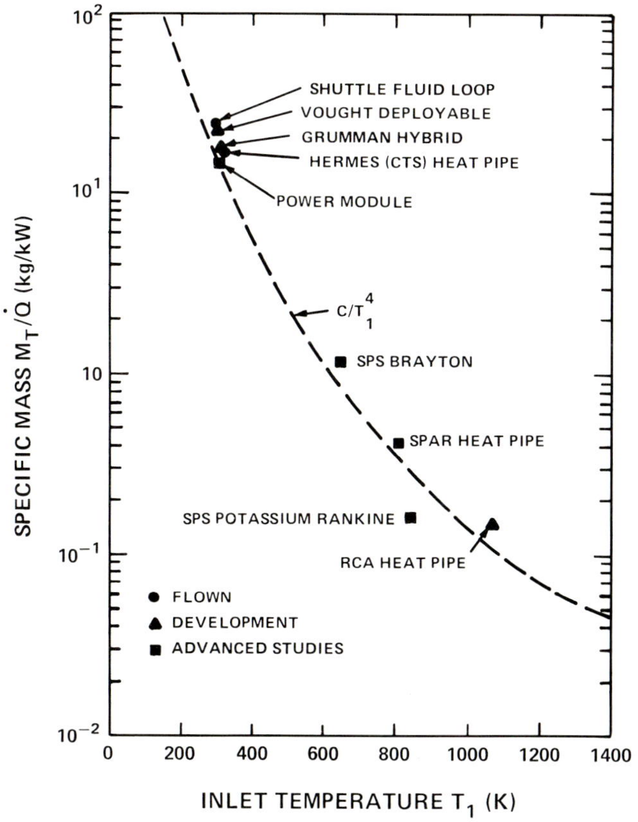

Fig. 13.17 Radiator performance [26] *expanded by authors.*

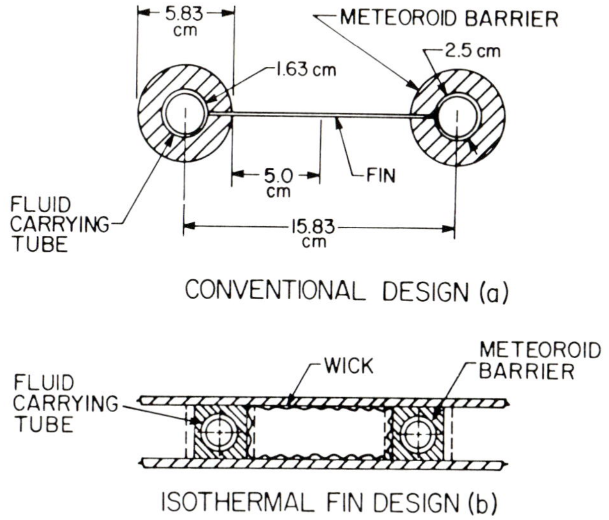

Fig. 13.18 Fin-tube radiator design: (top) armored fluid carrying tubes with crossed fins; (bottom) heat pipe fin configuration [28].

Mercury Rankine cycles were extensively developed as part of the SNAP-2 and SNAP-8 programs. For example, a ''brassboard'' mercury cycle was operated for 8,700 hours and all major components were demonstrated for over 10,000 hours. Turbine inlet temperatures were on the order of 900 K for a 35 kilowatt-electric converter unit.

A higher power and higher turbine inlet temperature Rankine system was under development, using potassium as the working fluid. This system had a primary power level of 300 kilowatts-electric and operated at 1,420 K. Its components were successfully tested for 10,000 hours and were ready for systems-level demonstration. Engineering problems remaining to be investigated included: (1) a demonstration of the jet condenser in microgravity; (2) potential problems with seals in turbomachinery; and (3) the need for an improved understanding of turbine erosion mechanisms.

Stirling cycle engines offer the potential of highly efficient converters (i.e., over 30%) for space power systems, while still permitting high temperature heat rejection to minimize radiator system size and mass. The free piston Stirling engine (FPSE) with gas-bearings in hermetically sealed units appears to be the most interesting for space power applications. These engines have already been demonstrated at a few kilowatts for 1,000 hours. The hermetic seal and the use of gas-bearings eliminates the lubrication, friction, and rod-seal problems encountered with kinematic Stirling engines [24]. FPSE technology, taking advantage of gas-bearings and non-contacting, narrow-clearance seals, represents a significant step toward higher efficiency space power plants, since traditional mechanical losses are greatly reduced. Isothermalizers are also being developed to reduce non-isothermal expansion losses and silicon carbide material is being used to reduce heat leaks [25]. Consequently, the FPSE promises to become the lowest specific mass conversion system, offering perhaps 100,000 hours of maintenance-free, highly reliable operation.

Heat Rejection Technologies

Up until now, space flight experience with radiators has

Table 13.7 Rankine cycle technology status for space power plants.

Working Fluid	Programs	Power (kW_e)	Temperature (K)	Component Technology (hr)	System Endurance (hr)
Organic	DIPS	1.3	615	10,000	2,000
Mercury	SNAP-8	35	900	10,000	8,700
Potassium	SNAP-50/ Advanced Rankine	300	1,420	10,000	—

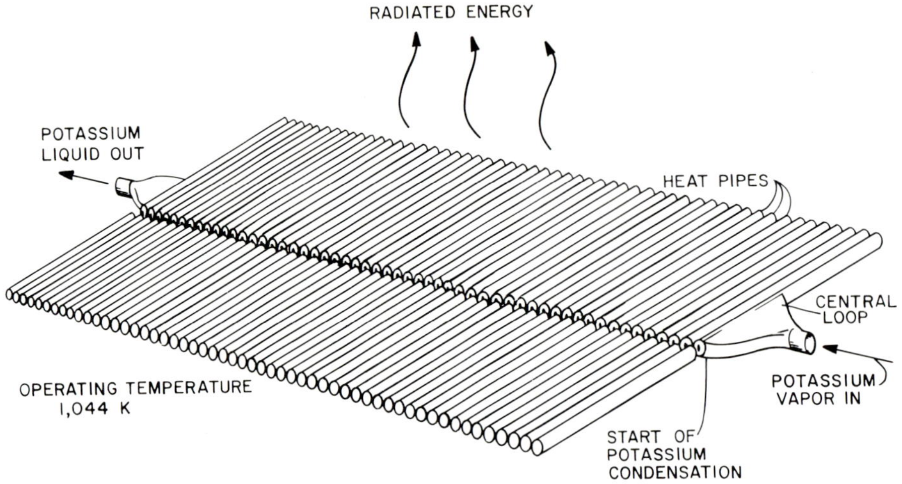

Fig. 13.19 A 50 KW_t heat pipe radiator.

been at relatively low temperatures and at relatively low power levels—at least when compared to the heat rejection demands of many future space reactor systems. In fact, this experience has centered around a heat rejection temperature of 300 K. Figure 13.17 summarizes contemporary radiator development [26].

As discussed in Chapter 6, conventional radiators generally employ thin, solid fins that are extended surfaces between fluid-carrying tubes [27]. These fluid transporting tubes must be protected against possible meteoroid damage. Armoring against meteoroids tends to make such radiator systems heavy; but they have also proven to be highly reliable and satisfactory for the spacecraft flown to date. An isothermal fin–heat pipe arrangement (see Fig. 13.18) can be used to improve performance, but meteoroid protection requirements limit performance gains.

Redundant heat pipe concepts, as shown in Fig. 13.19, use a central manifold with circulating coolant and heat pipes as fins. The most significant hardware program to date involved a 50 kilowatt-thermal radiator that had a mass of 7.7 kg, including a 2 kg section of pumped loop. Its operating temperature was 1,044 K. This radiator contained over 100 stainless steel heat pipes that used sodium as the working fluid. These heat pipes were bonded to a central loop section in such a manner as to protect the coolant loop from meteoroid bombardment. In radiator tests, the heat pipes performed properly, but variations in the quality of the thermal bonds resulted in variations in heat pipe operating temperatures [28, 29].

Heat pipe radiator components were also under development for the SPAR [30] and NEP [31] programs. For SPAR, titanium heat pipes of 0.3, 2.0 and 5.5 meter lengths were fabricated and tested with potassium as the working fluid. These particular pipes were semicircular in shape and had screen arteries in the corners. A distribution wick was formed on the titanium surface by steel shot-blasting. In tests, such heat pipes transported 1.86 kilowatts-thermal at an evaporator exit tempera-

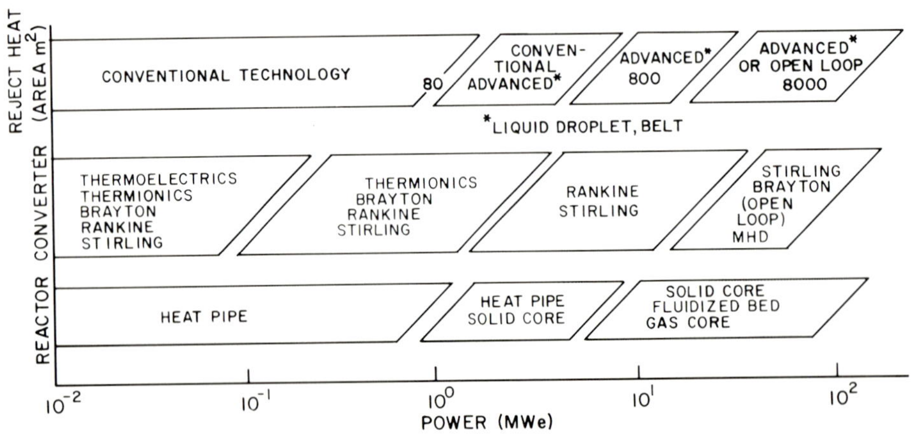

Fig. 13.20 Space nuclear reactor system technologies as a function of power level.

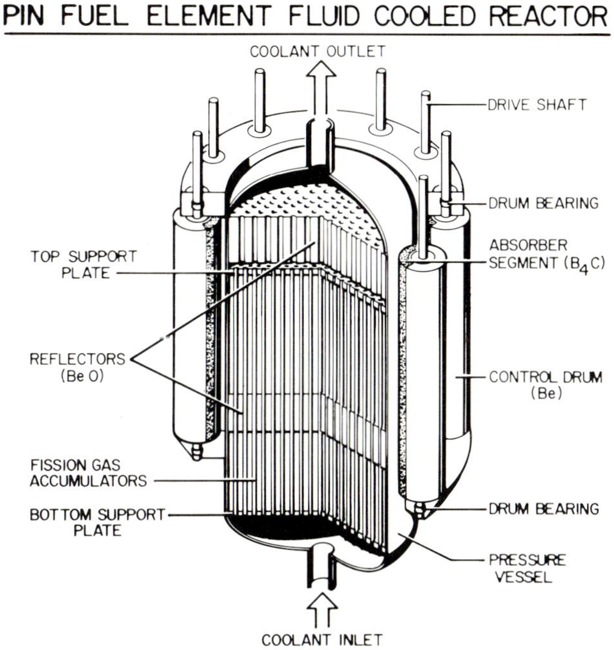

Fig. 13.21 Pin fuel space reactor (liquid metal or gas-cooled configuration). *Courtesy of Los Alamos National Laboratory.*

ture of 753 K. However, limitations in the shot-blasted distribution wick prevented the devices from reaching their design level of 2.9 kilowatts-thermal. (This design deficiency can be overcome through the use of a wire screen distribution wick.) Although a change in power plant configuration eliminated further development of this radiator concept, a seven year lifetime radiator with a specific mass of 0.38 kg/kW and operating at 775 K appears achievable.

Advanced radiator concepts, involving belts or particle streams to eliminate the need for the majority of meteoroid protection mass, are being considered (see Chapter 6). Although none of these concepts have yet been demonstrated, if successfully developed, they could reduce the general values in the specific mass curve appearing in Fig. 13.17 by a decade or more.

Technology Options as a Function of Reactor Power Level

Component selection for a particular reactor power level requires a careful consideration of the total functional requirements of the system. The overall design process includes an evaluation of such factors as lifetime, reliability, mass, launch loads and environments, and safety. A program manager must also consider the development time allowed and the fiscal resources available to meet a given set of requirements. These factors are frequently unique to a particular mission or power application. Consequently, the material presented here should be regarded as a technical generalization of the selection process. Figure 13.20 provides an approximate classification of component technology as a function of power level.

Technology Options for the 25-500 kW_e Regime

If we assume a goal of 100 kilowatts-electric, a specific mass of 30 kg/kW_e, and a lifetime of seven years, a large number of options exist for both the reactor and the power conversion system. Reactor choices include solid core, in-core thermionic, and distributed heat transport systems. Power conversion options include thermoelectrics, thermionics, Brayton cycle, Rankine cycle and Stirling cycle systems. For these power converters, reactor outlet temperatures on the order of 1,200 to 1,500 K are needed to meet the specific mass goal of 30 kg/kW_e.

As used in this book, the term "solid core" refers to that class of reactors for which the fuel is in the solid state and the coolant fluid is pumped through the core. Liquid-metal-cooled, solid core reactors normally use pin type fuel elements that have the fuel encased in a metal clad. The coolant passages are made up of the interstices between the stacked fuel elements (see Fig. 13.21). Because of the operating temperature levels, the cladding material will most likely be a refractory metal.

Solid core reactors can also be gas cooled. In such reactors, the fuel is a matrix type and the coolant passages are formed as an integral part of the fuel elements (see Fig. 13.22). The fuel can be coated with a

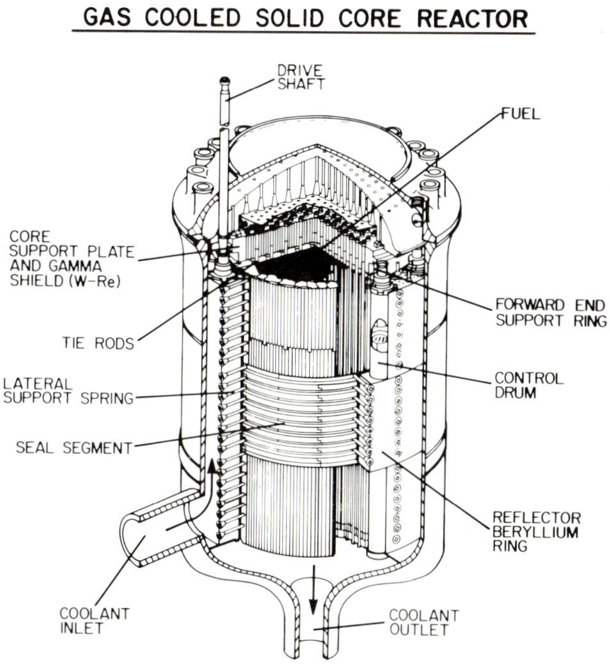

Fig. 13.22 Matrix fuel (gas-cooled) space reactor. *Courtesy of Los Alamos National Laboratory.*

thin metal carbide to prevent attack by a reactive coolant.

Because of the temperature range (1,200–1,500 K), candidate solid core reactor fuels include UO_2, UC, and UN in matrix, coated particle, or cermet forms. These reactors can be designed for core hot-end support, cold-end support, or both. However, cold-end core support designs are easier to achieve. A pressure vessel or core containment vessel is required for coolant containment and ducting through the core region. Gas-cooled reactors typically operate at pressures of 3 to 4 MPa (30–40 atmospheres). This requires the use of a more massive and thicker pressure vessel than required for liquid-metal-cooled reactors which typically operate at pressures nearer 0.1 MPa (1 atmosphere). While some savings in mass can be achieved for a gas-cooled reactor by locating the pressure vessel inside the reflector, this design approach may also require separate ducting for reflector cooling.

In typical solid core reactor designs, the working fluid cools the reflector prior to entering one end of the core through a flow distribution baffle or orifice plate. The working fluid then passes through the core coolant passages and is mixed in the exit plenum prior to distribution to the power conversion equipment.

Thermionic reactors are built up from basic thermionic converter elements that are stacked together to form fuel modules. As discussed in Chapters 5 and 11, the basic thermionic diode contains the fuel in the interior followed by an emitter, an interelectrode gap, a collector, an insulator, and finally an outer sheath. The diode is cooled by liquid metal that passes between the outer sheath and the insulator on the collector. The emitter operates at temperatures in the range of 1,600–2,000 K, whereas the collector and outer structure of the diode operate near 1,000 K. Electrons flowing from the emitter surface must travel across the interelectrode gap (about 0.13 mm wide) and deposit on the collector. A neutral space charge is maintained in the interelectrode gap by the continuous introduction of cesium atoms. These cesium atoms are ionized by collisions with electrons and by direct contact with the hot emitter surface. The cesium also attaches itself directly to both the emitter and collector surfaces, thereby lowering the work functions needed to achieve high efficiency.

Thermionic diodes are stacked—as many as six in a series—to form thermionic fuel elements (TFEs). These elements are nominally 2.5 cm outside diameter by 41 cm long and can produce several hundred to several thousand watts. To support megawatt-level power needs, these fuel elements would be assembled in configurations similar to a liquid-metal-cooled solid core reactor.

Some design advantages of a thermionic reactor include the modularity of the fuel elements and the fact that the reactor coolant temperature does not have to approach the fuel temperature. The core cooling can be accomplished with conventional liquid metal systems. For example, both NaK eutectic and K are candidate liquid metal coolants for use with the thermionic reactor. Insulators operate at a relatively low temperature and the "flashlight" thermionic concept can be scaled over a large power range.

The major difficulties encountered with thermionic diodes have been fuel swelling (that shorts out the diode) and insulator breakdown due to progressive neutron damage. Other problems include an output voltage that is limited by arcing and fast neutron damage to other construction materials. The flashlight concept can be used in either a fast reactor or in a moderated reactor. Use of a moderator, such as zirconium hydride, reduces the critical dimensions of the core, but also decreases the electric power output of a given volume core.

The externally fueled diode represents a second major thermionic reactor concept. It involves a thermionic diode with the fuel on the outside. Reject heat from the collector is removed from the center of the diode either by flowing liquid metal or by a heat pipe. The diodes are connected in a series-parallel network in order to build up the reactor output voltage and to improve system reliability. The higher fuel volume fractions characteristic of these externally fueled diodes result in smaller core sizes than encountered with internally fueled diode designs. Other advantages of the externally fueled diode are that fuel swelling is predominantly away from the interelectrode gap and that redundant cooling in the case of a failed element might also be

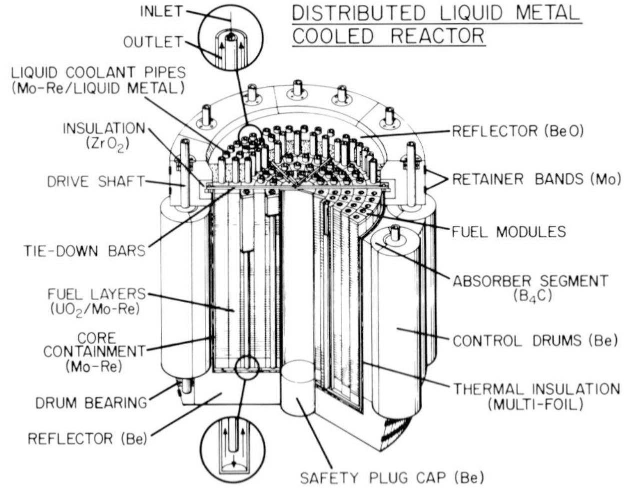

Fig. 13.23 Distributed cooled (liquid metal) space reactor. *Courtesy of Los Alamos National Laboratory.*

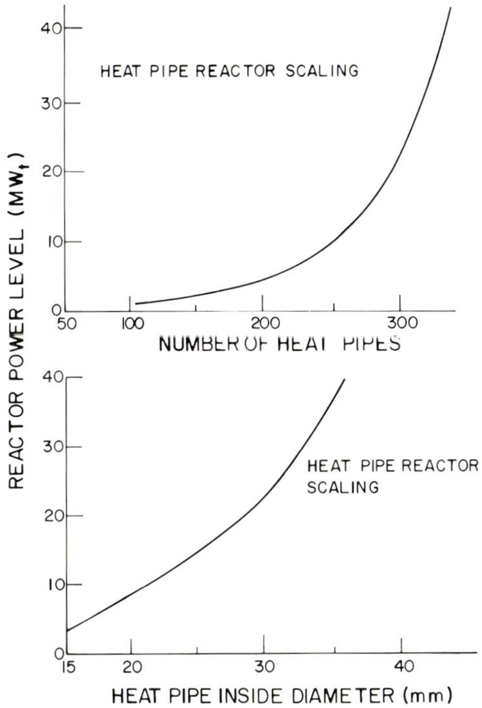

Fig. 13.24 Heat pipe reactor scaling. *Courtesy of Los Alamos National Laboratory.*

possible. Disadvantages include the fact that the addition of a moderator is difficult; electrode spacing must be accurately maintained over extensive lengths; fuel elements must be electrically isolated from each other (but the temperature is too high to use an effective insulator); and the larger diode size gives rise to very high currents.

Distributed heat transport reactors have separate, independently sealed coolant paths in the core. These coolant paths can take the form of either heat pipes or pumped liquid metal loops. In the latter design approach, several electromagnetic pumps are arranged in series to provide redundancy and independent coolant paths are maintained through these pumps. Figure 13.23 depicts a liquid-metal-cooled, distributed heat transport reactor concept.

A typical core arrangement is essentially a mirror image design of the solid core reactors previously described. In the distributed reactor, the fuel is placed external to the coolant channel and is composed of thin wafers inserted between radial fins machined on the coolant channels. Alternate fuel geometries, such as coated particles in a supporting matrix, are also possible. Fuel swelling in the wafer design is accommodated by voids that surround the fuel or by porosity in the fuel. For coated particle fuel, the swelling is absorbed in a porous graphite coating and the fission gases are contained by a metal carbide coating, such as zirconium carbide (ZrC). If heat pipes are used, the optimum number of heat pipes depends on the reactor power level. For example, about 120 heat pipes are desirable for a 1.5 megawatt-thermal reactor. As the reactor power level increases, the vapor flow area necessary for core cooling also increases and the optimum number of heat pipes increases as shown in Fig. 13.24, reaching about 340 at a reactor power level of 40 megawatts-thermal. The heat pipe inside diameter also increases with reactor power level as shown in Fig. 13.24. Table 13.8 provides some additional representative data for heat pipe cooled reactors. The peak fuel temperature for the design described is approximately 200 K higher than the heat pipe working temperature of 1,500 K.

One desirable design feature of the distributed cooled reactor is the absence of a core pressure vessel. This design feature, when coupled with a segmented reflec-

Table 13.8 Representative parameters for heat pipe reactors.

PARAMETER	REACTOR THERMAL POWER (megawatts) 1.5	4	20
Average Fuel Temperature (K)	1,580	1,580	1,580
Maximum Fuel Temperature (K)	1,685	1,700	1,715
Burnup Fraction (7 Yr Life)	0.036	0.076	0.162
Fuel Swelling (7 Yr Life) %	8	12	27
Number of Heat Pipes	120	180	290
O.D. of Heat Pipes (cm)	1.59	1.86	3.13
Vapor Space Diameter (cm)	1.43	1.56	2.80
Core Diameter (cm)	33.1	39.8	67.6
Core Height (cm)	33.1	39.8	67.6
Total Reactor Mass of Reactor Plus 1 m of Heat Pipes (kg)	500	860	2,310

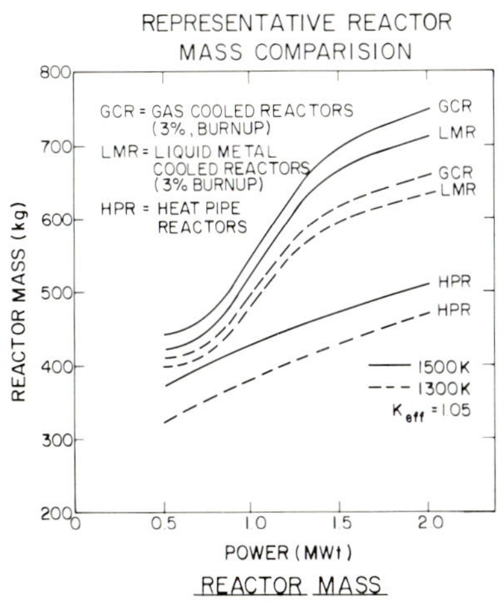

Fig. 13.25 Mass comparison of various solid core reactor concepts. *Courtesy of Los Alamos National Laboratory.*

tor, can be used to enhance the probability of core burnup and dispersal during atmospheric reentries. The distributed cooled cores are surrounded by a thin containment can and several layers of multifoil insulation that reduce heat leakage to the reflector. The reflector itself is made of Be and BeO, depending on temperature, and contains cylindrical control drums with B_4C as a neutron absorber. The core radial power profile can be flattened by varying the amount of fuel surrounding each heat pipe. A candidate coolant channel and core liner material for a distributed cooled reactor operating at 1,500 K is Mo-13% Re. Here, lithium would be used as the coolant and UO_2 in wafer form would be used as the fuel. Another prime design feature of the distributed cooled reactor is that it permits several independent coolant channel failures to occur without initiating overall core failure.

Figure 13.25 provides a mass comparison of representative solid core reactors, while Tables 13.9 and 13.10 provide an overall assessment of these type reactors from an aerospace nuclear safety perspective.

Table 13.9 Solid core reactor assessment.

INTEGRAL HEAT TRANSPORT REACTORS					
Description:	Matrix Fuel		Pin Fuel		In-Core Thermionics
Temperature	1,200 K	1,500 K	1,200 K	1,500 K	
Fuel Form	Matrix		Pin		Cylindrical
Fuel Cladding	Super alloy or refractory	Mo-Re or Ta alloys	Super alloy or refractory	Mo-Re or Ta alloys	W
Heat Transport	Turbopumps		EM Pumps		EM Pumps
Coolant	He	He	Na	Li	NaK
Reliability:	Lacks redundancy in lines and distribution elements. Redundant turbopumps add valves that need to operate at 1500 K. Complicated periphery design for launch loads.*		Lacks redundancy in lines and distribution elements. Large amount of experience with pin fuel design (lower temperature).*		Lacks redundancy in lines and distribution elements (operate at 1000 K). Highly interactive fuel and converter elements at 1700–2000 K lead to tolerance questions.
Mass:	Heaviest NSS because: —high pressures (3-4 MPa) —larger core diameter because of poor heat transfer —additional length for flow distribution —increase shield because of larger diameter		Heavier than distributed reactors because: —added length from flow distribution system —pressure containment vessel (0.1 MPa)		Not comparable because NSS includes electric converters. Needs pressure vessel and shield; tends to be bigger and heavier because reactor diameter larger.
Safety: Subcritical On Liquid Immersion	Heavy pressure vessel plus long lines with bellows makes meeting this requirement difficult. Can use scram rods in core or design pressure vessel around removable plug.		Pressure vessel with long lines makes configuration difficult. Can use scram rods in core or design pressure vessel around removable plug.		Pressure vessel with long lines makes configuration difficult. Can use scram rods in core or design pressure vessel around removable poison plug.
Significant Negative Temp. Coefficient of Reactivity	—All configurations appear to possess a negative temperature coefficient of reactivity—				

*See Table 13.10. NSS = nuclear subsystem

Table 13.9 continued on next page

Table 13.9 Solid core reactor assessment—*Continued.*

INTEGRAL HEAT TRANSPORT REACTORS—*Continued*

Safety:	Matrix Fuel	Pin Fuel	In-Core Thermionics
Subcritical On Land Impact	Heavy pressure vessel will probably absorb shock of impact, preventing core compression.	Pressure vessel will probably absorb shock of impact, preventing core compression. Design features should make meeting requirement feasible.	Pressure vessel will probably absorb shock of impact, preventing core compression. Design features should make meeting requirement feasible.
Independent Reactivity Subcriticality	—All designs meet this requirement by using separate scram and reactivity actuators—		
Independent Decay Heat Removal (After Shutdown)	Very difficult. Double jacketed lines and flow elements plus redundant pumping systems needed.	Very difficult. Double jacketed lines and flow elements plus redundant pumping systems needed.	Very difficult. Double jacketed lines and flow elements plus redundant pumping systems needed.
Unirradiated Fuel Safe	—All designs use ^{235}U and will not be operated at significant power until a safe orbit is reached—		
Restart After Shutdown	Needs battery for control drums. No apparent problem.	Needs battery for control drums. Needs heating system for coolant outside core.	Needs battery for control drums. (NaK is a liquid at 292 K.)
Core Dispersal On Reentry	Very difficult because of heavy pressure vessel and fuel form.	Very difficult because of pressure vessel and coated fuel form.	Very difficult because of pressure vessel and encapsulated fuel arrangement (W acts as a barrier).
Launch Environment:	Rover reactor designed for launch loads. No barrier anticipated for successful design.	SNAPSHOT was a liquid metal cooled reactor. Design solutions appear to exist for launch environment.	Design solutions appear to exist for launch environment.

DISTRIBUTED HEAT TRANSPORT REACTORS

Description:	Heat Pipe		Liquid Metal		In-Core Thermionics
Temperature	1,200 K	1,500 K	1,200 K	1,500 K	
Fuel Form	Wafer or Coated Particles		Wafer or Coated Particles		Wafer or Coated Particles
Fuel Cladding	Nb-Zr	Mo-Re	Nb-Zr	Mo-Re	W
Heat Transport	Heat Pipes		EM Pumps		EM Pumps or Heat Pipes
Coolant	Na	Li	Na	Li	NaK
Reliability:	No single failure points. Heat pipes have proven to be highly reliable (at lower temperatures).*		No single failure points. EM pumps have tended to be highly reliable if properly cooled.*		No single failure points. Both heat pipes and EM pumps have tended to be highly reliable.
Mass:	Lightest mass: —no pressure vessel or pumps —built-in redundancy		Heavier than heat pipe reactor: —core itself about same mass —EM pumps, lines, etc. heavier than heat pipe extensions		Not comparable because NSS includes electric power converters. Larger core size will probably result in a larger shield.

*See Table 13.10. NSS = nuclear subsystem

Table 13.9 continued on next page

Table 13.9 Solid core reactor assessment—*Continued.*

DISTRIBUTED HEAT TRANSPORT REACTORS—*Continued*

Safety:	Heat Pipe	Liquid Metal	In-Core Thermionics
Subcritical On Liquid Immersion	Removable plug or poison elements designed into current configurations.	Removable plug or poison elements designed into current configurations.	Removable plug or poison elements designed into current configurations.
Significant Negative Temp. Coefficient of Reactivity	—All configurations appear to possess a negative temperature coefficient of reactivity—		
Subcritical On Land Impact	Banded reflector and core should result in core splitting on major impact. Curved structure of reflector also helps avoid core compression.	Banded reflector and core should result in core splitting on major impact. Curved structure of reflector also helps avoid core compression.	Banded reflector and core should result in core splitting on major impact. Curved structure of reflector helps avoid core compression.
Independent Reactivity Subcriticality	—All designs meet this requirement by using separate scram and reactivity actuators—		
Independent Decay Heat Removal (After Shutdown)	Independent parallel flow paths provided.	Independent parallel flow paths provided.	Independent parallel flow paths provided.
Unirradiated Fuel Safe	—All designs use ^{235}U and will not be operated at significant power until a safe orbit is reached—		
Restart After Shutdown	Needs battery for control drums. Heat pipes have demonstrated restart capability.	Needs battery for control drums. Needs heating system for coolant outside core.	Needs battery for control drums. Restart not a problem with NaK or heat pipes.
Core Dispersal On Reentry	—Segmented/banded core and reflector configuration; bands ablate easily under aerodynamic heating of reentry; use of uncoated fuel also helps meet this requirement—		
Launch Environment:	—Use ductile materials; preliminary studies indicate NSS can be designed to meet launch loads—		

Table 13.10 Solid core reactor single failure point modes.

LIQUID METAL COOLED (PIN FUEL) REACTOR

Component and Function	Potential Failure Modes	Failure Effect
FUEL ELEMENT Acts as heat source for power plant.	Fuel corrosion by lithium.	Loss of criticality. Premature shutdown of system.
CORE SUPPORT STRUCTURE Maintains core integrity.	Radiation-induced material damage, flow-induced vibration, wear, weld, or fastening failure.	Premature shutdown, reactor disintegration.
PRESSURE VESSEL Provides containment for lithium fluid.	Rupture of vessel from weld failure or radiation-induced material damage.	Loss of coolant to reactor terminating operation.
REFLECTOR Improves reactor power distribution, lower critical mass	With conduction and radiation cooling, no foreseen failure modes.	

Table 13.10 continued on next page

Table 13.10 Solid core reactor single failure point modes—*Continued.*

LIQUID METAL COOLED (PIN FUEL) REACTOR—*Continued*

Component and Function	Potential Failure Modes	Failure Effect
CONTROL DRUMS Establishes power level by varying their position.	Multiple failure of redundant elements. Drums could bind from warped reflector or failure of mechanical linkage.	None for single element. Several drums fail-in would cause reactor shutdown; fail-out would cause supercritical condition leading to reactor mechanical disintegration; fail-in-place would lead to premature shutdown.
ACCUMULATOR Provides pressurized working fluid to fluid loops at startup.	Internal leakage through bellows due to manufacturing defect.	Fluid in bellows, gas in fluid loop. System may not start or reduction in power capability.
START VALVE Valve opens upon receipt of electric start command to allow pressurized fluid flow from accumulator to fluid loops. Start valve remains open when electric signal is removed and remains open during system operation.	Valve fails to open at start.	System does not start.
BOILER Transfer heat from reactor loop to power conversion loop.	Cracks due to material defect.	External leakage results in loss of working fluid and power loss. Over-temperature shut-down.
ELECTROMAGNETIC PUMP Provides pressurization working fluid for cooling reactor.	Loss of pump coolant loop. Wall rupture.	With redundant pumps, lose one pump. External leakage results in loss of working fluid, loss of power output.

GAS-COOLED (MATRIX FUEL) REACTOR

Component and Function	Potential Failure Modes	Failure Effect
FUEL ELEMENT Acts as heat source for power plant.	Xe-He gas not reactive with fuel but could react with fuel impurities	Loss of criticality. Premature shutdown of system (low probability of occurrence).
CORE SUPPORT STRUCTURE Maintains core integrity.	Radiation-induced material damage, flow-induced vibration, wear, weld, or fastening failure.	Premature shutdown; reactor disintegration.
PRESSURE VESSEL Provides containment for Xe-He gas.	Rupture of vessel from weld failure or radiation-induced material damage.	Loss of coolant to reactor terminating operation.
REFLECTOR Improves reactor power distribution, lower critical mass.	With conduction and radiation cooling, no forseen failure modes.	

Table 13.10 continued on next page

Table 13.10 Solid core reactor single failure point modes—*Continued.*

GAS-COOLED (MATRIX FUEL) REACTOR—*Continued*

Component and Function	Potential Failure Modes	Failure Effect
CONTROL DRUMS Establishes power level by varying their positions.	Multiple failure of redundant elements. Drums could bind from warped reflector or failure of mechanical linkage.	None for single element. Several drums failing would cause reactor shutdown; fail-out would cause supercritical condition leading to reactor mechanical disintegration; fail-in-place would lead to premature shutdown.
TURBO-COMPRESSOR Turbine spins compressor. Compressor provides pressurized working gas to converter loop.	Turbine wheel rubs due to mechanical interference.	Possible damage to rotating hardware and bearings and loss or decrease in power output.
	Bearing failure due to loss of gas film or mechanical failure.	Loss or decrease in power output.
	Turbine housing cracks due to thermal stresses, material defect.	External leakage results in loss of gas, loss of power output and over-temperature shutdown.
	Compressor rubs due to misalignment or out-of-tolerance.	Decrease system flow resulting in reduced power output.
	Turbine seal fails.	Loss of efficiency and performance.
	Blade failures in turbine or compressor.	Shutdown.

HEAT PIPE REACTOR

Component and Function	Potential Failure Modes	Failure Effect
FUEL ELEMENT Acts as heat source for power plant.	120 heat pipes used to remove heat and isolated fuel from fluid system. Loss of heat pipe will not affect system output.	
PERIPHERY COMPRESSION BANDS Maintain core in compression.	Redundant bands to bind core together.	None for single failures. Multiple failures affect core compression and integrity.
REFLECTOR Improves reactor power distribution, lower critical mass.	With conduction and radiation cooling, no foreseen failure modes.	
CONTROL DRUMS Establish power level by varying their position.	Redundant elements require multiple failures. Drums could bind from reflector warpage or failure of mechanical linkage.	None for single element. Several drums fail-in would cause reactor shutdown; fail-out would cause supercritical condition leading to reactor mechanical disintegration; fail-in-place would lead to premature shutdown.

Table 13.10 continued on next page

Table 13.10 Solid core reactor single failure point modes—*Continued.*

LIQUID METAL COOLED DISTRIBUTED REACTOR

Component and Function	Potential Failure Modes	Failure Effect
FUEL ELEMENT Acts as heat source for power plant.	120 lines used to remove heat and isolated fuel from fluid system. Loss of line will not affect system output.	
PERIPHERY COMPRESSION BANDS Maintain core in compression.	Redundant bands to bind core together.	None for single failures. Multiple failures affect core compression and integrity.
REFLECTOR Improves reactor power distribution, lower critical mass.	With conduction and radiation cooling, no foreseen failure modes.	
CONTROL DRUMS Establish power level by varying their position.	Redundant elements require multiple failures. Drums could bind from reflector warpage or failure of mechanical linkage.	None for single element. Several drums fail-in would cause reactor shutdown; fail-out would cause supercritical condition leading to reactor mechanical disintegration; fail-in-place would lead to premature shutdown.
ELECTROMAGNETIC PUMP Provides pressurization working fluid for cooling reactor.	Loss of pump coolant loop.	With redundant pumps, lose pump but system output not affected.

Table 13.11 Performance goals for megawatt-class space nuclear power systems.

POWER SYSTEM PARAMETER	GOALS	COMMENT
Power Level	0.5–5.0 MW_e	
Lifetime	7 years at full power 10 year total system	
Mass	Less than 15,000 kg @ 5 MW_e Less than 8,000 kg @ 1 MW_e	
Size	Packageable in the Shuttle bay as a single payload	
Power Distribution	100–400 VDC	Available at power plant/payload interface
Startup	Less than 1 hour Less than 2 minutes	To 1% power 1% to 100% power
Reliability	95%	Probability of success of delivering rated power at end of life (EOL)
Payload Exposure Dose (at 25 m)	10^{12} neutrons/cm^2 10^6 RAD	For reactor operating at 7 years at rated power

Technology Options for the 0.5–5 Megawatt-electric Regime

Table 13.11 describes a set of performance goals for space nuclear reactors in the MW_e regime. In addition to the solid core reactor designs just described, these goals might also be satisfied by pellet bed reactors or fluidized bed reactors.

In the pelletized reactor concept, the fuel is in the form of spherical pellets that are 0.5 to 2.0 cm in diameter. These fuel pellets are contained peripherally by a cylindrical liner and on the ends by perforated plates. Highly enriched (93%) uranium in an oxide, nitride, or carbide fuel is encapsulated to minimize erosion and corrosion. A conceptual pellet bed reactor

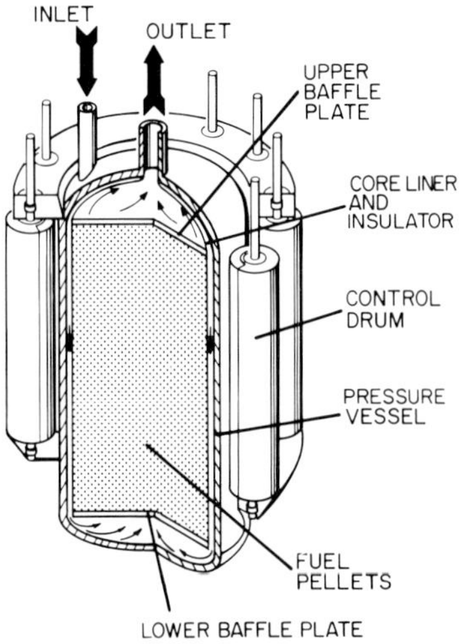

Fig. 13.26 Pellet bed reactor concept.

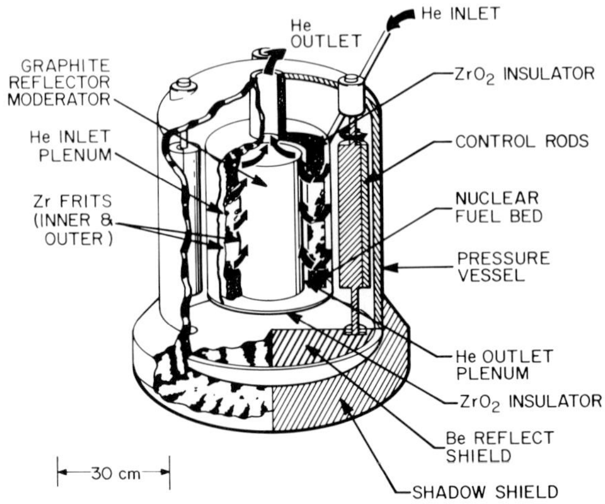

Fig. 13.27 Fixed bed reactor concept. *Courtesy of Brookhaven National Laboratory.*

for space power is shown in Fig. 13.26. The reflector and control drums are configured similar to the arrangements for liquid-metal-cooled and gas-cooled solid core reactors. Reflector cooling is achieved by radiation heat transfer to space. Engineering questions such as the exact geometry of the core periphery, the cooling of the reflector, and the location of the pressure vessel (i.e., inside or outside the reflector) must be addressed.

The pellet bed reactor has the potential for greatly simplifying core design and it supports the consideration of refueling. For very longlife systems (as might be used in lunar or Martian settlements), the ability to refuel would also support low burnup operations and therefore make fuel swelling and fission gas release problems more manageable.

Fuel for a helium gas-cooled pellet bed reactor might take the form of UC spheres coated with 0.02 to 0.05 mm of pyrographite, while lithium-cooled cores might use UC coated with molybdenum or tungsten. Core periphery construction materials could include refractory metals for the core liner, molybdenum multifoil insulation, beryllium or BeO for the reflector and control drums, B_4C for the neutron absorber, and a nickel alloy or stainless steel for the pressure vessel.

The current concept for a fluidized bed reactor involves 100–500 micrometer diameter fuel particles contained in an annular shell. The fuel particle bed is contained on its outside boundary by a perforated cylinder called a frit. The fuel and its containment frit are rotated and the particle bed is fluidized by passing the coolant gas rapidly inward.

A lower power version of this concept is commonly called a fixed bed reactor (FBR) (see Fig. 13.27). It operates without fuel particle rotation and fluidization. In this concept, the fuel is bounded by frits on the inside and outside. The inner porous frit, operating at essentially the fuel temperature, would be made of graphite or a ceramic material. The center of the core is constructed of graphite that serves as both a neutron moderator and reflector. One of the advantages of this reactor design concept is the high heat transfer rate it supports to the working fluid, thereby lowering temperature differences between the fuel and coolant. These temperature differences would typically be on the order of 10 to 20 K. The fuel bed itself is about 3 to 6 cm thick. The FBR core is surrounded by a Be reflector that incorporates cylindrical control drums with a neutron poison.

Uranium dicarbide (UC_2) is typically considered as the candidate fuel for FBR systems. It would be coated with a layer of porous pyrographite for retention of

Table 13.12 Potentially achievable performance characteristics for megawatt-class nuclear power plants.

	POWER LEVEL	
Power Level (MW_t)	4	20
Efficiency (%)	25	25
Converter Inlet Temperature (K)	1,500	1,500
Reject Heat Temperature (K)	1,000	1,000
Radiator Surface Area (m^2)	62	310
Power Plant Deployed Length (m)	5.6	14.5
POWER SYSTEM MASS (kg)		
Reactor*	860	2,310
Primary Loop	290	330
Shield	880	1,300
Converter**	875	1,400
Radiator	621	3,105
Structure (10%) of Total	394	935
TOTAL	3,920	9,380

*Based on heat pipe reactor

**Based on improved Stirling cycle

Table 13.13 Initial assessment of megawatt-class (0.5–5.0 MW_e) power plants.

	SOLID CORE	DISTRIBUTED CORE REACTOR	FLUIDIZED BED	PELLET	IN-CORE THERMIONICS
MASS (kg)					
Reactor	Heavier mass than distributed core reactors due to added length of flow distribution system and pressure containment vessel.	Lowest mass due to absence of pressure vessel and inherent redundancy.	Heavier mass than distributed core reactors due to large moderator and reflectors and need for pressure vessel.	Heavier than distributed core reactors due to pressure vessel.	Not directly comparable due to inclusion of conversion. Shields tend to be heavier due to larger cores. Pressure vessel required.
SAFETY					
Subcritical Liquid Immersion	Safety rods in core	Removable elements	Insert fuel in space	Insert fuel in space	Safety rods in core
Significant Negative Temperature Coefficient	Design	Design	Design	Design	Design
Subcritical on Land Impact	Pressure vessel absorbs shock	Banded reflector and core; arch design	Separate fuel from core	Separate fuel from core	Pressure vessel absorbs shock
Independent Shutdown Heat Removal	Separate coolant and pump channels	Separate parallel paths inherent	Questionable	Questionable	Questionable
Reentry Core Dispersal	Questionable	Separate band arrangement	Remove fuel to augment dispersal	Remove fuel to augment dispersal	Questionable
STARTUP AND RESTART	Requires auxiliary heating system for liquid metal cooled reactors and components.	Heat pipe cooled OK but liquid metal cooled system requires auxiliary power.	Not a problem for gas cooled reactors.	Limited problem for gas cooled reactors.	Limited problem for NaK cooled systems.
LIFETIME	Need to develop longlife containment. Swelling and fission gas released are problems.	Fuel swelling an issue. Fuel is not contained.	Burnup is a problem Frits not developed. High density fuel particles need development.	Fuel development needed. Fuel breakup and dispersal may be a problem.	Dimensional stability problems limit TFE lifetime. Insulator damage also is a problem.
RELIABILITY GOAL 95% (No Single Failure Points)	Single failure points difficult to eliminate.	Can be designed for many independent and redundant coolant channels.	Single failure points difficult to eliminate.	Single failure points difficult to eliminate.	Single failure points difficult to eliminate.

Table 13.14 Candidate parameters for 10 and 100 MW_e power plants.

	Nearer-Term Technology		Advanced Technology	
	10 MW_e	100 MW_e	40 MW_e	100 MW_e
Power Level (MW_e)	40	400	40	303
Efficiency (%)	25	25	25	33
Converter Inlet Temperature (K)	1,500	1,500	1,500	1,650
Reject Heat Temperature (K)	1,000	1,000	1,000	1,000
Emissivity	0.85	0.85	0.20	0.20
Radiator Area (m^2)	625	6,225	2,645	17,900
Radiator Specific Mass (kg/kW_t)	0.15	0.15	0.017	0.017
Mass (kg)				
Reactor	3,620	4,000	3,620	4,000
Primary Loop	390	2,000	390	2,000
Shield	1,800	3,200	1,800	2,760
Converter	15,000	40,000	2,200	11,000
Radiator	4,500	45,000	510	3,450
Structure (10% of total)	2,530	9,420	850	2,320
Total	27,840	103,620	9,370	25,530

fission gases and with an outer layer of ZrC to contain the pressure. The critical mass for a representative FBR design ranges from 20 to 40 kg of uranium-235.

Table 13.12 summarizes what might be achievable in the 1 to 5 megawatt-electric regime in space power systems. Improved power conversion technology will be needed. Conventional radiator technology is satisfactory until power levels approach the 5 MW_e regime. At this point, an advanced technology heat rejection system would be highly advantageous. A qualitative assessment of power plant technologies in the 0.5 to 5.0 megawatt-electric range appears in Table 13.13.

Technology Options for the 5–100 Megawatt-electric Regime

At present, there is a genuine paucity of definitive engineering work concerning multimegawatt-electric reactors for space applications. Table 13.14 suggests representative design parameters for such power systems in the context of both nearer term (i.e., 1995–2005) and advanced technologies (i.e., beyond 2005). Improvements in power converter systems and heat rejection systems are essential if decreased specific mass objectives are to be realized for these systems. An advanced-design, heremetically sealed, free piston Stirling engine has the potential of reducing the converter system mass to 11,000 kg for a 100 megawatt-electric power plant [32]. Similar innovative developments in advanced radiator systems might also be anticipated.

References

1. George P. Dix, "Nuclear Safety of Space Nuclear Systems," American Nuclear Society Proceedings, Huntsville, Ala., April 1969.
2. Gary Bennett, "Overview of the U.S. Flight Safety Process for Space Nuclear Power," *Nuclear Safety*, Vol. 22, No. 4, July–August 1981, pp. 423–434.
3. D. Buden, "The Acceptability of Reactors in Space," Proceedings of the 16th Intersociety Energy Engineering Conference, August 9–14, 1981, Atlanta, Ga.
4. U.S. Department of Energy, "Nuclear Safety Criteria and Specifications for Space Nuclear Reactors," OSNP-1, Rev 0, August 1982.
5. D. Buden and G. Bennett, "On the Use of Nuclear Reactors in Space," *Physics Bulletin*, Vol. 33, 1982, pp. 432–434.
6. "Studies on Technical Aspects and Safety Measures of Nuclear Power Sources in Outer Space," United States of America working paper to Working Group on the Use of Nuclear Power Sources in Outer Space, U.N. Committee on Peaceful Uses of Outer Space, 23 January 1980.
7. E. Cortwright, "Apollo Expeditions to the Moon," NASA SP-350, National Aeronautics and Space Administration, Washington, D.C., 1975.
8. "Questions Relating to the Use of Nuclear Power in Outer Space," U.S.S.R.: Working Paper Studies on Technical Aspects and Safety Measures of Nuclear Power Sources in Outer Space, U.N. General Assembly A/AC.105/C.1/WG.V/L.10, dated 25 January 1980.
9. U.S. Department of Energy, "Operation Morning Light," Nevada Operations Office, NV-198, Las Vegas, Nev., September 1978.
10. W. R. Gummer et al., "COSMOS 954, the Occurrence and Nature of Recovered Debris," Canadian Government Publishing Center, Catalogue No. CC 172-2/1980E, May 1980.
11. United Nations, "Report of the Scientific and Technical Sub-Committee on the Work of Its Fifteenth Session," U.N. Report A/AC/105/216, dated 6 March 1978.
12. "Use of Radioactive (Nuclear) Materials by the United States of America for Space Power Generation," Working Paper Submitted by U.S.A. to Committee On Peaceful Uses of Outer Space, U.N. General Assembly Report No. A/AC.105/L.102, dated 15 March 1978.
13. "Questions Relating to the Use of Nuclear Power Sources in Outer Space," U.S. Working Paper: Studies on Technical Aspects and Safety Measures of Nuclear Power Sources in Outer Space, U.N. General Assembly Paper No. A/AC.105/C.1/WG.V/L.8, dated 25 January 1980.

14. Draft Report of the Working Group on the Use of Nuclear Power Sources in Outer Space On the Work of Its Fourth Session, U.N. Committee On the Peaceful Uses of Outer Space, A/AC.105/C.1/WG.5/L.16, dated 16 February 1984.
15. U.S. Department of Energy, "Overall Safety Manual," 1 December 1981 issue.
16. C. M. Cox et al., "Fuel Systems for Compact Fast Space Reactors," 1st Symposium On Space Nuclear Power Systems, The University of New Mexico, January 8–13, 1984.
17. A. W. Cronenberg and W. A. Ranken, "Irradiation Effects on Fuels for Space Reactors," 1st Symposium On Space Nuclear Power Systems, The University of New Mexico, January 8–13, 1984.
18. P. H. Horton and W. J. Kurzeka, "Zirconium Hydride Reactor Control Reflector Systems Summary Report," Atomics International Report AI-AEC-13078, Canoga Park, Calif., 30 June 1982.
19. L. E. Donelan, "Zirconium Hydride Reactor Control Drive Actuator Development," Rockwell International Report No. AI-AEC-13080, 30 June 1973.
20. V. Kesheshian et al., "Radiation Shielding for Zirconium Hydride Reactor Systems," Rockwell International Report No. AI-AEC-13081, 26 June 1973.
21. "Proceedings of the AFOSR Special Conference on Prime Power for High Energy Space Systems," Space Prime-Power Conference, Norfolk, Va., 22–25 February 1982.
22. F. Hoffman et al., "Thermionic Infrastructure for Space Power," 17th Intersociety Energy Conversion Engineering Conference, August 1982, pp. 1908–1912.
23. L. Yang and G. Fitzpatrick, "Thermionic Conversion for Space Power Application," AFOSR Special Conference on Prime-Power for High Energy Space Systems, Norfolk, Va, February 22–25, 1982.
24. G. Benson, W. Rifkin, and R. Vincent, "An Advanced 15 kW Solar Powered Free-Piston Stirling Engine," 15th Intersociety Energy Conversion Engineering Conference, Paper No. 809414, Seattle, Wash., 18–20 August 1980.
25. R. J. Vincent, W. D. Rifkin and G. Benson, "Test Results of High Efficiency Stirling Machine Components," 17th Intersociety Energy Conversion Engineering Conference, Paper No. 829362, August 1982, pp. 1867–1874.
26. Earl Knapp, "Final Report Lightweight Moving Radiators for Heat Rejection in Space," Astro Research Corp., Report ARC-TN-1107, Carpenteria, Calif., 10 November 1981.
27. R. W. Werner and G. A. Carlson, "Heat Pipe Radiator for a 50-MWth Space Power Plant," Lawrence Livermore National Laboratory, Report No. UCRL-50294, 30 June 1967.
28. F. C. Prenger and J. A. Sullivan, "Conceptual Designs for 100 MW Space Radiators," National Academy of Science, Washington, D.C., 18 November 1982.
29. G. M. Kikin and M. L. Peelgren, "Test of a 50 kW Heat Pipe Radiator," Jet Propulsion Laboratory, Pasadena, Calif.
30. Steven P. Girrens, "Design and Development of a Titanium Heat Pipe Space Radiator," Los Alamos National Laboratory, Report LA-9251-MS, March 1982.
31. "A Study of the Application of Advanced Heat Pipe Technology Radiators for Nuclear Spacecraft," Thermacore, Inc., Leola, Pa., 3 December 1975.
32. G. M. Benson, "Free-Piston Stirling Alternator Conversion System for SP-100 Nuclear Reactor," Private Communication, 82-ERG-1282, 23 November 1982.

Appendix A. The Elements

ALPHABETICAL LIST OF ELEMENTS AND SYMBOLS

Element	Symbol	Atomic Number	Atomic Weight*
Actinium	Ac	89	227
Aluminum	Al	13	26.98
Americium	Am	95	243
Antimony	Sb	51	121.75
Argon	Ar	18	39.948
Arsenic	As	33	74.92
Astatine	At	85	210
Barium	Ba	56	137.34
Berkelium	Bk	97	249
Beryllium	Be	4	9.012
Bismuth	Bi	83	208.98
Boron	B	5	10.81
Bromine	Br	35	79.913
Cadmium	Cd	46	112.40
Calcium	Ca	20	40.08
Californium	Cf	98	251
Carbon	C	6	12.010
Cerium	Ce	58	140.12
Cesium	Cs	55	132.90
Chlorine	Cl	17	35.453
Chromium	Cr	24	51.996
Cobalt	Co	27	58.93
Copper	Cu	29	63.54
Curium	Cm	96	247
Dysprosium	Dy	66	162.50
Einsteinium	Es	99	254
Erbium	Er	68	167.26
Europium	Eu	63	152.0
Fermium	Fm	100	253
Fluorine	F	9	19.00
Francium	Fr	87	223
Gadolinium	Gd	64	157.25
Gallium	Ga	31	69.72
Germanium	Ge	32	72.60
Gold	Au	79	197.0
Hafnium	Hf	72	178.49
Helium	He	2	4.003
Holmium	Ho	67	164.93
Hydrogen	H	1	1.0079
Indium	In	49	114.82
Iodine	I	53	126.90
Iridium	Ir	77	192.2
Iron	Fe	26	55.85
Krypton	Kr	36	83.80
Lanthanum	La	57	138.91

Element	Symbol	Atomic Number	Atomic Weight*
Lawrencium	Lr	103	257
Lead	Pb	82	207.19
Lithium	Li	3	6.942
Lutetium	Lu	71	174.97
Magnesium	Mg	12	24.31
Manganese	Mn	25	54.94
Mendelevium	Md	101	256
Mercury	Hg	80	200.59
Molybdenum	Mo	42	95.94
Neodymium	Nd	60	144.24
Neon	Ne	10	20.182
Neptunium	Np	93	237
Nickel	Ni	28	58.71
Niobium (Columbium)	Nb	41	92.91
Nitrogen	N	7	14.007
Nobelium	No	102	254
Osmium	Os	76	190.2
Oxygen	O	8	15.999
Palladium	Pd	46	106.4
Phosphorus	P	15	30.973
Platinum	Pt	78	195.09
Plutonium	Pu	94	242
Polonium	Po	84	210
Potassium	K	19	39.10
Praseodymium	Pr	59	140.91
Promethium	Pm	61	147
Protactinium	Pa	91	231
Radium	Ra	88	226
Radon	Rn	86	222
Rhenium	Re	75	186.21
Rhodium	Rh	45	102.90
Rubidium	Rb	37	85.47
Ruthenium	Ru	44	101.1
Samarium	Sm	62	150.35
Scandium	Sc	21	44.96
Selenium	Se	34	78.96
Silicon	Si	14	28.09
Silver	Ag	47	107.87
Sodium	Na	11	22.990
Strontium	Sr	38	87.62
Sulfur	S	16	32.064
Tantalum	Ta	73	180.94
Technetium	Tc	43	99
Tellurium	Te	52	127.60

ALPHABETICAL LIST – *Continued*

Element	Symbol	Atomic Number	Atomic Weight*
Terbium	Tb	65	158.92
Thallium	Tl	81	204.37
Thorium	Th	90	232.04
Thulium	Tm	69	168.93
Tin	Sn	50	118.69
Titanium	Ti	22	47.90
Tungsten (Wolfram)	W	74	183.85
Uranium	U	92	238.03
Vanadium	V	23	50.94
Xenon	Xe	54	131.30
Ytterbium	Yb	70	173.04
Yttrium	Y	39	88.91
Zinc	Zn	30	65.38
Zirconium	Zr	40	91.22

*Atomic weight of the most abundant or best known isotope, or (in the case of radioactive isotopes) the isotope with the longest half-life, relative to atomic weight of Carbon-12 = 12.

THE TRANSURANIUM ELEMENTS

Atomic Number	Element	Symbol	Atomic Weight*
93	Neptunium	Np	237
94	Plutonium	Pu	242
95	Americium	Am	243
96	Curium	Cm	248
97	Berkelium	Bk	249
98	Californium	Cf	249
99	Einsteinium	Es	254
100	Fermium	Fm	253
101	Mendelevium	Md	256
102	Nobelium	No	254
103	Lawrencium	Lr	257

*Mass number of longest-lived or more available isotope.

PERIODIC TABLE OF THE ELEMENTS

1 H																	2 He
3 Li	4 Be											5 B	6 C	7 N	8 O	9 F	10 Ne
11 Na	12 Mg											13 Al	14 Si	15 P	16 S	17 Cl	18 Ar
19 K	20 Ca	21 Sc	22 Ti	23 V	24 Cr	25 Mn	26 Fe	27 Co	28 Ni	29 Cu	30 Zn	31 Ga	32 Ge	33 As	34 Se	35 Br	36 Kr
37 Rb	38 Sr	39 Y	40 Zr	41 Nb	42 Mo	43 Tc	44 Ru	45 Rh	46 Pd	47 Ag	48 Cd	49 In	50 Sn	51 Sb	52 Te	53 I	54 Xe
55 Cs	56 Ba	57–71 La* Series	72 Hf	73 Ta	74 W	75 Re	76 Os	77 Ir	78 Pt	79 Au	80 Hg	81 Tl	82 Pb	83 Bi	84 Po	85 At	86 Rn
87 Fr	88 Ra	89–103 Ac† Series															

*Lanthanide Series	57 La	58 Ce	59 Pr	60 Nd	61 Pm	62 Sm	63 Eu	64 Gd	65 Tb	66 Dy	67 Ho	68 Er	69 Tm	70 Yb	71 Lu
†Actinide Series	89 Ac	90 Th	91 Pa	92 U	93 Np	94 Pu	95 Am	96 Cm	97 Bk	98 Cf	99 Es	100 Fm	101 Md	102 No	(103) Lr

Transuranium elements start with element number 93, Neptunium. The actinide series of elements as a group occupies a single square in the main figure. The lanthanide series of elements also occupies a single square in the large chart.
Courtesy of U.S. Department of Energy

Appendix B. Glossary

AEC (U.S.) Atomic Energy Commission
ALARA as low as reasonably achievable
ALSEP Apollo Lunar Scientific Experiment Package
amu atomic mass unit
ARC asteroid resource complex
BOL beginning of life
BOM beginning of mission
CCV cooldown control valve
CDA control drum actuator
COPUOS (UN) Committee On Peaceful Uses of Outer Space
CRU combined rotating unit
DARPA Defense Advanced Research Project Agency
DEC direct energy conversion
DIPS Dynamic (Radio)isotope Power System
DOE (U.S.) Department of Energy
EASEP Early Apollo Scientific Experiment Package
EM electromagnetic
E-MAD Engine Maintenance, Assembly and Disassembly
EMP electromagnetic pump
EOL end of life
EOM end of mission
EOT Earth orbital terminal
EPA (U.S.) Environmental Protection Agency
EST-1 Engine Test Stand Number One
eV electron volt
EVA extravehicular activity
FBR fixed bed reactor
FP fission product(s)
FPSE free piston Stirling engine
FSAR Final Safety Analysis Report
GEO geostationary earth orbit
geosynchronous orbit
GIS graphite impact shell
GLFC graphite lunar module fuel cask
GSOC geosynchronous orbit complex
GW gigawatt
HEO high Earth orbit
HTGR high temperature gas (cooled) reactor
ICRP International Commission on Radiological Protection
INSRP Interagency Nuclear Safety Review Panel
IUS Inertial Upper Stage
keV thousand electron volts
kT kiloton
kW kilowatt
LCP large communications platform
LEO low Earth orbit
LES Lincoln Experimental Satellite
LET linear energy transfer
LH_2 liquid hydrogen
LiH lithium hydride
LOT lunar orbital terminal
LRC lunar resources complex
L2 Lagrangian libration point 2
L5 Lagrangian libration point 5
MeV million electron volts
MHW multihundred watt (RTG)
MPD maximum permissible dose
MPRE Medium Power Reactor Experiment
MW megawatt
NaK sodium-potassium eutectic mixture
NASA (U.S.) National Aeronautics and Space Administration
NCV nozzle control valve
NEP nuclear electric propulsion
NERVA nuclear engine for rocket vehicle application
NF nuclear furnace
NRDS Nuclear Rocket Development Station
NTS Nevada Test Site
nvt neutron velocity time
OTV orbital transfer vehicle
PbTe lead telluride
PCS power conversion system
PPO pressed plutonium (-238) oxide
PSM power system module
PSOV propellant shutoff valve
PV pressure vessel
QF quality factor

rad	radiation absorbed dose
RBE	relative biological effectiveness
RBR	rotating bed reactor
rem	roentgen equivalent man
RIFT	reactor in-flight test
R-MAD	Reactor Maintenance, Assembly and Disassembly
RTG	radioisotope thermoelectric generator
SER	Safety Evaluation Report SNAP-2 Experimental Reactor
SETI	search for extraterrestrial intelligence
SiGi	silicon germanium
SMF	space manufacturing facility
SNAP	System(s) for Nuclear Auxiliary Power
SPAR	Space Power Advanced Reactor
SPR	Advanced Space Nuclear Power Program
SPS	Satellite Power System
STS	Space Transportation System (Space Shuttle)
S2DR	SNAP-2 Developmental Reactor
S8DR	SNAP-8 Developmental Reactor
S8ER	SNAP-8 Experimental Reactor
TAGS	silver antimony germanium telluride
TBCV	turbine bypass control valve
TE	thermoelectric
TFE	thermionic fuel element
TI	thermionic
TITR	thermionic test reactor
TSCV	turbine shutoff control valve
TVA	turbine valve actuator
TZM	molybdenum alloy (Mo-0.6, Ti-0.1, Zr-0.035 C)
T-111	tantalum alloy (Ta-8W-2Hf)
UO_2	uranium dioxide
U-ZrH	uranium zirconium hydride
XE	experimental engine
ZrH	zirconium hydride

Illustrations

Illustrations Continued

Illustrations Continued

Illustrations Continued

Illustrations *Continued*

Tables

Tables continued

Index

NOTES

NOTES

NOTES

NOTES

NOTES

NOTES